S0-BLQ-019

ELECTRONICS IN OUR WORLD

A SURVEY

PRENTICE-HALL, INTERNATIONAL, INC., *London*
PRENTICE-HALL OF AUSTRALIA, PTY. LTD., *Sydney*
PRENTICE-HALL OF CANADA LTD., *Toronto*
PRENTICE-HALL OF INDIA PRIVATE LTD., *New Delhi*
PRENTICE-HALL OF JAPAN, INC., *Tokyo*

ELECTRONICS IN OUR WORLD

A SURVEY

GREGORY J. NUNZ
Electronics Department
Los Angeles Pierce College

Prentice-Hall, Inc., Englewood Cliffs, N. J.

Current printing (last digit):

10 9 8 7 6 5 4 3 2 1

13-252288-8

Library of Congress Catalog Card Number: 70-146682
Printed in the United States of America

To my dearest wife,
Georgia

PREFACE

If one were to check the list of books in print each year for the past decade, he might quickly be convinced that electronics books proliferate at a rate second only to that of rabbits. The number of such books—thick and thin, erudite and elementary, specialized and general—grows annually at an ever-increasing pace. Why, then, another entry into an already crowded field? The answer is that this book represents the first step in the expression of a *new pedagogical idea.*

In this age of technology most educators are convinced that every person should have some science background in his educational makeup. This usually takes the form of a course, typically called General Science, which comprises a "once over lightly" of chemistry, physics, biology, etc. Yet the day-to-day experience of the average person, the non-technologist, is not readily assimilable in terms of these pure sciences. Rather, he is confronted by an array of gadgets which, in at least nine of ten instances, are electrical or electronic devices: toaster, telephone, hi-fi, dictaphone, television, vacuum cleaner, etc. It is, therefore, this author's position that an elementary conceptual knowledge of electronics is as indispensible to every educated person as "reading, 'riting and 'rithmetic."

Thus, this book is intended as the basis for a conceptual survey course in electronics for *everyone.* No mathematical demand is made upon the reader beyond his ordinary abilities in arithmetic. The few additional ideas of quantitative relationship utilized in this book are introduced in Chapter 2. In specific application, this book may also provide, for the electronics technician, an introductory course (to maintain motivation) while he is acquiring the mathematical skills prerequisite to his subsequent theory courses.

The author wishes to express his appreciation to Miss Melanie Martin, Mrs. Amy Nakatani, Mrs. Claudia O'Connor, and Mrs. Sally Patton who typed the manuscript. Heartfelt thanks is also due to the many companies, too numerous to mention here individually, who generously provided illustrative material, and to Mr. Thomas Vlachos for the ancillary photography.

It is the author's earnest desire that this book be readable—never overly pedantic, frequently entertaining, and always instructive. Any recommendations for its improvement are wholeheartedly welcomed.

Gregory J. Nunz

CONTENTS

1

LOOK AROUND YOU

1-1 WHAT IS ELECTRONICS?

Perhaps the expected way to begin this chapter is to state at the outset the subject of this book. That seems simple enough. It concerns electronics. Such a statement inexorably leads to some attempt at defining *electronics*, a task that the author would have preferred to dodge at this point. The reason for shilly-shallying is that a suitable definition of electronics requires the use of other words (e.g., *electron*, *semiconductor material*, *vacuum tube*, etc.) which would also require definition. Two extreme results are possible: One is a veritable encyclopedic article on electronics, which is certainly out of place, since that is the function of this book as a whole. The other alternative is the *circular definition* approach used in some (poor) dictionaries, wherein the reader looks up the word *squog*, finds the definition "to fliggle," then patiently looks up *fliggle* and finds the definition "a squogging action." Here we attempt to steer a course between "squogging" and the encyclopedia.

In modern society, most people have some intuitive feeling for the abstractions we call *electricity* and *magnetism*. Ask any ten people to define either term and you will get ten different definitions. Despite the differences, however, there would be sufficient agreement for three persons to communicate meaningfully about topics relating to electricity and magnetism.

Relying temporarily on this degree of intuitive understanding, we offer the following working definition for the purposes of this book:

> **Electronics is that body of scientific knowledge and technology devoted to the control of electricity and magnetism in producing useful work.**

This definition would not satisfy the purist, who would have us distinguish between *electromagnetic theory* and *electronics* proper, but when the reader has reached the end of this book, he will be able to make the distinction of his own accord. It must also be pointed out that *control* here, of necessity, implies "understanding" as a prerequisite.

1-2 LOOK AROUND YOU

It is intriguing to try to imagine the state of our civilization had electronics (as we have broadly defined it) never come into being. The title of this chapter is meant as an invitation. After you read this sentence, stop for a few moments, look around your own small corner of the world, and imagine that every article you see whose existence depends *in any way* on electronics suddenly vanishes. If you have let your imagination go far enough, you should, in your mind's eye, be seated on a roughhewn bench in a rather primitive hovel (or at best, and at the whim of ancestry, on a handcrafted chair in a very medieval castle), probably in a different part of the world! You didn't quite reach that conclusion? Let's see if we can work from the obvious to the not so obvious.

For the sake of common understanding at this point, we restrict our discussion to the household environment and everyday experience. As you paused to consider the annihilation of electronics-dependent items, the first articles that suggested themselves to you were probably in the communications category: the television set, radios, hi-fi equipment, tape recorders, and, perhaps, the telephone—those things we usually associate with the word *electronics.* Naturally, all appliances with electric motors are also gone: the washer, dryer, refrigerator, vacuum cleaner, electric knife, electric toothbrush, electric can opener, electric clocks, forced air heater, air conditioner, and so on. Furthermore, all devices using electricity to produce heat and/or light have disappeared; there are no toasters, irons, electric ranges, space heaters, electric skillets or coffeemakers, and, of course, no electric lights. Thus far, only the obvious direct applications of controlled electricity and magnetism have been eliminated. How about articles that have been electroplated, electropolished, or welded by electrical means?

Gone is much of your jewelry and many other valuable household items! Consider that certain metals are produced by electrolytic means. All traces of aluminum and magnesium have now vanished. Is life becoming very primitive? The worst is yet to come! Many of the other household possessions you take for granted would be either unobtainable or priced far beyond your means if they had to be handcrafted rather than produced by electrically automated machines; so they, too, have vanished. Finally, if we diligently annihilate all traces of electrical or magnetic technology, the compass has never been invented and thus Columbus, Magellan, et al. never made their celebrated voyages. This leaves America to those few of us descended from its original Indian inhabitants, and the rest of us still in other lands whence our ancestors came.

1-3 ELECTRONICS FOR EVERYMAN

From the brief discussion of Section 1-2, the reader is probably now convinced (if he was not previously) of the tremendous impact of electronic technology on civilization. What is even more remarkable is the pace of such progress. Although man has been on this planet for a million years or more, magnetic phenomena have only been put to use during the last 500 years, and electronics has really been exploited only in the last hundred years. Research is producing new discoveries at an ever-increasing rate, and several novel and useful electronic devices will certainly go into production even in the brief interval from the moment these words are first set on paper to their appearance in print.

Everyone is, at least dimly, aware of the revolution in industry effected by the availability of the electronic computer. Research and design engineers tell us that the fully computerized home, with a robot staff, is no longer a science fiction writer's dream but an inevitable actuality only decades from realization. The tentacles of electronic technology embrace areas as diverse as medicine, law, politics, art, defense, and, of late, even romance. Indeed, there is no aspect of human experience to which electronics either has not already been or will not soon be applied profitably (in all senses of that word). Therefore every man and woman—doctor, lawyer, Indian chief, or housewife—should have at least a passing familiarity with electronic principles. In the following pages, the reader will be conducted on a tour of the expanding vistas of electronics in our world.

REVIEW QUESTIONS

1. Would the study of the phenomenon called *lightning* properly belong to the body of knowledge we have termed *electronics* in Section 1-1?
2. Continuing the "disappearance" game of Section 1-2, cite at least ten non-obvious examples of household items we should have to do without in the absence of electronic technology. Justify each selection.
3. Can you think of areas of human activity that have not yet been *directly* affected by electronics? Can you envision how electronics might have applications thereto in the foreseeable future?

2

ON SIZE AND RELATION

2-1 NUMBER AND SIZE

You may be surprised to learn that there are people in this world who can manage life nicely with almost no concept of numbers. There is, for example, a tribe of natives called Hottentots who, it is said, have no words in their language for numbers greater than three. In describing any larger quantity, they simply say *many*. Thus a Hottentot chieftain may tell you that he has one wife, three sons, or two belts, but many beads and many warriors. If you don't happen to be a Hottentot, however, such a limitation in your concept of numbers would prove a trifle inconvenient. Imagine telling a real estate agent only that you required a home with "many" rooms, or trying to sew a size "many" dress. Would you accept a job with an employer who would only contract to pay you "many" dollars per week? From these few examples it is clear that, no matter what your calling or profession, a certain degree of sophistication in dealing with numbers is indispensable. In particular, to persons involved in banking, commerce, science, and technology, facility with numbers means "bread and butter."

Generally speaking, numbers are required to specify the *size* of anything measurable—numbers of *dollars* for budgets and deficits, numbers of *yards* for cement foundations, numbers of *megatons* (millions of tons of TNT) for nuclear explosions, and so on.

Our concern in this book is electronics. Electronics is a branch of physical science, and one of the major tasks of all science is to *measure*, that is, to determine the size of certain quantities involved in the phenomena being investigated. In the following chapters, we will have occasion to talk about numbers that are very large and very small. We will also want to discuss relationships between numerical quantities. For these reasons, we shall spend some time in this chapter developing ideas that will enable us to communicate more effectively about numerical quantities and the relationships between them.

As the reader finished the preceding paragraph, he may have moaned, either mentally or actually. In particular, the nontechnical student, or the person who "absolutely hates math," may have been upset by the path he thinks this chapter (or perhaps the whole book) is going to take. If so, let us set the record straight right now:

> **THIS IS NOT A MATHEMATICAL TEXTBOOK!!!**
> **YOU ARE NOT GOING TO BE TAUGHT A LOT OF MATHEMATICS!!!**

Feel better? Okay, then! Listen carefully once more to the purpose of this chapter: it is to develop ideas that will enable us to *communicate about* numerical quantities and their relationships effectively (and *not* to do calculations, as the reader may have feared). Furthermore, the concepts to be developed in the succeeding paragraphs are not limited in scope to electronics but are of general value to anyone who tries to follow scientific articles in the newspapers or popular magazines. Thus the author sincerely hopes that the reader will make these concepts a part of his mental "tool box."

2-2 A NEGATIVE ATTITUDE

One of the first novel concepts that we encounter in dealing with physical quantities is the negative number. A negative number is *a number less than zero* and is written as an ordinary positive number *preceded by a minus sign* ($-$), for instance, -5, -100, -0.71, etc. Since we are committed to talk about physical quantities, the so-called practical person may ridicule the very idea of a negative quantity. "Who can conceive of something 'less than nothing'?" he may ask. Such a comment illustrates why it is important to have the "right kind of a negative attitude."

The important idea relative to negative physical quantities is that *zero* represents an *arbitrary reference point* on some scale. On such a scale, we may count away from zero in two directions, one of them being (arbitrarily) designated *positive* and the other *negative*. In our sheme of ordering numbers, we say that 10 is *greater* (i.e., more positive) than 5. We must therefore also say that -10 is *less* (more negative) than -5, which is the same as saying that -5 is greater than -10. This suggests a way of drawing a sort of picture of all the so-called *real* numbers, both negative and positive. Draw a straight line, as shown in Fig. 2-1, arbitrarily pick a point on this

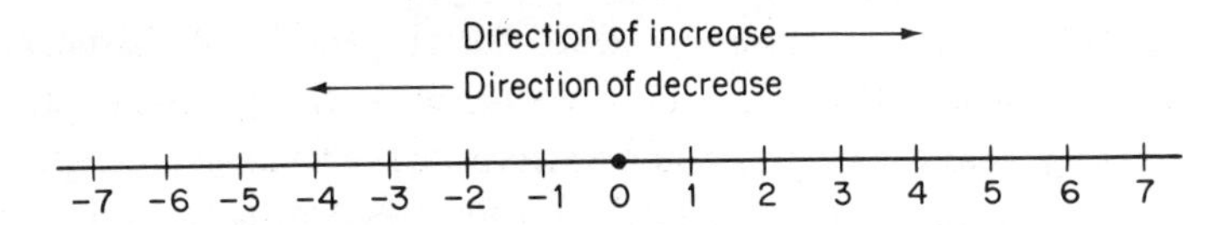

Fig. 2-1. A line (or scale or axis) of numbers.

line, and call it zero (0). Next, mark off a number of equal spaces on both sides of the zero point. Working from 0 to the right, number the marks consecutively 1, 2, 3, 4, etc., and working from 0 to the left, number the marks consecutively -1, -2, -3, -4, etc. Such a picture is called a *line* (or *axis* or *scale*) of *real numbers* and must be imagined to extend indefinitely to both the right and left, since the sequence of numbers continues indefinitely.

On a real-number scale of this type, negative numbers can be given many physical meanings. When Mr. Fahrenheit devised the temperature scale named after him, he first placed his thermometer in the coldest available standard he knew (a mixture of salt and melting ice) and marked his thermometer scale 0 where the mercury came to rest. He then placed his thermometer in the hottest available standard he knew (boiling water[1]) and marked his scale 212 where the mercury stopped. On this arbitrary scale, then, as shown in Fig. 2-2, there are 180 units (degrees) of temperature between the point at which water freezes (32°) and the point where it boils. Mr. Fahrenheit's thermometer now has a scale that is beginning to resemble

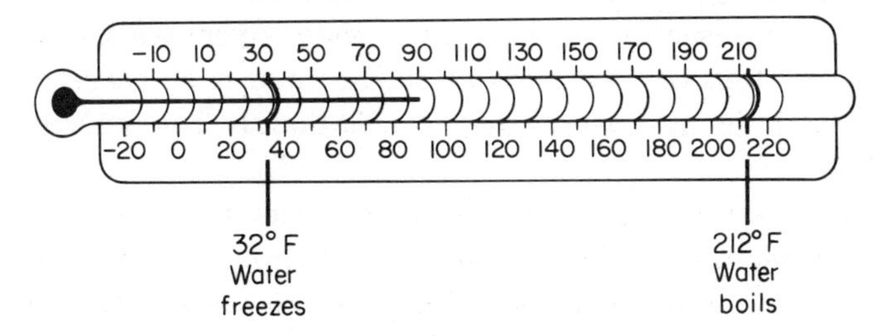

Fig. 2-2. The Fahrenheit thermometer.

[1] Actually, he first used human body temperature for his "hot" reference point and called it 96°. Later he chose to set the boiling point of water at exactly 212°.

at least the right half of Fig. 2-1. If Mr. Fahrenheit had pressed his thermometer against a piece of dry ice (frozen carbon dioxide gas), or had taken his thermometer to the polar regions in winter, he would have found that the mercury stopped somewhere to the left of the zero point on his scale. Therefore he would have had to continue his scale in the manner of the left half of Fig. 2-1 and would have associated *negative temperatures* (temperatures less than 0 °F) with polar winters and dry ice.

Similarly, taking the reference date for the birth[2] of Christ as 0, an historian could consider Fig. 2-1 a time scale, with the positive part of the scale representing years A.D. and the negative part of the scale representing years B.C.. An accountant could consider it a financial scale, with the positive numbers representing credits (money received) and the negative numbers representing debits (money owed).

In summary, the main ideas that the reader should have about negative numbers are

1. The negative numbers, together with the positive numbers, can be placed in an orderly sequence on a scale like the one shown in Fig. 2-1, on which numbers increase from left to right;
2. Negative numbers have physical meaning when 0 is understood as an arbitrary reference point on such a scale.

2-3 LILLIPUT AND BROBDINGNAG

When one stops to consider, it is rather surprising, how narrow the range of "sizes" is that describes the everyday experience of mankind. If a range of numbers were to be specified, it is true that most of the daily experiences of the average person can be described by numbers between one hundredth ($\frac{1}{100}$ or 0.01) and one hundred thousand (100,000). Thus the smallest division on the common ruler is $\frac{1}{16}$ of an inch and the smallest unit of our currency is $\frac{1}{100}$ of a dollar (one cent). At the other extreme, the biggest single financial transaction of the average man is perhaps the purchase of a home for a few tens of thousands of dollars; and if he travels all the way around the world in one business trip or vacation, he has covered only about 25,000 miles.

In his entertaining satires, Swift[3] exploits this narrow scale of our experience by sending his hero, Gulliver, to Lilliput, where all life is microscopic in scale compared to man's ordinary experience, and later to Brobdingnag, where the gigantic proportions of civilization make ordinary man doll-

[2] Because of certain corrections to our calendar, it is believed that the birth of Christ actually took place somewhere between 4 B.C. and 7 B.C.

[3] Jonathan Swift, **Gulliver's Travels.**

size by comparison. Comfortable in our scale of experience, we are amused at Gulliver's predicaments. Similarly, we tend to apathy when the Congress approves the decrease, or (more likely) the increase, of a budget measured in billions of dollars by a few millions, but we react with alarm at the seemingly unnecessary expenditure of a few tens of dollars by a husband or wife.

In the world of science, and specifically in the world of electronics, we will frequently encounter both Lilliputian- and Brobdingnagian-sized quantities. This means talking about numbers that, in our ordinary way of writing them, contain many zeros either before or after the decimal point. Try to read quickly and appreciate the meaning of

$$65{,}200{,}000{,}000{,}000$$

or

$$0.000000000184$$

and similar very large or very small numbers. Do you find our notation for such numbers cumbersome? You are not alone, and it is for this reason that what is called *scientific notation* has been devised.

2–4 EXPONENTS OR POWERS OF NUMBERS

In his work, *The Sand Reckoner*, Archimedes tries to convince King Gelon that the number of grains of sand in all the world is neither infinite nor bigger than any number that can be written. Such an argument might not seem a formidable undertaking today, but the Greeks of Archimedes' day did not have our convenient numerical notation. We may write a number as large as we please by merely adding zeros at the right of a number. The ancient Greeks, on the other hand, had individual symbols for distinct numbers, using letters of their alphabet; the largest single number they could write was a *myriad* (10,000). In order to write the much larger numbers he required, Archimedes was forced to make perhaps the first significant use of *exponents*.

What is an exponent? It is a shorthand way of showing that a number is to be multiplied *by itself* several times. The quantity to be multiplied by itself is called the *base* and the number of times it is to be multiplied is called the *exponent* or *power*. Thus

$$2^5 \text{ means } 2 \times 2 \times 2 \times 2 \times 2 \ (= 32)$$

and

$$10^4 \text{ means } 10 \times 10 \times 10 \times 10 \ (= 10{,}000)$$

The first expression, 2^5, is read as "two-to-the-fifth-power," where 2 is the base and 5 is the exponent or power; the second, 10^4, is read as "ten-to-the-fourth-power," with the 10 as the base and 4 the exponent. Expressions like 2^5 and 10^4 are called *exponentials*. In exponentials, the exponent (power) is always written as a small number above and to the right of the base. By definition, the first power of any base is simply the base itself, and by convention the exponent "one" is never written, except for illustrative purposes, thus

$$2^1 = 2$$
$$4^1 = 4$$
$$10^1 = 10$$

Also, by definition, the zero*th* power of any nonzero base is equal to one, thus

$$2^0 = 1$$
$$4^0 = 1$$
$$10^0 = 1$$

Negative exponents are interpreted as *one divided by the corresponding positive exponential;* for instance,

$$2^{-1} = 1 \div 2^1 = \tfrac{1}{2} = 0.5$$
$$10^{-2} = 1 \div 10^2 = 1 \div (10 \times 10) = \tfrac{1}{100} = 0.01$$

It is helpful for our purposes, to be familiar with the powers of ten, and we needn't bother with any other exponentials. A brief list of powers of ten is given in Table 2-1.

Table 2-1 Powers of Ten

Exponential Form	Value	Name*
10^{12}	1,000,000,000,000.	One trillion
10^9	1,000,000,000.	One billion
10^6	1,000,000.	One million
10^3	1,000.	One thousand
10^2	100.	One hundred
10^1	10.	Ten
10^0	1.	One
10^{-1}	0.1	One tenth
10^{-2}	0.01	One hundredth
10^{-3}	0.001	One thousandth
10^{-6}	0.000001	One millionth
10^{-9}	0.000000001	One billionth
10^{-12}	0.000000000001	One trillionth

*American system of numeration.

Reference to this table will help in following the discussion of scientific notation in Section 2-5. Notice that the value of the exponent in the exponential form is the number you would obtain if you were to place an imaginary decimal point after the first *nonzero* figure (in all these cases, 1) and count the number of places you must move to arrive at the real decimal point, considering moving to the right as positive and to the left as negative.

2–5 SCIENTIFIC NOTATION

As we have noted in Section 2-3, figures containing a large number of zeros, either before or after a decimal point, are tedious to read or write and conducive to making errors. What is worse, with respect to scientific measurements, such figures may give a false impression of accuracy. For example, an astronomer might (under pressure from a news reporter) quote the distance of a newly discovered star as 12 quintillion miles. If we write this out as 12,000,000,000,000,000,000 miles, we give the impression that it has been measured down to the last mile. In actuality, there are many errors and uncertainties in measuring such stellar distances and the astronomer's best estimate is that this particular star lies "somewhere between" 11,900,000,000,000,000,000 and 12,300,000,000,000,000,000 miles distant. In electronics, a similar situation prevails with respect to measurement errors and manufacturing tolerances. Because of such measurement errors and apparatus tolerances, only a certain number of digits can be considered meaningful in a figure representing a physical quantity. These digits are called the *significant figures* of a number.

A most useful notation, which obviates the need for long strings of zeros in very large or small numbers and simultaneously has the advantage of showing just the significant figures in the number, is *scientific notation.* In scientific notation, any given number is written as some decimal between 1 and 10 multiplied by a suitable power of ten. For example, to write 265,000 in scientific notation, assuming the first four figures were significant figures:

$$265{,}000 = 2.650 \times 100{,}000 = \underline{2.650 \times 10^5}$$

Similarly, the distance to the star mentioned earlier should be written in scientific notation as 1.2×10^{19} miles.

Thus a number in scientific notation consists of two parts: a decimal multiplier, which gives the significant figures and sign (positive or negative) of the number; and an exponential, separated by the multiplication sign ($\times$). The decimal multiplier is always a number between 1.000... and 9.999... and may be positive or negative, depending on whether the number to be represented is positive or negative. The exponential is that power of

10 required to give the desired number when it is multiplied by the decimal multiplier. Here are additional examples of scientific notation (three significant figures assumed in all cases)

$$7{,}260{,}000 = 7.26 \times 1{,}000{,}000 = \underline{7.26 \times 10^{6}}$$

$$0.0000493 = 4.93 \times 0.00001 = \underline{4.93 \times 10^{-5}}$$

$$-856 = -8.56 \times 100 = \underline{-8.56 \times 10^{2}}$$

$$-0.00300 = -3.00 \times 0.001 = \underline{-3.00 \times 10^{-3}}$$

Another benefit of scientific notation is that it avoids the confusion caused by the differences in the American and European names for very large and very small numbers. For instance, to an American, a "billion" is 1,000,000,000 ($= 10^{9}$), while to a European a "billion" is 1,000,000,000,000 ($= 10^{12}$), but neither can misunderstand the other if he says 10^{9} or 10^{12} instead of billion.

2-6 PHYSICAL QUANTITIES

Part of the challenge to the scientist in studying any phenomenon is to decide (1) what to measure and (2) how to measure it. With regard to *what* is to be measured, science requires that it be sufficiently well defined so that another scientist observing the same phenomenon (perhaps at a different place or time) could make the same measurement. The early work in what is now called *physical* science involved studying relatively large bodies of matter that were falling, rolling, and otherwise moving. Useful concepts derived from this area (mechanics) of physical science were carried over into other areas (such as electronics) as they developed. Consequently, it is worth our while to study a few of these concepts briefly.

2-7 SCALAR AND VECTOR QUANTITIES

It is convenient to distinguish between two types of physical quantities: *scalars* and *vectors.*

A scalar quantity is one that has only magnitude.
A vector quantity is one that has both a magnitude and an associated direction.

We can consider a scalar any quantity whose value is given merely by a reading (i.e., a number) on a scale. A vector, on the other hand, must be defined symbolically by an arrow; the length of the arrow showing the magnitude, to some scale, and the direction of the arrow showing the direction associated with the quantity.

The temperature of an oven is a typical scalar quantity. No matter which may the thermometer is turned, the reading is still, say, 350°F, and the value 350°F completely specifies the temperature.

A force is a good example of a vector quantity. When the rocket engine ignites under the spaceship resting on the launch pad, a force is generated whose magnitude must be greater than the magnitude of the force of gravity (i.e., the "weight" of the ship) and whose direction must be opposite (upward) to that of gravity (downward).

2-8 UNITS OF PHYSICAL QUANTITIES

Having decided what one wished to measure, one must decide *how* to measure it. Usually this involves comparison with some predetermined standard. To determine the length of a rod, we lay a ruler or measuring tape alongside it. The ruler or tape is already graduated with markings that refer back to the length of a bar of precious metal kept in a museum, which is taken to be the standard, or *unit*, of length. Similarly, the average time required for our planet to turn once around its axis is taken as the basis for a unit of time measurement. Our watches are graduated so that each tick marks 1/86,400 of a complete rotation of our planet, and this unit of time is called a *second*.

Unfortunately, there are many different standards (units) for measuring the same thing, varying with the period in history, country, and profession. For physical quantities pertinent to scientific and engineering work, there are two fundamental systems of units: the *metric* system and the *English* system. The English system is associated with the English people and is currently used in Great Britain, the United States, and their territories. The metric system, named for its unit of length (the meter), is used in all other civilized parts of the world.

Both the English and metric systems use the same unit of time, the second. The units of force, mass, length, and temperature are different, however, in the two systems. A comparison, with approximate equivalents, is given in Table 2-2.

Perhaps the concept of *mass* requires a little clarification, for it is often confused with weight. Weight is the force that a mass experiences because of the pull of gravity. A body in space, far away from any star or other source of gravitational attraction, has no weight but still has mass.

Table 2-2 Common Metric and English Units

Fundamental Quantity	English System (Common Unit)	Metric System (Common Unit)	(Approx. English Equiv.)
Force	pound (lb)	newton (N)	0.22 lb
Length	foot (ft)	meter (m)	3.3 ft
Time	second (sec)	second (sec)	Same
Temperature	degree Fahrenheit (°F)	degree Celsius* (°C)	1.8°F
Mass	slug	kilogram (kg)	0.068 slug

*Formerly called Centigrade.

Mass is a scalar quantity; weight, a force, is a vector quantity. Thus a body with a mass of one slug, located at the surface of the earth, experiences a force (weight) of about 32 pounds in the direction of the earth's center (downward).

2-9 WORK, ENERGY, AND POWER

When a force acts on a body, either producing or tending to resist motion through a certain distance, the quantity obtained by multiplying that force by the distance through which it acted is called *work*.

In order for work to be done there must be motion.

Work is the measure of what we get out of a device, and is a scalar quantity. Since

$$\text{Work} = \text{force} \times \text{distance}$$

it has the units pound-feet (lb-ft) in the English system and newton-meters (more usually called *joules*) in the metric system.

Energy is usually defined as the "capacity to do work" and is subdivided into two broad categories: potential energy and kinetic energy. Potential energy represents a stored energy of some kind, and kinetic energy represents energy of motion. Many specific types of energy are included within these two categories: potential energy due to location, chemical energy, nuclear energy, heat energy, radiant energy, electric energy, mechanical energy, etc. Energy cannot be created or destroyed, but, through various natural and man-made processes, any form of energy can be converted to any other and made to do some useful work in the transformation. Energy, like work, is a scalar quantity with the unit of pound-foot (English) or joule (metric).

Often we are interested not only in getting a job done but also in doing it within a certain limited period of time. From this standpoint, a useful physical quantity is the *time rate* of doing work (or time rate of consuming energy), which is called *power*. Thus power is a scalar quantity defined by

Power = work done (or energy consumed) ÷ time required to do it

and has the unit pound-foot per second (lb-ft/sec) in the English system or joule per second (called a *watt*) in the metric system. Sometimes it is desired to compare the power developed in a machine with the capability of a "standard horse" as defined by James Watt, who invented the steam engine. This standard horse can do 550 lb-ft of work in one second; therefore 550 lb-ft/sec (or 746 watts) equals one *horsepower*.

We will find these notions of work, energy, and power useful in discussing the capabilities of various electrical devices. One other quantity, which is useful in telling us how economically a device may be operated, is *efficiency*. It is defined by the statement

Efficiency = useful power output ÷ total power input

No process or device is a perfect converter of energy, and there is always some part that is not available for doing useful work in the transformation process. Consequently, the efficiency of any device is always a number less than one. It is usually given as a percentage and is thus always less than 100 percent. Since efficiency represents power divided by power, it has no units associated with it.

2-10 RELATIONSHIPS BETWEEN QUANTITIES

In studying the phenomenon, once the scientist has decided which quantities to measure and has measured them, his next task is to determine the relationship among them. The importance of knowing the relationship among quantities is that, once known, the scientist can *predict* what will happen to the other quantities when he does something to change one of them.

Now, the relationship between any two physical quantities will fall into one of three categories: *independence*, *statistical correlation*, and *functional dependence*. Two quantities are said to be mutually independent if changing one of them has no effect on the other. For instance, the number of bird species living on Gibraltar is independent of the price of tea in China.

(Perhaps it would be safer to say apparently independent, for the author is not aware of anyone's having studied this particular relationship in great detail.)

Two quantities are statistically correlated if the one generally varies with the other, but not in any exact and predictable fashion. Thus the number of copies of *Life* magazine sold in New York City in any given year is statistically correlated with the city's recorded population for that year. By studying the rate of population growth, the magazine's marketing manager can estimate "about" (but not exactly) how many additional copies to print for the next year.

The third type of relationship, functional dependence, is the one that concerns us most in electronics (and in science, in general). A quantity B is said to be *a function of* the quantity A, if for any given value of A there is a precise value for B determined by some fixed (and usually mathematical) rule. For example, the length of time required for a radar pulse to reach the moon and return is a function of the distance between the earth and moon. Since functional relationships are so important in electronics, we need a way in which to show them for discussion purposes. In general, there are two ways in which a function can be described easily: one way is by writing a mathematical equation, which the author has promised *not* to do in this book; the other way is by drawing a picture of the function. As our final topic in this chapter, we will discuss the art of drawing and interpreting such pictures.

2-11 PICTURES OF FUNCTIONAL RELATIONSHIPS

Imagine yourself in the following (admittedly, a bit farfetched) situation. You have spent a day out with a friend and are walking home. You have come to a street corner where you go separate ways. Having said "So long!" you stand and watch him or her walk away from you. Then, a helicopter, hovering motionless overhead, begins taking aerial photographs of this scene. The situation is depicted in Fig. 2-3. The helicopter is at a sufficiently high altitude so that you and your friend only appear as dots on the photographs taken. A typical photograph is shown in Fig. 2-4. Now suppose that one picture was taken every 10 seconds for 3 minutes. Suppose, further, that you obtained a set of these pictures and cut out and saved only the narrow strips that showed you and your friend, as indicated in Fig. 2-4.

If you place all the strips side by side in sequence from left to right, with the tiny images of yourself lined up in a horizontal row, you obtain something that looks like Fig. 2-5. You can draw in a vertical scale against

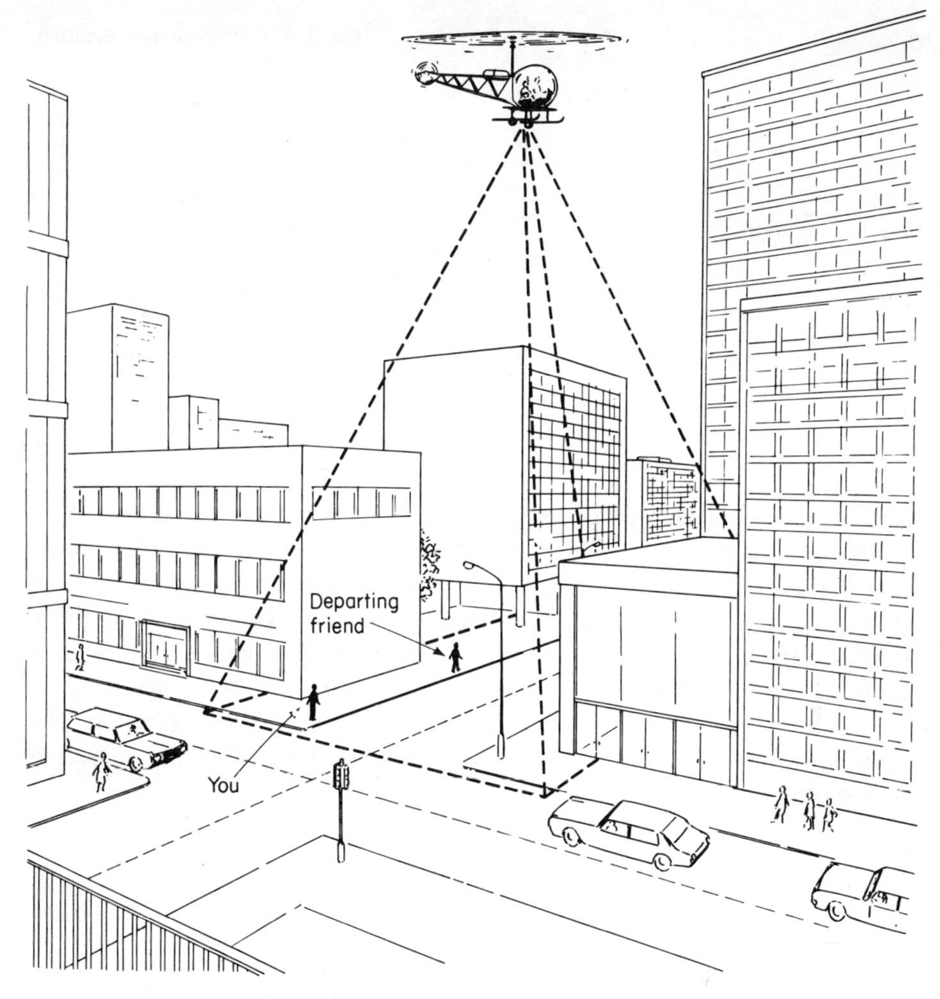

Fig. 2-3. Helicopter photographing departure scene.

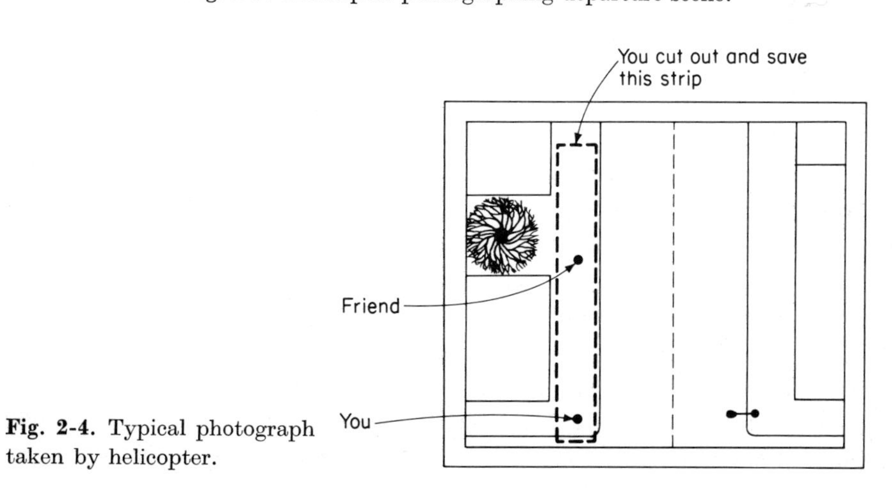

Fig. 2-4. Typical photograph taken by helicopter.

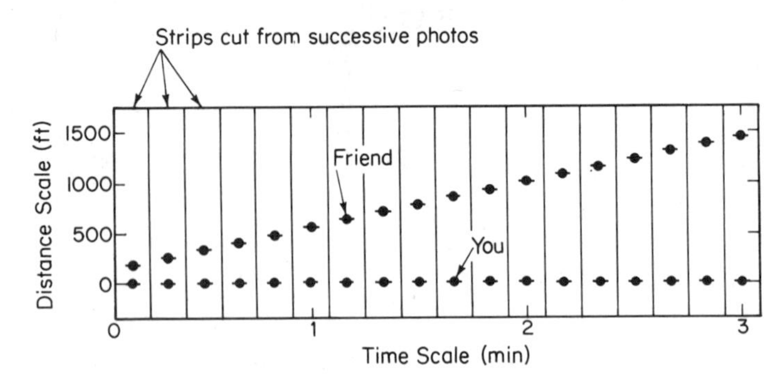

Fig. 2-5. Assembling the sequence of strips from photographs taken ten seconds apart.

which you can measure the distance between you and your friend. You can also mark a time scale under the strips to show when each was taken. These scales are shown in Fig. 2-5. What you now have is the beginning of a *picture history* of the distance separating you and your friend in those 3 minutes. If, instead of every 10 seconds, the helicopter could have taken a picture every second, the resulting history would have been made up of many more (180 instead of 18) strips, with much smaller "gaps" in the successive positions of you and your friend, and it might look something like Fig. 2-6.

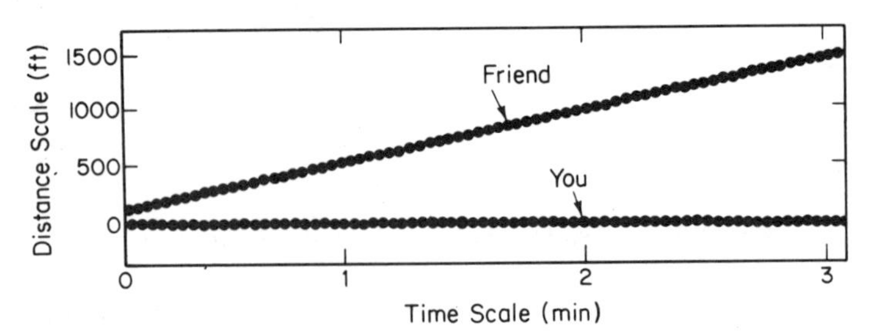

Fig. 2-6. Assembly of strips from sequence of photographs taken only one second apart.

It is not a far jump now, mentally, to imagine that the helicopter was taking pictures infinitely fast, that is, one picture every instant, during those 3 minutes. (Well, we did admit that this situation was farfetched right in the beginning, didn't we?) The result would be an infinite number of photos, which when cut and assembled in sequence as before would look like Fig. 2-7. Notice that the individual dots, representing the successive positions of you and your friend, have "smeared out" into continuous lines.

A picture like Fig. 2-7 is called a *graph* (more strictly, a Cartesian graph, but since it is the type with which we work almost exclusively, we

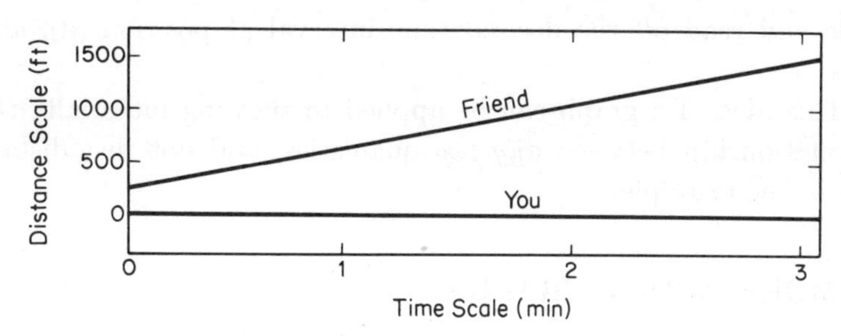

Fig. 2-7. Continuous history resulting from photographs taken every instant.

simply say graph). It is the picture of the functional relationship between two quantities, in this case the *distance* from your friend to you and the *time* since he began walking away.

Figure 2-7 is redrawn a bit more conventionally in Fig. 2-8. Here

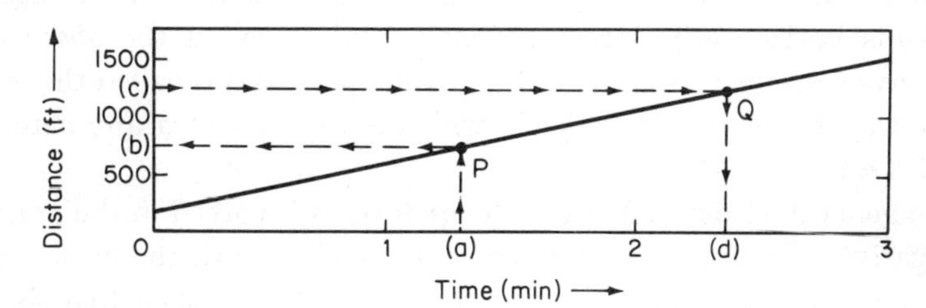

Fig. 2-8. Graph of the distance between you and friend as a function of time.

your position has become the horizontal axis. Also, a gridwork of lines has been drawn in to make the graph easier to read. In this example, distance is a *function* of time because at any given time your friend is at a definite, precise distance from you. Hence we have succeeded in drawing a picture of a functional relationship—the graph. From this graph, we can find the distance your friend has gone at any time (within the limits of the horizontal scale of the graph). Suppose, for instance, that we wanted to know how far your friend had walked in $1\frac{1}{2}$ minutes. We enter the "picture" on the horizontal axis at 1.5 minutes, position (a), and proceed vertically upward until we intersect the graph (the line representing your friend's position) at point P. We then proceed horizontally to the left and read the distance we were looking for on the vertical scale at position (b), 750 feet. Alternatively, we may wish to know how much time it took your friend to walk a certain distance, say 1200 feet. We then enter the picture on the vertical scale at 1200 feet, position (c), and move horizontally to the right until we contact the graph at point Q. We then proceed vertically downward to the horizontal scale and read off the desired time interval at position (d), about 2.4 minutes.

This idea of a graph can be applied to showing pictorially the functional relationship between *any two* quantities, and not just distance and time as in the example.

2-12 MORE ABOUT GRAPHS

In Section 2-11, you were introduced to the graph. That wasn't a very terrifying experience, was it? Now, with a bit of useful terminology (which, at least, can always be used to impress one's nontechnical friends) and a few more remarks about the general shapes of graphs, we will be in position to discuss all kinds of mathematical relationships in the chapters to come without ever mentioning a mathematical equation.

The graph is said to be drawn on a *set of axes*. One of these axes always runs horizontally across the page and is called the *abscissa*. The other always runs vertically up and down the page and is called the *ordinate*. (An easy way to remember which is which is that *a*bscissa and *a*cross both begin with *a*.)

In general, either or both of the quantities involved in the graph may have negative values as well as positive ones. Thus both the horizontal and vertical axes may look like Fig. 2-1. They are always placed so that their *zero points coincide*, as shown in Fig. 2-9, with the positive direction of the

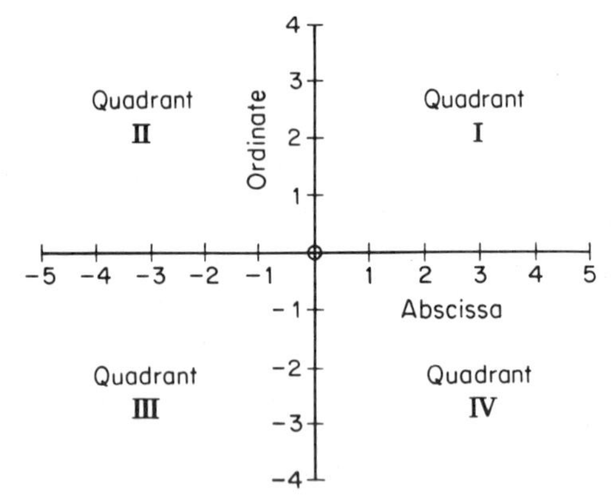

Fig. 2-9. Placement of the axes.

abscissa to the right and that of the ordinate upward. The common zero point is called the *origin*. Notice that the two intersecting axes divide the page in four regions called *quadrants*, numbered counterclockwise I, II, III, and IV. Notice, further, that both the ordinate and abscissa values are positive in Quadrant I; both are negative in Quadrant III; while only the ordinate values are positive in Quadrant II and only the abscissa values are positive in Quadrant IV. In many physical situations, such as our distance-time example of Section 2-11, only positive values of the quantities involved are meaningful and so only the Quandrant I part of the graph needs to be shown. When it is important to be able to read a graph accurately, a grid of fainter lines, parallel to the axes, is also drawn in.

Once the axes are drawn, two numbers (the abscissa and ordinate values) can be associated with any point. These numbers are called the *coordinates* of the point and are determined by proceeding horizontally and vertically from the point in question to the two axes and reading the value from the scales. Figure 2-10 shows six points, labeled A, B, C, D, E, and F.

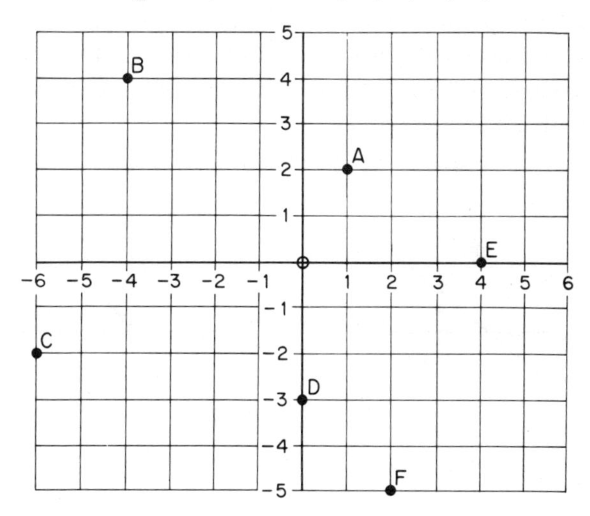

Fig. 2-10. Coordinates of points.

See if you can determine their abscissa and ordinate coordinates; then check your answers against those in Table 2-3.

Table 2–3 Coordinates of Points in Figure 2-10

Point	*Abscissa Coordinate*	*Ordinate Coordinate*
A	1	2
B	-4	4
C	-6	-2
D	0	-3
E	4	0
F	2	-5

When two quantities are functionally related, one of them is usually considered to be "free" or "independent"—that is, to have its value specified at will—and this quantity is called (unimaginatively) the *independent variable*. The other quantity, whose value is then functionally determined by the value assigned to the independent variable, is called the *dependent variable*. In drawing a graph, the independent variable is always associated with the abscissa and the dependent variable with the ordinate. Thus, in our distance-time example of Section 2-11, the independent variable was time. It is often a moot point as to which of two physical quantities is the independent variable, and the distinction is made arbitrarily. If time is one of the quantities involved, it is usually the independent variable, for man can do little to control its passage.

We next consider the graphs themselves and their general significance. The graph proper is the line (usually curved), which passes through all the points whose coordinates represent the corresponding values of the two variables. (In the examples that follow, no numbers are shown on the axes, for we are interested in discussing only the general trends of the graphs and not exact values.)

Consider the graphs shown in Fig. 2-11. The arrowheads on the axes

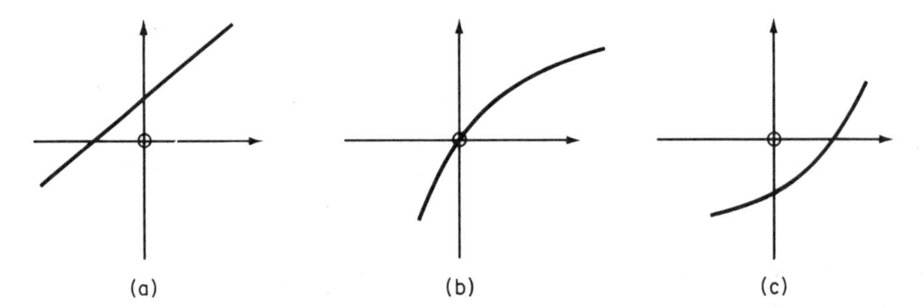

Fig. 2-11. Increasing functions.

are to remind us in which direction numbers increase (i.e., become more positive) on those axes. In each case, as we follow along the curves from left to right, we find that as the independent variable's value (abscissa) increases so does the dependent variable's value (ordinate) increase. Similarly, when we follow the curves from right to left, the independent and dependent variables both decrease. When the dependent variable increases as the independent variable increases, it is said to be an *increasing function* of the independent variable and can be seen pictorially as a graph whose trend is *upward to the right*. For example, the gasoline consumed in an automobile trip is an increasing function of the distance traveled.

By contrast, the graphs in Fig. 2-12 all have the property that the dependent variable decreases as the independent variable increases, and vice versa. The pictorial effect of this property is that the trend of the curves is *downward to the right*. Such relationships are called *decreasing functions*.

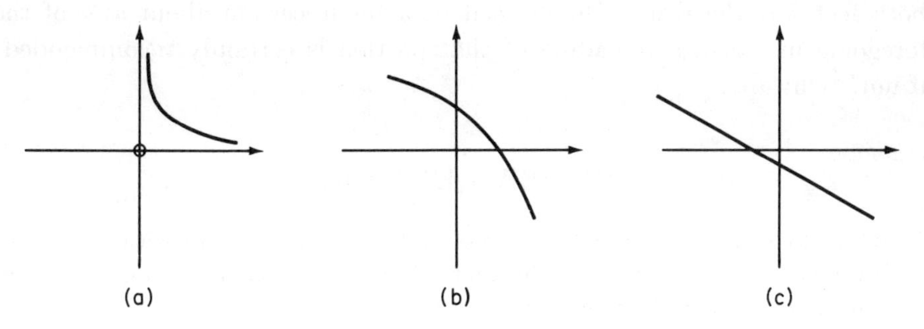

Fig. 2-12. Decreasing functions.

For instance, the length of time a given bowl of party punch will last is a decreasing function of the number of drinking guests at the party.

Finally, consider the graph of Fig. 2-13, a horizontal line. No matter

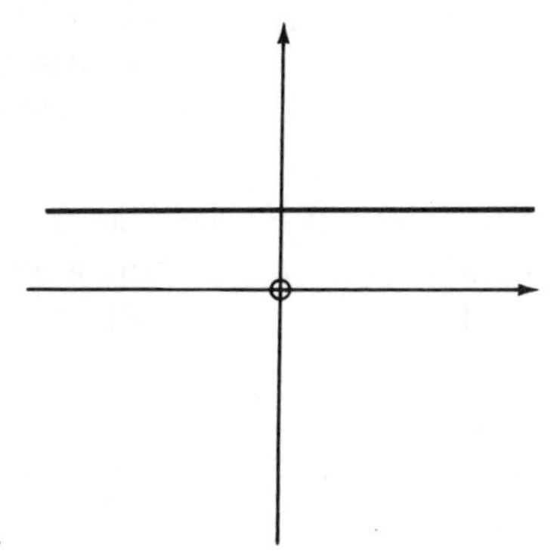

Fig. 2-13. Graph of a constant.

what value we select for the independent variable, the dependent variable has one, and only one, value. Thus the dependent variable does not depend at all on the independent variable and, in fact, is not a variable but rather a constant. If one were to plot the area (dependent variable) of each page in this book against its page number and connect the points, the graph would look like Fig. 2-13.

2-13 AGAIN, THE "WHY"

To the reader who has encountered the material in this chapter for the first time and has persevered to this point, the writer wishes to offer one more reassurance as to where this chapter is leading. With the means now at our disposal to discuss (1) numbers, positive or negative, large or small, (2) physical quantities and their units of measurement, and (3) relationships between quantities, we are ready to jump into the world of electronics "with both feet." If the reader should still be a bit uncertain about any of the foregoing material, a rereading of that portion is certainly recommended; if not, "Onward!"

REVIEW QUESTIONS

1. What do you believe are the largest and smallest numbers with which the "average" person must deal in his lifetime? Have these limits changed, in your opinion, over the last one hundred years?
2. Give some examples, other than those in this text, of the usefulness of negative numbers in describing some physical thing or process.
3. Express the following numbers (all considered to have *three* significant figures) in scientific notation:
   (a) 2840 (b) 0.00735
   (c) 4,000,000,000 (d) −54,000
   (e) −0.00000437 (f) 0.961
   (g) −2 (h) 0.000000000133
   (i) −0.081 (j) 93,700,000,000,000,000
4. Which of the following are vector quantities and which are scalar quantities?
   (a) height of a mountain (b) weight of a sack of potatoes
   (c) velocity of an airplane (d) atmospheric pressure
   (e) relative humidity (f) drag on a bullet
5. What is your approximate
   (a) height in meters?
   (b) weight in newtons?
6. A light bulb is rated at 100 watts. Is this a measure of the total light energy the bulb will deliver? How do you know?
7. Your neighbor tells you that he has just completely overhauled his automobile engine, and his engine is now running at an efficiency of 117 percent. What would you tell him?

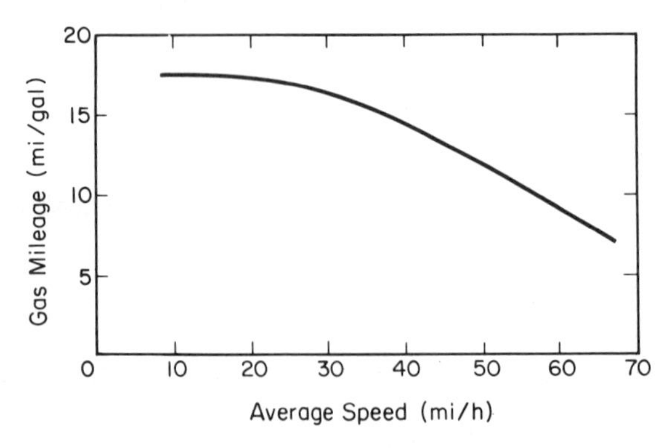

Fig. 2-14. Average speed (mi/h).

8. Figure 2-14 shows the gasoline mileage delivered by a certain automobile as a function of the average speed at which it is driven. Is mileage an increasing or decreasing function of speed? What kind of mileage can be expected at 30 miles per hour? At 60 mph? If the driver says he consistently gets about 16 miles per gallon, at what average speed would you say he drives?

3

OF CATS AND KITES AND SEALING WAX

3-1 THE PATCHWORK QUILT OF ELECTRICAL SCIENCE

"The time has come," the walrus said,
"To talk of many things:
Of shoes—and ships—and sealing wax—
Of cabbages—and kings—"

So run the most often-quoted lines of Lewis Carroll's "The Walrus and the Carpenter." The Walrus might well have been addressing us on the history of electronic science. Progress in electronics, like every other branch of science, is the happy combination of the fruits of directed research and pure accident. The pace of progress becomes ever more brisk, as each new discovery elucidates several already known phenomena while simultaneously pointing out many new areas of investigation. The history of electronics spans perhaps millions of years, and men of every race and nationality have been among its contributors.

3–2 FOREVER AMBER

The first recognition of electricity, as a distinct phenomenon in nature, is lost in antiquity. Perhaps somewhere in the Pleistocene period (i.e., in the middle part of the last two million years or so), some unnamed and forgotten *Homo erectus*[1] was the first manlike creature to be frightened out of his few wits by a bolt of lightning striking a nearby tree. For the greater part of man's history, lightning was by and large his only electrical experience. To be sure, in certain latitudes, ancient man observed the *aurora borealis* (the Great Northern Lights) and its southern counterpart, the *aurora australis,* but this was not a universal experience. Later, particularly when man became the sailor, he occasionally noted an eerie glow about the tops of tall spires, such as a ship's mast, which came to be called St. Elmo's fire. These phenomena, however, were beyond the control of man and most ancient people believed them to be of divine rather than natural origin.

It is the ancient Greeks who must be credited with the first recorded experiments in electricity. They found that when a piece of amber (the fossilized resin from an ancient species of tree, translucent and golden in color) was rubbed briskly, it exhibited an attraction for tiny bits of light material. Almost every child at some time discovers that a comb run through his hair will pick up bits of paper, which is another instance of the same effect. The ancient Greeks, however, had neither hard rubber nor polystyrene combs; therefore they believed that this mysterious attractive force could only be generated in amber. Although the Romans apparently knew of at least one other substance, a kind of lignite, in which electric forces could be generated by friction, this fact does not seem to have been common knowledge. Thus the Greek view of amber as the only electrifiable substance seems to have persisted until the late sixteenth century.

William Gilbert experimented with the electrification by friction of various substances. He showed that amber was not the only such material and, circa 1600, wrote a book discussing the substances with which he had experimented. From the latin word *electrum,* which is, in turn, derived from the Greek word ἤλεκτρου (*elektron*) meaning amber, Gilbert coined the word *electrics* to classify the substances that could be electrified by rubbing. For this work Gilbert is regarded as the "Father of Electricity." Note that the ancient Greek concept of the uniqueness of amber is thus, albeit unwittingly, perpetuated in the modern words electricity, electron, electronics, and all related words beginning with electr____.

[1] Earlier species of man, preceding *Homo sapiens.*

3-3 THE BEGINNINGS OF ELECTRICAL THEORY

The ancient Greeks, in the person of Thales of Miletus, hypothesized that the charged amber has a sort of "spirit" or "soul" and explained the observed attraction for bits of certain substances on the basis of amber's "likes" and "dislikes."

In 1551 Girolamo Cardan, an Italian physician, was the first to distinguish between the attractions exhibited by amber and by lodestone, thus separating the sciences of electricity and magnetism.

In 1733 a Frenchman named Charles François de Cisternay DuFay discovered that there were apparently *two* kinds of electrification, which he called vitreous and resinous, and which have subsequently come to be called *positive* and *negative*, respectively. In those days it was common scientific practice to assume the existence of "subtle fluids" to explain physical phenomena. DuFay postulated that these electric fluids permeated matter and were of two types, corresponding to vitreous and resinous electrification. When mixed in equal amounts, they were supposed to neutralize each other, but they could evidently be separated by friction (rubbing). He also discovered that *any* substance could be electrified if it were isolated on a glass stand.

Benjamin Franklin, a notable figure in the political arena of colonial America, was also an extremely competent scientist. Quite independently, he duplicated DuFay's discovery.

3-4 AN EXPERIMENT WITH CHARGE

The essentials of Franklin's classical experiment are illustrated in Fig. 3-1. A glass rod was rubbed with a piece of silk, as shown in Fig. 3-1 *a*, and brought near a pith ball suspended by a thread. (Pith is the soft, spongy, woody substance in the core of plant stems. When dry, it has a consistency not unlike cork.) Immediately, the pith ball was attracted to the rod and swung toward it, as in Fig. 3-1*b*. Upon contacting the rod, however, the pith ball apparently "changed its mind" and was repelled by the glass rod as shown in Fig. 3-1*c*. This experiment was repeated using a bar of sealing wax that had been rubbed on a cat's back instead of a glass rod rubbed with silk cloth. As shown in Fig. 3-1*d*, *e*, and *f*, the pith ball behaved just as it did under the influence of the glass rod. However, when the pith ball, already influenced by contact with the fur-rubbed sealing wax, was approached by a silk-rubbed glass rod, it was also attracted, as shown in Fig. 3-1*g*.

What is the significance of this experiment? In the first place, it shows that ordinarily neutral objects, like glass rods and bars of sealing wax,

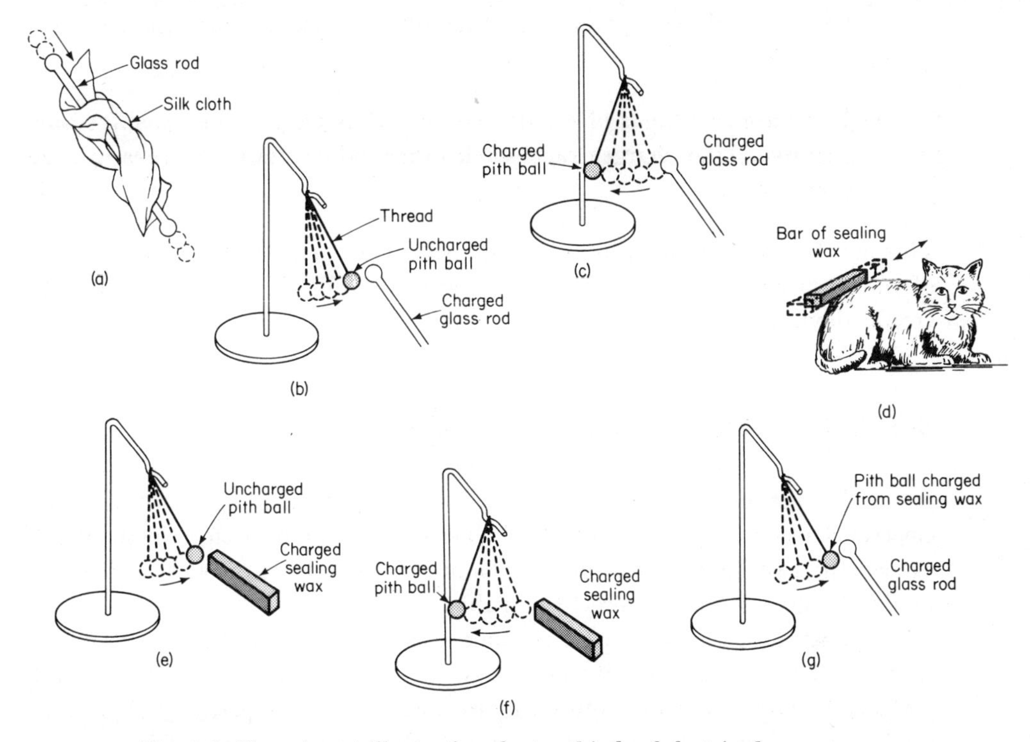

Fig. 3-1. Experiment illustrating the two kinds of electric charge.

can be electrically charged by rubbing. Secondly, we learn that a charged body can then influence a neutral one, either at a distance (with one effect) or on contact (with the opposite effect). Note that the neutral pith ball was attracted by both the glass rod and the sealing wax, but that after contact it was repelled in both cases. Thirdly, from the foregoing points, plus the observation that a pith ball charged by contact with charged wax is attracted to a charged glass rod, we learn that there are two kinds of charges. Unlike charges evidently attract each other, while like charges repel each other. Thus, when the glass rod was rubbed with the silk, it acquired one kind of charge (positive). When brought near to (but not touching) the pith ball, it somehow caused (we will hereafter say *induced*) the opposite kind of charge (negative) on the pith ball, and the resulting attraction caused the ball to move over and touch the rod. At the instant of contact, some of the rod's charge was transferred to the pith ball so that both then had the same kind of charge, resulting in a repulsive force that causes them to fly apart. The same behavior is observed for a neutral pith ball and a charged bar of sealing wax, but the attraction between the pith ball charged from fur-rubbed sealing wax and a silk-rubbed glass rod indicates that the charge on the wax is opposite (i.e. negative) to the charge on the glass.

From the observed results of this experiment, we state the *first law of electricity:*

> **There are two kinds of electric charge, arbitrarily termed positive and negative. Like charges repel each other, while unlike charges attract each other.**

This fact has been verified from the microcosm of the atom to the macrocosm of outer space and is essential to an understanding of electrical and electronic devices.

3-5 THE "SINGLE FLUID" THEORY

The reader must be aware that although it was clear that two kinds of electric charge exist, it was not at all clear to those pioneer electricians what caused the two kinds of charge. It was originally believed that two distinct kinds of "electric fluid" were responsible.

A major step forward was made when Ben Franklin postulated a *single fluid theory* in 1747. He proposed the existence of "electric fire" in all matter. A body that had acquired more than its normal quantity of electric fire was thus *positively* charged, while one that had lost some of its normal amount was *negatively* charged.

Franklin's theory was enlarged upon by a German scholar, Franz Ulrich Theodor Aepinus, born in Rostock, Saxony, in 1724. Aepinus took up residence in St. Petersburg as a member of the Imperial Academy of Sciences and professor of physics, devoting himself to research until his retirement in 1798. In 1759 he published his principal work on electricity and magnetism, entitled *Tentamen Theoriae Electricitatis et Magnetismi*. In it Aepinus writes that

> . . . . the particles of electric fluid repel each other and attract and are attracted by the particles of all bodies with a force that decreases in proportion as the distance increases; the electric fluid exists in the pores of bodies; it moves unobstructedly through nonelectric [i.e., conductors] but moves with difficulty in insulators; the manifestations of electricity are due to unequal distribution of the fluid in body, or to the approach of bodies unequally charged with the fluid . . . .

Aepinus is also credited with being the first to realize the interrelationship between electricity and magnetism.

3-6 ELECTROSTATICS, CHARGE, AND COULOMB'S LAW

The electrical charges we have been discussing are called *static* electric charges, or simply *static electricity*. The word static means "at rest" and refers to the fact that the position of the charges in space does not change with time. The study of such charges and the laws governing them constitutes the field of *electrostatics*.

For a long time the study of electrostatics remained purely qualitative because the equipment necessary for making systematic electrical measurements was not available. The earliest investigators of the quantitative aspects of electrostatics were the English scientist Henry Cavendish and the French physicist Charles Coulomb. Cavendish used his body as a measuring device. He is alleged to have become extremely proficient in assigning a numerical value to a quantity of electricity based on the quality of the shock he received from it. He was the first to discover the inverse square law, to be discussed presently, but this fact was not known until after his death.

Coulomb was a military engineer who, having retired for reasons of ill health and not wanting to become entangled in the seething politics of pre-Revolutionary France, devoted himself to scientific research. He invented the *torsion balance*, a sensitive force-measuring device in which the force to be measured is opposed by the torsion (*twist*) of a wire. Using this device, he was able to establish that the magnitude of the force of repulsion (or attraction) between two like (or unlike) charges depends on three factors: (1) the size of the charges, (2) the distance separating them, and (3) the nature of the material separating them; the direction of the force depends, of course, on whether they are like or unlike charges, as illustrated in Fig. 3-2.

Historically, several units have been used to describe a quantity of electric charge, and we need not go into all of them. The unit of electric charge most commonly employed today is that of the so-called "practical system" of electrical units, and it is called the *coulomb* in honor of the French virtuoso. For the time being, it is sufficient to say that a coulomb is approximately the quantity of electrical charge that flows through an ordinary (110 volt) household hundred-watt bulb in one second. Thus we may speak of so and so many coulombs of electricity analogous to the way we would speak of so and so many pounds of butter.

Charles Coulomb found that the force between two charges is proportional, independently, to the size of both charges. An example of this fact is shown in the graphs of Fig. 3-3. If the first charge is 1 coulomb (lower graph), the magnitude of force acting on both charges is: 1 pound, if the

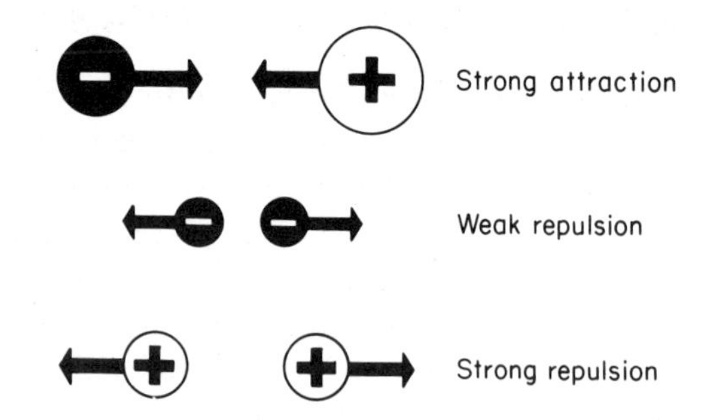

(a) Effect of size and kind of charge.

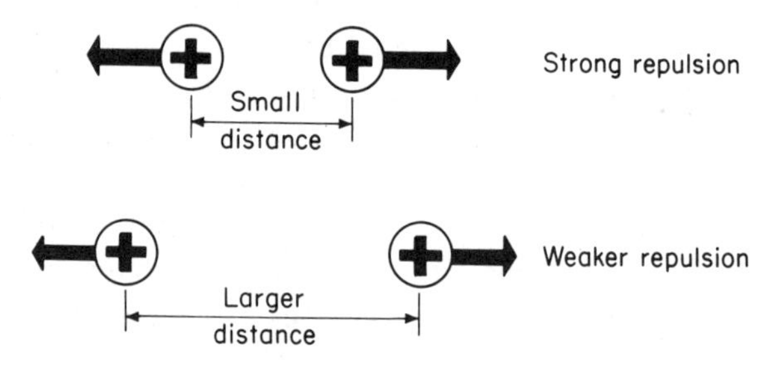

(b) Effect of separation distance.

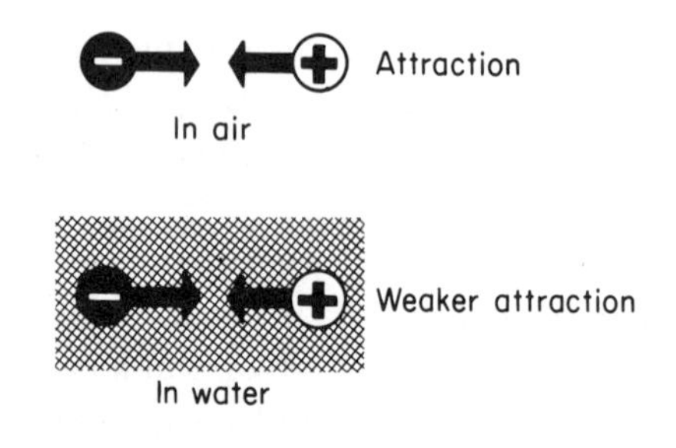

(c) Effect of separating material (medium).

Fig. 3-2. Factors affecting the force between two charges.

second charge is also 1 coulomb; 2 pounds, if the second charge is 2 coulombs; 3 pounds, if the second charge is 3 coulombs, etc. If the first charge is increased to 2 coulombs (middle graph), all these forces are doubled, and if the first charge is tripled (upper graph), these forces are similarly tripled. Mathematically, the effect of the sizes of both charges is represented by the product (i.e., multiplying together) of their values in coulombs.

The effect of the distance separating the charges might be anticipated. We expect that increasing the distance should result in a weakening of the

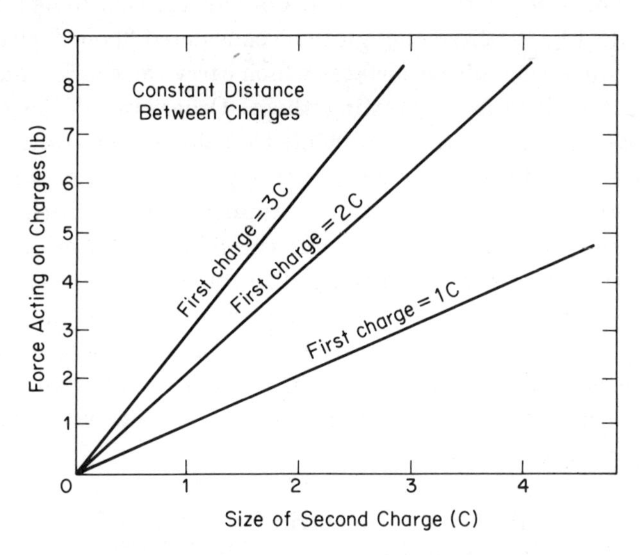

Fig. 3-3. Dependence of electrostatic force upon charge.

force, and indeed it does. The magnitude of the force, irrespective of its direction, decreases very rapidly with increasing distance, as shown in Fig. 3-4. If the charges are such that the force is 1 pound at 1 foot separation, then it is one-fourth ($= \frac{1}{2 \times 2}$) pound at 2 feet, one-ninth ($= \frac{1}{3 \times 3}$) pound at 3 feet, one-sixteenth ($= \frac{1}{4 \times 4}$) pound at 4 feet, etc. A decreasing function of this kind is called an *inverse square* relationship; hence the force is an inverse square function of distance. We will encounter another inverse square

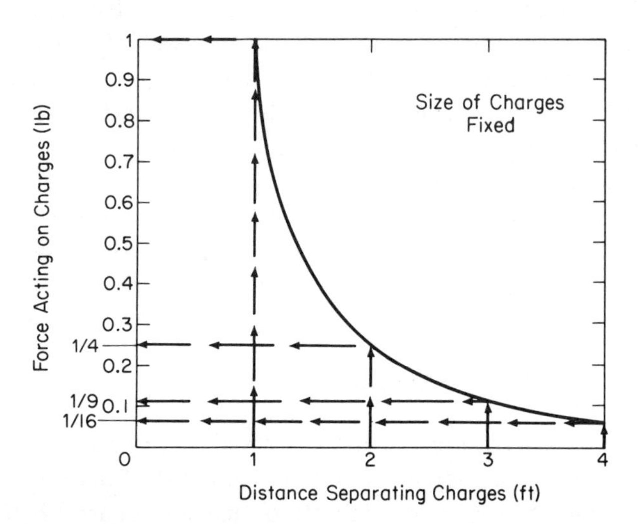

Fig. 3-4. Dependence of electrostatic force upon distance.

law later in connection with magnetism. For this relation to be strictly true, the charges must be localized enough to be considered "point" charges—that is, the largest dimension of the surfaces which carry the charge must be small compared to the distance separating them. Otherwise, if the charges are spread out over a large area, it is difficult to define the separating distance.

Finally, the effect of the substance separating the charges, which is called the *medium*, is characterized by a number called the *dielectric constant* of that medium. This number represents the shielding effect of the medium compared with empty space (vacuum). The force acting on two charges in any given medium is equal to the force that would exist if the charges were in a vacuum, divided by the dielectric constant of that medium. Some representative dielectric constants are given in Table 3-1. Note that the dielectric constant of air at ordinary atmospheric conditions is not very different from that of empty space.

Table 3–1 Dielectric Constants of Various Substances at 20 °C (68 °F)

Air	1.0006	Olive oil	3.11
Ammonia (liquid)	15.5	Paraffin	2.0–2.5
Asphalt	2.68	Polyethylene	2.3
Beeswax	2.75–3.0	Porcelain	6–8
Glass, Pyrex	3.8–6.0	Rubber	2.8
Glass, Corning 8870	9.5	Sulfur	4.0
Mica	7–9	(Vacuum)	1
Nylon	3.5	Water	80

In all the foregoing examples, we treated the effect of one variable at a time on the electrostatic force between charges and left the other variables fixed at some unspecified values that made the forces come out at reasonable numbers. Although this was a good practice for illustrating the points being made, it probably left the reader without a true appreciation of the tremendous strength of electrostatic forces. It is therefore well to point out that two point (contained in a small volume) charges of one coulomb each, having the same sign (i.e., both positive or both negative) and separated by a distance of one foot, repel each other with a force of over *ten million tons*!

3–7 CURRENT AND POTENTIAL

Although electrostatics concerns itself with charges at rest, some terminology relating to charges in motion is not out of place here. According to the single fluid theory, in order for an electrically neutral body to become a charged body, some electric charge had to be moved either to or away from that body. The *rate* at which charge is transported past a given point is called the *current* at that point. Current is, in other words, how much electricity *per unit time* passes a given point. The unit of current represents the transport of one coulomb of electricity in one second and is called one *ampere*, after another distinguished Frenchman. The reader is cautioned to remember that although not evident from the name, an ampere is a rate of one coulomb per second.

Another concept of great value is that of *electrostatic potential*, or simply *potential*. In everyday speech, when we speak of a person's potential we mean his capacity or capability to accomplish something good and useful, whether he achieves it or not. Similarly, in electrical terminology, potential is defined as the work done upon a unit of charge to move it to where it is located, and therefore the useful work that could be done by that unit charge if it were allowed to return (whether achieved or not). Since work is measured in joules, the unit of potential is *one joule per coulomb*, which is called one *volt*, in honor of Alessandro Volta. Potential is, in most electrostatic situations, considerably easier to measure than charge. Yet charge is simply related to potential; hence the latter is an extremely useful parameter.

The potential existing at any given point is of significance only with relation to that at some other point. Therefore we speak of the *potential difference* between points A and B which represents the amount of work done on (or by) a unit of charge in going from point A to point B. Since this potential difference is measured in volts, it is often called *voltage* as well, and we shall frequently use this shorter terminology although it is less precise. Furthermore, since it is this potential difference that will cause charges to move, it is also called *electromotive force*, or *emf* for short. This name is misleading, for it really represents work per unit charge and *not* force per unit charge. Nevertheless, emf is also established terminology. Thus the reader will have to bear in mind that **potential difference = electromotive "force" = voltage = work/unit charge.**

3–8 CONDUCTORS AND NONCONDUCTORS

Early in the history of electrical science, it was discovered that electricity would apparently pass through certain substances like water through a sieve. These substances, notably metals, are called *conductors*. Du Fay had discovered that the most easily electrified (by friction) substances were the poorest conductors. The reason is that when opposite charges are connected by a conductor, the charges will move (i.e., there will be a current in the conductor) to cause neutralization. If the same connection were made with other substances (not conductors), there would be little motion of the existing charges and so these substances are termed *insulators*. Stephen Gray, in 1729, first evolved the concept of conductors and nonconductors, which led to the discovery of electrical insulation. This terminology supplanted Gilbert's classification as electrics and nonelectrics.

There is no sharp line between conductors and insulators, but rather a varying degree of conduction for all substances from very good to very poor. More is said on this subject in Chapter 4.

3-9 THE ELECTRIC FIELD

The way in which a charge acted to produce a force on another charge at a distance was quite mysterious. Attempts to reformulate the problem led to the concept of an *electric field* around every charge.

If, in a given region of space, one can insert a test device at *every* point and determine both a magnitude and a direction of some vector (see Chapter 2, Section 2-7) quantity P, a P-field may be said to exist in that region. It is to be noted that the test device must not interfere with the value of the parameter it is intended to measure. In the case of electrostatics, the agreed-upon test device is a very, very small positive charge. Placed near a positive charge, the little test charge will be repelled away along the line connecting the two; and placed near a negative charge, the test charge will be attracted along the line connecting them. A "map" showing the strength and direction of the field all around an isolated positive or negative charge is given in Fig. 3-5. The direction of the field is represented by the direction

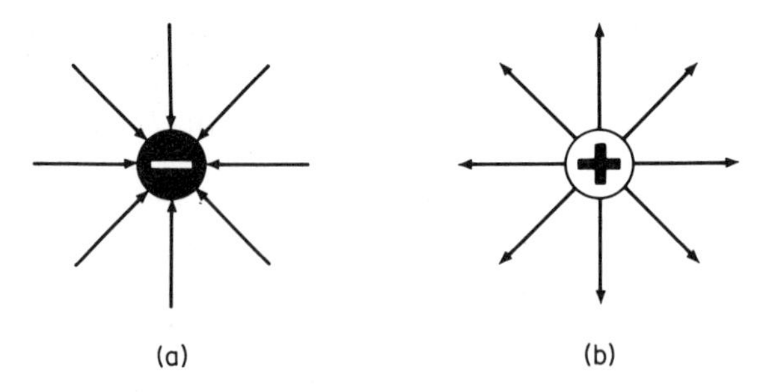

Fig. 3-5. Electric field around an isolated charge. (a) Negative charge. (b) Positive charge.

of the lines, and the strength of the field by the density of the lines (i.e., how close they are to each other). From Fig. 3-5 we see that the field around a negative charge is directed radially inward and that the field around a positive charge is radially outward. Of course, these radial fields must be imagined in three dimensions rather than just in the plane of the paper as in Fig. 3-5.

The magnitude associated with the field at each point is called either the *electric field intensity* or the *electric field strength* by various writers. In this text, the term *intensity* will be used exclusively. The value of the electric field intensity is the force that would be experienced by a unit positive charge (i.e., a positive charge of one coulomb) placed at that point in the field *if it did not alter the existing field.* Of course, a real one-coulomb charge would affect any field, so experimentally one would use a very small positive charge, as indicated previously, and calculate what the force would be on a one-coulomb charge. The natural unit of electric field intensity is therefore force per unit charge or newtons/coulomb. It must be noted that the unit newtons/coulomb is exactly equivalent to the unit volts/meter, and it is this latter unit that is most frequently used.

Having mapped the electric field around isolated charges, we now consider the field around two charges in close proximity. Figure 3-6 shows a map of the field around two equal unlike charges. Similarly, Fig. 3-7 shows the field around two equal like (positive) charges. Remembering that a little

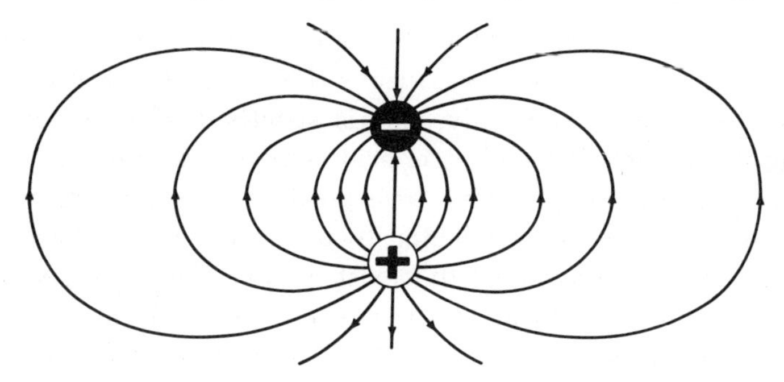

Fig. 3-6. Electric field around two equal opposite charges.

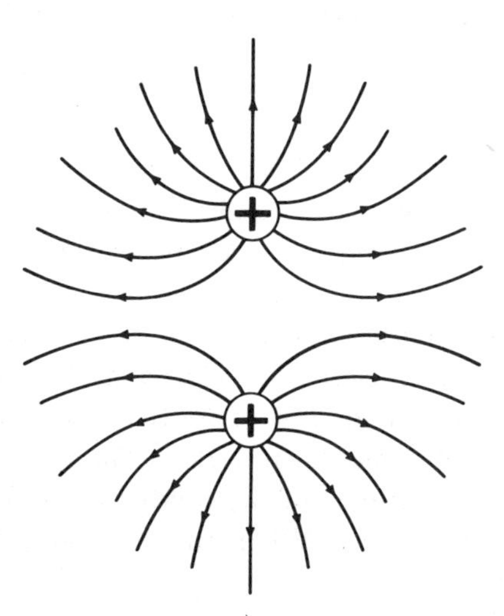

Fig. 3-7. Electric field around two equal positive charges.

positive test charge will be repelled by a positive charge and attracted by a negative charge along the line joining them, the reader should be able to convince himself that the lines drawn in Fig. 3-6 and 3-7 do in fact show the direction of the field. It is left as an exercise for the reader to draw the field around two equal negative charges.

It is convenient to treat those lines in Fig. 3-5 through 3-7 as though they represented a reality—that is, we speak of *electric lines of force* that emanate from positive charges and terminate on negative charges. Whether or not such lines of force do "really exist" is a moot point and irrelevant to the fact that they are a useful concept in understanding electronic devices. Two important properties of electric lines of force are

1. They never cross (intersect) because the field cannot have two directions at the same time.
2. Near the surface of a metal conductor, they must be perpendicular to its surface.

There is no electric field within a conductor—that is, the electric field intensity is everywhere zero within the conductor *under electrostatic conditions* (no current). (Actually, this is strictly true only for a perfect conductor, and a very feeble electric field can be maintained in a real conductor, but the statement is true to a good degree of approximation for metals.) The reason is that a conductor, by definition, allows charges to move easily. Thus, if an electric field did exist at any time, charges under the influence of the field would move to neutralize themselves, thereby annihilating the field. Consequently, electric lines of force always terminate at the surface of a conductor.

3-10 ELECTROSCOPE TO ELECTROMETER

One of the first instruments used to indicate the presence of an electric charge was the *electroscope*. The first electroscope, devised by William Gilbert, consisted simply of a light needle mounted on an insulated pivot, as shown in Fig. 3-8*a*. Either end of the needle was attracted by a charged body. The next step was to make use of the repulsion of similarly charged bodies. Thread, straw, and pith balls were used for this purpose with excellent results, but the most successful and widely used design employed gold leaves and was introduced by Abraham Bennett in 1787. The gold leaf electroscope is shown in Fig. 3-8*b*. It consists of a glass jar into which an L-shaped conductor has been sealed. The gold leaves are cemented at one end to the horizontal part of the conductor, and they hang freely from that end alongside each other. When the knob atop the conductor is influenced

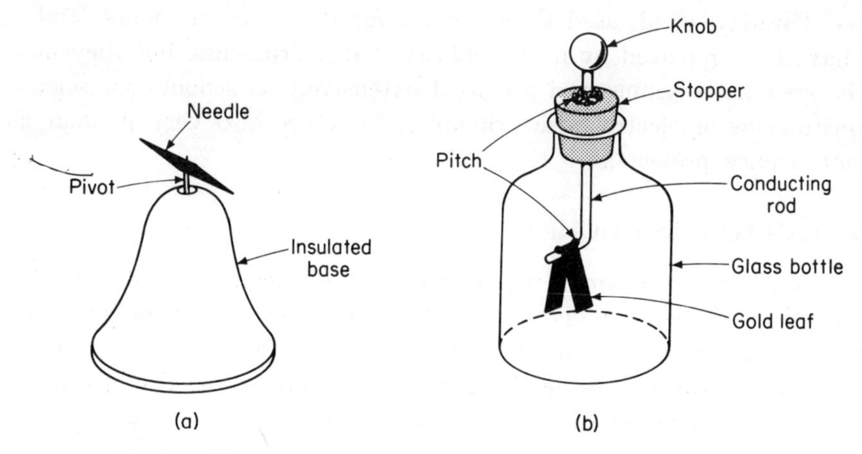

Fig. 3-8. Electroscopes. (a) Gilbert's needle. (b) Gold leaf electroscope.

by an electric charge, charge is conducted down the rod to the gold leaves, which, being similarly charged, fly apart. The angle at which the leaves are separated represents the balance achieved between the electrostatic force of repulsion and the force of gravity (weight of the leaves).

The next successful innovation was to replace one of the leaves with a rigid blade, actually part of the conducting rod, and to add a numerical scale against which the deflection of the movable leaf could be read. This device, shown in Fig. 3-9, is now called an *electrometer* and can be used as a quantitative measuring instrument. Our early experimenters, Franklin,

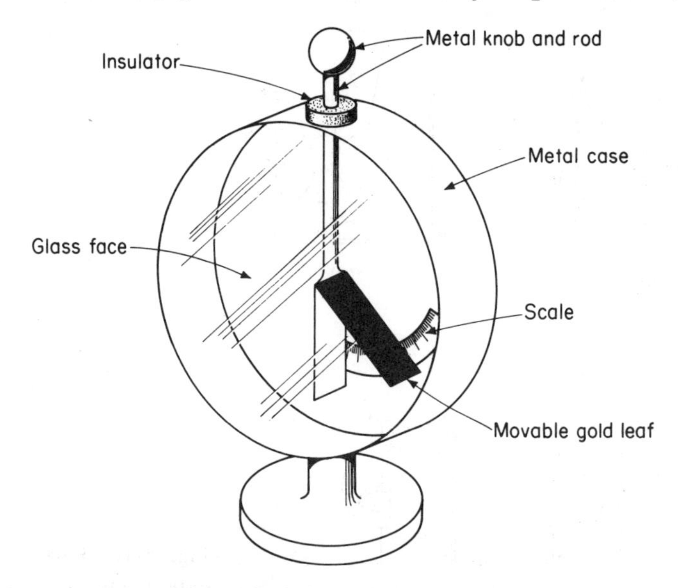

Fig. 3-9. Gold leaf electrometer.

DuFay, Faraday, et al. used these devices for their measurements. Today they have been replaced by more sophisticated instruments, but they may still be seen in museums and are used extensively in schoolroom science demonstrations of electrostatic principles. They are also very popular as student science projects.

3-11 FARADAY'S BUCKET

We shall now examine the distribution of electric charges at each step in an experiment involving metallic conductors, this time using an electrometer to indicate the level of charge. Michael Faraday performed this experiment in 1843 with an ice bucket and an electroscope. The operations are illustrated schematically in Fig. 3-10.

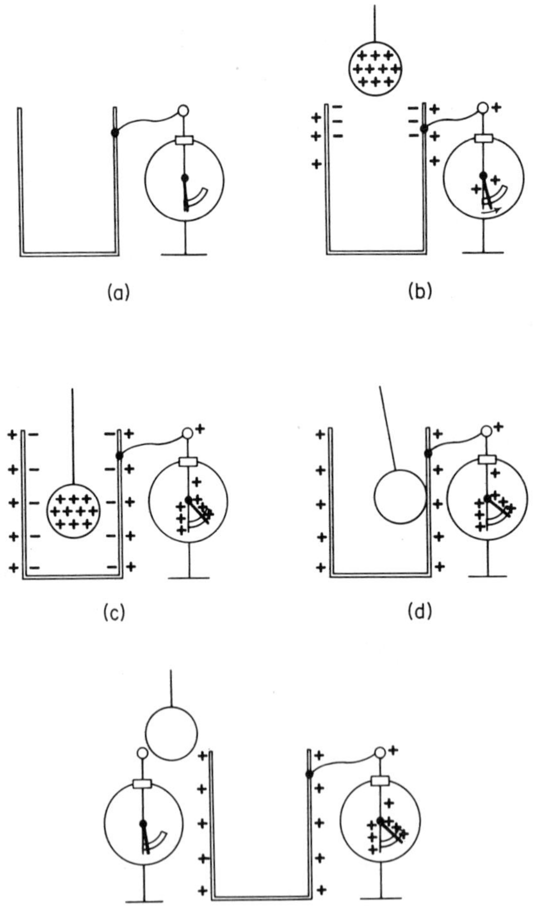

Fig. 3-10. Faraday's bucket experiment.

The experiment begins with a metal cylinder connected to the electrometer, which indicates an uncharged condition, as shown in Fig. 3-10*a*. As a positively charged metal ball suspended from a silk thread is brought near the opening in the cylinder (see Fig. 3-10*b*), negative charges in the metal of the cylinder are attracted and migrate toward the area nearest the charged ball and the electrometer begins to deflect. When the charged ball is entirely contained within the cylinder (Figure 3-10*c*), a negative charge exists on the inside surface of the cylinder, equal to the positive charge on the ball. Thus there is also an equal positive charge on the outside surface of the cylinder, as indicated by the deflection of the electrometer. When the charged ball is allowed to touch the wall of the cylinder, as in Fig. 3-10 *d*, the internal negative charge on the cylinder exactly neutralizes the positive charge on the ball, leaving the ball uncharged and the original positive charge now on the cylinder. This last situation is illustrated in Fig. 3-10*e*, where a second electrometer shows the ball to be indeed neutral. This experiment demonstrates two points:

1. Placing a charged object inside a neutral conductor induces an equal opposite charge on the inside surface of the conductor, and an equal charge of like sign on its outside surface.
2. All of the charge on a conductor is concentrated at its surface.

3–12 CONDUCTION VERSUS INDUCTION

Let us now take a closer look at the two processes by which a charge can be placed upon an insulated neutral conductor by the influence of a previously charged body. The first process involves contact between the two bodies, and part of the charge on the originally charged body is transferred directly to the neutral conductor, as shown in Fig. 3-11. In this case the neutral conductor was *charged by conduction.*

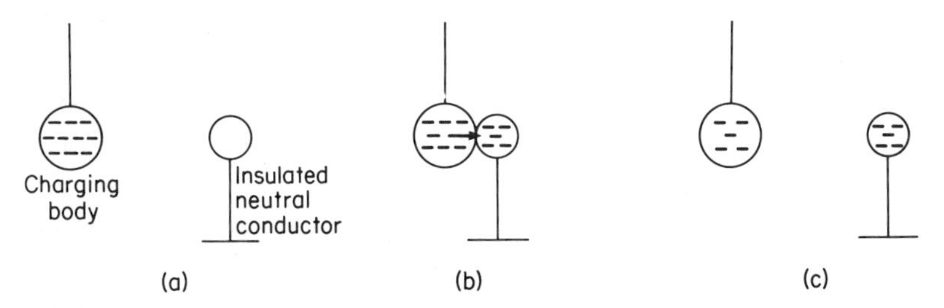

Fig. 3-11. Charging by conduction. (a) Before charging. (b) Negative charges transferred on contact. (c) After charging.

To understand the second process, we must first understand what is meant by *grounding*. Relative to the size of most objects that man uses in his experiments, our planet Earth may be considered effectively infinite in extent. It has an (almost) infinite capacity to "absorb" electric charge without itself showing any (significant) charge. Thus, if we provide a conducting path from a charged object to *ground* (i.e., to earth), the charged object will "lose" its charge to earth. We indicate grounding, that is, providing a path to ground, by the symbol ⏚.

Now, let us return to the second charging process, shown schematically in Fig. 3-12. In this case, the charged body is merely brought near the

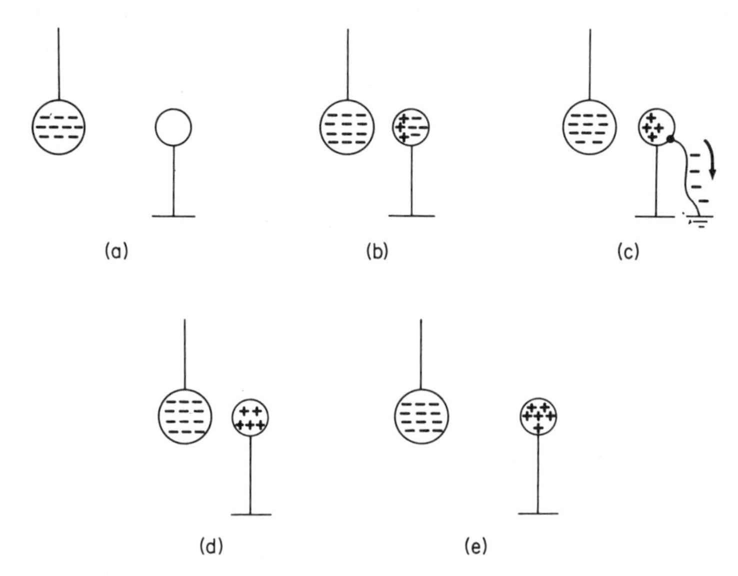

Fig. 3-12. Charging by induction. (a) Before charging. (b) Charging body brought near neutral conductor. (c) Neutral conductor momentarily grounded. (d) Ground path removed. (e) After charging.

neutral conductor. In so doing, as in the Faraday bucket experiment, the like charges (here shown as negative) within the neutral conductor are repelled to the side opposite the charged body. If the conductor is now momentarily grounded, the like charges within it have the opportunity to escape even further from the influence of the charged body, all the way to the earth. Upon removing the ground path, we find that the insulated conductor is no longer neutral but has acquired a charge opposite to that of the charging body, and it retains this charge when the charging body is removed. A conductor charged in this way is said to have been *charged by electrostatic induction*.

Referring back to Fig. 3-11 and 3-12, note that a conductor charged by conduction receives a charge of the same sign as that of the charging body, while a conductor charged by induction is charged opposite to the charging body.

3–13 A JUG OF ELECTRICITY

Having discovered that substances could be electrified by friction, the early electrical scientist set about finding more effective ways of producing "bigger and better" charges. Electrostatic machines that would produce impressive spark discharges were developed. Yet electricity would have remained a curiosity had a means not been found to store up a charge for some time and deliver it at will: this means was the *Leyden jar*. It permitted the electrician to "can" electric charge just as the housewife cans fruits and pickles.

The Leyden jar principle was discovered accidentally by both Professor P. van Musschenbroek at the University of Leyden in the Netherlands and by E. G. von Kleist of Kammin, Pomerania (Germany), in 1745. Von Kleist's story of the discovery was not well documented, so the Dutchman was originally given the only credit. The term Leyden jar first appears in the writings of the Abbé Nollet, whose name we shall encounter again in Chapter 5.

The original form of the device, as described by van Musschenbroek and shown in Fig. 3-13*a*, consisted simply of a vial (narrow bottle) containing water, which was closed with a cork through which a wire or nail protruded into the water. The experimenter held the vial with the protruding nail touching the main conductor of some charging device, such as one of the previously mentioned electrostatic machines. His body, of course, served as a grounding path. The water, not being perfectly pure, served as a conductor and distributed the charge over its outer surface, at the inside surface of the vial. On the other side of the glass, an equal opposite charge was induced on the surface of the other conductor (the experimenter's hand!). When the experimenter touched the protruding nail, he received a considerable shock. Van Musschenbroek described the intensity ". . . as if struck by lightning." (We doubt if he really had any first-hand experience of this nature to compare with, but scientists of those days were much more picturesque in their writings.)

The final form of the Leyden jar, as such, is shown in Fig. 3-13*b*. Here the water is replaced by a metal foil coating on the inner surface of the jar, connected by a wire, chain, or other flexible conductor to the center rod. The experimenter's hand is similarly replaced by a metal foil coating of the outer surface of the jar.

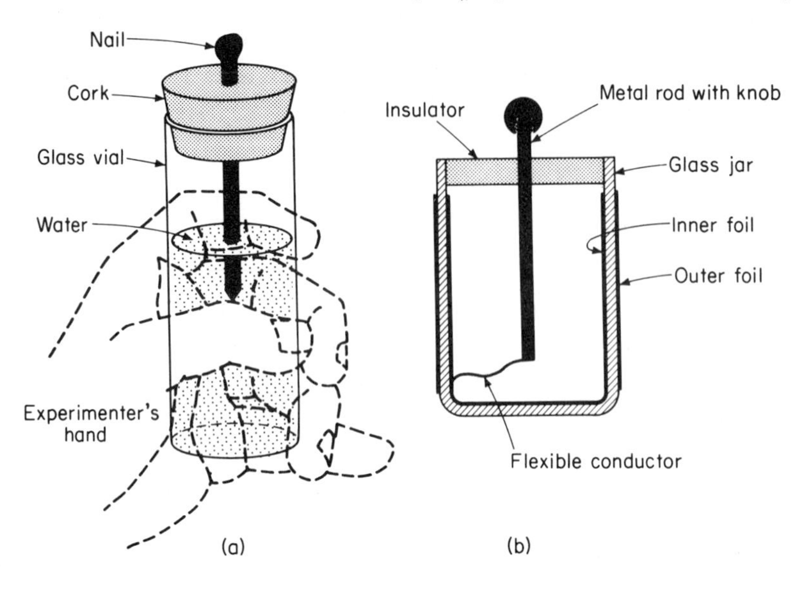

Fig. 3-13. The Leyden jar. (a) Original version. (b) Final configuration (cutaway view).

Like the electroscope, the Leyden jar is now a mere museum piece and schoolroom demonstration device. Unlike the electroscope, however, a modern version of the Leyden jar is in widespread use today, and without which most modern electronic devices could not function—the *capacitor.* The capacitor (formerly called the electrical condenser) is in reality nothing more than an efficiently designed Leyden jar that will fit an equivalent charge-storing capability into a much smaller space. As shown in Fig. 3-14, the capacitor consists of two conductors of comparatively large surface area called the *plates,* separated by an insulator called the *dielectric.* The capacitor may actually be built in several ways, such as in Fig. 3-15. If the required plate size is small, the plates may be of sheet metal bonded to opposite sides of a thin sheet of some suitable dielectric, such as mica or a ceramic. If a large plate size is required, the plates will be of metal foil in long, thin strips that are separated by a thin flexible dielectric, such as paper or mylar, and this "long, skinny sandwich" is rolled up into a tubular capacitor. If being able to vary the opposing plate area at will is desirable, the capacitor is built in the form of a frame with each "plate" actually consisting of several semicircular pieces, the pieces of one plate being rigidly fixed and the pieces of the second plate being movable, in a rotary fashion, in and out from between the pieces of the first plate. In the latter case, the dielectric is usually air, and this type of capacitor is called an air-dielectric variable capacitor.

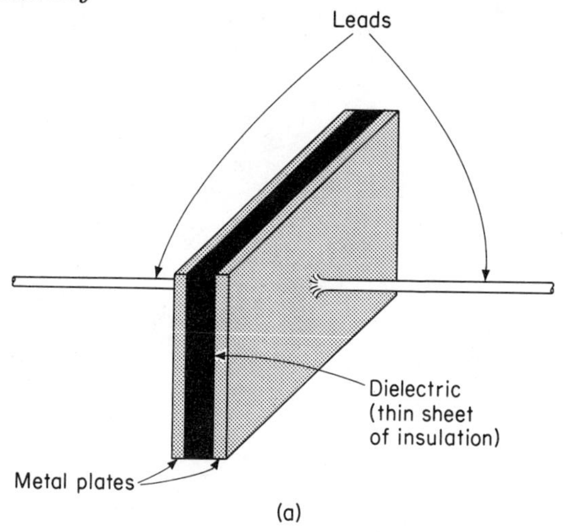

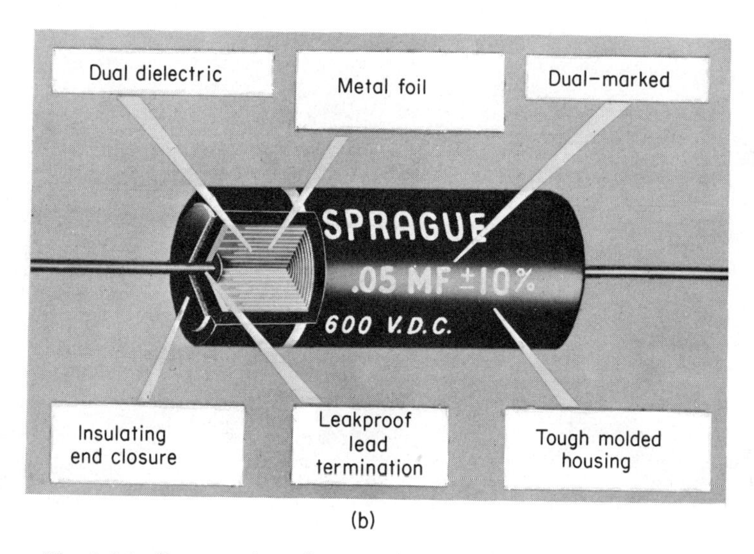

(b)

Fig. 3-14. Construction of a capacitor. (a) Conceptual. (b) Actual construction of tubular capacitor. (Courtesy of Sprague Electric Co.)

The basic quantitative law of capacitors (and also Leyden jars) is a simple one and relates the charge contained to the difference of potential between the plates. The law is

$$\textbf{(Contained charge)} = \textbf{(capacitance)} \times \textbf{(potential difference)}$$

The capacitance is a characteristic of each capacitor and depends on its size, shape, and type of dielectric. In general, the larger the plates, the *thinner* the

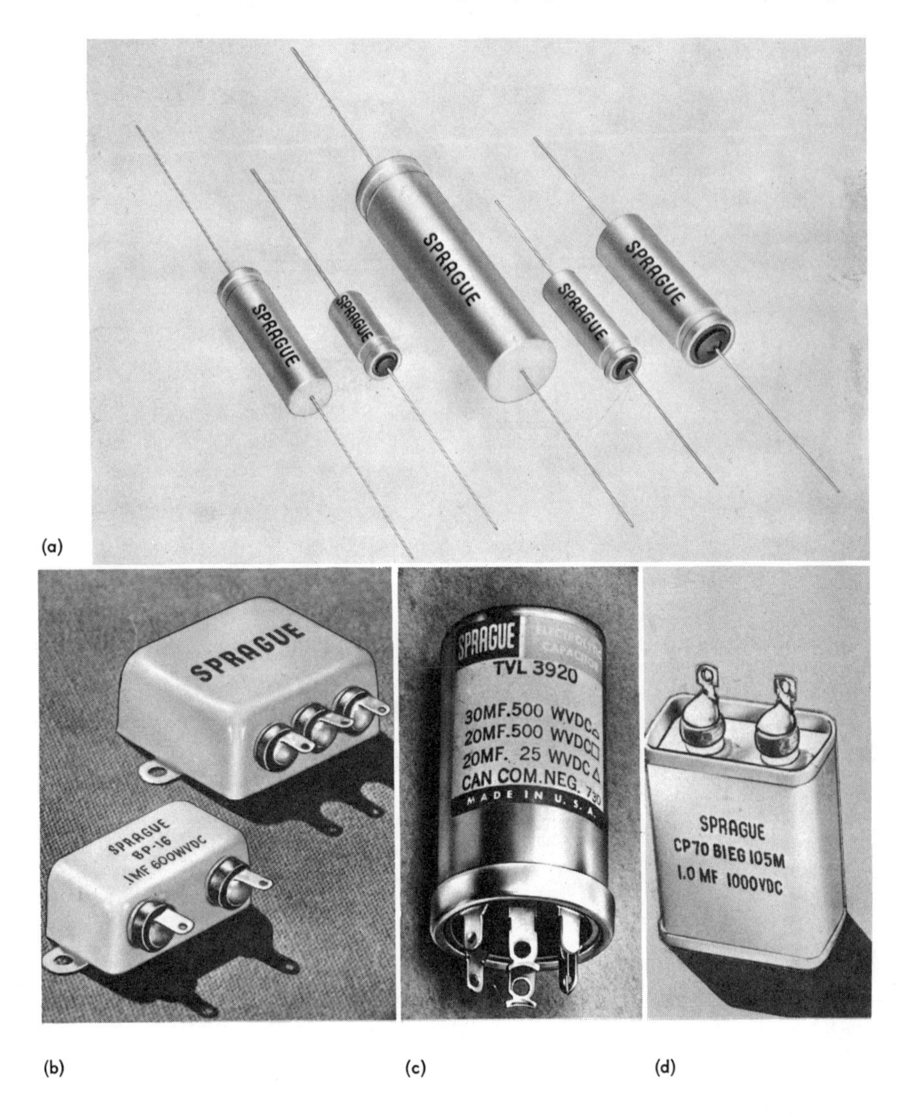

Fig. 3-15. Various types of capacitors. (a) Tubular. (b) Bathtub. (c) 3-in-1 electrolytic. (d) Oil-filled high voltage. (Courtesy of Sprague Electric Co.)

dielectric, and the higher its dielectric constant, the greater is the capacitance. The unit of capacitance is the *farad*, named after Michael Faraday. The farad is the capacitance of a capacitor that would hold a charge of one coulomb when the potential difference between its plates was one volt. A one-farad capacitor would be enormous in size! The practical unit of capacitance is the *microfarad*, which is one millionth of a farad. (*Note:* The prefix *micro* means one millionth of a . . . , or $\times 10^{-6}$. Later we shall have occasion to speak of microcoulombs, microvolts, microamperes, and other such units, when it is convenient to work with millionths of the basic unit.) Commerical capacitors are ordinarily sold in sizes ranging from as large as thousands of microfarads down to millionths of one microfarad.

The charge held by a capacitor, illustrated by the graphs of Fig. 3-16, varies with the potential difference produced across the plates, much

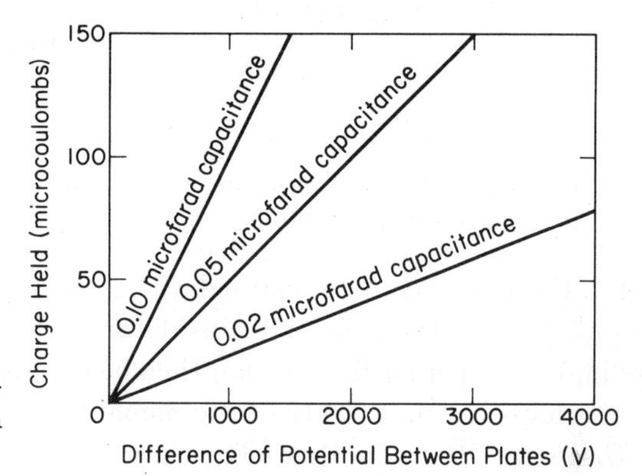

Fig. 3-16. Dependence of capacitor's charge upon capacitance and potential between plates.

as the weight of air in a basketball varies with the inflation pressure. Furthermore, just as there is a certain pressure at which the basketball will burst and hold no air, there is also an electric "pressure" (i.e., voltage) at which the dielectric will break down and the capacitor will no longer hold a charge.

As we progress through the discussion of various electronic circuits and devices, the reader will find that capacitors have universal application, perhaps second only to the resistor (to be discussed later). Their need is particularly evident whenever it is required to store a charge for a short time and then make it rapidly available for use.

3-14 ELECTROSTATIC MACHINES

As stated in the beginning of the previous section, the early electricians sought to devise machines that would produce larger charges with less effort on the part of the experimenter. The earliest models were the *frictional* machines. The principle of all frictional machines is illustrated by the manner of charging the *electrophorus*, shown in Fig. 3-17*a*. It consists of a metal cup that is filled with a disk of insulator (such as pitch, resin, or sealing wax) and is called the sole plate, together with a second metal plate to which a glass handle is attached, which is called the proof plane. The sole plate is held in the hand, and the upper surface of the insulator is rubbed with cat fur, wool,

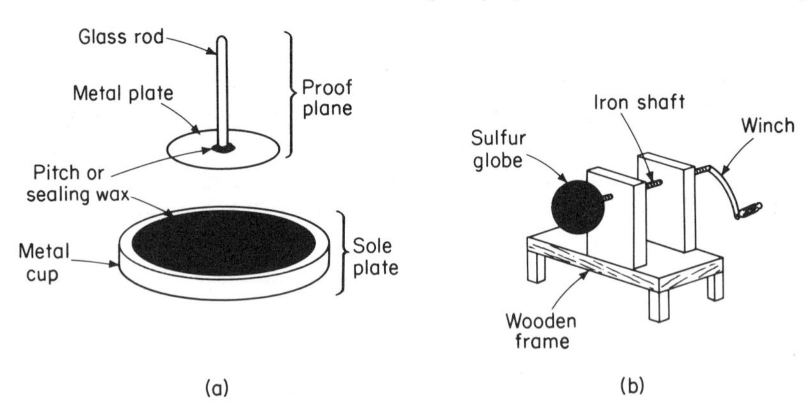

Fig. 3-17. Frictional electrostatic generators. (a) Electrophorus. (b) Von Guericke machine.

silk, etc., thereby becoming charged. [The proof plane is charged by placing it in contact with the sole plate insulator and momentarily touching (i.e., grounding) its upper surface. It can then be used to apply a charge elsewhere.]

Perhaps the first electrostatic machine was a friction machine devised by a German, Otto von Guericke, circa 1663. This device, shown in Fig. 3-17*b*, was merely a globe of sulfur mounted on an axle and rotated by a winch. The sulfur globe was electrified by holding one hand against it while the crank was turned briskly.

The frictional machine went through a brief period of evolution. A glass globe was used in place of the sulfur; leather and rubber friction pads took the place of the hand; the glass globe became a glass disk or plate. Many other refinements were added until frictional machines were developed which could produce both positive and negative charges at a potential of several hundred thousand volts. These machines would yield sparks several feet long.

The frictional machine, however, was soon supplanted by a more efficient generator of electrostatic charge, the *influence* machine. The operating principle of this type of machine is illustrated in Fig. 3-18. Consider two Leyden jars, 1 and 2, whose outer coatings are grounded and whose inner coatings carry small opposite charges, together with two suspended neutral conducting spheres, A and B, as shown in Fig. 3-18*a*. Let sphere A be brought near (but not touching) jar 1, and let B be brought near jar 2. The negative charge on 1 induces (recall Sections 3-10 and 3-11) positive charge on the side of sphere A nearest 1 and a negative charge on the remote side of A. Since jar 2 is positively charged, the reverse situation occurs on sphere B, as in Fig. 3-18*b*. Now let A and B be momentarily connected as in Fig. 3-18*c*, by a wire, called the *neutralizing conductor*, which is subsequently

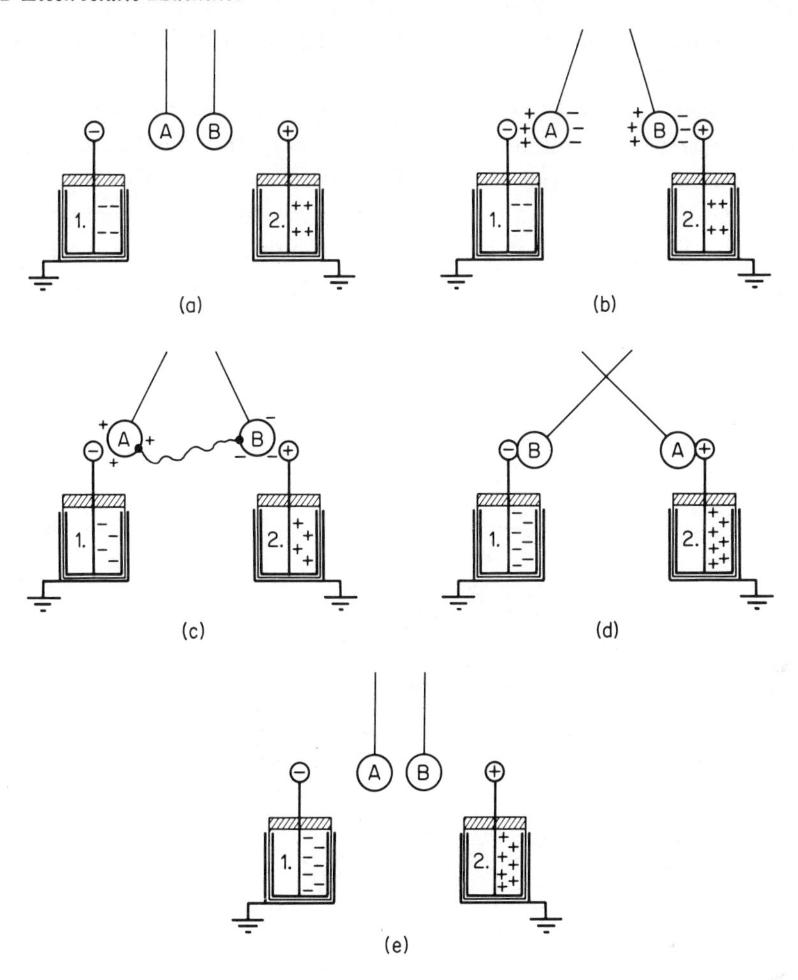

Fig. 3-18. Principle of the influence machine.

removed. The neutralizing conductor gives the negative charge from A and the positive charge from B an opportunity to neutralize each other, leaving A with a positive charge and B with a negative charge. Next B is moved so that it touches the knob of jar 1 and A is moved over to touch the knob of jar 2. This is the situation in Fig. 3-18*d*. Note that energy has to be expended (i.e., work done against the electrostatic attraction) to get B away from the vicinity of 2 and A away from the vicinity of 1. Upon contact, B gives up its negative charge to jar 1 and A gives up its positive charge to 2, as in the Faraday bucket experiment, because the capacitance of the jars is much greater than that of the spheres. Finally, with the spheres, which are now (almost) completely discharged, back in their original positions, we have the situation shown in Fig. 3-18*e*. The charges on the Leyden jars have been increased, and the cycle may be repeated again to increase the charges on the jars still further.

Machines using this principle are called influence machines because they use electrostatic "influence" (i.e., induction and conduction) rather than friction to produce the charges. Beginning with a small charge, they effectively multiply it over and over, parlaying it into as large a charge as the Leyden jars will hold without discharging through the air as a spark.

Influence machines also went through an evolution, beginning with simple devices called doublers and culminating in a machine constructed by James Wimshurst, circa 1878. This type of machine is illustrated in Fig. 3-19.

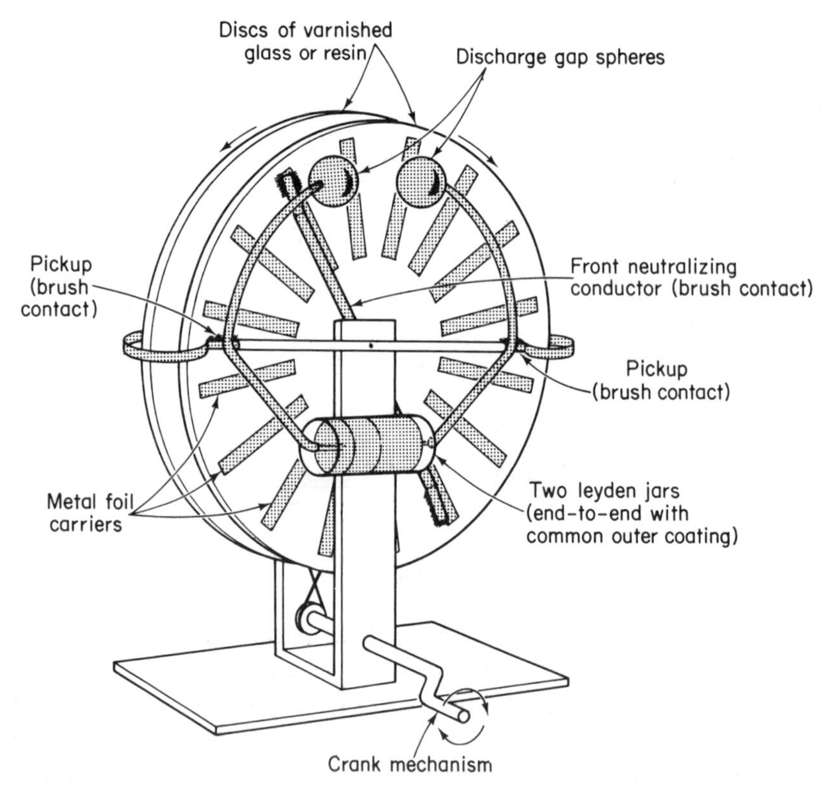

Fig. 3-19. Wimshurst influence machine.

A winch and pulley mechanism is used to turn two disks of varnished glass, resin, wax, or other insulating material in opposite directions. Attached to the disks are metal foil carriers, which take the place of the spheres in Fig. 3-18. Across the diameter of both disks are placed neutralizing conductors, which are rigidly attached to the frame and fitted with flexible or brush contacts where they touch their respective rotating disks. The neutralizing conductor on the rear disk is mounted at right angles to (i.e., a quarter turn away from) the neutralizing conductor on the front disk, although this fact is not apparent in the illustration. Also, at opposite ends of a diameter are two sets of pickups, flexible or brush contacts that connect opposing pairs of carriers on the front and rear disks. Conductors connect the pickups to the center rods of two Leyden jars, which share a common (grounded) outer coating and which serve the same purpose as the two separate jars in Fig. 3-18. The pickups are also connected to a pair of knobs; the air gap between the knobs is adjustable. The gap between the knobs determines how high the potential difference between the Leyden jars will get and therefore how large a charge will be stored before a spark discharge occurs.

To understand the operation of the Wimshurst machine, recall the simplified experiment described in Fig. 3-18 and suppose that one of the carriers on the rear disk is positively charged, while one diametrically opposite on the same disk is negatively charged. Simultaneously, the opposing carriers on the front disk are momentarily connected by the front neutralizing conductor. The positive carrier on the rear disk then induces a negative charge in the opposing carrier on the front disk, while the reverse occurs at the other end of the diameter. As the disks rotate, the charged carriers encounter the pickups and give up their charges to the Leyden jars. Furthermore, as soon as a pair of carriers on the front disk become charged, they, in turn, can induce charges on other carriers on the rear disk when the latter are connected by the rear neutralizing conductor. Hence, after a few turns, this mutual influence causes half the carriers on each disk to be charged negatively and the other half to be charged positively, with the neutralizing conductors forming the boundary between them. The location of the pickups and the opposite rotation of the two disks result in only positive charges being fed to one Leyden jar and only negative charges to the other. Thus, beginning with a very small charge (which might have been left from previous use of the machine, or from something brushing against a disk, or even from air friction as the disks began to turn), a very large potential is rapidly generated across the two Leyden jars.

Wimshurst-type influence machines have gone the way of other electrostatic devices discussed, that is, into the museum and the school science lab. There was a time in the late nineteenth century when every prominent physician had one in his office, where its awesome display was supposed to convince the sick of the tremendous powers that the practitioner ostensibly had at his command.

3-15 THE VAN DE GRAAFF MACHINE

Electrostatic machines might have lost all interest for man had the atomic age not arrived. In nuclear research, supplies of extremely high voltage are necessary to accelerate particles to high velocity. An interesting device that can produce a potential difference of as much as 8 million volts is the *Van de Graaff generator*. This type of machine, designed by Robert J. van de Graaff in 1931 and still used today, is represented conceptually in Fig. 3-20. The major components are a large hollow conductor, either

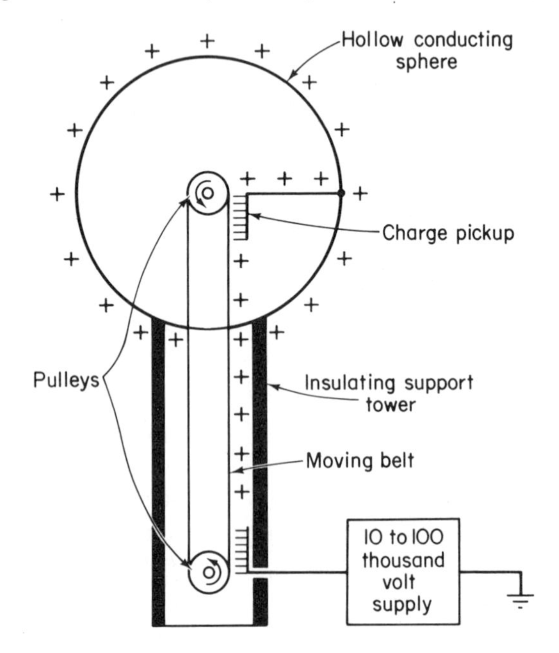

Fig. 3-20. The Van de Graaff generator.

spherical, as shown, or cylindrical with curved ends; a continuous belt of some insulator, such as nylon, rubber, silk, linen, or even paper; and a supply (transformer-rectifier set) of 10 to 100 thousand volts. The belt runs over two pulleys. The lower pulley is grounded and is driven by a motor. The upper pulley is mounted within, but is electrically insulated from, the hollow conducting sphere. Charge is fed onto the belt by a "comb" of sharp points connected to the supply source and is collected from the belt by a similar set of sharp points mounted within and connected to the hollow sphere. In Van de Graaff's original machine, the charge was generated by the friction between the belt and pulleys, but better and more consistent performance is attained in modern machines by using the separate supply. The maximum attainable voltage is limited by how well the sphere is isolated from its surroundings, by its size, and the breakdown of the air surrounding it due to the intense electric field. In practice, for a desired potential difference, the required size of the generator can be minimized by enclosing the whole structure within a large steel tank in which the environment can be controlled. By using a mixture of dry gases rather than ordinary moist air, and a pressure of about ten times atmospheric pressure, the breakdown potential is increased to several times its value at atmospheric pressure.

3–16 IS LIGHTNING ELECTRICITY?

As soon as the early electrical experimenters could produce sparks large enough to observe, and especially after they had perfected the friction machines, they should have realized the identity of these discharges with nature's lightning bolts. It is probable that several of them at least suspected it, but man is always slow to overthrow prevailing ideas, and the accepted theory was that lightning bolts were explosions of clouds of combustible gases.

It was the nimble-minded Benjamin Franklin who first publicly debunked the "fire-in-the-sky" theory in his writings, suggesting instead that lightning bolts were enormous electric sparks. He was not the first to test his own theory, however, because his intended experiment called for placing a metal rod atop the spire of the then uncompleted Christ Church in Philadelphia. Consequently, several Europeans had tested and proven Franklin's theory before he did himself, among them the Frenchman d' Alibard, who was able to produce sparks inductively, using charges drawn from the atmosphere during a storm by an insulated 40-foot metal rod.

However, news traveled slowly in those days before the advent of the telegraph, telephone, radio, and television and meanwhile Franklin had another idea. Instead of waiting for the spire to be completed, he would achieve the desired altitude for his conducting wire by flying it on a kite. So, in the midst of a thunderstorm, Franklin trudged out to the pasture and lofted his celebrated kite, oblivious to the personal danger of electrocution. Charges from the air were drawn into the wire that protruded from the kite frame and, true to Franklin's theory, were conducted down the wet silken cord to the dangling key and to Ben. He was able to charge a Leyden jar and thereby demonstrate the identity of lightning and electricity.

The curious reader is cautioned not to tempt the fates by trying this experiment himself. Several of Franklin's contemporaries who performed similar experiments were not so lucky—they were electrocuted.

REVIEW QUESTIONS

1. Can you cite examples, not mentioned in the text, of electricity in nature? Were these common knowledge before the nineteenth century?

2. The Greek word for sulfur is *theion*, which, as an English prefix, becomes *thio*. Suppose that amber had not existed in ancient Greece and the Greeks had discovered instead that sulfur can be electrified by friction. What English words would be different today and how might they be spelled?

3. What role(s) did the objects named in the title of this chapter play in the development of electrostatics?

4. If Franklin had named the kinds of charge in the reverse way (i.e., if what we now call a positive charge were called negative, and vice versa), what differences would there have to be in the wording of the first law of electricity? In Coulomb's law?

5. A positive point charge and a negative point charge of the same size are being held a fixed distance apart in a vacuum. How would the force acting on each of the charges be affected if
   (a) half of one charge was removed?
   (b) the positive charge was replaced by an equal negative charge?
   (c) the charges were placed five times as far apart?
   (d) the charges were buried in sulfur instead of in a vacuum?

6. How much charge passes a given point in one second if there is a current of one microampere at that point?

7. If forces can be measured in pounds, why can't electromotive force be measured in pounds? Why not in "pounds per coulomb"?

8. Gravity can also be considered a field. Why? What is the test device? What would the gravitational field around an isolated ball of matter look like?

9. Sketch the electric field around two equal negative charges.

10. If you were going to design a Leyden jar of predetermined dimensions, for maximum charge capacity, would you
    (a) make the jar wall thick or thin?
    (b) make the jar from polyethylene or glass?

11. How much charge does a 0.02-microfarad capacitor store when it is charged to 2500 volts? To 5000 volts?

12. A frequently performed demonstration involves having a person stand on an insulated platform and place his hand on the dome (main conducting sphere) of a Van de Graaff generator. When the generator is turned on, the person's hair stands on end. Why? What would happen if the person did not stand on an insulated platform?

4

ATOM AND *e*

4-1 EARLY IDEAS ABOUT MATTER

To us, Benjamin Franklin's "electric fluid" would seem a remarkable "fluid" indeed! It would flow through solid metal wire just as water flows through a hollow pipe. The scientists of the eighteenth and early nineteenth centuries, however, had only worked in the realm of tangible things. They felt comfortable with fluids and were aware of many of the principles of what we now call fluid dynamics. It was natural for them to approach electricity as a flux of a homogeneous electric fluid (or fluids). This concept of electricity held sway until around the turn of the century, when some revolutionary facts concerning the nature of matter itself were being discovered.

Have you ever considered the question of the continuity of matter? Suppose that you had an extremely sharp knife, one so sharp that it could cut through the tiniest bit of substance. If you then began to cut up some object, say a cube of sugar, first in half, then cut one of the halves in half, then cut one of these pieces in half again, and so on, could you continue to cut indefinitely and always obtain a still smaller piece of sugar? Alternatively, would you at some point in the procedure reach a "smallest size piece" that refused to be further divided?

The nature of matter and its continuity were pondered by the early Greek philosophers. They sought the simplest substances, or *elements*, from which all matter was constituted. From the speculations of the Ionian philosophers came the concept of four fundamental elements: earth, water, air, and fire. All types of matter were then supposed to be composed of proper combinations of these four elements. The picture was extended somewhat later by Democritus of Abdera, who postulated that all matter was made of indivisible particles called *atoms*. The word atom comes from two Greek roots meaning "not divisible." Democritus' atoms were all of a single "elementary" substance, but differed from each other in size and shape, and were supposed to combine with each other in appropriate ways to produce the known forms of matter.

4-2 ELEMENTS, COMPOUNDS, AND MIXTURES

Early work in the field now called chemistry—by such pioneers as John Dalton and Antoine Lavoisier—supported the atomic concept of matter. However, the elements earth, air, fire, and water, so dear to the hearts of the Ionians, had to be discarded in favor of substances like tin, copper, iron, sulfur, and oxygen. (The ancient concept of elements is still perpetuated in the literary sense, as when we speak of a man who "pitted himself against the elements" or of an edifice "weathered by the elements.") Substances that could not be broken down into simpler substances by any means then known (i.e., by any physical process or chemical reaction) were now termed elements. Elements were found to combine with each other in *definite proportions* to form new substances, which do not resemble or behave like any of their constituent elements and which cannot *easily* (i.e., by means *other* than chemical reaction) be broken down into those constituent elements. Such combined substances are called *chemical compounds,* or more simply, *compounds.* Elements and/or compounds can be intimately mingled with each other in such a manner that the separate constituents are individually recognizable while in the mingled condition, and they can be relatively easily separated by physical means. Such a composite of intermingled elements and/or compounds is called a *mixture.* The distinguishing characteristics of a mixture are (1) the constituents can be combined in *any* proportions, (2) the individual components can be recognized within the mixture, and (3) the components can be separated by purely physical processes.

The classical experiment illustrating the difference between elements, compounds, and mixtures is the combination of iron and sulfur. Iron and sulfur are both elements, which is to say that they cannot be changed to simpler substances by ordinary (nonnuclear) processes. Iron is a fairly heavy, silver-colored metal. Sulfur, as commonly encountered, is a light yellow powder. If we take a quantity of fine particles of iron, called *filings*, and physically mix it with a quantity of powdered sulfur, the result is merely a mixture of iron and sulfur in which both the iron and sulfur particles may be easily distinguished, as shown in Fig. 4-1. This mixture is readily separable

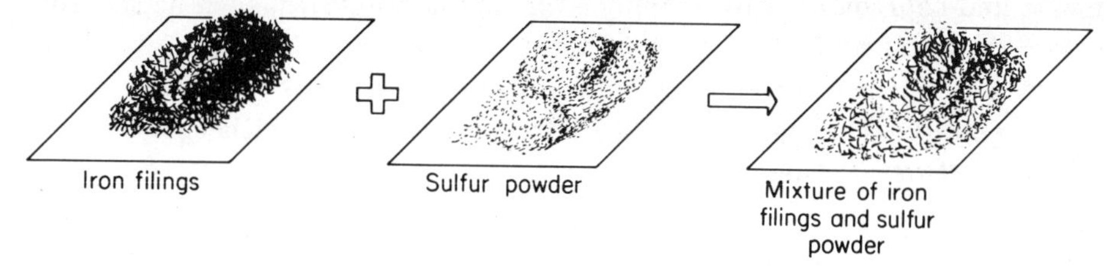

Fig. 4-1. An iron–sulfur mixture.

into its original constituents. If a magnet is brought near the mixture, the iron filings will be attracted to, and attach themselves to, the magnet, while the sulfur remains behind. Alternatively, if the mixture is poured into a jar of water, the heavier (denser) iron will sink to the bottom, while the lighter (less dense) sulfur will float to the surface. The separation is illustrated in Fig. 4-2. If, now, some of the iron-sulfur mixture is placed in a test tube

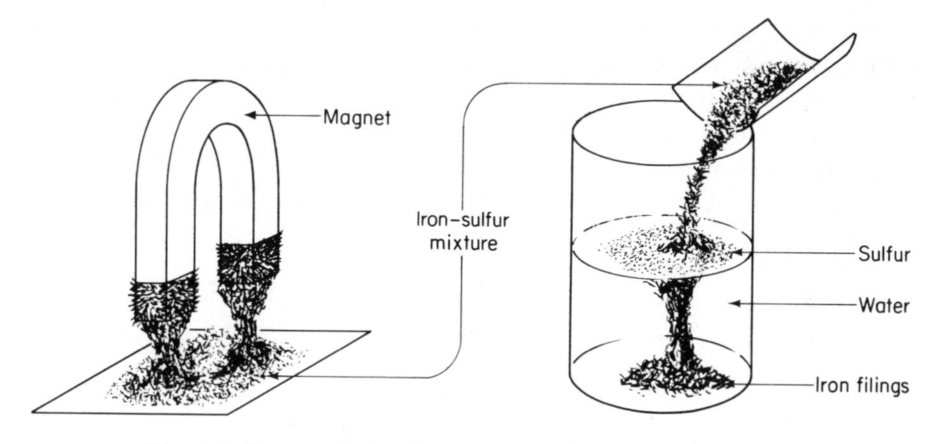

Fig. 4-2. Two methods of separating an iron–sulfur mixture.

and heated beyond the point at which sulfur melts, a temperature is reached at which a drastic change takes place. Examination of the resulting substance, upon cooling, reveals a brownish-black material that resembles neither iron nor sulfur, together with some surplus of either iron or sulfur. If the original mixture had been "just right," comprising approximately 56 parts by weight of iron and 32 parts by weight of sulfur, there would be no excess of either iron or sulfur; only the brownish-black substance would remain. This substance is the chemical compound iron sulfide, more properly called ferrous sulfide, from the Latin word for iron, *ferrum*. It is not metallic in appearance, nor magnetic, like iron. Neither is it yellow, nor will it float on water, like sulfur. We thus see how elements, like iron and sulfur, can combine to form compounds, like ferrous sulfide. We also see how elements and compounds can be mingled in simple mixtures, such as the iron-sulfur mixture, or the mixture of ferrous sulfide with the excess iron or sulfur. Both elements and compounds are considered *pure* materials, in contrast to mixtures. Thus all substances found in nature are either pure materials or mixtures of them.

4-3 MOLECULES VERSUS ATOMS

The work of the early chemists revealed that matter is *not* continuous—that is, that there *is* a "smallest size" particle of any pure material. The smallest particle of any pure substance that displays the characteristic physical and chemical properties of that substance is called a *molecule*, from Latin roots meaning "little mass." The smallest particle of an element is called an *atom*, retaining the name originated by Democritus. Since compounds are formed from elements, a molecule of a compound must contain at least one atom of each of its constituent elements, and will therefore contain at least two atoms. A molecule of an element, on the other hand, may consist merely of one atom of an element—that is, the terms atom and molecule may be synonymous in the case of pure elements. (This fact is generally true of elements in the liquid or solid condition.) The molecule concept is illustrated in Fig. 4-3 for iron, sulfur, and iron (ferrous) sulfide. Note

Fig. 4-3. Molecules of iron, sulfur, and ferrous sulfide.

that an iron molecule consists merely of one iron atom the sulfur molecule is just one sulfur atom, while the ferrous sulfide molecule comprises two atoms, one of iron joined to one of sulfur.

4-4 CHEMICAL SYMBOLISM

Chemists represent the atoms of elements symbolically by the first letter or two of their names, usually taken from the Latin form. Thus an atom of sulfur is represented by S, oxygen by O, nickel by Ni, but iron by Fe (from *fe*rrum), copper by Cu (from *cu*prum), and tin by Sn (from *s*ta*n*num).

A chemist represents the molecule by writing, next to each other, the symbols for the atoms composing that molecule. Thus the molecules of iron, sulfur, and ferrous sulfide, as illustrated in Fig. 4-3, are represented by the symbols Fe, S, and FeS, respectively. If more than one atom of the same element appears in a given molecule, this is indicated by writing the number of such atoms as a little numeral to the right of, and slightly below, the symbol for the atom. For example, iron forms a second compound with sulfur, called ferr*ic* sulfide, the molecule of which contains two atoms of iron and three of sulfur. Instead of writing something like FeFeSSS, the chemist writes Fe_2S_3 as the symbol for the molecule. The little numerals, 2 and 3, are called *subscripts*, and indicate how many atoms of each kind are included in the molecule. In the same way, water is a compound whose molecule consists of two atoms of hydrogen (H) and one atom of oxygen (O). The symbol for the water molecule is therefore H_2O. The subscript 2 indicates two hydrogen atoms, and the lack of any subscript on the O (*Note:* the subscript 1 is never used.) indicates only one oxygen atom.

Over one hundred elements have been isolated and named to date. A complete list of these elements, together with their chemical symbols, is given in Table 4-1. Some have only been man-made (in nuclear reactions) in recent times, and are not found in nature. Many of those occurring naturally are very rare. Only some forty-odd elements are used commonly enough so that their names would even be recognized by persons not actively engaged in the fields of science and technology. Yet what a far cry from the four "elements" of the early Greeks, and those, we now know, are not elements at all! Air is a mixture of nitrogen, oxygen, and other gases; earth is a loose term for a mixture of the compound silicon dioxide (sand) with various other compounds and elements; water is a compound; and fire is not a substance at all but a phenomenon that accompanies the process of rapid combustion.

The chemical reaction of two (or more) elements to form a compound is represented by the symbols for the participating molecules, together with the + sign and the = sign. The plus sign is read as *and*, understood in the sense of ". . . together with . . . ." The equals sign is read as *yields*, and is understood to mean ". . . react(s) chemically to form . . . ." Thus

$$Fe + S = FeS$$

means "one molecule of iron and one molecule of sulfur react chemically to form one molecule of ferrous sulfide." Similarly, the expression

Table 4-1 The Elements

Atomic Number	Name	Symbol	Atomic Number	Name	Symbol	Atomic Number	Name	Symbol
1	Hydrogen	H	36	Krypton	Kr	71	Lutetium	Lu
2	Helium	He	37	Rubidium	Rb	72	Hafnium	Hf
3	Lithium	Li	38	Strontium	Sr	73	Tantalum	Ta
4	Beryllium	Be	39	Yttrium	Y	74	Tungsten (Wolframium*)	W
5	Boron	B	40	Zirconium	Zr	75	Rhenium	Re
6	Carbon	C	41	Niobium	Nb	76	Osmium	Os
7	Nitrogen	N	42	Molybdenum	Mo	77	Iridium	Ir
8	Oxygen	O	43	Technetium	Tc	78	Platinum	Pt
9	Fluorine	F	44	Ruthenium	Ru	79	Gold (Aurum*)	Au
10	Neon	Ne	45	Rhodium	Rh	80	Mercury (Hydrargyrum*)	Hg
11	Sodium (Natrium*)	Na	46	Palladium	Pd	81	Thallium	Th
12	Magnesium	Mg	47	Silver (Argentum*)	Ag	82	Lead (Plumbum*)	Pb
13	Aluminum	Al	48	Cadmium	Cd	83	Bismuth	Bi
14	Silicon	Si	49	Indium	In	84	Polonium	Po
15	Phosphorus	P	50	Tin (Stannum*)	Sn	85	Astatine	At
16	Sulfur	S	51	Antimony (Stibium*)	Sb	86	Radon	Rn
17	Chlorine	Cl	52	Tellurium	Te	87	Francium	Fr
18	Argon	Ar	53	Iodine	I	88	Radium	Ra
19	Potassium (Kalium*)	K	54	Xenon	Xe	89	Actinium	Ac
20	Calcium	Ca	55	Cesium	Cs	90	Thorium	Th
21	Scandium	Sc	56	Barium	Ba	91	Protactinium	Pa
22	Titanium	Ti	57	Lanthanum	La	92	Uranium	U

Table 4–1 (Cont'd.) The Elements

Atomic Number	*Name*	*Symbol*	*Atomic Number*	*Name*	*Symbol*	*Atomic Number*	*Name*	*Symbol*
23	Vanadium	V	58	Cerium	Ce	93	Neptunium	Np
24	Chromium	Cr	59	Praseodymium	Pr	94	Plutonium	Pu
25	Manganese	Mn	60	Neodymium	Nd	95	Americium	Am
26	Iron (Ferrum*)	Fe	61	Promethium	Pm	96	Curium	Cm
27	Cobalt	Co	62	Samarium	Sm	97	Berkelium	Bk
28	Nickel	Ni	63	Europium	Eu	98	Californium	Cf
29	Copper (Cuprum*)	Cu	64	Gadolinium	Gd	99	Einsteinium	Es
30	Zinc	Zn	65	Terbium	Tb	100	Fermium	Fm
31	Gallium	Ga	66	Dysprosium	Dy	101	Mendelevium	Md
32	Germanium	Ge	67	Holmium	Ho	102	Nobelium	No
33	Arsenic	As	68	Erbium	Er	103	Lawrencium	Lw
34	Selenium	Se	69	Thulium	Tm			
35	Bromine	Br	70	Ytterbium	Yb			

$$2H_2 + O_2 = 2H_2O$$

means "two molecules of hydrogen (gas) and one molecule of oxygen react chemically to form two molucules of water." Such expressions are called *chemical equations.*

4-5 THE QUEST FOR THE ATOMIC GLUE

We return now to the question of what happens as we continue to subdivide the sugar cube with our imaginary infinitely sharp knife. Eventually we reach the smallest particle having the properties of sugar, a single molecule of sucrose (the proper chemical name for cane sugar). This molecule consists of 22 hydrogen atoms and 11 oxygen atoms joined to 12 carbon atoms and therefore is represented by the symbol $C_{12}H_{22}O_{11}$. If we cut this molecule with our theoretical knife, the sweet white sugar is no more, and in its place we have tiny bits of black solid and clear liquid. The solid comprises molecules (i.e., atoms) of carbon and the liquid is water. In particular, we have 12 molecules of carbon and 11 molecules of water. The carbon can be subdivided until one molecule (=one atom) of carbon remains, and this cannot be divided further. The water can still be divided until only one molecule remains, and if *this* molecule is cut, the water disappears (figuratively) in a wisp of gas that consists of two hydrogen atoms and one oxygen atom. Like the carbon atom, the hydrogen and oxygen atoms can be divided no further. Thus, in subdividing a particle of sugar, the end of the line is reached when atoms of carbon, hydrogen, and oxygen remain.

What has all this to do with electronics? you may well inquire at this point. The answer lies in the nature of the forces that bind atoms together. The early chemists speculated as to the reason atoms of certain elements combined, while others did not, and hypothesized that chemical combination was associated with their physical shape. In a compound, the atoms of the constituent elements were thought to interlock, like the pieces of a jigsaw puzzle. For example, it was once believed that the oxygen atom was a torus (shaped like a doughnut) and that hydrogen atoms were small spheres which could nestle neatly in the "hole" of the oxygen "doughnut" to form a water molecule, as shown in Fig. 4-4. Other atoms were thought to have various kinds of hooks, eyelets, and holes to account for the reactions in which they would take part. As the list of known elements grew longer, the number of mathematically possible combinations (i.e., compounds) of them increased even faster. It took more and more imagination to devise unique shapes for all the atoms, which would explain why only certain compounds

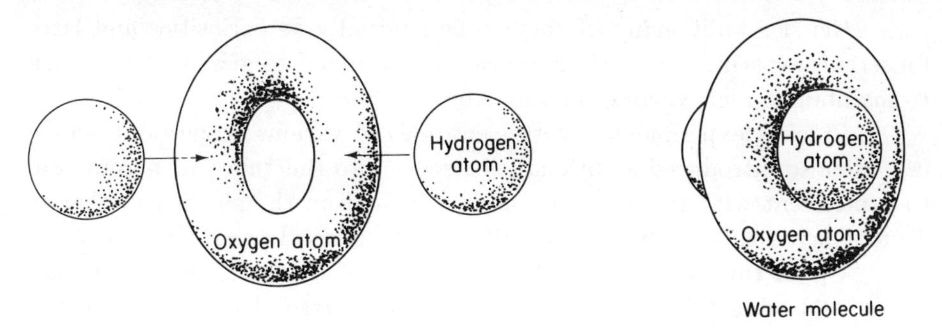

Fig. 4-4. Early conception of the formation of a water molecule.

could be formed. All in all, the jigsaw puzzle theory of atomic structure was not very satisfactory.

Finally, J. J. Thomson, a noted English scientist who won a Nobel Prize for his work in the field of electricity, made the discovery that led to both the modern theory of the atom and to the science of electronics. He alone cannot receive all the credit, however. Two other Englishmen, William Crookes and Michael Faraday, as well as a German, Heinrich Geissler, paved the way for Thomson's historic discovery.

4-6 THE CROOKES TUBE

Faraday originated the experiment that led to the ultimate discovery. He built a glass tube with metal electrical connectors sealed into the opposite ends. Such a tube is shown in Fig. 4-5. When the wires attached to the elec-

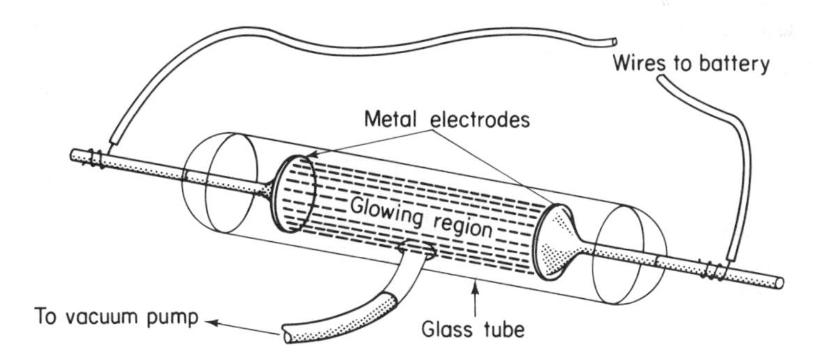

Fig. 4-5. A Faraday tube.

trodes (the metal connectors) were connected to a battery and most of the air was pumped out of the tube, the remaining air in the region between the electrodes would glow, indicating the passage of electricity.

Geissler built many of these tubes, initially as curiosities and later for serious experimental work. He perfected the metal-to-glass seal necessary to maintain a high vacuum in the tube.

Crookes experimented with electrodes of various shapes and materials. He also introduced additional electrodes into the tubes to permit him to experiment with the effect of an electric field on the glowing discharge. To this day, a tube of this type is called by the general name *Crookes' Tube*.

Up to this point it had been determined that, within the Crookes Tube, a glowing beam emanated from the negative electrode, called the *cathode*, and moved through the tube to the positive electrode, the *anode*. Because of its point of origin, the beam was said to consist of *cathode rays*. Experimentation had shown that this beam of cathode rays could be deflected (i.e., bent) by the influence of a magnet. Attempts to produce deflection of the beam with an *electric* field had failed, however. It was the consensus of European scientific opinion that the pretty colored glow consisted of "waves in the ether," although most British scientists felt that the cathode rays were beams of negatively charged atoms of the metal from which the cathode was made. The stage was set for J. J. Thomson's appearance.

4-7 THE ELECTRON UNVEILED

Thomson believed that the cathode rays could indeed be deflected by an electric field, provided that a sufficiently high vacuum could be created within the Crookes Tube. With a tube having the configuration shown in Fig. 4-6, in which he had obtained an unusually good vacuum after much painstaking effort, Thomson was able to deflect the beam with an electric field. He used a circular anode with a small hole in its center, which permitted

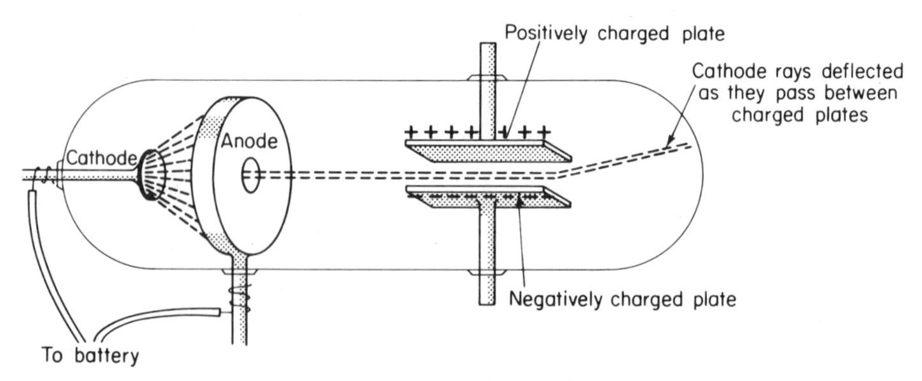

Fig. 4-6. Thomson's experiment.

some of the cathode rays produced to pass on through the anode in a narrow beam. With no charge applied to the deflection plates, the beam continued in a straight line until it struck the opposite end of the tube. When a second battery was connected to the deflection plates, the beam was bent toward the positively charged plate. He therefore hypothesized that the beam was constituted of negatively charged particles. In 1897, from careful measurements of both magnetic and electric deflections of the beam, Thomson was able to calculate the ratio of the charge to the mass of an individual particle. Faraday had already determined the value of the smallest known electrical charge, which we represent today by the symbol e. Thomson assumed that the charge on his newly discovered particle was e, and from its charge-to-mass ratio, calculated its mass. This inconceivably small particle, which has been named the *electron*, has a mass of only about one billionth of one billionth of one billionth of a gram (more correctly, it has been determined to be 9.108×10^{-28} gram). To get a better appreciation of this tiny mass, imagine that the masses of all objects in the universe were amplified so that the average human being weighed as much as our own star, the sun, normally weighs; then on this same scale, an electron would weigh only as much as a large grape!

4–8 INTO THE HEART OF THE ATOM

In experimenting with various materials, Thomson found that the cathode rays always consisted of particles of the same charge and mass, regardless of the metal from which the cathode was made or the gas with which the tube was filled. Consequently, he concluded that the particles (i.e., electrons) were contained *within the atoms* of gas or metal. In other words, the "indivisible" atoms of Democritus, Dalton, et al. could be further subdivided into electrons and "something-elses." This concept was further confirmed when it was found that among the products of the radioactive disintegration of the element radium, which had just been discovered by Pierre and Marie Curie, was a stream of electrons. These electrons emitted in the disintegration process, which are called *beta particles*, must have been part of the radium atoms. For his discovery, Thomson received the Nobel Prize in 1906 and was subsequently knighted.

Following Thomson, in 1909 an American physicist, Robert Millikan, made a very accurate determination of the electron's charge by observing the deflection of tiny charged droplets of oil falling in an electric field. The presently accepted value of this important constant, e, is 1.602×10^{-19}

coulomb. In other words, our unit of charge, the coulomb, is 6.24×10^{18} (i.e., about 6,240,000,000,000,000,000) times as big as the charge on one electron.

We pause for a moment to consider an implication of Thomson's discovery. Since lumps of matter, metallic or otherwise, do not normally attract or repel each other, they are evidently electrically *neutral*—that is, ordinary matter carries no electric charge. It must therefore be true that atoms contain positive charges as well, in order to balance the electrons' negative charges.

Next to play important roles in the drama of unveiling the atom were two other Englishmen (The English seemed to have had a virtual monopoly on discoveries in atomic physics in those days!), Ernest Rutherford and H. G. J. Mosely. Rutherford had studied under Thomson at Cambridge, and the latter had directed him into the study of radioactivity. It had been found that among the emanations of disintegrating radium was a stream of positive particles, now called *alpha particles*, in addition to the previously mentioned negative beta particles and other products. Rutherford bombarded sheets of gold foil with these alpha particles. He found that a large percentage of the particles passed through quite readily, but a few rebounded like rubber balls off a wall. Electrons were too light, and of the wrong charge (negative rather than positive as required for mutual repulsion) to repel the relatively heavy alpha particles. From the energetic rebounds, Rutherford reasoned that the gold atom must contain something with a positive charge much greater in magnitude than the electron's negative charge, and with a considerable mass. Moreover, the ease with which most of the alpha particle beam penetrated the foil indicated that the atom had to be mostly *empty space*, and the positively charged large mass had to be localized within a small portion of the total volume of the atom. Rutherford had, in short, discovered the *nucleus*! He theorized that the atom comprised a small, heavy, positively charged core (the nucleus) surrounded by negatively charged electrons. After many experiments, he concluded that the hydrogen nucleus, 1846 times as massive as the electron, had the smallest positive charge and was actually a fundamental particle, which he later named the *proton*. In 1908 he received a Nobel Prize for his contribution to atomic theory.

4-9 STRUCTURE OF THE ATOM

Mosely, one of Rutherford's assistants, enunciated the principle of balanced atomic charge: that the number of protons in an atom always equals the number of electrons. The number is the fundamental characteristic of the atom, which determines its chemical behavior as an element, and is called the *atomic number*. In a manner of speaking, it is the atomic number of iron which makes it "iron" and not, say, "gold." Thus an iron atom has 26 protons and 26 electrons, while a gold atom has 79 of each.

Niels Bohr, a Dane, added further sophistication to the evolving picture of the atom by grouping the electrons into *shells* that have differing electron capacities, depending on how close the given shell lies to the nucleus. The innermost shell can contain only 2 electrons; the second shell can contain 8 electrons; the third and subsequent shells may contain 18 or more electrons, but the *outermost shell always contains 8 or less.*

The picture was not yet complete, however, in that the mass contributed to the atom by protons, as dictated by the atomic number, did not account for the total mass of the atom. The additional mass contributed by the atomic number of electrons is far too small to make up the difference. The answer came in 1932 when James Cavendish, another assistant of Rutherford, discovered the *neutron*, a particle approximately equal to the proton in mass but carrying *no charge*. Neutrons contribute the required additional mass to the nucleus of the atom without affecting its electrical balance. The total mass of the nucleus, relative to hydrogen, is called its *atomic weight* and is numerically equal to the total number of protons plus neutrons in that nucleus.

Since 1932, the picture of the atom has changed in many respects. Today the question of what an atom "really is" is a question of metaphysics more than of physics, if it has any meaning at all. Depending on the type of experiment conducted, the subatomic "particles" sometimes behave as "particles" and sometimes as "waves." Perhaps they are both, if such a statement is meaningful. In any case, one only speaks now of a *model* of the atom. From the standpoint of physical and chemical behavior, particularly as applied to electronics, the Rutherford-Mosely-Bohr model is more than sufficient to explain the observed phenomena. Hence, in this text, we shall content ourselves with this model, realizing that we are dealing with an idealization and nothing more.

4-10 ISOTOPES

Early measurements of atomic weight were fairly crude and, to the accuracy of these measurements, the relative masses of all atoms seemed to be some whole number times the mass of the hydrogen atom. This made for a very nice, simple picture, particularly for those who theorized that all elements were made up of some number of hydrogen atoms, somehow "fused together." As more refined measuring techniques became available, it became quite clear that the relative atomic weights were not whole numbers at all. A striking example is chlorine, whose atomic weight is about $35\frac{1}{2}$ and is definitely not 35 or 36.

It was subsequently postulated, and proven with the advent of the *mass spectrograph*, that several different forms of the same element can, and do, exist. These different forms have the same atomic number (number of protons) but differ in the number of neutrons and therefore have different atomic weights. Atoms having the same atomic number but different atomic weights are called *isotopes*. Different isotopes of a given element have identical chemical properties but slightly different physical properties. Two isotopes of chlorine occur in nature: Both have 17 protons; hence the atomic number is 17. But one has 18 neutrons and the other has 20; hence their atomic weights or *mass numbers* are 35 and 37, respectively. These specific isotopes are represented by the symbolism, ${}_{17}Cl^{35}$ and ${}_{17}Cl^{37}$. Thus the chlorine that occurs in nature is a mixture of approximately 75 percent ${}_{17}Cl^{35}$ and 25 percent ${}_{17}Cl^{37}$, hence the natural "average" atomic weight of 35.457. The situation is similar with other elements. A brief list of isotopes of some elements is given in Table 4-2.

Table 4-2 Examples of Isotopes

Element	*Atomic Number*	*Atomic Weights of Natural Isotopes*
Hydrogen	1	1, 2, 3
Oxygen	8	16, 17, 18
Chlorine	17	35, 37
Iron	26	54, 56, 57, 58
Silver	47	107, 109
Uranium	92	234, 235, 238

An analogy may be drawn between the elements and the dog family. Atoms of different atomic number are as unlike as Great Danes and Chihuahuas. Isotopes of the same atomic number but different mass number are very similar, but differ from each other in mass like standard, toy, and miniature Poodles.

4-11 THE PERIODIC TABLE: METALS VERSUS NONMETALS

Our model atom, then, is a microscopic beehive of activity resembling somewhat a miniature solar system, with the nucleus as the "sun" and the the electrons as "planets." The analogy cannot be carried too far, however. The sun is a single body, while the nucleus is a composite of several protons and neutrons "stuck together." The electrons, unlike the planets, are all identical, and the outermost electrons can sometimes readily "escape" from the atom; in fact, it is this property that permits the formation of compounds and many other atomic phenomena. (We will discuss this property of the outer electrons in more detail later.) Furthermore, the solar system is essentially "flat" (i.e., almost lies in a plane) with the orbits of most planets forming nearly concentric circles. By contrast, even in our simple model, the electron orbits must be imagined in three dimensions, and many are highly elliptical. "Pictures" of the simplest elements, according to this model, are shown in Fig. 4-7. For simplicity, we will hereafter represent such atoms schematically, as shown in Fig. 4-8, rather than pictorially as in Fig. 4-7, because we will be interested principally in three facts concerning the

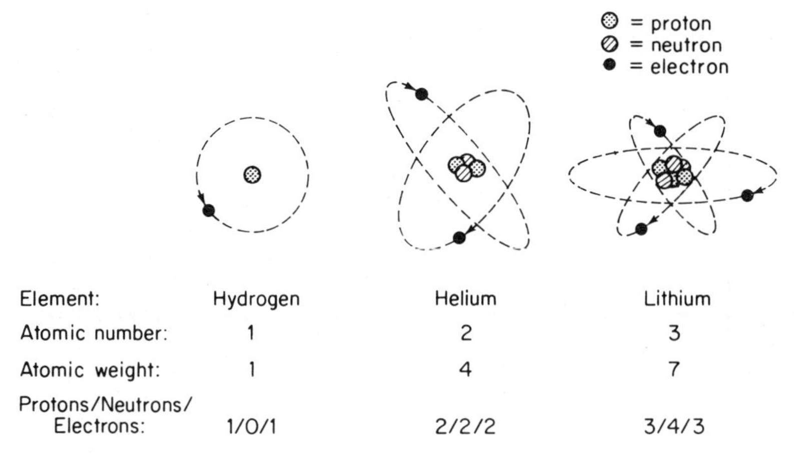

Fig. 4-7. The simplest atoms (particles and orbits not to scale).

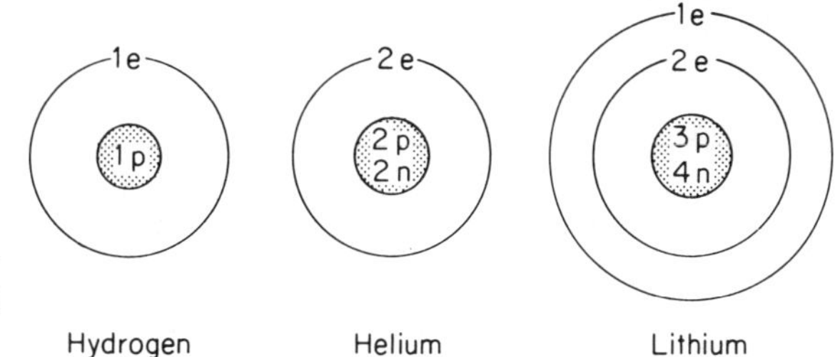

Fig. 4-8. Schematic representation of the atoms depicted in Fig. 4-7.

atom: its atomic number, atomic weight, and the number of electrons in its outer shell. The outer shell electrons are called the *valence* electrons.

The valence electrons, in general, govern the chemical reactivity and electrical properties of the element. This fact, coupled with the fact that the outer shell of any atom can contain a maximum of eight electrons, causes a sort of periodic similarity in properties among the elements as we climb

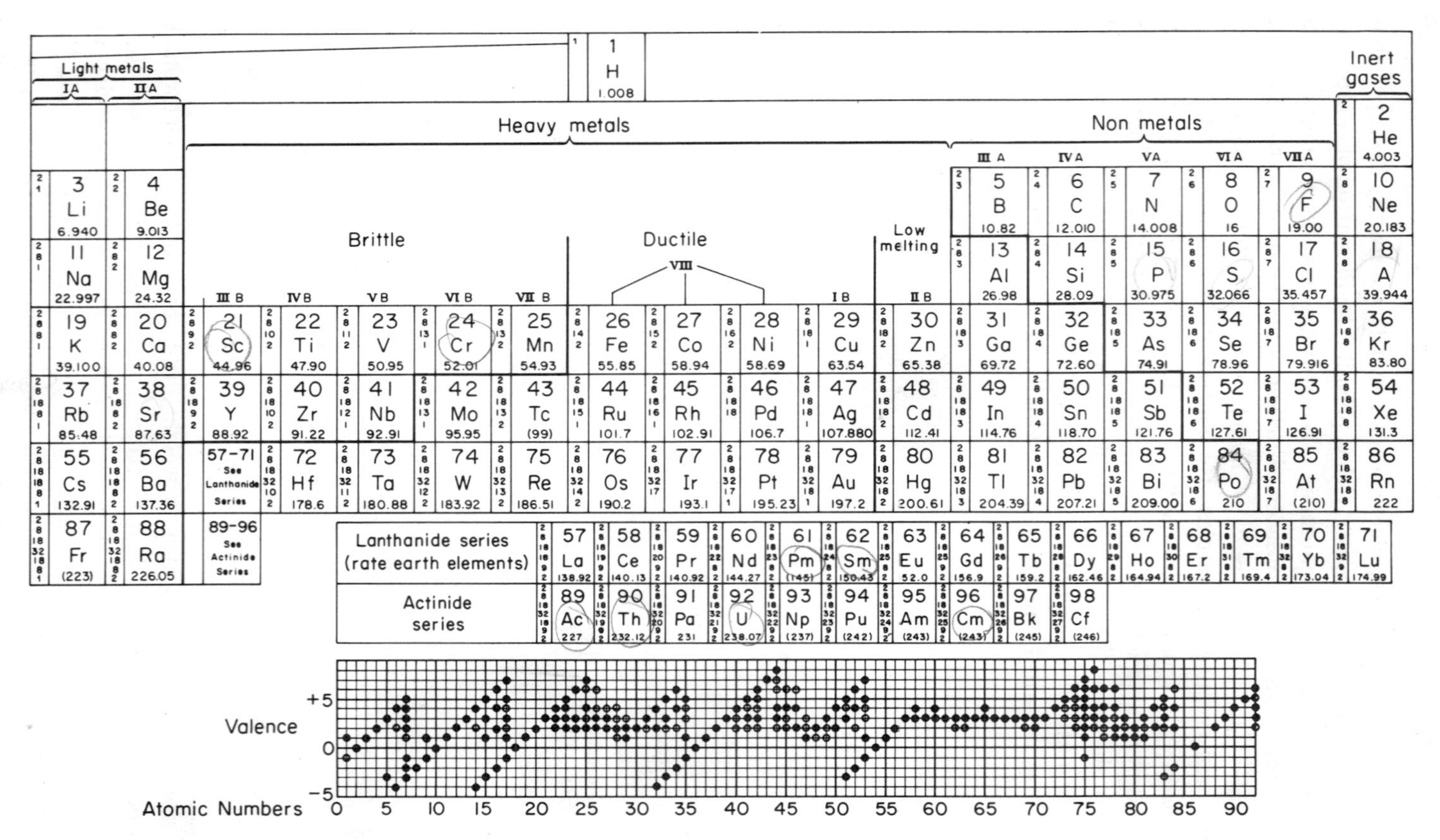

Fig. 4-9. Periodic table of the elements.

the atomic number scale. The periodic repetition of properties was first noted by the Russian chemist, Dmitri Mendeleev, who grouped the elements in a chart that has come to be called the *periodic table* of the elements. A modern form of this table is presented in Fig. 4-9. The meaning of the data contained in each box in the table is explained by Fig. 4-10, which is

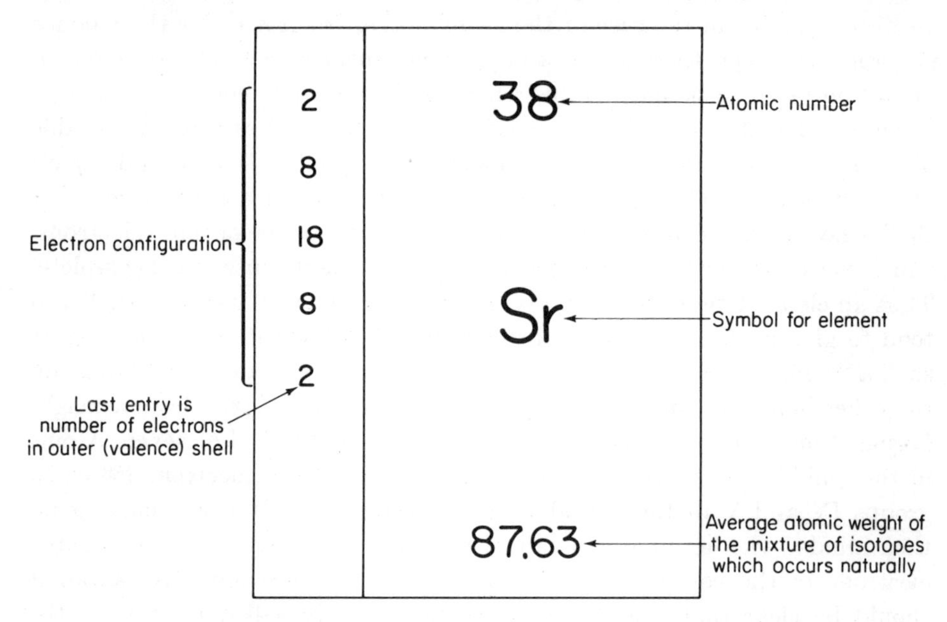

Fig. 4-10. Explanation of symbolism used in periodic table entries, Fig. 4-9.

an enlargement of one entry from Fig. 4-9. Families of elements with similar (but not identical) properties are represented by vertical columns. Some very strange groups of elements appear in the table, such as the Lanthanide and Actinide groups given separately at the bottom. The members of each of these groups are so similar to each other in behavior that the whole group must be "filed" as one element in the table. We will look further into the grouping of elements later in the text. For the moment, we will simply make this one very coarse distinction. If the atom has four or less valence electrons, it is (loosely speaking) a *metal:* if it has five or more valence electrons, it is a *nonmetal.*

4–12 IONS

An atom may become electrically charged by the gain or loss of one or more electrons. In such a charged condition, the atom is referred to as an *ion* and the process in which the charge is acquired is called *ionization.* It is only the valence electrons that are affected in the ionization of the atom; that is, electrons are either added to or lost from the outermost shell. The force that binds the valence electrons to the atom is the electrical attraction of the positively charged nucleus for those negatively charged electrons, just as in the case of the atom's other electrons. For the valence electrons, however, there is also an opposing consideration. The atomic configuration tends to be most stable when its outer shell is complete—that is, when its outer shell contains the maximum number of electrons it is capable of containing. For the first two elements, this number is two, and for all others it is eight. Given the proper circumstances, an atom whose outer shell is not already complete will either acquire or lose one or more electrons, whichever is "easier" (requires less energy), to make its outer shell complete. Thus an element that contains, say, only one electron in its outer shell will tend to give up this electron rather readily, thereby making its next inner shell a "complete outer shell." An atom that has seven valence electrons, on the other hand, requires less energy to "borrow" an electron from a neighboring atom to complete its true outer shell, rather than to lose seven. Atoms in the "middle" position of having four or five valence electrons (those in groups IV and V in the periodic table) have a difficult time "making up their minds" whether to lose or gain electrons and in many cases just "share" electrons in the compounds they form. From our previous discussion, it should be clear that the elements we have loosely called metals are the ones that will tend to lose electrons, while nonmetals will tend to acquire them.

Consider the ionization of a metal atom having more than one valence electron. Initially it is a neutral atom. Given a small amount of energy, the atom will readily lose one of its valence electrons, becoming a positive ion with a charge of plus one (+1). This net charge is due to the fact that there is now one proton in the nucleus which does not have a charge-compensating electron in orbit around the atom. Removal of a second valence electron requires a good deal more energy because the ion, now positively charged, is very strongly attracting its negative electrons. Removal of a third valence electron is still more difficult, and so on. Similarly, in the ionization of a nonmetal, the neutral atom readily accepts the first electron, becoming a negative ion with a net charge of minus one (—1). Successive electrons are progressively more difficult to add, however, because the negative ion exerts a repulsive electrical force on them.

4-13 MODERN VIEW OF THE ELECTRIC CURRENT

According to our present understanding, an electric current comprises a flow of electrons; thus Ben Franklin guessed wrong when he postulated that current comprised a flux of positively charged fluid. This should be visualized not as a flow of fluid through a pipe, as in Ben Franklin's concept, but more like the passage of a large number of charged marbles or ball bearings through a stack of large mesh screens. A bar of metal lying inert on a table possesses some electron motion. This is due to thermal energy, as measured by the temperature of the bar, which, although small, is sufficient to cause some of the metal atoms to lose one valence electron. Such a loose electron will drift around until it is captured by a neighboring atom that has also lost an electron and therefore is (at least temporarily) positively charged. As individual atoms gain sufficient energy, the process repeats over and over again on a statistical basis. There is no net electric current, however, for the electrons drift randomly and the effect "averages out" over the enormous number of atoms in the bar. In the previous chapter we saw how a momentary electric current was produced when a charged body was discharged. In the next chapter we will see how a continuing electric current can be maintained.

4-14 CONDUCTORS AND INSULATORS REVISITED

From the foregoing discussion, it should be evident that certain substances should readily admit the passage of an electron flow, since the atoms of those substances are "eager" to give up their own electrons, while other substances, which tend to capture electrons, would impede the passage of an electric current. As indicated in Section 3-7, substances that easily pass an electric current are called *conductors*; those that resist the passage of an electric current are called *nonconductors* or *insulators*; and those that fall somewhere in between in their capacity to conduct a current are called *semiconductors*. Once again the reader should realize that no fine line of distinction can be drawn, and the elements fall along a continuous scale that shades gradually from good conductors through fair conductors, poor conductors, and fair insulators to good insulators. On the basis of our definitions and what we have learned thus far about atomic structures, we can understand why in general, metals are conductors and nonmetals are insulators.

4-15 ELECTRON ELECTROSTATICS

All the phenomena of electrostatics, as described in Chapter 3, can be understood now in terms of accumulation or deficiency of electrons. Hence Franklin was on the right track with his single fluid theory despite his incorrect guess about the sign of the charge. [There may even be places in our universe where Franklin would have been correct—so-called *anti-matter*[1] worlds where the atoms have positively charged electrons (*positrons*) circling negatively charged nuclei.]

In any case, the reader should reconsider the experiments in Chapter 3 with the idea that a positive charge is caused by a loss of electrons and a negative charge by an excess. Thus, for example, the glass rod, when rubbed with a silk cloth, gives up some of the electrons from atoms at its surface to the cloth, thereby acquiring a positive charge.

REVIEW QUESTIONS

1. Can you tell which of the following are elements, which are compounds, and which are mixtures: glass, salt, pepper, brass, bourbon, carbon dioxide, diamond, nickel, oak, soap, and battery acid?
2. What do each of these chemical equations say, in words? (Refer to Table 4-1 for the meaning of unfamiliar symbols.)
   (a) $Na + Cl = NaCl$
   (b) $2Sb + 3S = Sb_2S_3$
   (c) $4Fe + 3O_2 = 2Fe_2O_3$
3. At the end of Section 4-6 it was stated that if the weight of all objects were amplified so that a man would weigh as much as the sun, then an electron would weigh about as much as a large grape (i.e., about $\frac{1}{18}$ of a pound). On this same scale, about what would a proton weigh?
4. Draw simplified schematic pictures, as in Fig. 4-8, of the atoms: ${}_{29}Cu^{63}$, ${}_{29}Cu^{65}$, ${}_{32}Ge^{72}$, ${}_{14}Si^{28}$, ${}_{92}U^{235}$. Refer to Fig. 4-9, as necessary.
5. Using the periodic table, classify the following either as metals or as nonmetals: strontium, osmium, arsenic, phosphorus, astatine, ytterbium.
6. Using the periodic table, group the following into the categories conductors, semiconductors, and insulators: titanium, germanium, silver, barium, iodine, sulfur, silicon, thorium, selenium, radon.

[1] The interested reader is referred to Hannes Alfven, *Worlds–Antiworlds Antimatter in Cosmology* (San Francisco: W. H. Freeman, 1966).

7. Which of the following elements all belong to one family: chromium, scandium, fluorine, samarium, polonium, actinium, uranium, potassium, thulium?
8. Discuss the experiments illustrated in Figs. 3-10, 3-11, 3-12, and 3-18 in terms of atomic theory.

5

LOCKED UP IN A CELL

5-1 HERDING ELECTRONS

Fortunately for us, the eighteenth-century electricians were not satisfied merely to make sparks with their electrostatic machines and attract bits of paper and pith. They were also interested in the phenomena associated with moving electric charges, that is, the science of electrons in motion called *electrodynamics* (as opposed to electrostatics). As stated earlier, the invention of the Leyden jar was the first step in providing more improved sources of electrons. The first electric currents to be studied involved discharging Leyden jars through conductors. Recall that the electrons in a conductor, at room temperature, are wandering around rather aimlessly, much like a herd of grazing cattle. When a difference of potential is created between the ends of the conductor, such as by connecting it between the inner and outer foils of a Leyden jar, the electrons respond like cattle to the rustlers' gunshots and move from the region of higher (i.e., more negative) potential toward the region of lower (i.e., less negative, or more positive) potential, resulting in an electric current.

One interesting group of early experiments with electric current was performed by our good friend, Ben Franklin, who entertained guests at a spring picnic in April, 1748 with some electrical demonstrations. First he fired several guns, using sparks from Leyden jars to ignite the charges. Next, he unwittingly anticipated the form of capital punishment used in places like New York State and electrocuted the turkeys for the picnic with a Leyden jar discharge current. Finally, he sent a Leyden jar current across the Schuylkill River to give a mild shock to a gentleman volunteer.

5–2 FRENCH "FRIED" MONKS

The Carthusians are a hermit-like order of monks founded in France in the eleventh century, whose name, as well as that of a delicious liqueur they prepare and sell, derives from the place of their establishment, Chartreuse. A notable Carthusian was the Abbé, Jean Antoine Nollet, who is perhaps best known for his experiments with osmotic pressure in animal bladders. However, like most scientists of his day (they called themselves *natural philosophers*), he was interested in all aspects of emerging science, electricity among them. Perhaps because the Carthusians were ascetics, conditioned to disciplining themselves in expiation for the evils perpetrated by their fellowman, the glib Abbé Nollet was able to convince a number of them to participate in an experiment as *living conductors*. The monks formed a closed path for an electric current by holding wires between them in their hands, making a chain of the form monk-wire-monk-wire, etc. Ostensibly, enough monks were used to make a loop measuring a mile around, which would have required 500 to 1000 monks. In any case, regardless of the actual size of the loop and the number of monks involved, the Abbé (who was conspicuously not connected into the loop) did perform this experiment around 1748, with a number of the French gentry as witnesses. He connected several previously charged Leyden jars into the loop, and as the final connection completing the loop was made, Nollet notes that all the monks felt the shock equally (How did he judge that?) and at about the same instant. He had thus, in a qualitative way, demonstrated two basic facts about an electric current:

1. The speed at which an electric current is propagated in a conductor is very great.
2. If the various segments of the path for the electric current tend to "oppose" or "resist" that current equally (as we may suppose was approximately true of the monks' bodies), then those segments each get an equal share of the work done by the current as it traverses that path (witness the equal shocks).

5-3 A BETTER SOURCE IS NEEDED

Other experimenters performed less bizarre experiments than Nollet's in sending currents along lengths of conductor. Du Fay, whom we met in Chapter 3, discharged a friction machine through a piece of wet pack thread about 300 meters long. He noted that moisture aided conduction in the thread. Perhaps the long-distance record for a current from a man-made electrostatic source was set by the English physician, Watson, who discharged a Leyden jar through a wire about $2\frac{1}{3}$ miles long.

All these currents were too ephemeral for any useful purpose, lasting only a matter of a few seconds at most. Furthermore, the high voltages involved in the electrostatic devices meant that excellent insulation had to be provided along the route of the conductor to prevent inadvertent arcing to ground. The immediate practical goal of the late eighteenth-century experimenters was a long-distance electric signaling device that could be used in some way to transmit messages, and not just unpredictable sporadic pulses of energy. What was clearly needed was a device that could maintain a steady flow of charge (i.e., electrons) for some time, and at a lower potential.

5-4 FROGS AND A SANDWICH

Scientists are human, like anyone else, and an intellectual disagreement between two of them is often somewhat more than dispassionate. Such a disagreement erupted between two now-famous Italians: the physiologist Luigi Galvani and the physicist Alessandro Volta, from whom the unit of potential takes its name. As a result of this feud, two major contributions were made to modern science: the invention of the *battery* and the beginning of an understanding of the electrical nature of nerve actions in animals.

Galvani was born at Bologna in 1737 and his studies of anatomy resulted in his being appointed lecturer (professor) in anatomy at the University of Bologna when he was 25 years old. His reputation as a comparative anatomist grew out of his research on the auditory and genitourinary organs of birds, but his most renowned work began with one of those fortunate accidents mentioned in Section 3-1.

It happened in 1771[1] when he was dissecting a frog on a table near an electrostatic machine. It is alleged that when he touched a nerve in the dead frog with his scalpel while the nearby machine was in operation, the

[1] The same effect was discovered by Floriano Caldani in 1756 but apparently, was not pursued further.

frog's legs kicked spasmodically. The effect so intrigued Galvani that he repeated similar experiments over and over in an attempt to understand. He discovered that the same response occurred, without the presence of a friction machine, but during a violent thunderstorm. Clearly, then, the effect was electrical and the frog was somehow a receiver of electricity. Then he made the most startling discovery of all; that without machine or thunderstorm, when a frog was suspended with a *copper* hook by its spinal cord from an *iron* stand, with its legs touching the iron base, the legs kicked violently anyway, as in Fig. 5-1. In this case, where did the electricity come

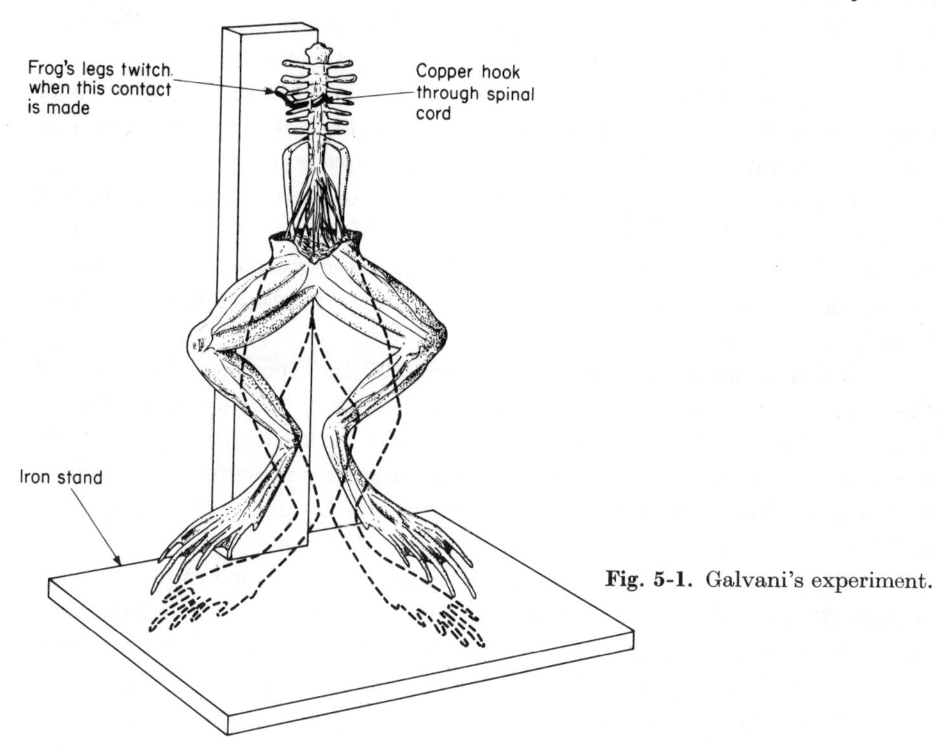

Fig. 5-1. Galvani's experiment.

from, the metals or the frog? Galvani was convinced that it originated in the frog. What a phenomenal discovery! The life force in living things must be electrical in nature! In 1786 he simplified the experiment, for demonstration purposes, to what he called "the metallic arc," which consisted of wires of two dissimilar metals being placed one into a frog's nerve and the other into an associated leg muscle. When the protruding ends of the wires were brought into contact, the muscle contracted. Galvani believed that the motions of the muscle were due to the interaction of a negative charge on the outside of the muscle and a positive charge generated in its inner substance, which flowed along the nerve and through the metallic arc when contact was made. So in 1791 Galvani announced his theory of *animal electricity.*

Meanwhile Volta had heard of Galvani's experiments. Born at Como in 1745, Volta became an eminent physicist and was first appointed professor of physics at the Como Gymnasium, and subsequently to the chair of physics at Pavia in 1774. He traveled through Europe making the acquaintance of many scientific celebrities. Repeating Galvani's experiments, Volta at first accepted the older man's theory of animal electricity. But when he performed the same experiment with both pieces of wire in the metallic arc made from the *same* material, the muscle did not move. Evidently the wires had a more important role than merely being conductors.

Volta recalled earlier experiments of his in which small quantities of electricity had been generated simply by bringing together pieces of two dissimilar metals, and he had called it *contact electricity.*

Galvani was offended by the brashness of this young Lombardian upstart, for he felt he had proved his point. After some experimentation, he was able to get a feeble twitch out of the frog's muscle with two pieces of the same material and saw that this result was transmitted to Volta forthwith with an attitude of *vendetta.*

Volta immediately repeated Galvani's latest experiment and verified the result. In addition, he tried enough variations of his own to be able to point out that even though the experiment will work with two pieces of the same substance, the pieces still had to be different in some way, such as by being at drastically different temperatures or by one piece being shiny and clean while the other was corroded (with the metal's oxide).

Before Galvani could take up the gauntlet again, one of his students performed what should have been the decisive experiment. Using *no metal* at all, he produced a twitch by merely touching the frog's leg muscle with the free end of the nerve itself. *Ecco la*! Animal electricity!

Undaunted, Volta claimed that nerve and muscle were merely two dissimilar conductors, like copper and zinc, or like clean copper and corroded copper. It was still contact electricity! Unfortunately, Galvani died before Volta proved his point (in 1800) by making history's most celebrated, although inedible, sandwich—the *voltaic cell.* This cell consisted of a disk of copper and a disk of zinc separated by a disk of moist paper. When the copper plate was connected to the zinc plate with a wire, a current flowed. From Volta's point of view, the significant thing was that the moist paper was just as good as frog's flesh in producing the current of contact electricity. From our point of view, however, the significant thing is that a current *persisted*, and at low voltage. Volta had invented the granddaddy of all "batteries."

The controversy as to whether Galvani or Volta was right raged on and off, until well into the nineteenth century, when Sir Humphry Davy and his even more famous assistant and successor, Michael Faraday, showed that both men were right and yet wrong, the energy source being chemical in nature.

Volta was well rewarded for his efforts. He received the Copley Medal of the Royal Society of London in 1791; was called to Paris by Napoleon for demonstrations of contact electricity and had a medal struck in his honor; was made a senator of Lombardy; and was appointed director of the philosophical faculty at Padua in 1815. Galvani, who died in 1798, did not achieve quite the degree of fame and fortune bestowed on his rival countryman, but he did carve himself a conspicuous niche in the hall of eternity by pointing the way to an understanding of the human nervous system in terms of electrical impulses. The medical applications of electronics are discussed later in the text.

5-5 A LITTLE MORE CHEMISTRY

In the next section we are going to study various types of "batteries" and how they work. What is going on inside a battery is basically a chemical process. The chemistry we learned in Chapter 4 will serve us well, but we need just a bit more to deal with electrode reactions.

First, recalling the meanings of compound and ion from Chapter 4, we distinguish between *ionic compounds* and *nonionic compounds*. In an ionic compound, the constituent elements actually exist as ions—that is, one element has lost one or more electrons and the other element has acquired the one(s) lost by the first. In Fig. 5-2, for instance, we see atoms of sodium and chlorine and a molecule of sodium chloride. Here the valence electrons are shown individually for clarity. Notice that, in the compound, the sodium atom has given up its one valence electron, leaving it a *sodium ion* with a charge of $+1$; while the chlorine atom has acquired that electron, leaving it a *chloride ion* with a charge of -1. It is, of course, these opposite charges that hold the molecule together. You will recall that we can write the reaction of sodium (Na) and chlorine to form sodium chloride (ordinary table salt) as

$$\mathrm{Na} + \mathrm{Cl} = \mathrm{NaCl}$$

If, however, we wish to show specifically the fact that the ions are formed, we can write this reaction in three steps:

$$\mathrm{Na} = \mathrm{Na}^{+} + ⓔ$$

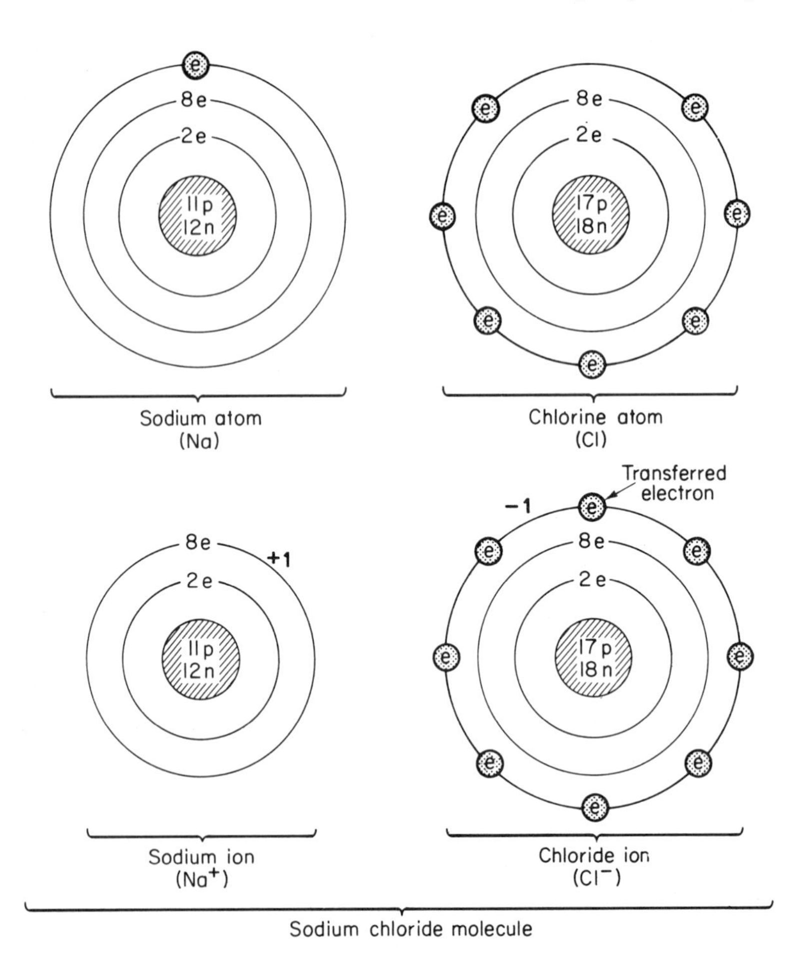

Fig. 5-2. Sodium, chlorine atoms and the sodium chloride molecule.

$$Cl + ⓔ = Cl^-$$

$$Na^+ + Cl^- = NaCl$$

where the little plus and minus signs at the upper right of the elements' symbol indicate that it is an ion and the charge on that ion. The symbol ⓔ is used to represent an electron, either lost or acquired. Because the individual atoms exist as ions within them, compounds like sodium chloride are called ionic.

In a nonionic compound, electrons are *shared* rather than transferred, and the constituent atoms do not exist as ions. The elements carbon (C) and hydrogen (H) and the molecule of methane (CH_4) are shown in Fig. 5-3. As a helpful visual model, one may imagine the shared electrons whirling

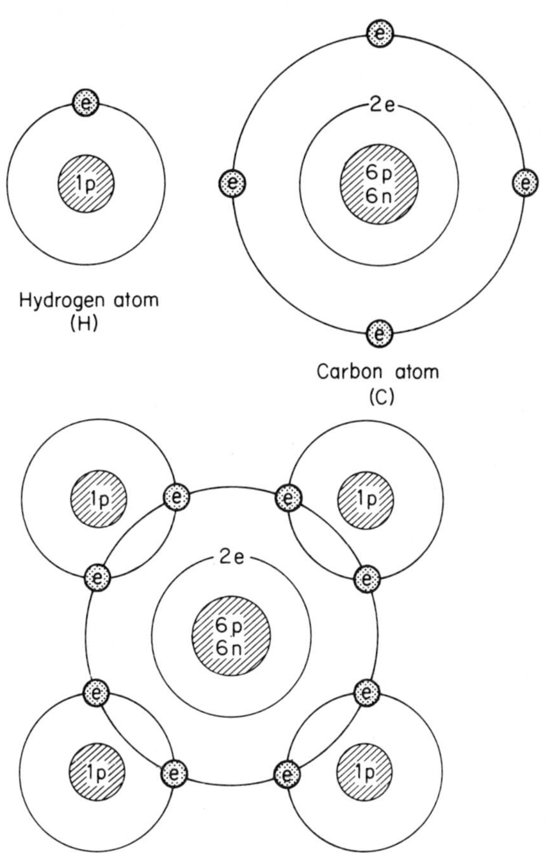

Fig. 5-3. Hydrogen, carbon atoms and methane molecule.

alternately around the hydrogen nuclei and then the carbon nucleus; thus each atom has a complete outer shell (at least on a part-time basis). In this case, it is the electron-sharing that holds the molecule together and compounds like methane are called nonionic. Being ionic or nonionic is not a black-white situation any more than being a conductor or insulator is. Electron bonds in compounds are of all degrees between 100 percent ionic and 100 percent covalent (shared).

Next, we present the concept of a *radical*, which is a group of elements within a compound that stay together and, as a group, act like a single element in many chemical reactions. We are particularly interested in ionic radicals—that is, radicals that occur in ionic compounds and behave as a single ion. Some important ionic radicals are listed in Table 5-1. Notice that there are both positive and negative radicals. The constituent elements of the radical may be joined nonionically although the radical as a whole

Table 5-1 Some Ionic Radicals

Name	*Symbol*	*Charge*
Ammonium	$(NH_4)^+$	+1
Hydroxide	$(OH)^-$	−1
Nitrate	$(NO_3)^-$	−1
Sulfate	$(SO_4)^=$	−2

forms ionic bonds. Thus, in sodium sulfate, Na_2SO_4, the two sodium atoms and the sulfate group form an ionic compound, but within the sulfate ion, the sulfur and oxygen atoms are bonded together nonionically.

Ionic compounds are grouped into three basic categories: *acids, bases* (or *alkalis*), and *salts.* An acid is a compound formed from one or more hydrogen, H^+, ions and some negative ion. Thus HCl, H_2SO_4, and HNO_3 are acids. A base is a compound formed from some positive ion (usually of a metal) and one or more hydroxide ions $(OH)^-$. Therefore NaOH, $Fe(OH)_3$, and NH_4OH are bases. A salt is a compound formed from some positive ion (usually metallic, and *not* hydrogen) and some negative ion (*not* hydroxide). For instance, NaCl, $Fe(NO_3)_3$, and $(NH_4)_2SO_4$ are salts. A salt is one of the products formed when an acid reacts with a base; the other is *always water.* A typical acid-base reaction, for example, is

$$H_2SO_4 + 2KOH = K_2SO_4 + 2H_2O$$

(sulfuric acid) + (potassium hydroxide)

= (potassium sulfate) + (water)

Water, itself, is a very interesting compound. If we write it as H(OH), instead of the usual H_2O, we see that under the previous definitions it seems to qualify as both an acid and a base. Actually, it is neither, and is only slightly ionic, but under the proper circumstances it can be made to act like a weak acid or a weak base. For our present discussion, two properties of water are important. First, it is an excellent solvent and will, in fact, dissolve almost anything to some extent (albeit slight). Secondly, when an ionic compound is dissolved in water (and they all do, quite readily), the water molecules act as a sort of shield to keep the oppositely charged ions apart. Remember the high dielectric constant of pure water? Thus, in a water solution, the constituent ions of an ionic compound are separated and free to move around in that solution; we say that the compound is *dissociated* in water. Of course, the solution as a whole has no electric charge, for the total charge on the positive ions balances that on the negative ions.

We are now prepared to return to a discussion of batteries.

5-6 PRIMARY CELLS

Volta had discovered that two plates of dissimilar metals, separated by a disk of moist paper, served as a continuous source of electrons. (He would have said "electric charge," since the electron was not yet discovered.) Fortunately, either the local drinking water was not too pure, or the moist paper picked up traces of ionic compounds from the perspiration of Volta's hands, or perhaps such compounds were left in the paper by the manufacturing process. Whatever the case, without ionic salts in the wet paper, the device would not have worked. Such a device is called a *voltaic cell*, or simply *cell*, if there is no chance of confusion with other meanings of the word. In popular speech, the word *battery* is often used instead of the correct term, cell, but a battery is, strictly speaking, a *group* of cells and we shall reserve the word battery solely for that purpose. The voltaic cell, then, consists of pieces of two dissimilar solid conductors, separated by an ionic solution that is called the *electrolyte*. The dissimilarity is required of the conductors so that one of them will give up a valence electron more readily than the other.

Volta soon discovered that he got more current from the cell if he moistened the paper with a salt or acid solution instead of with plain (impure) water. Let us consider a typical cell, such as that shown in Fig. 5-4, con-

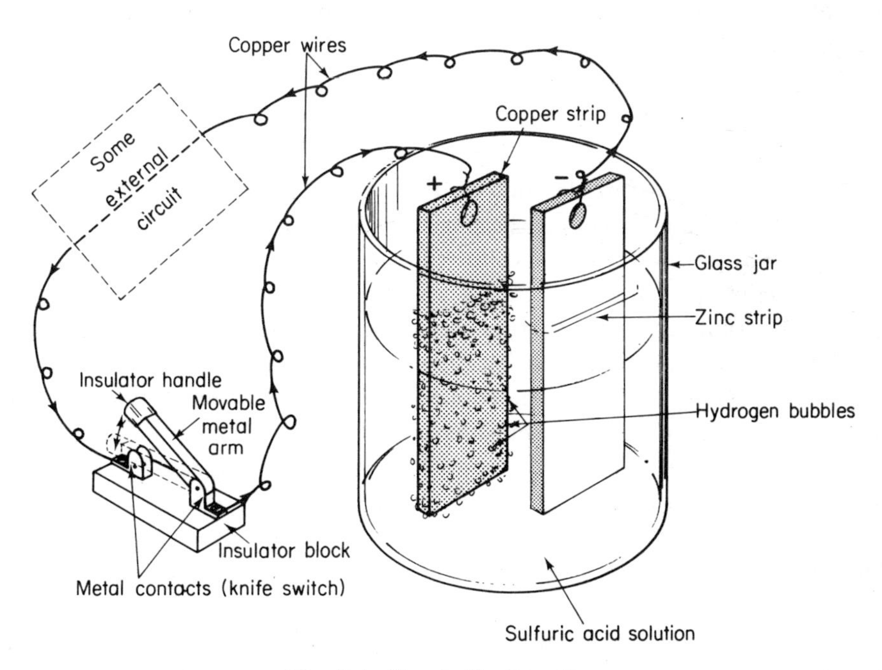

Fig. 5-4. Simple Cu–Zn cell.

sisting of a piece of copper, a piece of zinc, and a water solution of sulfuric acid (H_2SO_4). Wires connected to the copper and zinc provide an external path, or *circuit*, for electrons to follow outside the cell. We will not bother, at this time, about what the wires are connected to, except to note that a switch is provided for ease in completing or breaking the external circuit. (This type of switch is called a *knife switch.*)

Inside the cell, chemical reactions are taking place. Remember that, in the electrolyte, the H_2SO_4 is dissociated

$$H_2SO_4 = 2H^+ + SO_4^=$$

so that we really have a jar full of H^+ and $SO_4^=$ ions. Now, zinc gives up electrons much more easily than copper. Therefore the zinc electrode, under attack by $SO_4^=$ ions, is dissolving, and the zinc metal atoms are giving up electrons, becoming positive ions in solution, thus

$$Zn = Zn^{++} + 2ⓔ$$

Since electrons are made available at the zinc electrode, it becomes the negative electrode or *anode.* The zinc ions then force hydrogen ions to get out of the solution, and the H^+ ions do this by taking electrons from the copper electrode, the reaction being

$$2H^+ + 2ⓔ \text{ (from the copper)} = H_2 \text{ (gas)}$$

Thus hydrogen gas bubbles out at the copper plate, which, having been robbed of electrons, becomes the positive electrode or *cathode.*

With the handle of the knife switch in the position shown (solid lines), the circuit is open and there is no way for electrons to get from the zinc plate to the copper plate. In this condition, the zinc plate acquires enough of a negative charge to prevent any more Zn^{++} ions from forming and escaping into solution; similarly, the copper plate has acquired an equal positive charge, strong enough to prevent any more of its electrons from being stolen by H^+ ions. Hence the reactions quickly come to a screeching halt. If one were to connect a voltmeter (a device for measuring emf) across the two electrodes at this point, one would measure a characteristic potential called the *open-circuit potential,* which is about a volt and a half for the Cu–Zn cell.

Now let the knife switch be closed (position shown in phantom), providing an external circuit for electrons to follow between the copper and zinc. Electrons from the zinc anode will immediately move toward the copper cathode, as shown by the arrows, in an attempt to neutralize its positive charge. However, this permits H^+ ions to start "stealing" electrons from the cathode again, which, in turn, allows more zinc atoms to give up electrons to the anodes and go into solution as Zn^{++}; so the whole process continues and we get steady current. The emf now across the electrodes is near, but something less than, the 1.5 volts open-circuit potential, as will be discussed in Chapter 6.

Ideally this process would continue until either all the zinc or all the sulfuric acid was used up, whichever happened first. However, the design of our simple cell is a little too simple, and no matter how much zinc or acid is provided, it stops functioning very quickly. The problem is in the hydrogen gas formed. The bubbles coat the copper cathode, and since hydrogen gas is an insulator, the exposed area of the copper is reduced to almost nothing. This gas-film-coating problem is called *polarization.* (*Caution:* This is only one of several different meanings for the word, as used in a technical sense.)

The first successful solution to the polarization problem was the Daniell cell, or *crowsfoot* cell, invented by John Frederic Daniell in 1836. This cell, depicted in Fig. 5-5, made use of two electrolytes, one for the anode and

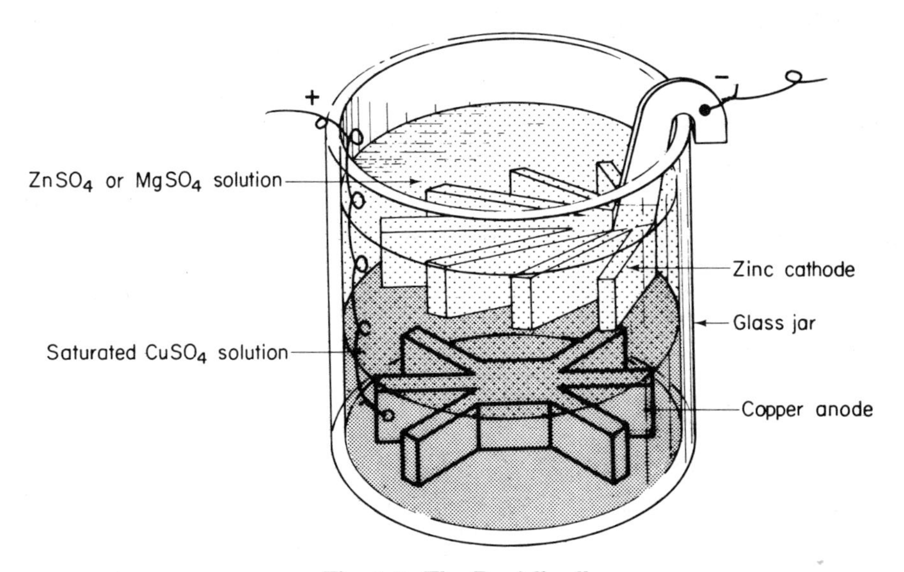

Fig. 5-5. The Daniell cell.

one for the cathode. The cell takes its nickname from the shape of the zinc anode and copper cathode. The anode electrolyte is a solution of zinc sulfate, magnesium sulfate or sulfuric acid, and the cathode electrolyte is a saturated (containing as much as will dissolve) solution of copper sulfate. The electrolytes are separated, in a few designs by a porous plate but usually just by gravity, because the copper sulfate solution is dense and will stay on the bottom. In operation, zinc dissolves from the anode while more copper (a conducting solid rather than an insulating gas) is plated out of the copper sulfate solution at the cathode. The reactions are

$$Zn = Zn^{++} + 2\textcircled{e}$$
$$Cu^{++} + 2\textcircled{e} = Cu$$

and the open-circuit potential is about 1.08 volts.

Several other cells, such as the Grove cell and the Bichromate cell, were subsequently developed, but all these cells suffered from one major disadvantage—they contained liquids—and since most of them had to have at least a small opening to let out gases formed in the electrode reactions, toting them about was a sloppy proposition at best, and dangerous with the acid-containing ones. What was sorely needed for many applications was a *dry* cell.

The first and most dramatic step in this direction was made by Georges Leclanché in 1865. In its earliest form, the Leclanché cell was a *semiwet* cell, using a zinc anode, a carbon cathode, a strong solution of sal ammoniac (ammonium chloride, NH_4Cl) as electrolyte, and a *depolarizer*, which consisted of manganese dioxide (MnO_2) mixed with carbon granules for better conductivity. The latter was packed in a porous cup around the anode. The MnO_2 depolarizer reacted with the hydrogen formed to produce water, and since provision for gas escape was therefore no longer necessary, the Leclanché cell could be a sealed unit. A further improvement was to add zinc chloride to the electrolyte, which gave longer consistent performance. The electrode reactions in the final form of the cell are somewhat more complex:

$$Zn\ (\text{from cathode}) = Zn^{++} + 2\textcircled{e}$$
$$2MnO_2 + Zn^{++}\ (\text{from } ZnCl_2 \text{ electrolyte}) + 2\textcircled{e} = ZnO \cdot Mn_2O_3$$

and the open-circuit potential is 1.5 volts.

The first sealed Leclanché cell appeared circa 1880. Thereafter, various materials were tried to thicken the electrolyte, including plaster of paris and cereal paste, as well as separating layers of absorbent paper. The first true dry cell (which should really be called a *damp* cell or *moist* cell) made its appearance about 1886, as represented by the design of Carl Gassner. The modern form of this cell is the classical "old number 6," shown in Fig. 5-6. This cell was, and still is, immensely popular, its 1.5-volt output being used to power field telephones, doorbell buzzers, ignitions, models, etc.

With the advent of the cylindrical hand flashlight in 1898, a demand was created for smaller versions of this cell, which evolved into the popular C, D, penlite, etc. cells now familiar to all of us. Some of these cells are shown

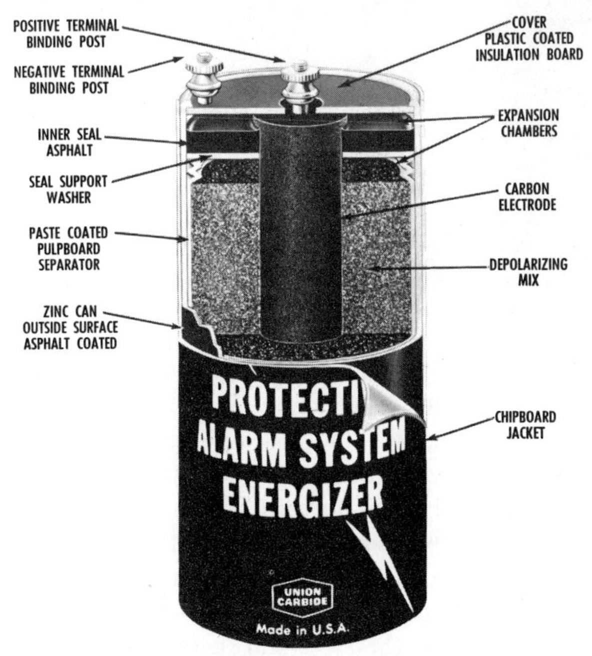

Fig. 5-6. Commercial 1.5-volt zinc–carbon dry cell. (Courtesy of Union Carbide Corp.)

in Fig. 5-7. This family of Leclanché-type cells illustrates an important fact about cells in general: the open-circuit potential of a given *type* of cell is essentially a fixed value, regardless of the size of the cell. All these cells, from the tiny penlite to "old number 6" are nominally 1.5-volt cells. The size of the cell determines how much total charge can be extracted from it, and the size, together with the configuration and composition of the components, further determines how fast that charge can be extracted (i.e., how much current can be drawn).

The cells discussed so far have had either salt- or acid-containing electrolytes. The electrolyte used may also be a base, and several such successful designs have been tried over the years. The first of these was the Lalande cell, introduced in 1880, using a zinc amalgam (alloy with mercury) anode, a copper oxide cathode, and a sodium hydroxide solution as elec-

Fig. 5-7. Assorted small dry cells.

trolyte. The most significant alkaline cell in present use is the *mercury cell*, which has come into wide use since World War II as a low-current source, primarily for use in hearing aids and other miniature applications. It contains a zinc amalgam anode, a mixed cathode comprising mostly red mercuric oxide together with a small amount of finely divided graphite, and an electrolyte that is a solution of caustic potash (potassium hydroxide, KOH) containing some zinc compound. The electrolyte is either gelatinized or absorbed in paper or cellulose cloth. The mercury cell produces an open-circuit emf of 1.35 volts and the electrode reactions are

$$\text{Zn} = \text{Zn}^{++} + 2\textcircled{e}$$
$$\text{Hg}^{++} + 2\textcircled{e} = \text{Hg}$$

All these cells behave alike in one respect. When the cathode material or the electrolyte is consumed, the cell is no longer usable. It is then said to be *discharged*, or more popularly, burnt out, dead, or shot. There is no way to rejuvenate these dead cells, short of cutting them open and replacing consumed material. Such cells are called *primary cells*, as distinguished from the type we meet next.

5-7 SECONDARY (STORAGE) CELLS

In contrast to the primary cell, we have the *secondary cell*, also called the *storage cell*, which is a device in which a useful quantity of electrical energy can be stored through a *reversible* chemical reaction. In a secondary cell, chemical reactions produce quantities of electrons at one electrode and a deficiency at the other, just as in a primary cell. The differences are that (1) no material is lost from the cell (like the hydrogen in some of the primary cells) and (2) passing an electric current from another source through the cell, in the reverse direction from that which the cell itself creates, causes the electrode reactions to reverse themselves and regenerate the starting materials. When electrical energy is being drawn from a storage cell, it is said to be *discharging,* and when electrical energy is being put into the cell to regenerate it, it is said to be *charging.* In theory, an ideal secondary cell could be completely discharged, and subsequently recharged, an indefinite number of times. In real cells, however, there are always slight losses of material. Regeneration is incomplete, for the chemical reactions are not 100 percent reversible in that side reactions (chemical reactions other than the desired ones) inevitably take place, consuming some active material. Hence, in due course, secondary cells also become permanently "dead." Two types of secondary cells are of great commercial importance: the *lead-acid cell* and the *nickel-cadmium cell.*

The principle of storage cells was discovered when it was found that if two identical platinum plates were used to pass a current through dilute sulfuric acid, and the current was then switched off, the plates and solution acted like a voltaic cell and would *supply* a small current of their own. It was subsequently found that if lead were used instead of platinum, the obtainable current was increased to the point where the cell offered promise as an electrical source, rather than merely a scientific curiosity. Perhaps the first useful lead-acid cell was constructed in 1859 by the Frenchman, Gaston Planté. Planté separated two long, narrow strips of lead by strips of flannel (allegedly torn from his wife's petticoat), rolled this "sandwich" up, and placed it in a jar of sulfuric acid. After charging (which converted part of the positive plate to lead oxide), he was able to obtain quite a respectable current from the cell.

Twenty-two years later, Planté's fellow countryman, Camille Faure, constructed an assembly (battery) of several lead-acid cells that was to be the prototype of the modern automobile battery. Faure's principal improvements were replacement of the simple sheet of lead with a heavy grid of lead, for greater active area, and, in the case of the positive electrodes, applying a lead oxide (PbO_2) powder directly to the grid.

The next major step in the evolution of these cells was the perfection of methods of construction that prevented the active materials from falling off the plates, as it did in the early Faure cells. In their present form, lead-acid batteries like the one shown in Fig. 5-8 find wide application in any situation where a large burst of electrical energy (i.e., high current) at low voltage is required for a short time and can subsequently be given back as

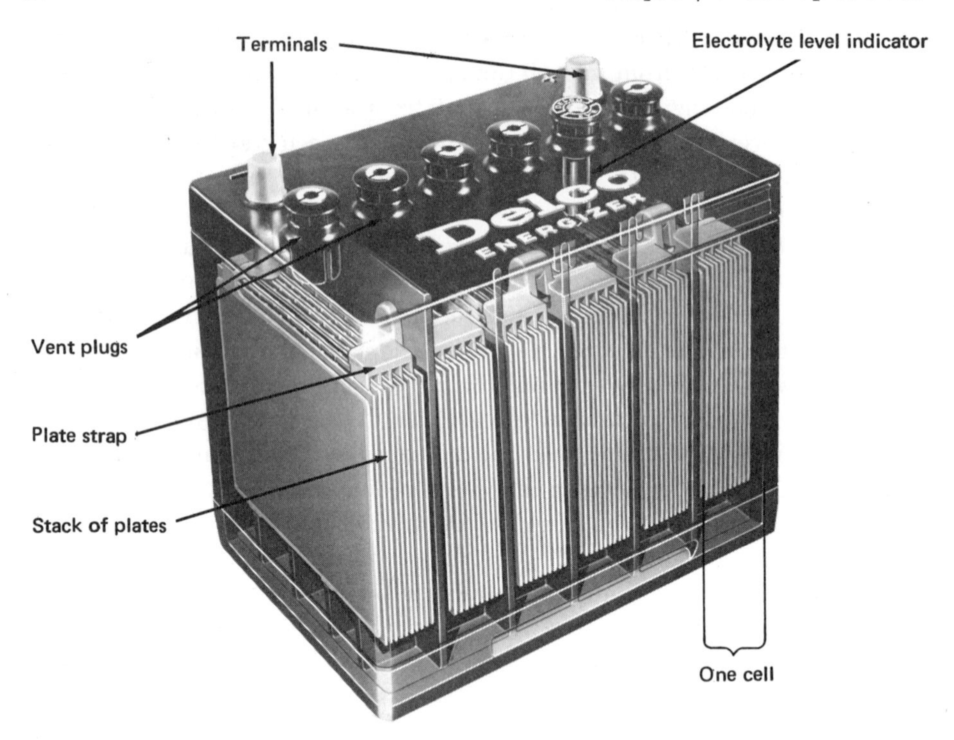

Fig. 5-8. Commercial 12-volt lead–acid automobile battery. (Courtesy of Delco-Remy, division of General Motors Corp.)

a "trickle" current over a much longer period of time, such as for starting car, truck, boat, and aircraft engines, and for emergency lights, as in hospitals. Over 30 million such batteries are produced in the United States each year.

The modern lead-acid cell is a nominal 2-volt cell, with a sponge lead (Pb) negative electrode and a lead dioxide (PbO_2) positive electrode. The electrode reactions are:

$$\text{charge} \longleftrightarrow \text{discharge}$$

$$Pb + SO_4^{=} = PbSO_4 + 2ⓔ$$

$$PbO_2 + SO_4^{=} + 4H^+ + 2ⓔ = PbSO_4 + 2H_2O$$

Notice that both the positive and negative plates are converted to lead sulfate during discharge, thus consuming the sulfuric acid and producing more water. Hence the electrolyte becomes more and more dilute as the cell discharges. This is how, merely by testing your auto battery's electrolyte concentration, the service station attendant is able to diagnose the state of discharge.

A storage cell that has come into more prominence in the space age is the nickel-cadmium cell. Thomas Edison laid the groundwork with his Edison cell, circa 1908, which used an iron cathode, a nickel oxide anode, and an alkaline electrolyte. Edison cells themselves are still used. They are inherently somewhat more rugged than lead-acid cells but give lower performance. The nickel-cadmium cell was invented in Sweden near the turn of the century by W. Jungner and K. Berg, who discovered that substitution of cadmium for the iron in an Edison cell improved its "shelf life" considerably. Thus, whereas an Edison cell will lose its entire charge in about three months if left idle, the cadmium cell will still remain half charged after a year on the shelf. For this reason, nickel-cadmium batteries are used in many of our satellites.

The active positive plate material of this 1.3-volt cell is a mixture of nickel hydroxide, $Ni(OH)_3$, and two nickel oxides, NiO and NiO_2, together with graphite for better electrical conductivity; the negative plate material is cadmium metal, Cd, mixed with some iron oxide; and the alkaline electrolyte is potassium hydroxide (KOH) solution. The electrode reactions during discharge are very complicated, but the net overall effect is that the cadmium is converted to oxide while some NiO_2 is reduced to NiO.

5-8 THE FUEL CELL

A most interesting type of primary cell is the *fuel cell,* within which the chemical oxidation of liquid or gaseous fuels, which would normally be called *combustion,* is conducted in a specially controlled fashion, the reaction energy being liberated as electrical energy. The principal virtue of this type of cell is that large quantities of reactants may be stored in tanks outside the cell and fed to the cell as needed. In theory, almost any liquid or gaseous fuel can be used, with either air or pure oxygen gas as the other reactant. Indeed, the feasibility of several types has been demonstrated. However, only one type of fuel cell has really been developed commercially to date—the *hydrogen-oxygen cell.* There are two major reasons for this. First, the hydrogen-oxygen cell operates near room temperature and pressure, while most other fuel cells operate at inconvenient temperatures, on the order of 900 °F or more, and at high pressure. Secondly, hydrogen and oxygen are available as rocket fuels on the big boosters used to launch our manned spacecraft, and it is these spacecraft in which the fuel cells are used as power sources because the weight of ordinary primary cells required to do the same job would be prohibitive.

A typical hydrogen-oxygen fuel cell is shown in Fig. 5-9.

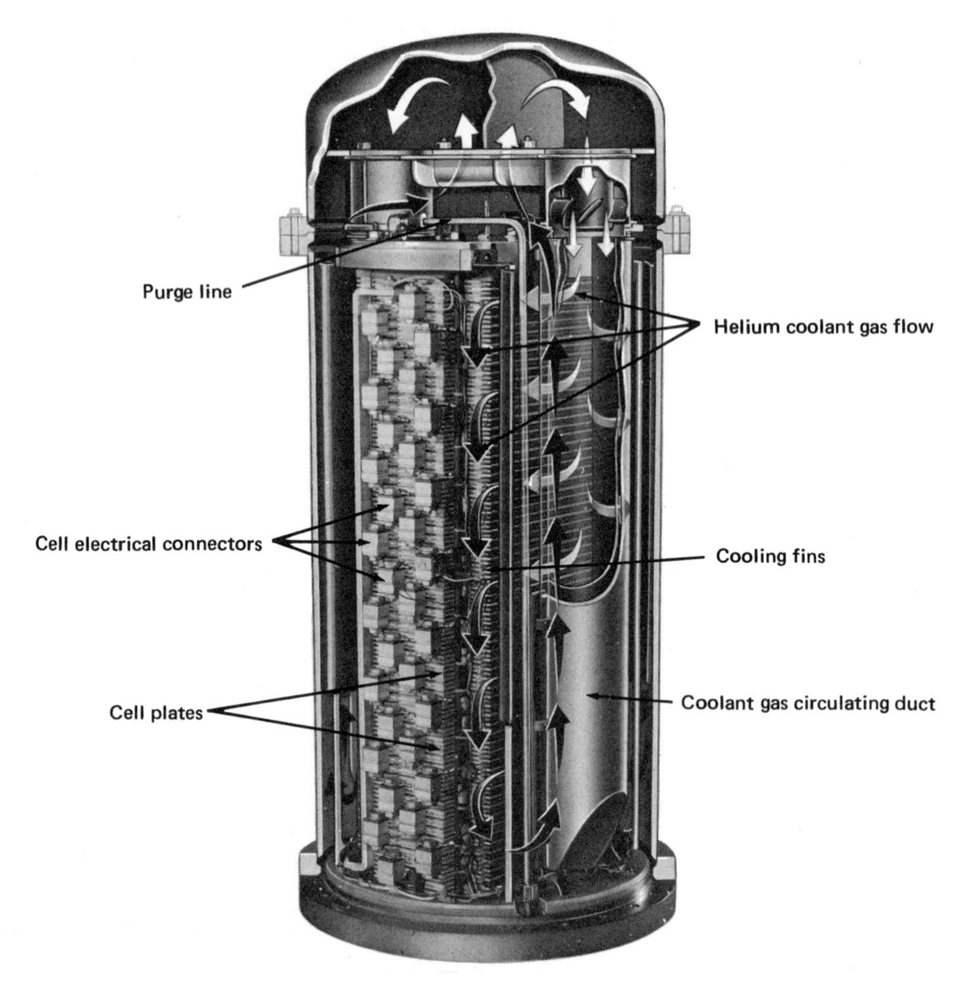

Fig. 5-9. Commercial H_2O_2 fuel cell power system developed for NASA. This 29-volt, 2000-watt unit contains 33 two-cell sections, each having an active area of 0.2 ft^2. (Courtesy of Allis-Chalmers Corp.)

In order to understand how the fuel cell works, we introduce two new concepts from chemistry.

A *catalyst* is a substance used to promote a chemical reaction, but which is not one of the reactants itself. The part played by a catalyst is not always well understood. One function of the catalyst is usually to provide an active surface upon which the reaction takes place. Sometimes, when the overall reaction takes place in a series of steps, the catalyst is actually consumed in an early step, only to be re-formed in a later step, so that the net effect is as though it did not participate.

Chemisorption, a term formed by combining the words *chemi*cal ad*sorption,* is distinguished from ordinary adsorption by the fact that activation, in the form of some sort of weak electron bond, occurs when the adsorbed substance contacts the surface upon which it is adsorbed. Thus chemisorption is the first step in many chemical reactions involving catalysts, particularly when one or more of the reactants is a gas.

We now return to the hydrogen-oxygen fuel cell. A simplified diagram of the typical configuration is given in Fig. 5-10. The major components

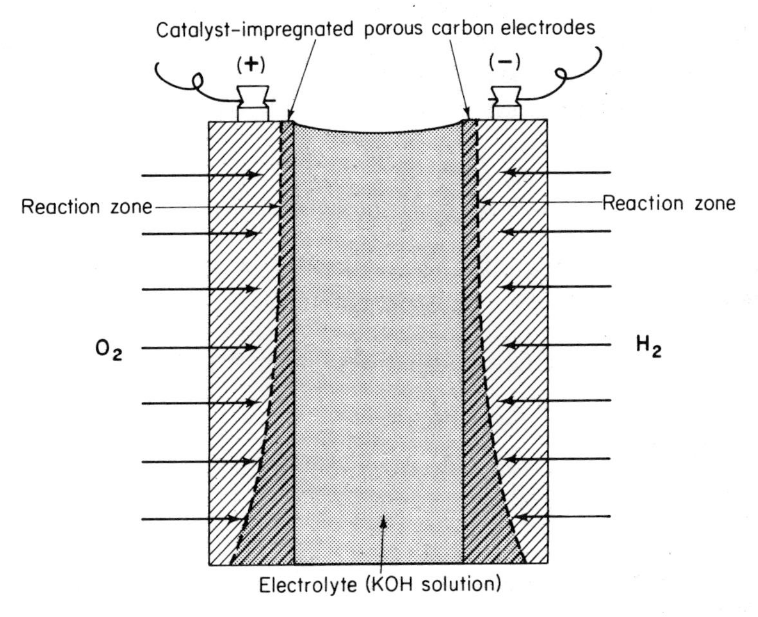

Fig. 5-10. Schematic representation of a hydrogen–oxygen fuel cell.

are two very special carbon electrodes, a container, an alkaline electrolyte, and the two gases. The carbon electrodes are very porous, the H_2 and O_2 being fed to the cell through these electrodes, and each is impregnated with a special catalyst. The H_2-electrode catalyst is one, or a mixture, of the so-called "noble" metals: platinum, rhodium, palladium, and ruthenium; a 40 to 60 percent mixture of palladium and rhodium gives good results. The O_2-electrode catalyst may be selected from among the oxides of copper, silver, gold, and nickel, mixed with some pure metal. The carbon from which the electrodes are made has no part in the chemistry of the cell but serves as a conductor to permit taking electrons from, and returning them to, the cell.

The electrolyte is a potassium hydroxide solution[2] and is not consumed by the cell reactions. The gas-feed pressures are carefully controlled so that the electrolyte permeates part way, but not all the way, through the electrodes. The reactions take place, on the catalysts, at the liquid-gas boundaries within the electrode. These regions are therefore called *reaction zones*. If gas pressures are too high, the electrodes are left "dry"; similarly, if gas pressures are too low, the electrolyte "floods" all the pores in the electrodes. In either case, there is no reaction zone and the cell will not function, hence the careful control of feed pressures.

In operation, a series of stepwise reactions occurs at each electrode. The hydrogen electrode becomes the negative electrode, and the oxygen electrode is positive. Both sets of reactions begin with chemisorption of the gas upon its respective catalyst. So, in the hydrogen electrode, the first step is

$$H_2 \text{ (gas)} = H_2 \text{ (chemisorbed)}$$

Next, under the activating influence of its catalyst, the two-atom hydrogen molecule breaks up into two individual free hydrogen atoms

$$H_2 \text{ (chemisorbed)} = 2H \text{ (chemisorbed)}$$

Two hydroxyl ions from the electrolyte then each give up an electron and combine with the hydrogen atoms to form water

$$2H \text{ (chemisorbed)} + 2(OH)^- \text{ (from electrolyte)} = 2H_2O + 2ⓔ$$

The overall effect is that hydrogen gas enters the electrode and combines with hydroxyl ions from the electrolyte to form water, leaving electrons available at that electrode.

Meanwhile, at the oxygen electrode, chemisorption is also occurring in the first step

$$O_2 \text{ (gas)} = O_2 \text{ (chemisorbed)}$$

In the next step, unlike the hydrogen electrode situation, the two-atom oxygen molecule does *not* break up but combines directly with a water molecule and two electrons to form a hydroxyl ion and a "perhydroxyl" ion (a new acquaintance, O_2H^-), thus

$$O_2 \text{ (chemisorbed)} + H_2O \text{ (from electrolyte)} + 2ⓔ = OH^- + O_2H^-$$

[2] Some modern O_2–H_2 fuel cells use a solid ion-exchange resin as electrolyte.

The OH^- goes into the electrolyte to replace one of those consumed at the H_2 electrode. The O_2H^- ion is an undesirable product, however, and if this were the last step of the O_2-electrode reactions, the cell would soon become polarized by the accumulation of these ions. Fortunately, the catalysts used at the O_2 electrode also catalyze the breakup of the O_2H^- ion into another hydroxyl ion, which can go back into the electrolyte, and a free oxygen atom, which can be used over in the oxygen reactions

$$O_2H^- = OH^- + O$$

The net effect at the oxygen electrode is therefore that the incoming oxygen gas combines with water to form hydroxyl ions, depleting the electrode of electrons in the process.

Looking at the net overall cell reaction now, we see that the OH^- ions consumed at the H_2 electrode are exactly replaced by those formed at the O_2 electrode but that the water formed at the H_2 electrode more than equals that consumed at the O_2 electrode. Hence the cell takes in H_2 and O_2 and produces water according to the reaction

$$2H_2\ (\text{gas}) + O_2\ (\text{gas}) = 2H_2O\ (\text{liquid})$$

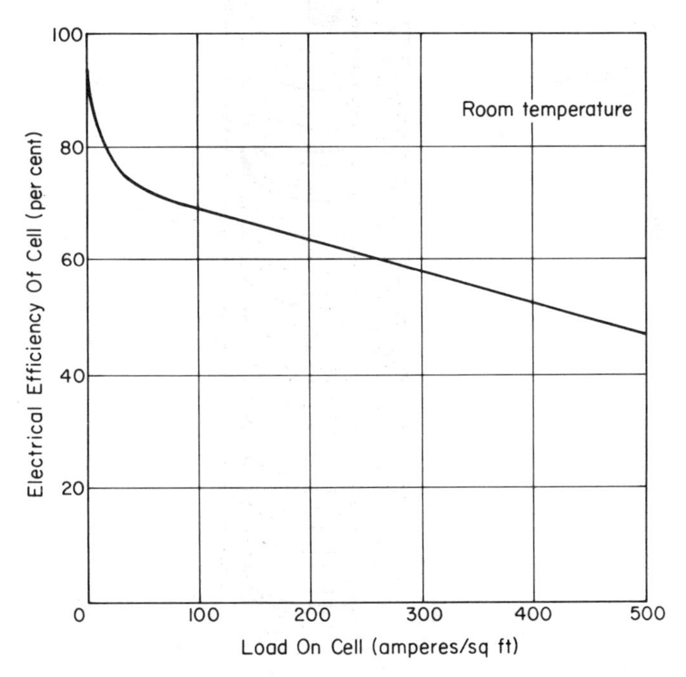

Fig. 5-11. Efficiency of a H_2O_2 fuel cell (typical).

which is the same reaction that would take place if hydrogen were burned in oxygen to produce heat energy instead of electrical energy.

This cell has an open-circuit potential of 1.12 volts and is theoretically capable of yielding 1.32×10^4 joules (enough electrical energy to light a 100-watt bulb for 2.2 minutes) for each pound of water produced. There are, of course, some losses. The determining factor in the efficiency of a fuel cell, as shown in Fig. 5-11, is the load it must bear—that is, how much current each square foot of plate area is forced to deliver.

5-9 BATTERIES

A group of cells, connected together, is called a *battery of cells*, or simply *battery*. As pointed out in Section 5-6, we often use the word battery

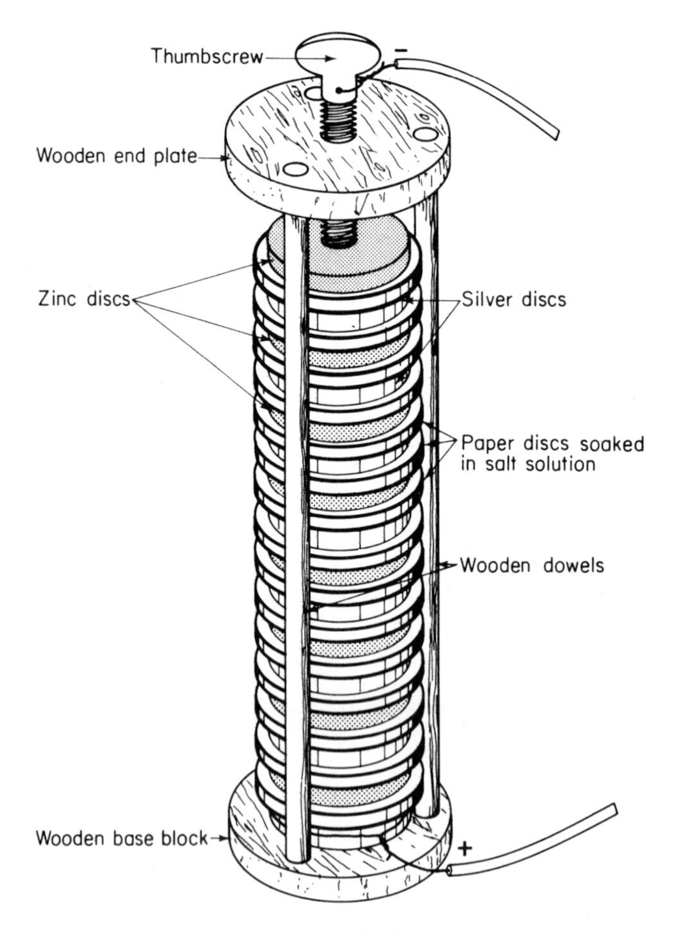

Fig. 5-12. A voltaic pile.

in everyday speech to mean just one cell, which is not strictly correct. On the other hand, in some cases, many little cells are packaged together in such a way that the assembly looks like one large cell when in reality it is a battery. For everyday purposes, it doesn't make too much difference, but the reader should be aware of the correct technical definitions of both words.

By 1800, soon after his discovery of the cell, Volta had combined a number of them to make the first operational battery. He called it a *pila* or "pile." The construction of the pile is indicated in Fig. 5-12. Disks of silver (or copper) alternated with disks of zinc, separated by paper disks that had been soaked in a salt solution. Volta supposed that it would function "... continuously, its charge being renewed after each discharge; it possesses in fine an inexhaustible charge." We know that the overenthusiastic Volta wasn't quite correct, for the pile would be exhausted when either the zinc or (more probably) the electrolyte was gone.

Another form of battery constructed by Volta is depicted in Fig. 5-13. He called it the *crown of cups*, each cell in the battery consisting of

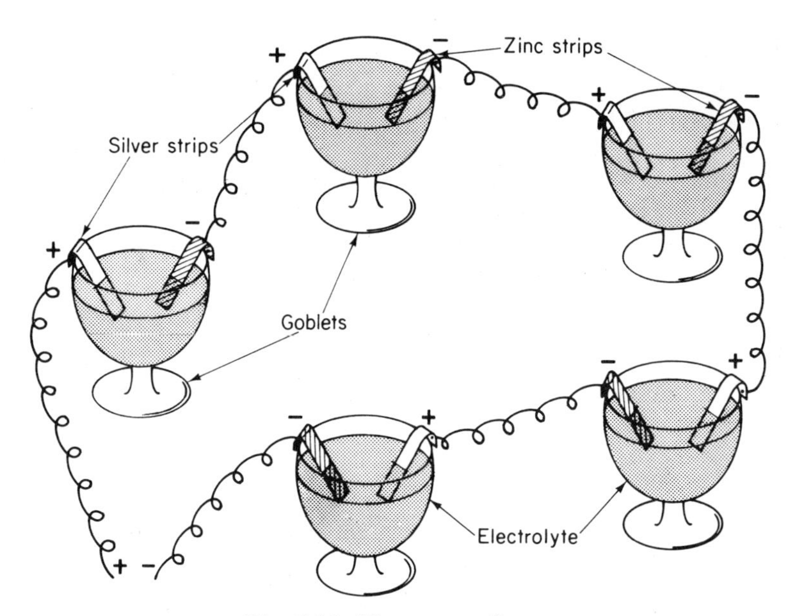

Fig. 5-13. The crown of cups.

a goblet partially filled with electrolyte into which a strip of zinc and a strip of silver or copper have been placed. The goblets are interconnected with wires.

Note that in both the pile and the crown of cups, the cells are connected in such a manner that electrons flow into a cell, out of it, into the next cell, out of it, and so on. Such an arrangement is called connection *in series*.

Furthermore, the order in which the electrodes are connected is negative-positive-negative-positive-etc. When a battery is connected in this way, the cells are said to be in series, *aiding*. A battery of cells in series, aiding, has an open-circuit voltage that is the *sum* of the individual open-circuit potentials of the cells. Thus a series-aiding battery of six 2-volt cells has an overall open-circuit potential of 12 volts as, for instance, in an automobile battery.

Another useful way to connect cells into a battery is to connect all the positive terminals together and all the negative terminals together. Cells wired together in this manner are said to be connected *in parallel*. The overall open-circuit voltage of a parallel-connected group of identical cells is the same as the open-circuit voltage of *one alone*. For example, the open-circuit voltage of six 2-volt cells in parallel is still 2 volts. The advantage is that six times as much current could be drawn from the parallel group as from one cell alone. Figure 5-14 *a* shows three "old number 6" 1.5-volt cells connected in series (aiding) and Fig. 5-14*b* shows the same three cells connected in parallel.

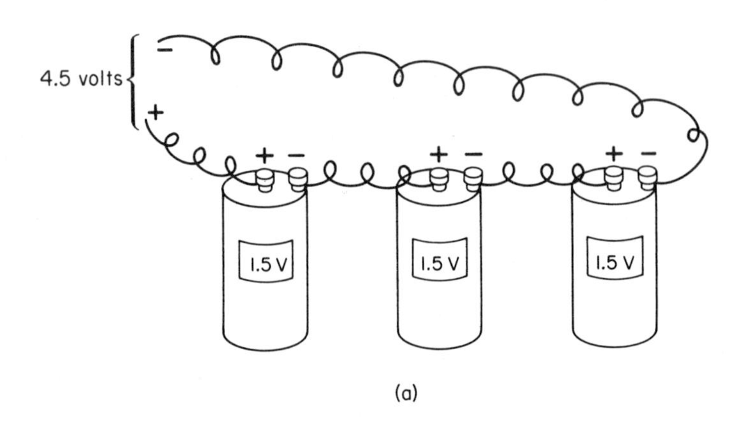

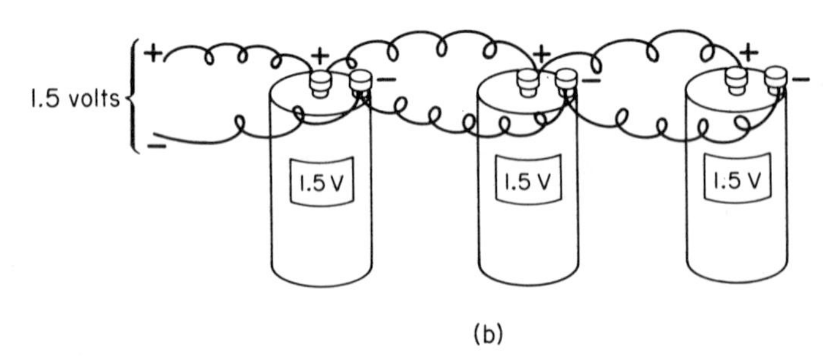

Fig. 5-14. Three 1.5-volt cells connected. (a) In series (aiding). (b) In parallel.

We may draw an analogy to fire-fighting in a primitive area where only men and buckets are used. If the burning house is on top of a hill and the source of water is at the base of the hill, the fire brigade must line up and pass the buckets up the hill so that they reach the fire one at a time. In the same way, cells in series provide a "hill's worth" of high potential but only a "bucket at a time" of current. If, on the other hand, there is a lake very near the burning house, each fire fighter can fill his own bucket and dash it on the flames. Similarly, cells in parallel provide only a "short distance" in potential but "many buckets at a time" of current.

5-10 OTHER SOURCES OF EMF

Almost all of this chapter has been devoted to various devices (voltaic cells) in which chemical energy is converted to electrical energy. Yet if it is true, as stated in Section 2-9, that *all* forms of energy are mutually interconvertible, then it follows that one should be able to obtain electrical energy from heat, radiant energy (light), and mechanical energy as well. Are there devices that do, in fact, accomplish these conversions? The answer is a definite Yes!. We have already seen one approach to the mechanical-to-electrical energy conversion in the electrostatic machines discussed in Chapter 3. Most of the electric current used today in our homes, office, factories, and automobiles comes from a mechanical-electrical conversion device called a *generator* (or sometimes dynamo), which is discussed in detail in a later chapter. A few words about devices producing electrical energy from heat and light are in order here.

In 1822 Thomas J. Seebeck found that when two different metals are used in a circuit and one junction point is hotter than the other, there is an electric current in the circuit. This effect is known as the *Seebeck effect*. A device consisting of wires or strips of two different metals, which are in contact at two junction points as in Fig. 5-15, is called a *thermocouple*. When there is a temperature difference between the two junctions, a difference of potential is generated in proportion to the temperature difference, which produces a current in the thermocouple. The voltages, and consequently the currents as well, produced by thermocouples are quite small. An indication of how small these voltages can be is given by Table 5-2, in which the voltage due to a temperature difference of 1°C is listed for several metals, with lead as the reference metal and one junction held at 0°C. Note that the voltage is measured in *microvolts* (millionths of a volt). A positive value indicates that current flows from the given metal to the reference metal at the cold junction. Because of their low output, thermocouples are rarely used as a primary source of emf. (You will find one such case in Chapter

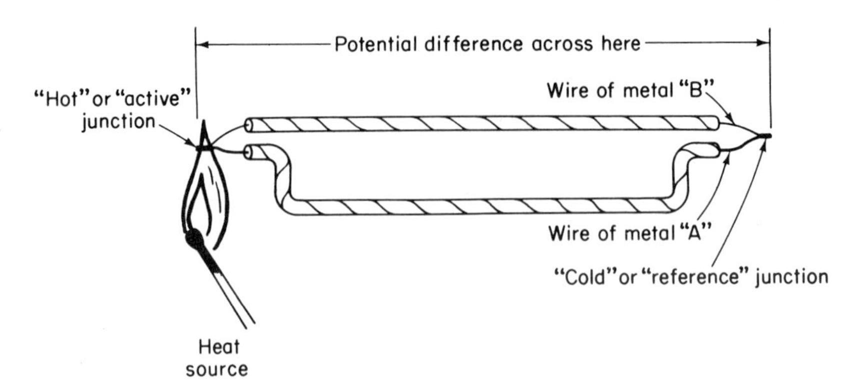

Fig. 5-15. The thermocouple.

Table 5-2 Thermoelectric Voltage Produced by Some Metals in Contact with Lead

Metal	*Voltage Produced by a Temperature Difference of 1°C (microvolts)*
Aluminum, 99% pure	−0.40
Brass, 66% Cu, 33% Zn	+0.70
Carbon, filament	+11.06
Constantan, 60% Cu, 40% Ni	−38.10
Iron	−51.34
Nickel	−19.07
Platinum	−3.04
Silver, annealed	+2.50
Zinc	+3.05

6.) They do, however, find wide application as temperature-measuring devices.

The practical conversion of light energy to electrical energy takes place in a device called a *photovoltaic cell, photoelectric cell, solar cell,* or sometimes simply a *photocell.* This device makes possible the photographer's lightmeter and the more sophisticated laboratory photometers, as well as providing power for today's satellites and other spacecraft.

As early as 1839, Alexandre Edmond Becquerel (father of the discoverer of radioactivity, Henri Becquerel) first noted and studied the effects of light on voltaic cells. It was thirty-seven years later, however, when W. Adams and R. Day found that a junction of selenium and platinum generated an emf when it was illuminated. This discovery is the foundation of photovoltaic cell design. A large number of semiconductors, such as selenium, germanium, silicon, and certain oxides, when impure, will show this effect, in conjunction with almost any metal, under strong illumination. Modern solar cells are constructed by forming a junction between two pieces of semiconductor containing different impurities. Ultraviolet light is much more effective than visible light in energizing photocells. The photoelectric effect was unexplained until 1905 when no less a scientist than Albert Einstein proposed the present theory. In brief, this theory states that light energy exists in discrete little "packages" called *photons*, which in some ways behave like particles in that they are capable of knocking electrons out of the semiconductor into the metal side of the junction.

Photocells, like thermocouples, have low current-producing capability although the photocell's output capability is somewhat greater than that of a thermocouple. Large numbers of them (solar cells) can, however, be connected in "strings" of series and parallel groupings to give any desired voltage and current capability. Typical solar cells might have an open-circuit voltage of 0.4 volts, be capable of delivering some 40 milliamperes (ma) of current for each square centimeter of junction area, and operate at an overall efficiency of about 12 percent. More is said about semiconductor photocells later in the text.

REVIEW QUESTIONS

1. Try to imagine a system of telegraphy, using Leyden jars charged from a friction machine as a source of energy. Describe how you would "send" a word. How long do you think it would take to send a 25-word telegram?
2. Suppose that you were to repeat Abbé Nollet's experiment with a very large Leyden jar, and with lamps in place of the monks. What results would you expect?
3. Can you guess which of the following are acids, bases, and salts: KNO_3, NH_4OH, $HCOOH$, H_3PO_4, $NaOH$, $CaCl_2$, $(NH_4)_2SO_4$, HF, HIO_3, $Mg(OH)_2$, $Pb(NO_3)_2$
4. What is the difference between a primary cell and a secondary cell? Suggest uses for each.
5. What is polarization? How is it avoided in modern cells?
6. A friend of yours has constructed an experimental Cu–Zn cell. Dissatisfied, he wants to obtain a higher voltage, so he is considering construction of a larger one. What would you tell him?
7. You have a box full of dry cells, each of which is capable of delivering up to 0.5 ampere at 2.0 volts. How could you connect several into a battery if you needed
   (a) 0.3 ampere at 6.0 volts
   (b) 0.9 ampere at 2.0 volts
   (c) 1.0 ampere at 4.0 volts
   (d) 0.1 ampere at 8.0 volts
   (e) 1.3 amperes at 2.0 volts
   (f) 1.5 amperes at 4.0 volts

8. How many millivolts potential difference would there be, when the cold junction is at 0°C and the hot junction is at 100°C, for
   (a) an aluminum-lead thermocouple?
   (b) a platinum-lead thermocouple?
9. What effect is visible in an operating O_2–H_2 fuel cell?

6

ELECTRONS GO OHM

6-1 ON CURRENT EVENTS

We are about to turn our attention to the "rules" that govern the flow of electric current. You will recall that an electric current is a *directed movement of a group of electrons* within a material (as opposed to the random wanderings of individual electrons, due to the material's temperature). The electrons constituting an electric current are somewhat akin to humans taking a vacation: they must be motivated to go where they are going, and they must have the means for getting there. The "motivation" that urges the electrons on their way is emf, the difference of potential between where they've been and where they're going. The "highway" they require in order to make the trip is the electric circuit, a *closed conduction path.* By a closed path, we mean that if you imagine yourself to be an electron and that you start to travel around the circuit from any given point, you will eventually return to the starting point without ever having encountered a "break" in your path (although you may have had to pass through several devices, besides wires, in making your tour). In describing a circuit, the word *closed* is not required because a prospective electric path simply is not a circuit if it is not closed.

The fact that a circuit is necessary for a current to exist marks one way in which the analogy between current in a wire and water flowing in a pipe fails badly. Thus water can be pumped from a well in one location, through a pipe, to (for instance) turn a water wheel, and the water leaving the water wheel *need not* be returned to the well. The electrons drawn from a source of emf, on the other hand, *must* be provided a way to return; otherwise a steady current cannot be established—hence the circuit.

It is therefore not strictly correct to say that we *consume* electricity, for all the electrons that we take from the electric company's source we also return. It is better to say that we *exploit* electricity. The situation in an electric circuit is something like the operation of a large commercial fruit farm: the farmworkers are electrons, the bunkhouse/refectory complex is the power source (the electric company's generators), and the orchards are the loads—the appliances we "plug in." The workers leave the bunkhouse/refectory area refreshed and well-fed (theoretically) and go to the orchards to pick fruit. When their energy is spent, at the end of the day, they return to the bunkhouse/refectory area to eat and restore their energy in rest. Similarly, the electrons leaving the source of emf at one terminal have *acquired* energy within that source, as measured by the potential difference across its terminals. In traversing the circuit, the electrons give up their energy by *doing work* of some kind, returning to the other terminal of the source depleted, and having their energy replenished again within the source. We pay the electric company for the electrons' "labor," just as the farmworkers are paid a salary for their labor.

6-2 DIRECT CURRENT AND OTHERWISE

For ease of study, electric currents are grouped into two categories: *transient* and *persistent.*

A transient current, like a transient resident of a city, is one that does not stay for long. Transient currents vanish in the proverbial "fleeting instant," which is to say that they exist for some period of time ranging from a tiny fraction of one second to perhaps a few seconds at most. Such transient currents, often called simply "transients," are usually associated with starting, stopping, or some other *sudden deliberate change* in an electrical process.

A persistent current, also called a *steady* current, is one that lasts for an indefinite time period and either (1) has a fixed value or (2) varies in some pattern that repeats over and over again.

When a circuit current flows only in one direction, that current is called a *direct current.* Here one may conveniently think in terms of the water flowing through the pipes in a home. Regardless of whether the water is turned on only momentarily (a transient) or left running for some time (a persistent current), the flow is only in one direction: from the water company's supply line, through the house pipes, and out through the drain to the sewer. Similarly, a direct current always flows from the negative terminal of the source of emf, around the circuit, and back to the positive terminal of the source of emf. A persistent current that has a fixed value *must* be a direct current, as we have defined it. Indeed, this idea is always conveyed by the term direct current, and if we mean anything other than a steady current of fixed value, we must say *fluctuating* direct current, *pulsating* direct current, *transient* direct current, etc. Figure 6-1 shows how the level

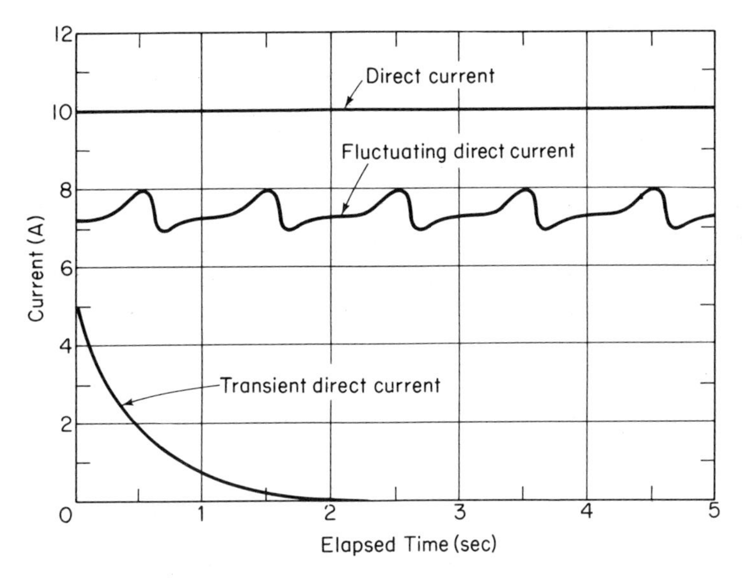

Fig. 6-1. Representative types of direct current.

of current changes with time for several examples of types of direct current. Note that what we call simply direct current is represented by a horizontal line, in this case at the constant value of 10 amp. The phrase direct current is used so often that it is abbreviated simply as *dc*, and we hereafter make use of this abbreviation. Furthermore, even the abbreviation has been in use for so long that people forget what the letters dc represent. We often hear the terminology dc current used, which if written out in full would be "direct current current," a ludicrously redundant expression. However, the phrase dc current is deeply embedded in the phraseology of most people who are involved with electronics (including the author) and has become accepted jargon.

Opposed to dc is *alternating current*, of course abbreviated as *ac* (and with the phrase ac current also accepted usage). An alternating current is said to exist in a circuit in which the direction of electron motion alternates, first in one direction, then in the other. In order for this to happen, the positive and negative terminals of the source of emf must frequently change places (i.e., change the sign of their charge, because of what is happening inside the source). Just as the water flowing through a hose was symbolic of dc, we may think of ac as the motion of many Ping-Pong balls in a number of games that are being played simultaneously, side by side, with all the players on the left hitting their respective balls at one instant and then all the players on the right returning them at the next instant. Thus the balls do not cover a distance of more than a few feet, but they are rapidly covering that same distance over and over again. In alternating current, the electrons, like the Ping-Pong balls, are constantly reversing their direction of travel. Nevertheless, there is "a directed movement of a group of electrons," which, under our definition, is legitimately called a "current." Some typical ac currents are shown in Fig. 6-3. Note that alternating currents must always vary from plus to minus about the zero line—plus and minus indicating opposite directions of flow in this case—and cannot have a fixed value. As with direct current, the words alternating current, or ac alone, describe a specific type of persistent alternating current, as indicated in Fig. 6-2. Any other type of alternating current carries some other qualifying term like transient or modulated.

We postpone further discussion of alternating currents of any kind, and of transient dc, until later in the text. For the present, we are concerned

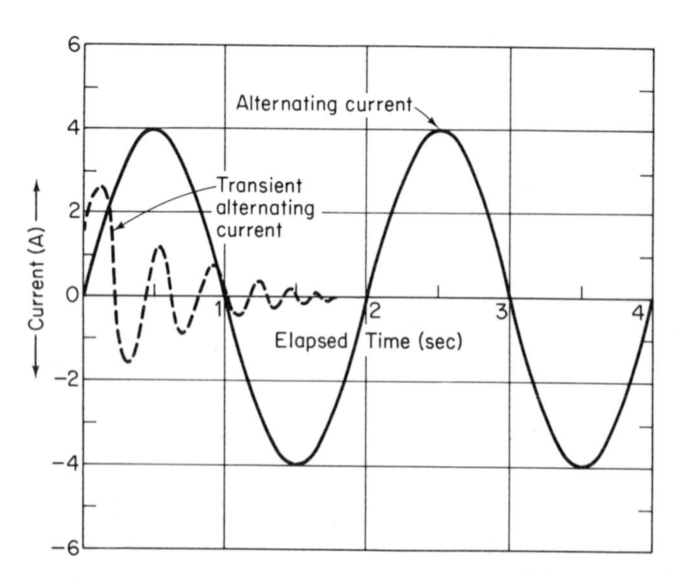

Fig. 6-2. Representative types of alternating current.

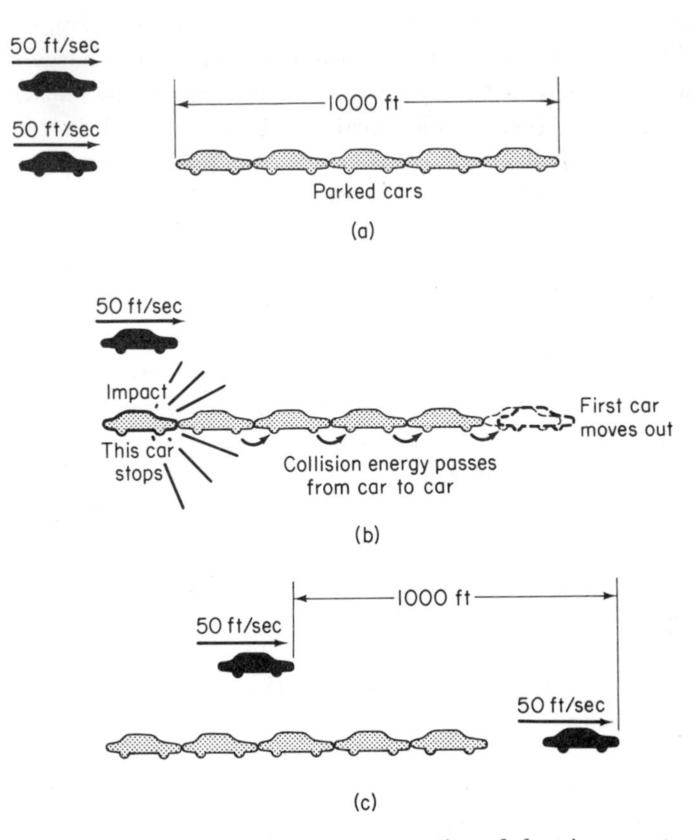

Fig. 6-3. Colliding car analogy to propagation of electric current in a conductor. (a) Before impact. (b) At impact. (c) After impact.

only with (persistent) direct current. All the sources of emf that we have studied thus far—electrostatic machines, voltaic cells and batteries, fuel cells, solar cells, and thermocouples—are dc sources. Furthermore, the natural occurrences of electric charge, such as lightning and electric eels, are also dc phenomena. Indeed, ac phenomena were virtually unknown on earth until man invented an ac generator.

6-3 WHICH WAY? HOW FAST?

Before electrons were discovered, Ben Franklin, as you may recall, had been more or less badgered into making a guess about which way the electric current flowed. He supposed that the current through the circuit was from plus to minus because, according to his single fluid theory, he believed the current to be a flow of positive charges. Today we know that Franklin was wrong—that the current consists of an enormous number of unimaginably small negative charges and actually flows from negative to positive in the external circuit. Yet the flow-of-positive-charge idea was so entrenched in electrical theory that the discovery of the electron did not overthrow it easily. Instead, electrical scientists found it less troublesome to keep all their old rules and conventions and relate them to *conventional current*, a *fictitious* current that was assumed to exist in a real circuit with the same magnitude as, but opposite flow direction to, that of the real (electron) current. Many older books on electricity still use conventional rather than electron current, and the reader is cautioned always to verify which concept a particular author is using. In this text, the term *current* is always to mean electron flux unless stated otherwise.

It would undoubtedly be easier to let the conventional current idea die out completely were it not for the fact that there is a very important area in modern technology where current is actually attributable to *positive* carriers. That area, to be discussed in some detail in a later chapter, is the theory of semiconductor devices, wherein we shall talk about positively charged "holes" as carriers.

Another interesting aspect of the electric current is the question of how fast it is propagated around a circuit. Remember Abbé Nollet's evaluation of the experiment with the monks? He claimed that all the monks in the chain felt the shock at about the same time. If they truly felt the shock at *exactly* the same instant, we would be forced to conclude that an electric current travels infinitely fast. In 1905, however, Albert Einstein showed, in his special theory of relativity, that no signal can be propagated faster than light travels in a vacuum. We would therefore expect that the *upper limit* on the speed of an electric current would be the speed of light in vacuum, about 3×10^8 meters per second (or about 186,000 miles per second).

The actual speed at which electrons move, on the other hand, is very slow. In a 16-gage wire (such as might be found on a lamp) carrying 0.9 amp of direct current, enough to light a 100-watt bulb, the speed of the average electron is about 1.7×10^{-4} feet per second (fps). If the length of wire between the wall switch and the bulb is 7 ft, and we assume that the electrons are "poised at the switch" like runners waiting for the starting gun, it would take about *half a day* for the first electrons to reach the filament. Yet we know that a lamp lights almost instantly when we snap the switch (and this includes the time required for the filament to heat up!). How can this be?

The answer lies in the fact that the phenomena we associate with an electric current are propagated as a *wave* effect although individual electrons move slowly. To understand this effect, consider two identical cars traveling alongside each other at a speed of 50 fps, as shown in Fig. 6-3*a*. (You can try this experiment with a group of toy autos or billiard balls.) Now let one of the moving cars collide with the last of a group of identical cars, parked bumper to bumper in a chain 1000 ft long, while the other moving car goes freely on its way. (Note that Fig. 6-3 is not to scale; it would take many more cars to make a chain 1000 ft long, but we will pretend that there are enough cars.) Physics tells us that, providing the cars are not deformed in the collision, the colliding car will come to a dead stop, while the first car in the chain will be knocked forward at the speed the colliding car originally had. The energy of the impact is passed down the chain of stopped cars at the speed of sound (traveling in steel). Thus, one second later, as per Fig. 6-3*c*, the noncolliding car has progressed 50 ft, while the moving car in the next lane is now just slightly less than 1000 ft ahead of it. If the cars were truly indistinguishable from each other, *as electrons are,* we would have to say that the car in the colliding lane apparently moved 1000 ft in one second although we know its speed was only 50 fps.

Similarly, although they do not really collide like the cars, electrons entering a conductor repel those "parked" in the valence shells of the nearest atoms. These electrons, in turn, repel those from the next group of atoms, and so on. This wave effect is propagated through the conductor at nearly the speed of light. Thus, instead of the entire day required by the electrons in our example to get to the bulb, the energy of the current arrives at the end of that 7-ft wire in just a bit more than seven billionths of a second (7×10^{-9} second). Hence, for many practical purposes, we can say that current traverses an electric circuit "almost instantaneously."

6-4 SHOWING THE WAY TO OHM

In every branch of science, there is a typical evolutionary pattern. First, the individual phenomena are *observed* and documented. Next, someone finds a convenient, and often clever, way to *classify* these observations into a number of useful categories, and it is at this point that the "science" is born. The scientists then mull over the classified observational data and propound some *qualitative theory* "explaining" those data (i.e., linking them to other already established scientific knowledge). The final phase in this evolution is *quantification*—being able to measure and predict "how much." To accomplish this last step, one must first decide (1) what is to be measured and (2) how to measure it. It is at this latter stage in the science of electrodynamics (i.e., electrical science involving persistent currents rather than static charges) that Georg Ohm entered the picture. Ohm's ideas were not plucked from a vacuum, however. The stage was set for him by others.

Henry Cavendish, the wealthy English recluse-scientist, actually anticipated Ohm by almost half a century. He introduced the idea of electric potential, but called it "degree of electrification," and studied how different materials conduct electrostatic discharges. Circa 1775, Cavendish completed a study whose results were tantamount to a statement of the law that now bears Ohm's name, but Cavendish did not publish.

By 1820 André Ampère, the Frenchman for whom the unit of current is named, had begun the development of a consistent system of electrical terminology. The first true current-measuring instrument, the *galvanometer*, was invented by a German physicist named Johann Schweigger in about 1823, the same year in which Seebeck discovered thermoelectricity (see Section 5-10). It is also worthy of note that, in 1822, Jean Fourier published his *Analytic Theory of Heat*, which was to influence Ohm's thinking profoundly. Without demeaning the greatness of Ohm's contribution, it could probably be said that his law was an "idea whose time had come."

6-5 THE LAW AND MR. OHM

Georg Simon Ohm made his entrance into the world in the quiet Bavarian university town of Erlangen on March 16, 1787. His father was a master mechanic and locksmith. Although the family's means were very limited, the father encouraged both Georg and his younger brother, Martin, to attend the university to study philosophy and mathematics. When his funds ran out, young Georg left the university to teach at a Swiss private school until his means permitted him to return to Erlangen once again, where he received his Ph.D. in 1811. The next six years were the difficult ones for Ohm, involving some rather unhappy teaching posts. Then, in 1817, he secured an appointment as instructor of mathematics and physics at the Jesuits' College of Cologne, where he enjoyed friends, a certain degree of prestige, access to a well-equipped laboratory, and—that most precious commodity—free time for research. He remained at Cologne for nearly ten years, and it was there that he made the discovery for which he is famous.

Influenced by Fourier's work with heat flow, Ohm believed that the electric current was under a "driving pressure" that was opposed by the circuit, the extent of this "resistance" depending on the geometry and composition of the wires. He conducted a series of experiments, using thermocouples as the source of emf and various lengths of copper wire as conductors. His current-measuring instrument was a Schweigger "galvanic multiplier" (i.e., a galvanometer) making use of a suspended compass needle. (The principle involved in the operation of this device will be made clear in

Chapters 8 and 9; for now, let it suffice to say that the angle through which the needle deflects is a measure of the current.)

In 1827 Ohm published a pamphlet entitled *Die galvanische Kette mathematisch bearbeitet* ("The Galvanic Circuit Investigated Mathematically"). In it he states the celebrated principle that has come to be called Ohm's Law of Constant Proportionality, or briefly, *Ohms law.* In modern terminology it may be stated as follows: In a conductor, maintained at a uniform temperature and containing no sources of emf:

The current in the conductor is simply proportional to the potential difference across it.

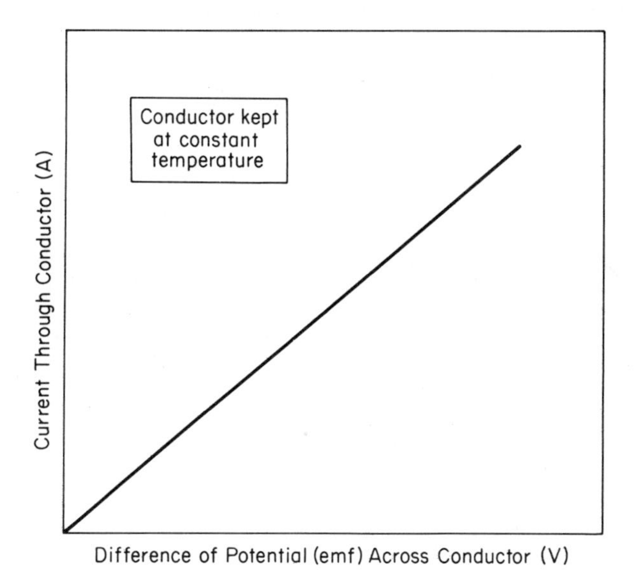

Fig. 6-4. Graphic representation of Ohm's law.

This beautifully simple relationship is shown by the graph of Fig. 6-4. In effect, Ohm's law states that if we double the potential difference across a piece of conductor, we double the current through it; and if we cut the potential difference in half, we also reduce the current by half (always assuming, of course, that we do not change the temperature of the conductor in question). The property of the conductor that opposes the current, and indeed determines how much current will flow for a given applied emf, is called the *resistance* of the conductor and is now measured in units called *ohms*, after Georg. Ohm also demonstrated that the resistance of the conductor depends on the material, temperature, and the shape of the conductor (as discussed in Section 6-6). The mathematics of Ohm's law is so simple that a child can do it:

$$\text{Current (in amperes)} = \text{applied emf (in volts)} \div \text{resistance (in ohms)}$$

Thus, if a conductor whose resistance is 2 ohms is connected across a 6-volt battery, the current in the conductor[1] will be $6 \div 2 = 3$ amperes. That's all there is to it!

The discovery of such a simple law for the electric circuit should have pleased Ohm's scientific contemporaries. It didn't! Worse, most of them refused to accept it and even denounced poor Georg as a madman. Part of their disbelief, if not their vehemence, can be understood if we recall that at that time most of the scientific world thought electricity to be

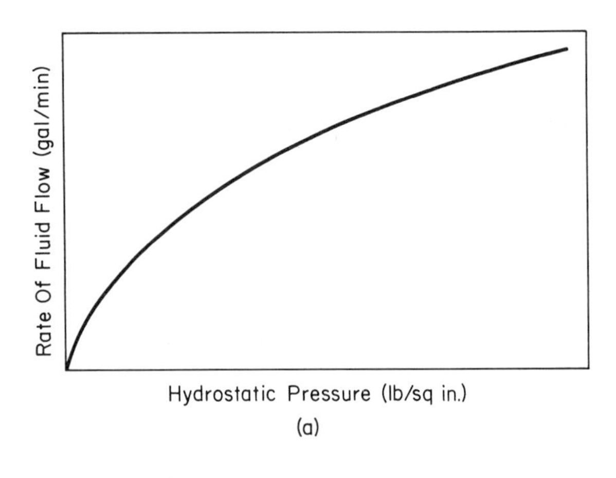

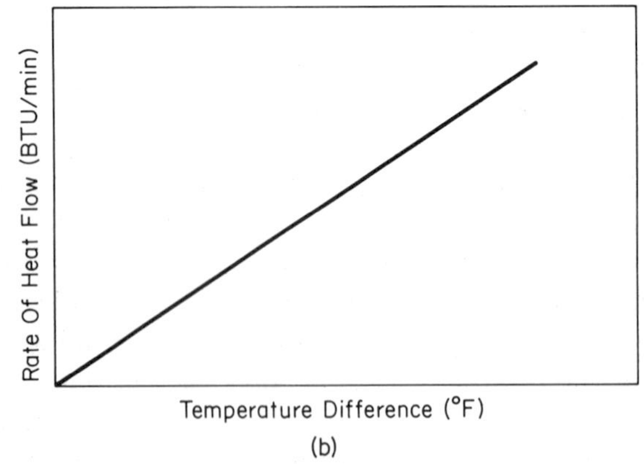

Fig. 6-5. Graphic representation of the laws for flow. (a) Of fluid through a pipe. (b) Of heat through a solid. Compare with Fig. 6-4.

[1] Neglecting the resistance "internal" to the battery and assuming that the temperature of the conductor does not change.

a fluid. In the newly discovered laws of fluid flow, the fluid "current" (volume of fluid flowing through a conduit in a unit of time) is *not* simply proportionial to the hydrostatic pressure, as shown in Fig. 6-5*a*. Ohm, on the other hand, had expected electric current to follow a law similar to that for heat flow, as expounded by Fourier and illustrated by Fig. 6-5*b*. Time eventually proved Ohm correct, but the initial wave of rejection and resentment caused him to resign his position in Cologne and spend 6 years in obscurity as a private tutor in Berlin. Gradually his work came to be recognized. In 1833 he was appointed to the Polytechnic School at Nürnberg. Real recognition was to come, not from his native Germany, however, but from the British. The coveted Copley Medal was awarded him by the Royal Society of London in 1841, and Ohm was made a member of that august society the following year. The eventual universal acceptance of his work, after a lapse of some 7 years, finally secured Ohm the post of professor of physics at the University of Munich, which he held until he died in 1854.

A fitting epilogue to the saga of Georg Ohm came in 1881 at the Paris meeting of the International Electrical Congress, where the ohm was officially adopted as the unit of resistance in honor of the perspicacious German. It was at this same congress that the units of potential and current, named respectively for the Italian, Volta, and the Frenchman, Ampère, were also adopted.

6–6 RESISTANCE AND CONDUCTANCE

In the previous section we learned from Georg Ohm that the amount of current (i.e., the rate of flow of electric charge) that will exist in any particular conductor, for a given dc potential difference maintained between the ends of the conductor, is determined by a property of that conductor called its *resistance*. The unit of resistance measurement is the ohm, and by international agreement one standard ohm has been defined to be the resistance of a column of pure liquid mercury, of specified diameter and length, and at a specified temperature. Thus we have a graduation to put upon a "yardstick" for measuring resistance. The word ohms is often replaced by the Greek capital letter omega (Ω), which looks something like a horseshoe. You may therefore see a particular resistance specified as 25 ohms or as 25 Ω, and they both mean the same thing.

It is often convenient to look at the tendency for a conductor to oppose the flow of current in another way—as the *tendency to pass current*—and in this way we speak of the *conductance* of a conductor. We hasten to emphasize that resistance and conductance are simply two different ways of looking at the *same* property of a conductor. It is something like discussing the relative "warmth" or the relative "coolness" of a group of objects; in either case, we are talking about ranking them according to temperature, but from two different points of view. As might be expected, an object having a high resistance has a low conductance (just as a very "cold" object has little "warmth") and vice versa. Numerically, the conductance of an object is defined as one-divided-by-its-resistance in ohms. The unit of conductance was, somewhat facetiously, called *mho* (ohm spelled backward), and this name stuck. This relationship between resistance and conductance is called a *reciprocal* relationship and is illustrated by the graph of Fig. 6-6 and Table 6-1. Note that a resistance of 4 ohms is equivalent to a conductance of $1 \div 4 = 0.25$ mho; a resistance of one ohm is the same as a conduc-

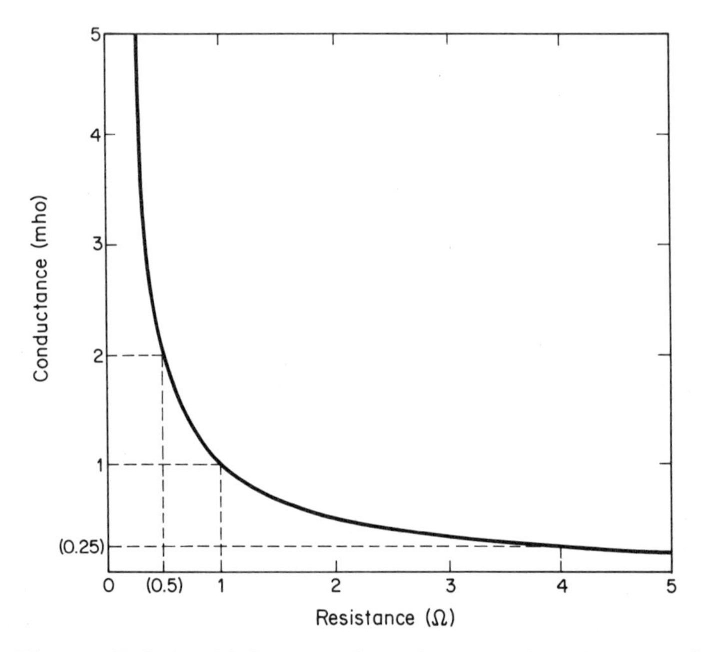

Fig. 6-6. Relationship between the resistance and conductance of an object.

Table 6–1 Some Equivalent Resistances and Conductances

Resistance (ohms)	*Conductance (mhos)*
1,000,000	0.000001
2,000	0.0005
10	0.1
1	1
0.05	20
0.0008	1250

tance of $1 \div 1 = 1$ mho; and a resistance of 0.5 ohm is equivalent to $1 \div 0.5 = 2$ mhos conductance. The factors that had to be specified in defining the standard ohm give us a clue as to what factors are important in determining the dc resistance or conductance of an object. These factors are

1. Nature of the material
2. Length in the direction of current flow
3. Cross-sectional area through which the current flows
4. Temperature of the material

The effect of each of these factors is illustrated in Fig. 6-7. The current in a length of copper wire of a given diameter at 70 °F varies with the emf across it as given by the graph A in Fig. 6-7. If we put the same emf across

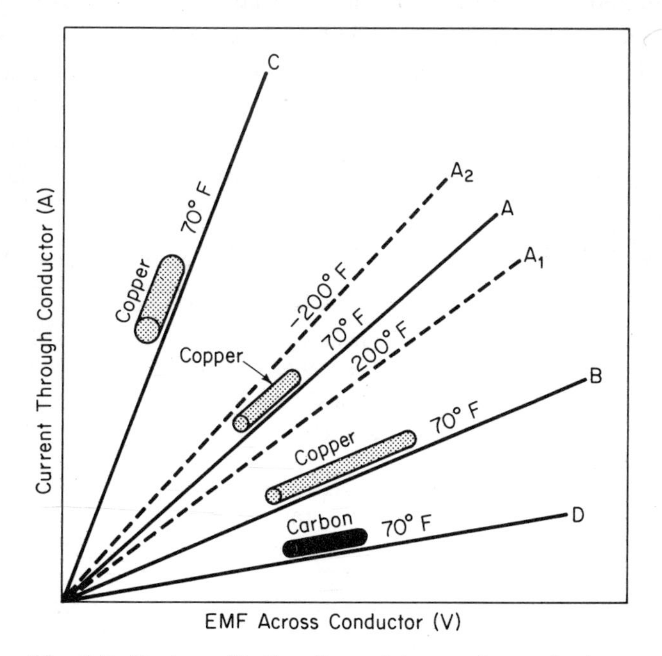

Fig. 6-7. Factors affecting the resistance of a conductor.

another copper wire twice as long, but with the same diameter (cross section) and also at 70 °F, the amount of current is one half as much, shown by graph B. So we see that *resistance increases* (conductance decreases) *with increasing length* of the conductor.

If next we put the same emf across a copper wire of the same length as the first, with a larger diameter (hence a larger cross-sectional area) than the first but still at 70°F, we see from graph C that the current increases, indicating a lower resistance. Therefore *resistance decreases* (conductance increases) *with increasing cross section.*

We return now to the original piece of copper wire and watch what happens to the current in it, for any given applied emf, if we heat the wire or cool it. A comparison of graphs A_1 and A_2 with A shows that less current flows in a warmer copper wire and more in a cooler one. This is the case for most metallic conductors, including all pure metals and most alloys. For nonmetallic conductors, however, the opposite is usually true. In this case, we must therefore be specific and say that *for metals, resistance increases* (conductance decreases) *with increasing temperature.*

Finally, if we apply the same emf to a copper wire and a carbon "wire" (filament), both at the same temperature and with the same dimensions as in graphs A and D, much less current flows in the carbon. This effect is related to the nature of the material itself and is usually expressed as a property of the material called *resistivity* (technically, volume resistivity), or the "reciprocal" property, *conductivity.* To get some appreciation for the broad variation among substances of the tendency to conduct an electric current, a brief table of relative resistivities and conductivities is presented as Table 6-2. In this table, the resistivity and conductivity of

Table 6–2 Relative Resistivities and Conductivities at 68°F

(Compared to Carbon = 1)

Substance	*Relative Resistivity*	*Relative Conductivity*
Aluminum	0.00081	1200
Beeswax	2×10^{17}	5×10^{-18}
Carbon (graphite)	1	1
Copper	0.0005	2000
Glass, plate	5.6×10^{15}	1.8×10^{-16}
Iron	0.0029	350
Mercury	0.027	37
Rubber	6×10^{17}	1.7×10^{-18}
Shellac	5×10^{11}	2×10^{-12}
Silver	0.00045	2200
Sulfuric acid (10%)	730	0.0014
Water, pure	1.4×10^{8}	7×10^{-9}
Wood, maple	10^{13}	10^{-13}

carbon are taken as unity. We can then see that copper is a 2000-times-better conductor than carbon, while wood only conducts about one ten-trillionth as well as carbon. Silver is seen to be the best conductor of all, but copper is such a close second and so much less expensive, that silver is used only in special applications.

The reasons for the effects of these various factors on resistance become evident when we recall the mechanism by which the electric current is propagated. Remember that each electron entering the conductor soon "bumps into" a free electron (i.e., an electron freed from the valence shell of one of the conductor's atoms), propelling it onward and, in effect, taking its place. The electron "pushed onward," in turn, bumps into another free electron from another atom, and so on; thus the "bump" is passed along the conductor. Now, when an electron is released from an atom, that atom becomes an ion (i.e., it is left with a positive charge), at least momentarily, until it acquires an electron and becomes a neutral atom again. Meanwhile, the electrostatic interaction between the moving electron and this positive ion robs energy from the electron and transfers it to the ionized atom, thereby causing that atom to vibrate. This energy loss by the electrons may be thought of as a kind of electrical "friction," and it is responsible for the resistance of the conductor.

If two wires are the same in all respects except that the cross section of one is larger than the other, a given current will pass more easily through the one with the larger cross section because the larger wire contains more free electrons for conduction. This is why resistance decreases with cross-sectional area.

As the length of a conductor is increased, the number of "collisions" the electrons will experience in traveling through it is also increased. Since the electron loses energy with each such collision, we can understand why resistance increases with length of conductor.

The variation of conductivity for different substances we can readily understand in the light of Chapter 4. We know that the atoms of the various substances differ in both the *number* of electrons available to be freed for conduction (i.e., those in the valence shell) and in the relative *energy required to release them.* We would therefore be surprised indeed if all substances conducted equally well!

The temperature effect is not so simple. Since the temperature of a crystalline solid is a measure of the vibration energy of the atoms comprising that solid, we can imagine the atoms vibrating back and forth in larger and larger swings about their "average" position as the temperature is raised. The greater this vibration, the greater the probability that "collisions" with electrons will take place. The result is that resistance increases with increasing temperature, at least for metals in the normal working temperature range. Later, we shall see that a strange thing happens to the resistance of many metals at very low temperatures—it vanishes!

It should also be noted at this point that the one *effect* of a current passing through a resisting conductor is to make the atoms of that conductor vibrate, which effect is observed as *an increase in temperature* of the conductor. In other words, the energy that the electrons lose in the conductor shows up as heat!

6-7 CONDUCTORS AND RESISTORS

In most electric circuits, there will be one or more specific devices in which virtually all the energy of the electrons is given up. Such a device is called a *load*. In dc circuits, the energy carried by the current is dissipated in the resistance of the load(s), while the energy lost as a result of both the resistance of connecting wires and the effective resistance internal to the source itself is *usually* negligibly small by comparison. For this reason, the connecting wires in a circuit are generally treated as if they had absolutely no resistance at all, that is, as very idealized conductors.

Very frequently, it is necessary to be able to place a certain definite amount of resistance into a circuit—so frequently, in fact, that devices to accomplish this purpose have become the most common circuit components. A circuit element whose sole purpose is to offer a controlled amount of resistance to the passage of an electric current is called a *resistor*. These little "bundles of ohms" come in a variety of types, sizes, and values.

Resistors can be broadly grouped into two classes: *fixed* resistors and *variable*, or adjustable, resistors. A fixed resistor is used when a single specific value of resistance is required at some point in a circuit. A variable resistor is used when, for one reason or another, it is desired to have some resistance within a specific range of resistance values, which can be selected at will, available at a point in a circuit.

A typical fixed resistor is shown in the foreground of Fig. 6-8, against a background circuit employing many such resistors. The four bands around the body of the resistor represent a coding system (to be discussed presently), which tells the value of its resistance. Basically there are three types of fixed resistors: *composition*, *metal film*, and *wirewound*. In a composition resistor, the resistance element is made of a composition material, as the name implies, in which the major conducting substance is carbon. Various filler materials are mixed with the carbon to give a particular resistivity and other desired physical properties. The carbon resistance element is formed as either a cylindrical slug or a hollow cylindrical film element. An example of the latter type of construction is shown in the cutaway view of Fig. 6-9. Here the carbon composition element has been bonded to a glass support. The copper conductors, called *lead wires* (or simply *leads*), which are used for making the connections to the circuit, also serve to carry away some of the heat generated in the resistor. These leads are electroplated with a tin/lead

Fig. 6-8. Typical fixed resistor (note size). (Courtesy of IRC, Inc.)

alloy for ease in making solder connections. Finally, the entire assembly is encapsulated in an insulative molding compound, with the color coding and other markings placed thereon.

The metal film and wirewound types of fixed resistor, shown in Figs. 6-10 and 6-11, respectively, are similar in construction to their composition counterparts except that the resistance element is a thin deposited film of high-resistance metal alloy in the former and a coil of alloy resistance wire in the latter. These types are used when more precise control of the resistance is required than is afforded by the composition type. In addition, the wirewound type can be sized to dissipate large amounts of power.

The physical size of a resistor relates not to its value in ohms but to the number of watts of power it can handle. Remember that the effect of resistance is to cause the moving electrons to give up energy as *heat*. The

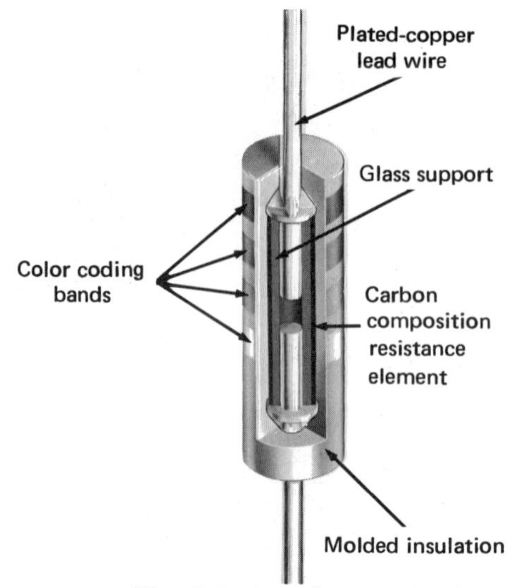

Fig. 6-9. A carbon composition resistor. (Courtesy of IRC, Inc.)

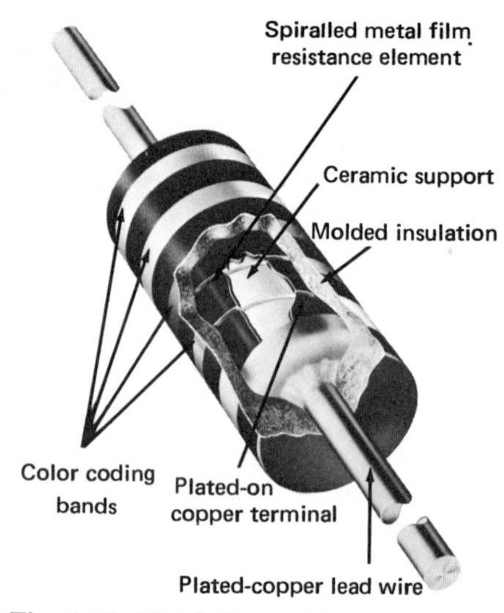

Fig. 6-10. Metal film resistor. (Courtesy of IRC, Inc.)

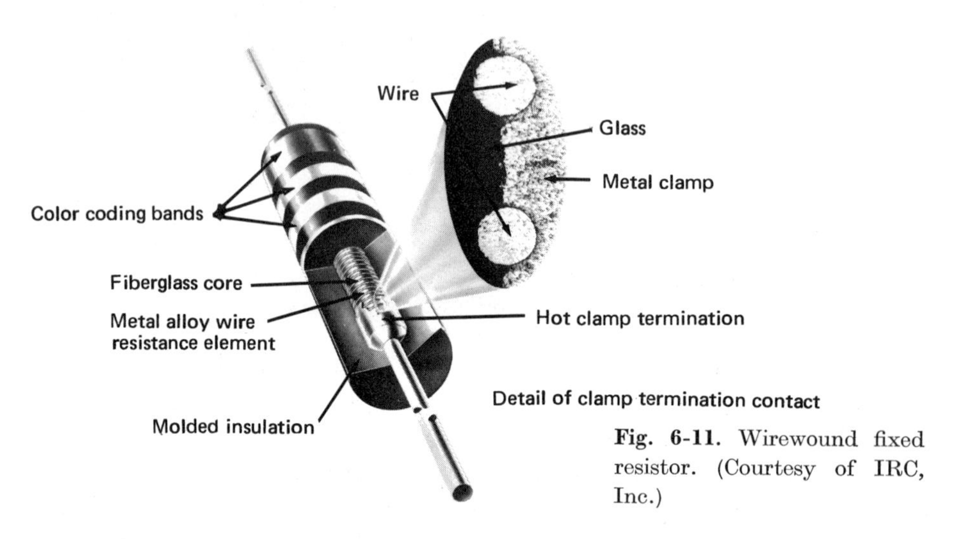

Fig. 6-11. Wirewound fixed resistor. (Courtesy of IRC, Inc.)

rate at which that heat energy is liberated—that is, the *power* (see Section 6-2) dissipated—determines how hot the resistor will get. If heat is generated faster than the resistor can get rid of it to the surroundings, the resistor's temperature will increase until it melts and "burns out." In choosing a resistor for an application, the circuit designer must therefore specify both the resistance value and the power-handling capacity of resistor. Fixed resistors, such as those we have seen in Figs. 6-8 through 6-11, come in various sizes, as shown in Fig. 6-12, that will safely dissipate up to 1 or 2 watts of

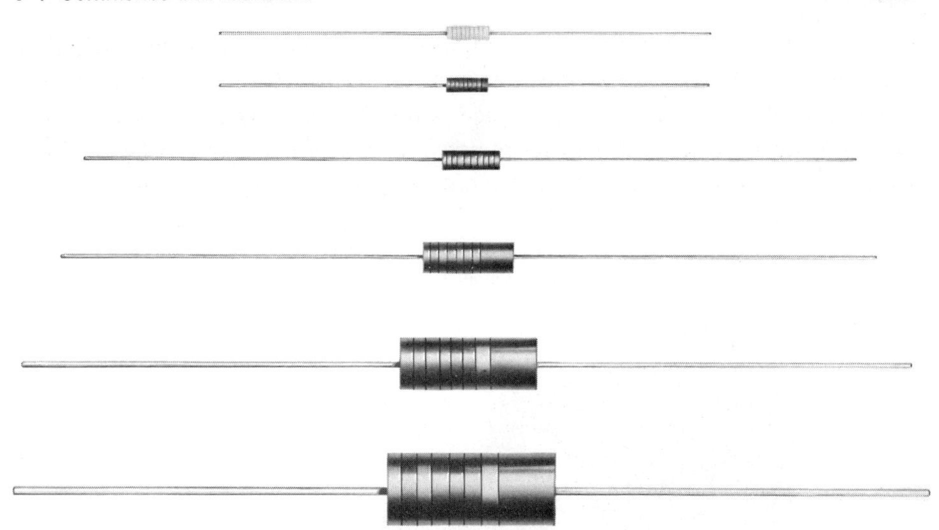

Fig. 6-12. Fixed resistors of various sizes, left to right: 0.1-watt, 0.125-watt, 0.25-watt, 0.5-watt, 1-watt, 2-watt. (Courtesy of the Allen-Bradley Co.)

power. For greater power-handling capacity, a power resistor of the type shown in Fig. 6-13 is used. Power resistors come in sizes rated up to 250 watts.

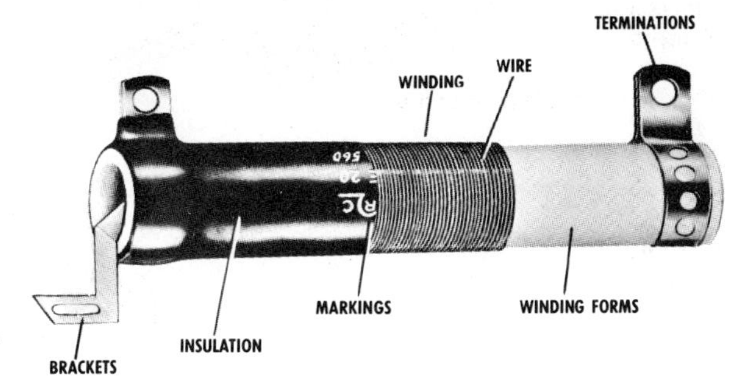

Fig. 6-13. Construction of a power resistor. (Courtesy of IRC, Inc.)

Variable resistors are of similar construction to fixed resistors, having either a composition, metal film, or wirewound resistance element. In the variable resistor, however, there is a movable contact called a *wiper*, which can be placed anywhere along the length of the resistance element by the adjustment mechanism, usually a rotational mechanism. The unit is fitted with one terminal that is connected to the movable contact, and either one or two terminals connected to the end(s) of the resistance element. The three-terminal version is called a *potentiometer* (i.e., a device to meter or dole out potential). The two-terminal version is called a *rheostat* (a device "to put current in static balance"). Actual rheostats are rarely seen items today, for a potentiometer can always be connected to function as a rheostat. The maximum resistance of the variable resistor is the total resistance of its element, while its minimum resistance is nearly zero. Its resistance at any given setting is the fraction of its maximum resistance represented by that portion of the total length of the element between the wiper contact and the end of the element connected to the circuit in question. Figure 6-14 shows

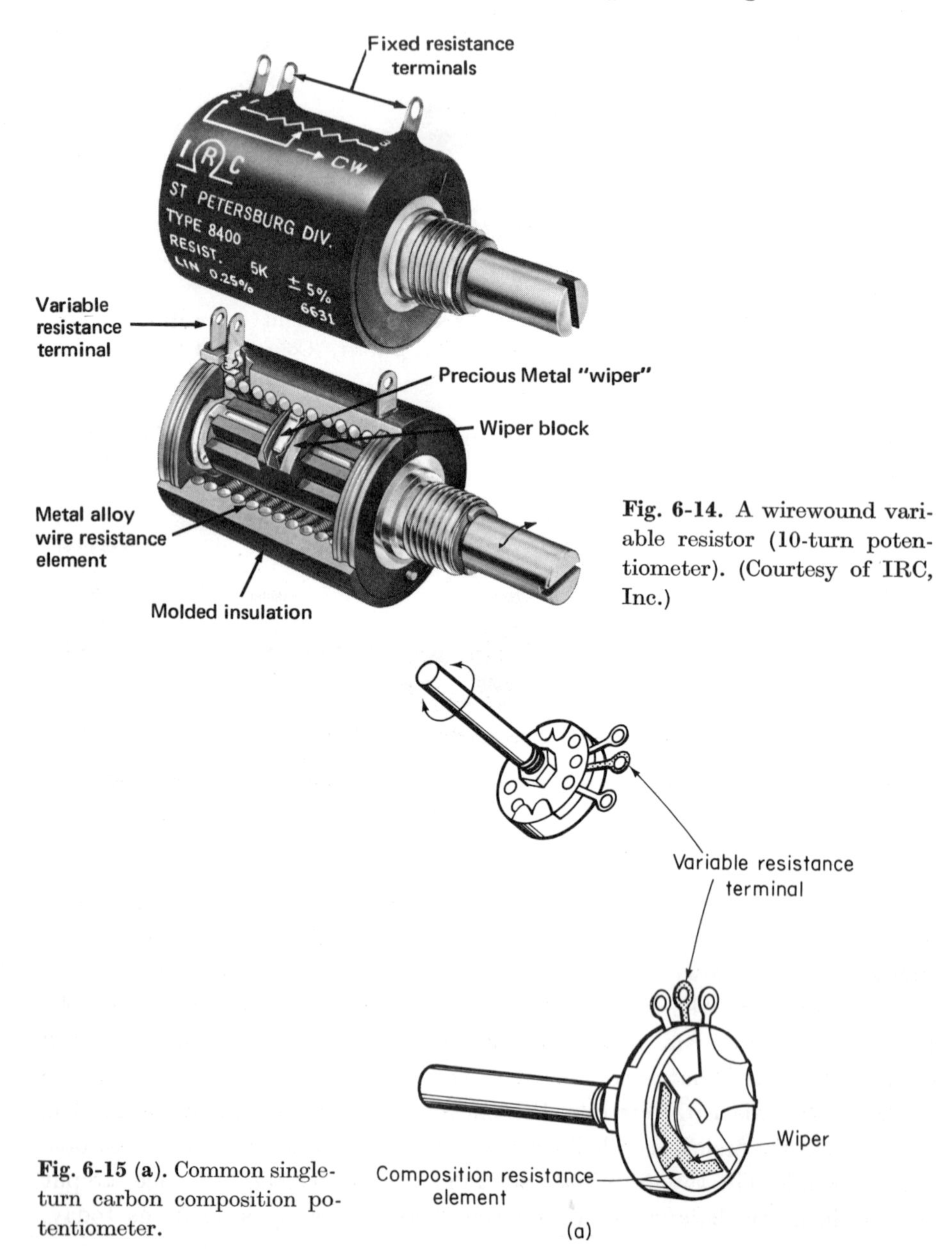

Fig. 6-14. A wirewound variable resistor (10-turn potentiometer). (Courtesy of IRC, Inc.)

Fig. 6-15 (a). Common single-turn carbon composition potentiometer.

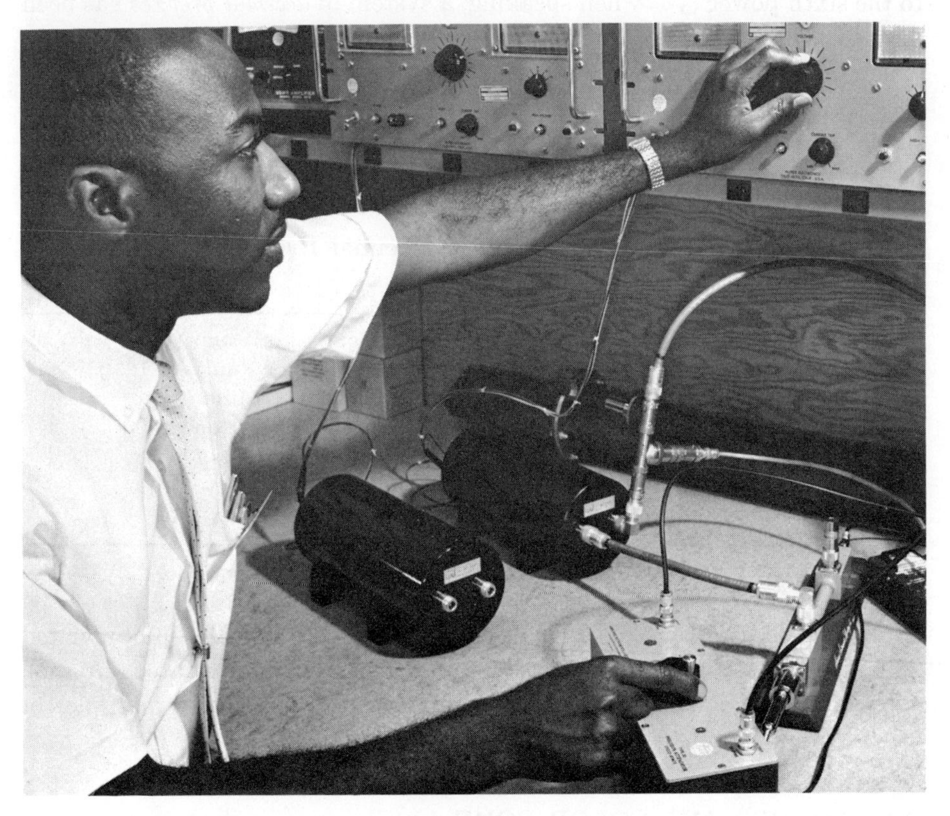

Fig. 6-15 (b). Using a potentiometer to adjust a voltage level.

a wirewound variable resistor (a potentiometer) that requires ten turns of the shaft for the full range of adjustment from zero to maximum resistance (in this illustration, 5000 Ω). Less-precise potentiometers employ 5 turns, 3 turns, and, most commonly, only one turn to cover the full range. The ordinary one-turn potentiometer is illustrated in Fig. 6-15*a*. It is commonly used in electronic devices as a means of voltage level adjustment, in the manner shown in Fig. 6-15*b*.

6-8 RESISTANCE VALUES; THE DECIMAL PREFIXES

As you may well imagine, the resistances of objects cover a wide range, from "almost nothing" in, say, a large bar of silver to a resistance so large (perhaps quintillions of ohms) that it may be considered practically infinite. Even commercial resistors, like those depicted in Fig. 6-9 through 6-11, are made in standard sizes ranging from $\frac{1}{10}$ to 100,000,000 ohms. To avoid the strings of zeros, and to avoid even having to say "... times ten to the sixth power . . ." when speaking, a system of *decimal prefixes* has been devised. These decimal prefixes represent some multiple of the basic unit, which in this case is the ohm, but they can be used with any physical unit. The decimal prefix is combined with the name of the basic unit to make a single word, which now represents a much smaller or larger unit. A brief[2] list of the more important decimal prefixes is given in Table 6-3. Thus

Table 6-3 Important Decimal Prefixes

Prefix	*Abbreviation*	*Pronounced*	*Means*
pico–	p	**pē**-kō	trillionths* of 1 unit ($\times 10^{-12}$)
nano–	n	**năn**-ō	billionths* of 1 unit ($\times 10^{-9}$)
micro–	μ (Greek mu)	**mȳ**-krō	millionths of 1 unit ($\times 10^{-6}$)
milli–	m	**mĭl**-lē	thousandths of 1 unit ($\times 10^{-3}$)
kilo–	k	**kē**-lō	thousands of units ($\times 10^{3}$)
mega–	M	**mĕg**-ā̆	millions of units ($\times 10^{6}$)
giga–	G	**jĭg**-ā̆	billions* of units ($\times 10^{9}$)

*American numeration system.

22,000,000 ohms is more commonly written as 22 *megohms* (abbreviated 22 MΩ), and 470,000 ohms is written as 470 *kilohms* (abbreviated 470 kΩ). We shall make constant use of these prefixes in dealing with multiples of other electrical units.

6-9 THE EIA-MIL COLOR CODE

On small fixed resistors, as noted in Figs. 6-8 through 6-12, the value of the resistance is indicated by a color coding scheme, which employs parallel bands of color placed nearer to one end of the resistor. This scheme has been standardized by both the Electronics Industries Association (EIA) and military (MIL) specifications; hence it is called the *EIA-MIL Color Code*. There are always three color bands to specify the value of the resistor, and sometimes a fourth to specify tolerance. The meaning of the colors is given in Table 6-4 and Fig. 6-16. As examples of coding, consider the four resistors shown in Fig. 6-17.

The resistor in Fig. 6-17*a* has four code bands, sequentially *brown, red, blue,* and *silver*. The first two bands give the significant digits, which from Table 6-4 are: brown = 1 and red = 2. The third band gives the multiplier; in this case, *blue* means add 6 zeros (or multiply by 10^6). Thus the resistance is 12,000,000 ohms = 12×10^6 ohms = 12 millions of ohms = 12 *meg*ohms. The silver fourth band tells us that the resistance has a ±

[2] See Appendix 2 for a complete list.

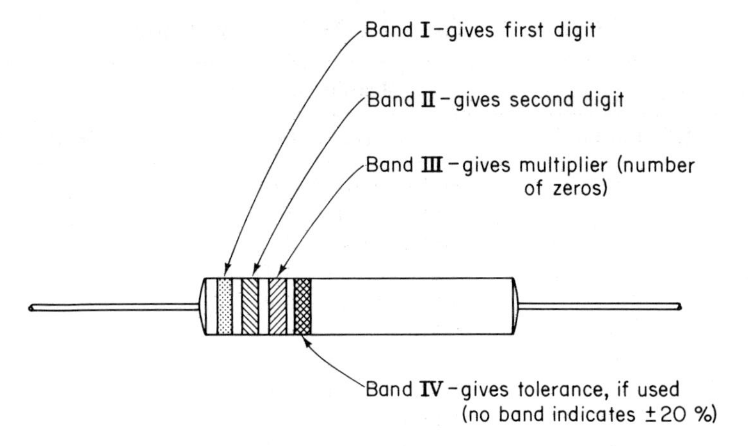

Fig. 6-16. The EIA-MIL color code.

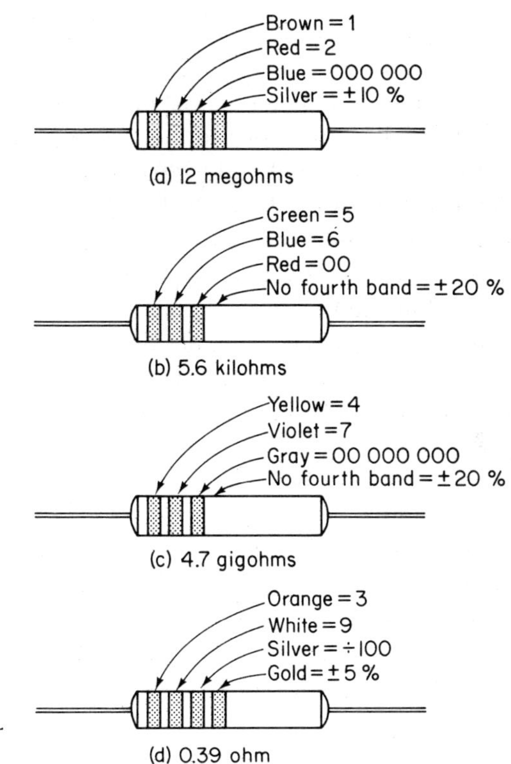

Fig. 6-17. Reading the EIA-MIL code.

10 percent tolerance. In other words, its measured resistance can be anywhere between 90 and 110 percent of the *nominal* 12 megohms, that is, between 10.8 and 13.2 megohms.

The second resistor (Fig. 6-17*b*) has the coding green (5), blue (6), red (2 zeros = $\times$ 100, or $\times$ 10^2). Its nominal value is therefore 5600 ohms

Table 6-4 The EIA-MIL Color Code

Color	Meaning of Color		
	In Band I or II	In Band III	In Band IV
Black	0	Add *no* zeros ($\times$ 1)	–
Brown	1	Add 1 zero ($\times$ 10)	–
Red	2	Add 2 zeros ($\times 10^2$)	–
Orange	3	Add 3 zeros ($\times 10^3$)	–
Yellow	4	Add 4 zeros ($\times 10^4$)	–
Green	5	Add 5 zeros ($\times 10^5$)	–
Blue	6	Add 6 zeros ($\times 10^6$)	–
Violet	7	Add 7 zeros ($\times 10^7$)	–
Gray	8	Add 8 zeros ($\times 10^8$)	–
White	9	Add 9 zeros ($\times 10^9$)	–
Gold	–	Divide by 10 ($\times 10^{-1}$)	$\pm 5\%$
Silver	–	Divide by 100 ($\times 10^{-2}$)	$\pm 10\%$

$= 5.6 \times 10^3$ ohms $=$ 5.6 thousands of ohms $=$ 5.6 *kil*ohms. The lack of a fourth band indicated a tolerance of 20 percent, so the actual resistance is some value between 4480 and 6720 ohms.

The third resistor, shown in Fig. 6-17*c*, also has no fourth band and is thus a 20 percent tolerance resistor. Its yellow-violet-gray markings show a nominal value of 4,700,000,000 ohms $= 4.7 \times 10^9$ ohms $=$ 4.7 billions of ohms $=$ 4.7 *gig*ohms.

The fourth resistor has four bands. The first three bands, orange-white-silver, indicate a nominal value of $39 \div 100 = 0.39$ ohm. A gold ± 5 percent tolerance band shows that its true value is between 0.37 and 0.41 ohms.

The reader may be somewhat surprised at the large tolerances permissible on commercial resistors. In most circuits, however, a 10 to 20 percent variation in resistance can be easily tolerated. There is usually this kind of variation, or more, in other components, such as tubes, transistors, and capacitors. In more critical circuits, either a final adjustment will be provided by using a variable resistor or special high-precision components will be used throughout when the additional expense is warranted.

REVIEW QUESTIONS

1. Recall the difference between potential (or emf, or voltage) and current. Why are there frequently signs warning "Danger—High Voltage" but *not* "Danger—High Current"?

2. A friend tells you that last month he consumed 360 kilowatt-hours of electricity. What does he really mean to say?
3. Give some examples of transient and steady currents.
4. A flashlight, a portable radio, an automobile horn, and an electric cigarette lighter are all operated from dc power sources. Can you tell which depend on transient currents and which depend on steady currents for their intended overall effect?
5. If the emf supplied to our homes is ac—that is, it alternates as described in Section 6-2—why do we not observe the lights flickering on and off at night?
6. We have stated that a closed circuit must be provided for a current to flow. Describe this path in a common flashlight.
7. Both my flashlight and transistor portable radio operate immediately when I turn the switch on. My console television set does not operate, however, until a few moments after the switch is turned on. Does this not show that electrons from a battery travel faster than those from a wall socket? Why?
8. When lit, an ordinary 60-watt bulb has a resistance of about 240 ohms. According to Ohm's law, how much current passes through it from the 120-volt supply?
9. To triple the current through a conductor, we could triple the voltage applied, but we decide to change the resistance instead. Will the new resistance be greater or smaller than the original value? By how much? How is the conductance changed?
10. What is the conductance of a circuit component whose resistance is 50 ohms? 100 ohms? 2 ohms? 4 microhms?
11. Which is the better conductor, aluminum or iron? Carbon or mercury? Wood or glass?
12. What is the nominal resistance of a resistor that is marked with the following color bands:
    (a) blue-black-green (b) white-orange-brown-gold
    (c) red-green-violet-silver (d) blue-white-orange
13. In a plant that manufactures resistors, the actual resistances of a sample of ten units, all color-banded brown-green-yellow-silver, were measured to be:

163,100	155,800
147,600	139,200
134,900	161,100
161,000	144,300
152,700	135,500

    Which, if any, should the inspector reject?
14. Compared to one watt of power, how much bigger or smaller is a microwatt? A kilowatt? A megawatt? A picowatt?

7

LET'S GO TO THE CIRCUITS

7-1 SAFARI INTO THE DC CIRCUITS

In this chapter we are going to begin to explore some dc circuits. The fundamentals gained here will then serve us well in understanding more sophisticated circuits later. We are also going to consider the energy expended in electric circuits, for this is what our monthly electric bill is all about. First, however, we need a clear and simple way to picture an electric circuit.

7-2 CIRCUIT CARTOONS

In describing an electric circuit, as for any other assembly of several component parts, a picture is worth at least a thousand words. It would, however, be cumbersome, indeed demanding of an artistic talent few of us possess, to draw a true-to-life picture of the components involved in every electric circuit we chose to discuss. Moreover, several supplementary cutaway drawings would be required to show the essential features of a circuit containing elements like transformers, vacuum tubes, and transistors. Fortunately, two convenient symbolisms have been developed through the years, for depicting the essential features of an electric circuit for the purposes of analysis. These are the *functional block diagram*, usually called simply *block diagram*, and the *electrical schematic diagram*, usually called simply *schematic.*

These diagrams bear about the same relationship to a picture of a circuit that a cartoon does to a photograph: the diagrams do not attempt to portray actual components; instead they show how the essential character of each component, or of a group of components, contributes to the behavior of the circuit.

In a block diagram, *a component or group of components is represented by a simple block* bearing a title that describes its function. The blocks are normally drawn as rectangles or squares although other shapes are often used. A single line is used to connect each block to any other block to which it sends, or from which it receives, one or more signals. The block diagram is used most frequently to illustrate the *logic* of a complex circuit, each block representing a major function and involving (usually) a group of components. The connecting lines, which show the flow of data from one function to another, may each represent a number of conductors. Figure 7-1, for example,

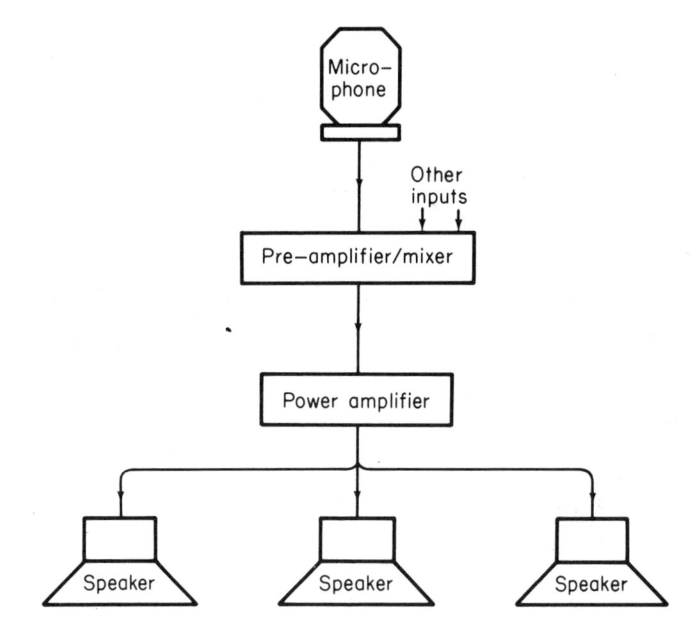

Fig. 7-1. Public address system. Example of a functional block diagram.

shows a block diagram of a public address (PA) system. Note that this type of diagram gives no information as to how an amplifier or a preamplifier/mixer is constructed. Block diagrams will play a prominent role in subsequent discussions of many electronic devices. For the moment, however, we are more interested in the second type of pictorial symbolism—the schematic diagram.

A schematic diagram is a line drawing representing an electric circuit, in which *a distinctive symbol is used to represent each circuit element* and a single line is used to represent *each interconnecting conductor*. Each symbol for a particular type of component is chosen to call to mind: (1) the functional nature of the component and (2) the appropriate number of connections that must be made to that component. Standard symbols have been evolved for all common circuit components. Some basic symbols are given in Fig. 7-2. A more complete listing can be found in Table 4 of the Appendix. The

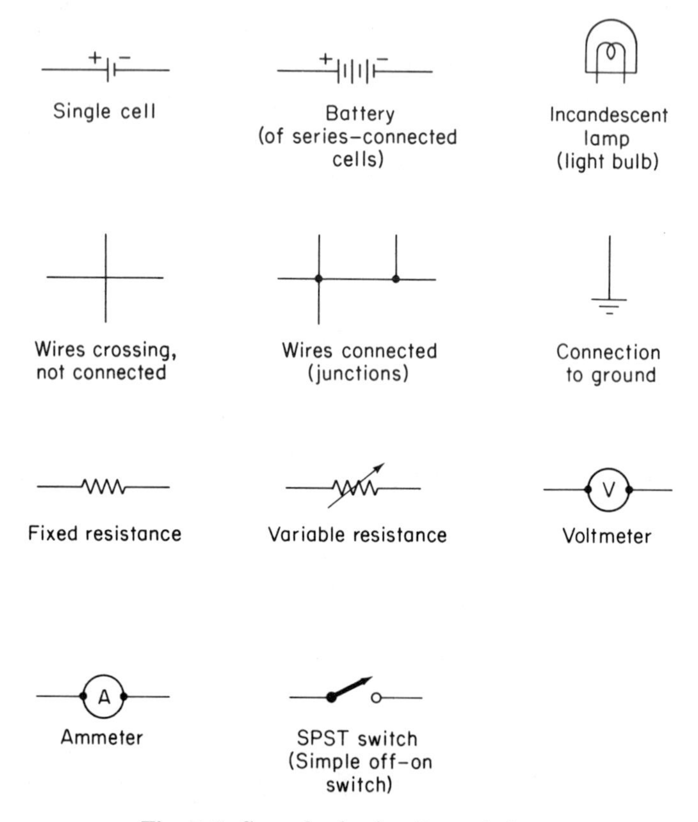

Fig. 7-2. Some basic circuit symbols.

lines representing connectors (i.e., the interconnecting conductors) are always drawn in such a way as to have only horizontal and vertical segments on the page (i.e., only "square corner" turns are allowed), except in the rare instance where a violation of this rule may be unavoidable. In any case, the objective of any schematic is to show how the circuit works—that is, to show both qualitatively and quantitatively how each component contributes to the behavior of the circuit—and maximum clarity is the goal. Appropriate numerical values, such as resistance of resistors, open-circuit potential of cells, and ranges of measuring instruments, are specified right on the diagram. Note that in Fig. 7-2 the symbols are given for two common electrical measuring instruments the ammeter and the voltmeter, whose principles of operation are discussed later in the text. For present purposes, it is sufficient to note that

> **an ammeter measures the current that flows *through* it**

and that

> **a voltmeter measures the potential difference *across* its terminals, having (ideally) *no current* at all through itself.**

Suppose now that we wished to consider a circuit in which a lamp (bulb) is connected to a dry cell through an off-on switch. Figure 7-3 shows a

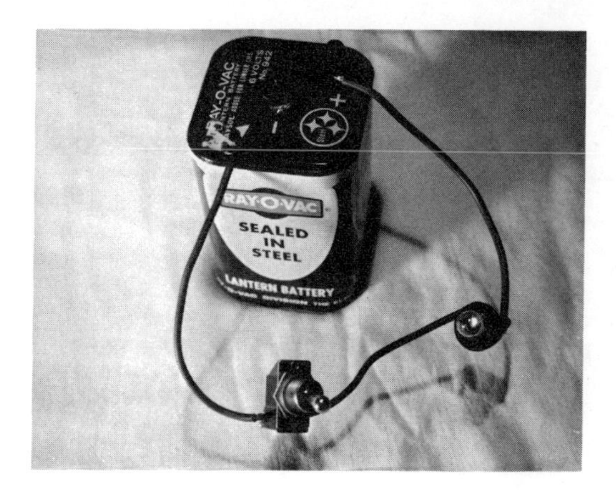

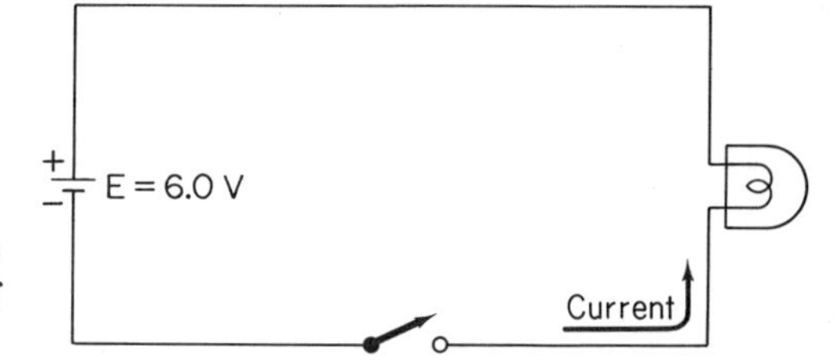

Fig. 7-3. Simple schematic diagram and photograph of actual circuit.

photograph of the circuit in the inset, plus the corresponding schematic diagram. See how the schematic shows all the essential features of the circuit! The cell symbol shows its *polarity* (i.e., which terminal is positive and which is negative); hence we can determine the direction of current (i.e., which way the electrons will move through the circuit). The notation $E = 1.5$ v next to the cell symbol indicates the magnitude in volts of the emf the cell produces. (The symbol E is always used for the emf produced by a dc source.) From Fig. 7-3 we can see that when the switch is closed there is a complete path for the electrons to move from the negative terminal of the cell, around through the lamp, and back to the positive terminal of the cell.

Our next circuit, shown in Fig. 7-4, consists of two resistors, one 15 kΩ and the other 10 kΩ, connected to a 9-volt battery (which, internally, is made of several cells). A very low-range ammeter (i.e., a *milli*ammeter) is inserted in the circuit to measure the tiny current. Again we can see how the schematic faithfully indicates the essential features of the circuit. In actual point of fact, the schematic often makes a clearer distinction between elements than a photo. For example, certain types of capacitors closely resemble, and could be taken for, resistors, whereas this confusion could never occur in a schematic.

Note also in Fig. 7-4 that the resistance value of each resistor is

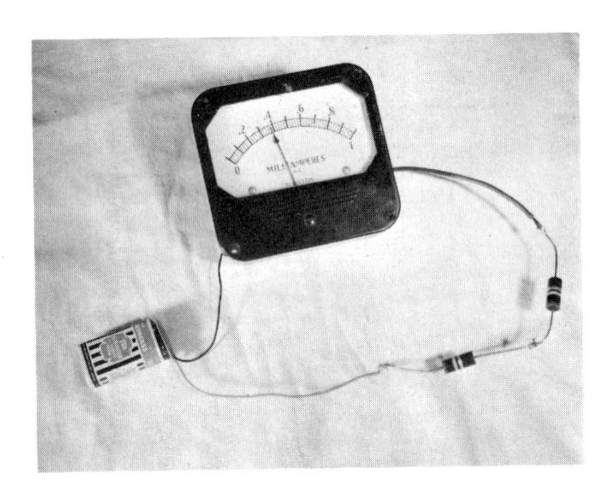

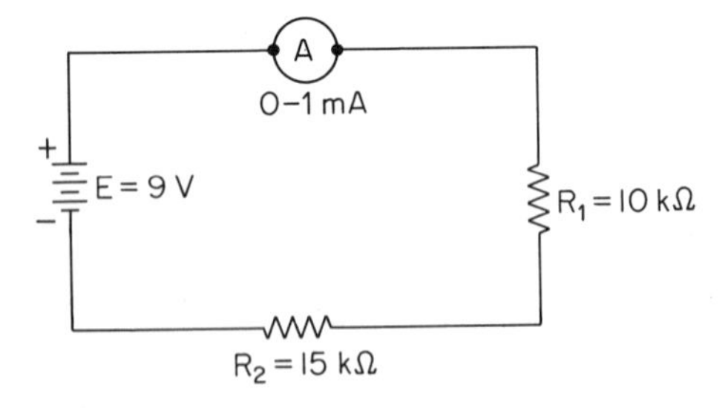

Fig. 7-4. Another simple schematic and photograph of actual circuit.

specified. The letter R is always used to represent resistance, and the various resistors in a circuit are usually distinguished by little numbers (subscripts) placed slightly below and to the right of the R. Thus our circuit contained two resistors, which were designated R_1 and R_2.

By now, you should be getting the idea of how a schematic is drawn. Hereafter we will show only the schematics, omitting the circuit photographs.

7-3 SERIES CIRCUITS

When an electric circuit is formed from a group of components in such a way that at every point in the circuit there is *only one path for current to follow,* the components are said to be connected *in series.* This type of circuit, shown conceptually in Fig. 7-5, is then called a *series circuit.* In such

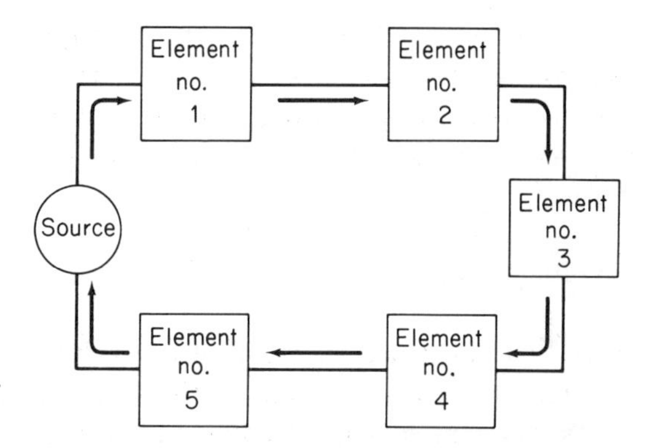

Fig. 7-5. Series circuit (conceptual).

a circuit, the components are connected end to end, and the electrons pass through each in turn. The circuits of Figs. 7-3 and 7-4 are both series dc circuits.

The most important characteristic of a series circuit is that the *current is the same at any point in the circuit;* in other words, an ammeter inserted anywhere in the circuit will indicate the same current. By analogy with water flowing in a pipe, where the flow rate is one gallon per minute whether we measure it at the entrance, middle, or exit of the pipe, so, too, a current of one coulomb per second (one ampere) in a series circuit is the same wherever we measure it.

The overall effective resistance of an entire series circuit is simply the total of all the individual resistances in the circuit (including the resistance of the wires and the source, if they are significantly large compared to the other resistances). Thus in the circuit of Fig. 7-4, assuming all other resistances are negligibly small compared to those of the two resistors, we may compute

$$\text{Total circuit resistance} = 15\ \text{k}\Omega + 10\ \text{k}\Omega = 25\ \text{k}\Omega$$

A useful idea in dealing with circuits is that of a *voltage drop.* Recall the discussion at the end of Section 6-1 and the analogy between the electrons and laborers. It was pointed out there that the source imparts a certain energy per unit of charge (i.e., voltage) to the electrons and that they give up this energy by doing work of some kind, returning to the source depleted. It is to be emphasized that this relationship is an *exact balance*. The energy imparted to each unit of charge by the source of emf is *exactly* equal to the energy it gives up in passing through the rest of the circuit. If this were not true, there could be some "voltage left over" but no "resistance left over" after the electrons have passed through the last resistance in the series group, and Ohm's law tells us that an unlimited (infinitely large!) current would exist in that part of the circuit, which simply does not happen. We may define a voltage drop across a resistance as the work *done*, or energy *loss*, by a unit charge as it passes through that resistance. Numerically, this voltage drop (measured in volts) can be calculated via Ohm's law, now arranged in the form

$$\text{Voltage drop} = (\text{resistance}) \times (\text{current through that resistance})$$

In a series circuit, then, the total of the voltage drops through all the resistances must be exactly equal to the net emf applied by the source.

As an example, consider the series circuit shown in Fig. 7-6. Since the

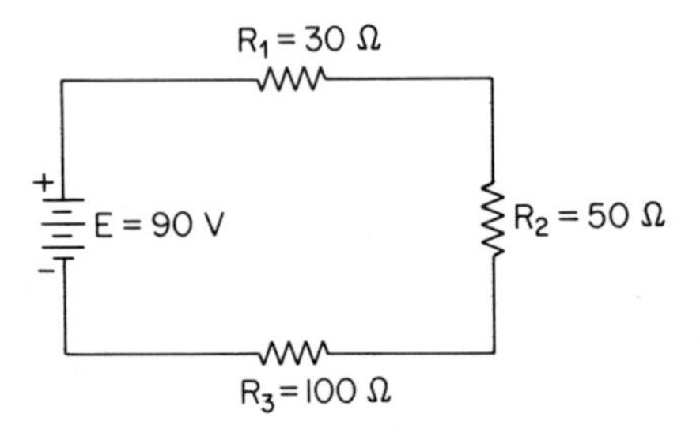

Figure 7-6

net resistance of the whole circuit is the total of the individual resistances we may compute

$$\text{Net resistance} = 30\ \Omega + 50\ \Omega + 100\ \Omega = 180\ \Omega$$

The current through this 180 ohms is given by Ohm's law as

$$\text{Current} = \text{emf} \div \text{resistance} = 90\text{ v} \div 180\ \Omega = \tfrac{1}{2}\text{ amp}$$

which is the same throughout the circuit. Then the respective voltage drops through the individual resistances are

$$\text{Drop through } R_1 = (30\ \Omega) \times (\tfrac{1}{2}\text{ amp}) = 15\text{ volts}$$
$$\text{Drop through } R_2 = (50\ \Omega) \times (\tfrac{1}{2}\text{ amp}) = 25\text{ volts}$$
$$\text{Drop through } R_3 = (100\ \Omega) \times (\tfrac{1}{2}\text{ amp}) = 50\text{ volts}$$
$$\text{Total drop} = 90\text{ volts}$$

verifying that the total drop is exactly equal to the source voltage.

When more than one voltage source is connected into a series circuit, as in Fig. 7-7, they may either be connected to *aid* or to *oppose* each other.

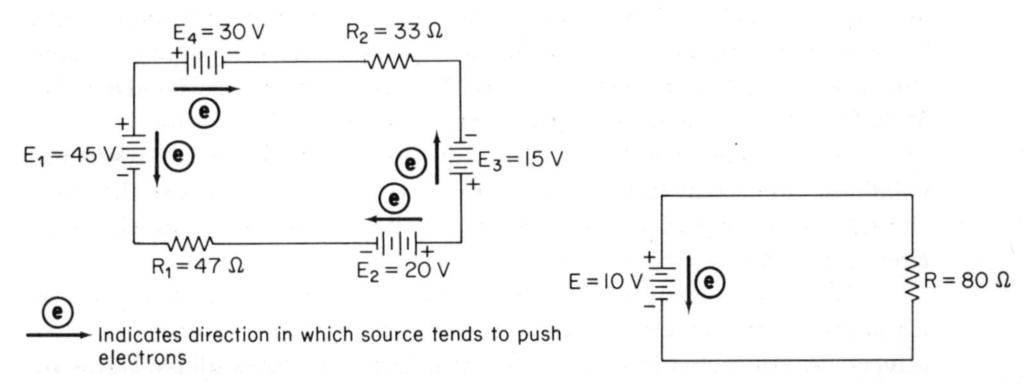

Fig. 7-7. Multiple sources of emf. (e) indicates direction in which source tends to push electrons

Fig. 7-8. Simpler circuit equivalent to that of Fig. 7-7.

Two sources connected in such a way that they tend to make electrons flow in the same direction are said to be *aiding* (each other), while two sources connected such that they tend to make electrons flow in opposite directions are said to be *opposing* (each other). The net effect of several sources is found by totaling the voltages of all those trying to make electrons go one way, totaling the voltages of all those trying to make electrons go the other way, and subtracting the smaller from the larger total. A battery having this value, and connected like those with the biggest total, will have the same net effect as all the separate sources. In Fig. 7-7, for instance, batteries E_1 and E_3 want to move electrons counterclockwise around the circuit, while E_2 and E_4 push for the clockwise direction. The total of E_1 and E_3 is $45 + 15 = 60$ volts, while that of E_2 and E_4 is $30 + 20 = 50$ volts. Thus E_1 and E_3 "win out" over E_2 and E_4 with a net effect of $60 - 50 = 10$ volts. Moreover, the resistors R_1 and R_2 could be replaced by one resistor whose resistance is $47 + 33 = 80$ ohms. Therefore the circuit of Fig. 7-7 could be replaced by the simpler one in Fig. 7-8.

7-4 THE OPEN CIRCUIT

Whenever a break occurs, or is deliberately made, in a circuit, that circuit is said to be open. The term *open circuit* is somewhat paradoxical in that an electrical path, once broken, is really no longer a "circuit." The terminology has stuck, however, so we shall continue to talk about open circuits. The break in the circuit is usually considered equivalent to an "infinitely large" resistance. This is not strictly true because if you take a pair of diagonal cutting pliers and snip a piece of wire out of a circuit, the gap is filled by air that has a large but finite resistance. For most practical circuits, however, the resistance of the air in the gap is so very, very large compared with any "normal" resistance in the circuit, that it may be considered infinite. To indicate an infinite quantity, the "lazy eight" symbol (∞) is used.

An open circuit is shown in Fig. 7-9. Since the total "circuit" resistance is infinite, no current flows. Thus there is no voltage drop across the finite (1.5 Ω) resistance and the total supply voltage appears across the open circuit, as read by the voltmeter.

The more inexpensive types of Christmas-tree lights, as well as the filaments (heating elements) of the tubes in many radio and television receivers, are connected in series as shown in Fig. 7-10. Since all ten lamps are

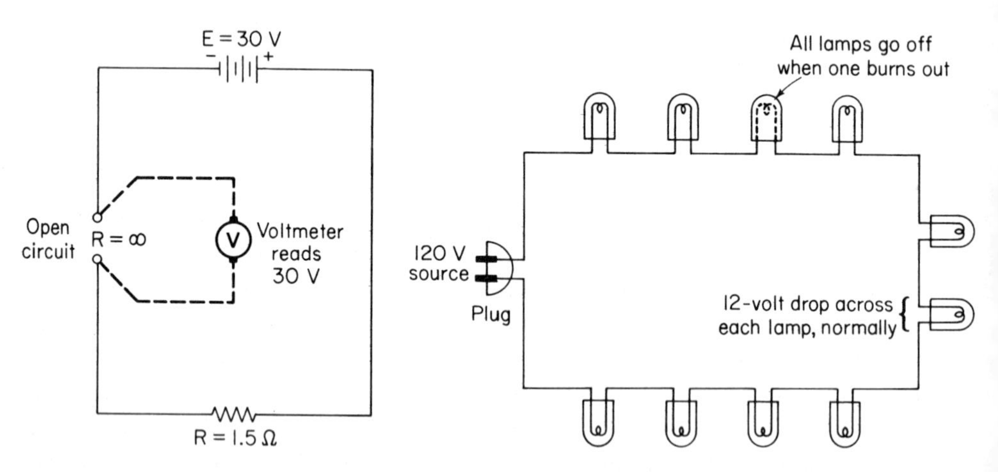

Fig. 7-9. Open-circuit behavior.

Fig. 7-10. Series-wired Christmas-tree lights.

essentially identical, their resistances are equal, and therefore the same amount of voltage is normally dropped across each lamp. The drop across each lamp is about 12 volts, because 10 lamps $\times$ 12 volts per lamp $=$ 120 volts. (The fact that household current is ac rather than dc is unimportant in this discussion.) If, now, any lamp burns out—that is, its filament breaks, as shown in Fig. 7-10—we have an open circuit. No current flows; no voltage is dropped across the other lamps, which, of course, don't light; and 120 volts appears across the burnt-out lamp. This is the primary disadvantage of series-wired Christmas-tree lights (and also of series TV-tube filaments)—when one burns out, they *all* go out! You can see that it would never do to wire the electric outlets of a home, or other building, in series.

7-5 PARALLEL CIRCUITS

A parallel circuit is one that contains several paths for the current to follow. In other words, at some point in a parallel circuit, such as the one represented conceptually by Fig. 7-11, the current divides itself between

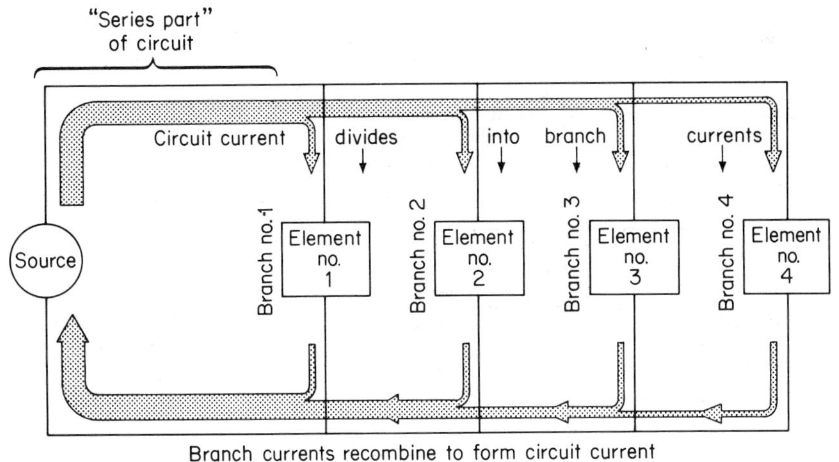

Fig. 7-11. Parallel circuit (conceptual). Branch currents recombine to form circuit current.

two or more alternative branches or the circuit. The components in that part of circuit having the multiple paths are said to be connected *in parallel.* A simple parallel circuit is depicted in Fig. 7-12. Note that the three resistors, R_1, R_2, and R_3, are connected in parallel with each other, making this a *parallel circuit,* but that the *group* of three resistors considered together is connected in series with the source. The key facts about a parallel circuit are as follows:

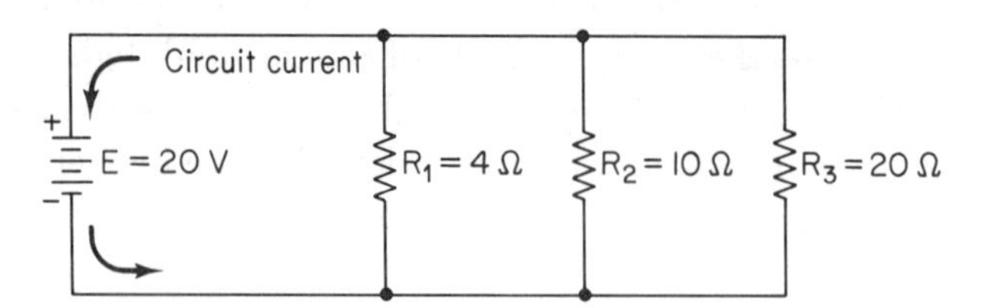

Fig. 7-12. Simple parallel circuit.

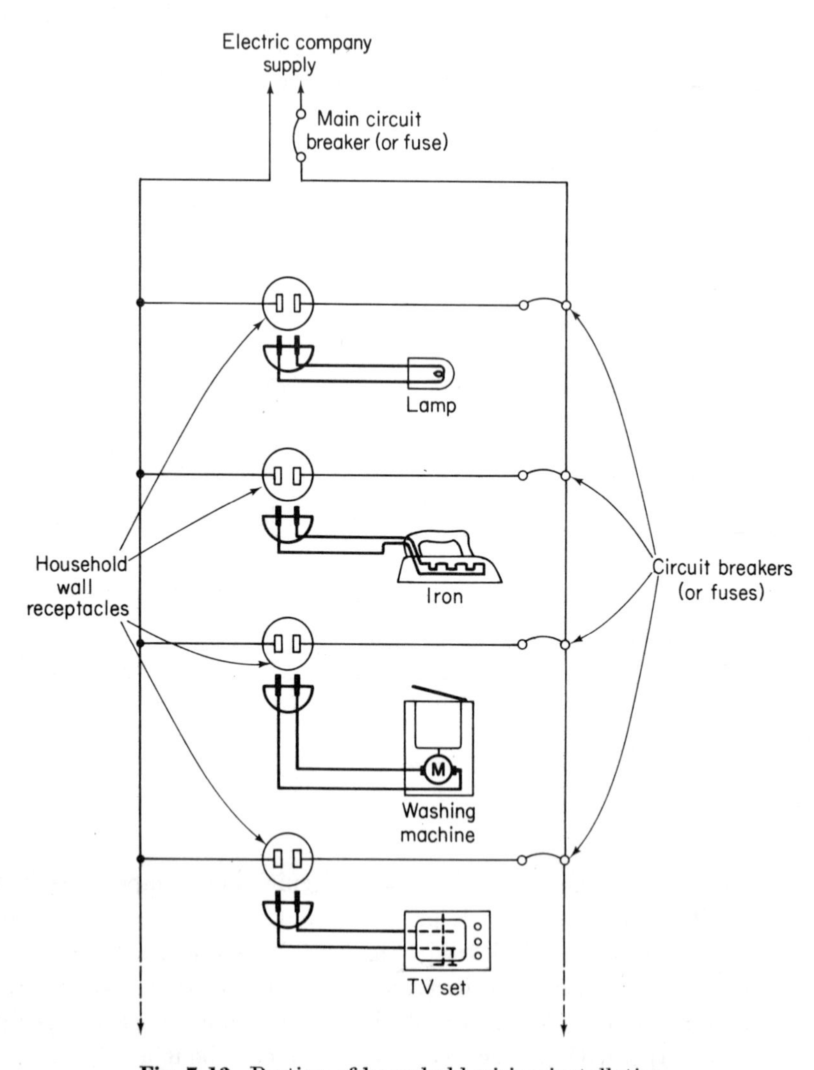

Fig. 7-13. Portion of household wiring installation.

1. Each parallel branch has its own current, and the total of these branch currents is equal to the *circuit* current which flows in the *series part* of the circuit.
2. The *same voltage* appears across each of the parallel branches.
3. The overall effective resistance of the whole circuit (i.e., the resistance that the source "sees") is less than the resistance of any one of the parallel-connected branches.

As a concrete example, let us analyze the circuit of Fig. 7-12. We note that the source voltage, 20 volts dc, appears across each of the branches, and Ohm's law permits us to calculate each of the branch currents, the total of which is the circuit current:

$$\begin{aligned}
\text{Current through } R_1 &= 20 \text{ v} \div 4\ \Omega = 5 \text{ amp} \\
\text{Current through } R_2 &= 20 \text{ v} \div 10\ \Omega = 2 \text{ amp} \\
\text{Current through } R_3 &= 20 \text{ v} \div 20\ \Omega = \underline{1 \text{ amp}} \\
\text{Total circuit current} &= 8 \text{ amp}
\end{aligned}$$

But for an 8-amp current to flow in a dc circuit to which 20 volts is applied, the overall effective resistance of the circuit must be $2\frac{1}{2}$ ohms (because, by Ohm's law, $20 \text{ v} \div 2.5\ \Omega = 8$ amp). Here we can see how the overall resistance is indeed less than any single resistance in the circuit. This fact leads us to point out another important difference between series and parallel connections: *Every element added in series decreases the total circuit current, whereas every element added in parallel increases the circuit current.*

All commercial electrical installations—homes, offices, factories, hospitals, etc—are parallel circuits. Figure 7-13 depicts a portion of a typical household wiring system with a few appliances indicated. The fuses or circuit breakers are installed to protect the installation from short circuits.

7-6 SOURCE AND CONNECTOR RESISTANCE; SHORT CIRCUITS

The connecting wires used in a real circuit actually have some resistance. All sources of emf also have an effective *internal* resistance. Normally these resistances (a few hundredths or thousandths of an ohm for typical connectors and perhaps a few tenths of an ohm for a battery) are very small compared to that of even the smallest resistive device in the circuit and may be neglected in circuit calculations. In some circuits, however, particularly in some parallel circuits (where the effective resistance of a large group of parallel elements may become very small), these source and connector resistances must be taken into account. This is usually accomplished by lumping all these small resistances together and considering the total as the resistance of a fictitious resistor connected *in series with the source.*

Consider, for example, the circuit of Fig. 7-14, in which the battery and connector resistance shown as the dashed resistance, R_{sc}, total 0.2 ohm.

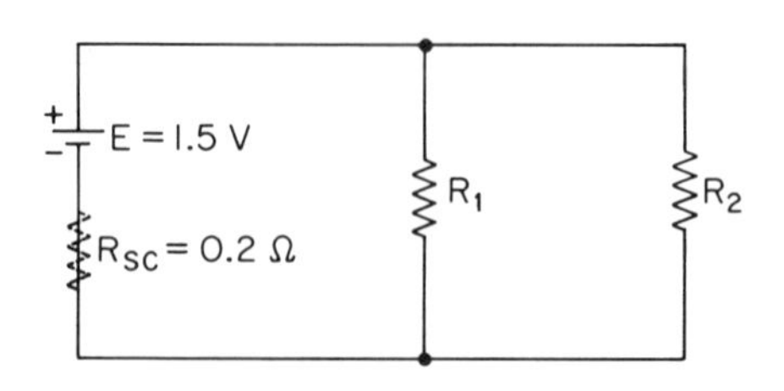

Fig. 7-14. Internal resistance of source and oonnector resistance.

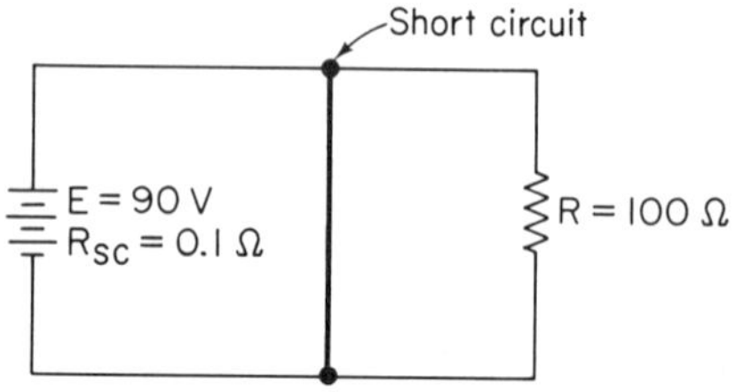

Short circuit
E = 90 V
RSC = 0.1 Ω
R = 100 Ω

Fig. 7-15. The short circuit.

Suppose, first that R_1 and R_2 are both 30-ohm resistors. Their combined effective resistance, in parallel, is 15 ohms. If we neglect R_{sc}, we calculate a circuit current of

$$\text{Current} = 1.5 \text{ v} \div 15\ \Omega = 0.1 \text{ amp} = 100 \text{ ma}$$

while if we include R_{sc}, the actual circuit resistance is 15.2 ohms and

$$\text{True current} = 1.5 \text{ v} \div 15.2\ \Omega = 0.0987 \text{ amp} = 98.7 \text{ ma}$$

The error incurred by neglecting R_{sc}, is about 2 percent, which is entirely negligible, since the tolerance on standard resistors is, at best, $\pm$ 5 percent, as you will recall.

Now suppose instead that R_1 and R_2 are both 0.30-ohm resistances. Their combined effective parallel resistance is then only 0.15 ohm. Calculating the circuit current, while neglecting R_{sc}, gives

$$\text{Current} = 1.5 \text{ v} \div 0.15\ \Omega = 10 \text{ amp}$$

whereas allowing for the value of R_{sc} yields a true circuit resistance of $0.15 + 0.2 = 0.35$ ohm, and

$$\text{True current} = 1.5 \text{ v} \div 0.35\ \Omega = 4.29 \text{ amp}$$

Here, neglecting R_{sc} results in an error of 133 percent! Thus, in doing circuit calculations, designers and circuit analysts must take the resistances of connectors and sources into account whenever the other circuit resistances are less than about ten times as great. (This doesn't happen too frequently!)

Let us consider the simple circuit shown in Fig. 7-15 without the short circuit connection. This is the way the circuit was intended to operate. The circuit current (we may neglect R_{sc}) is

$$\text{Current} = 90 \text{ v} \div 100\ \Omega = 0.9 \text{ amp}$$

and the connecting wires are chosen in a size to accomodate a current of about an ampere. Now suppose that through some accident, such as a breakage of the insulation covering a wire or a dropped screwdriver, a conductor of essentially zero resistance is connected directly across the source, as represented by the short circuit line in Fig. 7-15. Such a connection makes what is called a *short circuit.* If there were truly no source or connector resistances, the circuit current would be *infinite* and all of it would flow through the zero-resistance path with none through the resistance R. In a real circuit, like the one shown, R_{sc} has some finite value, in this instance 0.1 ohm. Instead of "infinity," the actual current then "only" climbs to

$$\text{Current} = 90 \text{ v} \div 0.1\ \Omega = 900 \text{ amps}$$

which is still almost a thousand times larger than what the wiring is designed for. Of course, this level of current, if ever reached at all, will not persist for long. We know that the energy dissipated in a resistance appears as heat and that the resistance of conductors increases as they get hot. The short-circuit current will decrease rapidly, but usually not rapidly enough. In the example given, the wires would probably get hot enough to set nearby combustibles afire, to melt their own insulation, and possibly to melt themselves. This is why a short circuit is such a dangerous occurrence in any circuit; the result is generally some kind of damage. To protect the source against short circuits, almost all electric-power distribution circuits incorporate one of two types of safety devices—the *fuse* or the *circuit breaker.* Both devices open the circuit, like an automatic switch, when too much current flows. Their construction is discussed presently.

7-7 MORE COMPLEX CIRCUITS

Many electric circuits are considerably more complicated than the simple series and parallel circuits discussed so far. A circuit containing cross-connections, which make it neither a series nor a parallel circuit, is called a *network.* Figure 7-16*a* shows one of the simplest networks, called a *bridge*

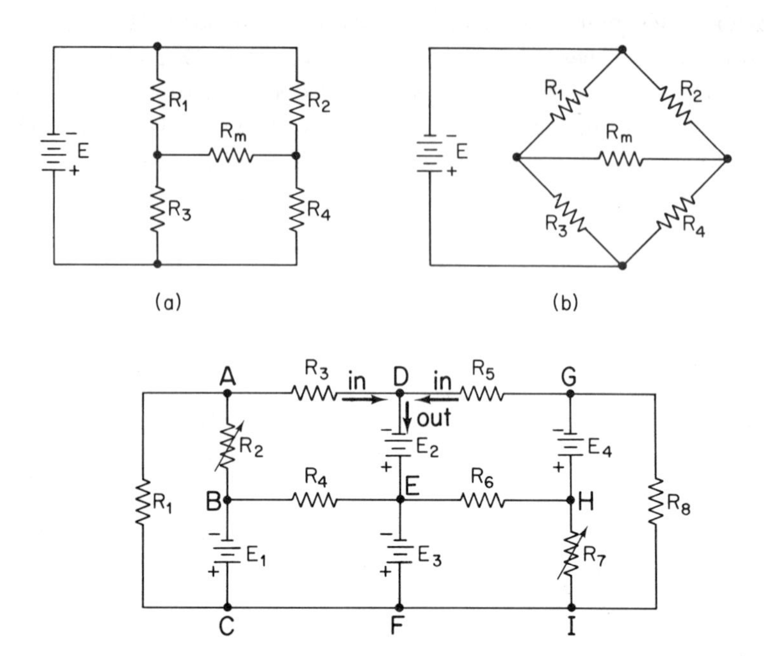

Fig. 7-16. Networks. (a) Simple (bridge) network. (b) Conventional representation of bridge network. (c) More complex network.

network. Note that if the resistor R_m were removed, this network would become a simple parallel circuit. The more conventional way to draw this particular network is shown is Fig. 7-16*b*. A more complex network is given in Fig. 7-16*c*.

Networks cannot be analyzed mathematically by using Ohm's law directly. Their solution requires application of more generalized methods. These generalized principles were first clearly enunciated by the noted German physicist, Gustav Robert Kirchhoff.

Kirchhoff, better known for his later pioneering work in the field of spectroscopy, had done much interesting work in electrical theory early in his career. A strong mathematician, he had received his training at the University of Königsberg, and was thoroughly acquainted with the work of the great Newton. In the midnineteenth century, after a thorough study of Ohm's work, Kirchhoff formulated the general rules that now bear his name. These rules, or laws are so simple that at first glance you may wonder why all the fuss. Their very simplicity is what makes for their general applicability to any network.

The first, now called *Kirchhoff's Point Law*, states that at any point in a circuit, such as point *D* in Fig. 7-16*c*, the total current entering the point is equal to the total current leaving that point. In other words, electrons do not accumulate or pile up at any point in the circuit.

The second principle, now called *Kirchhoff's Loop Law*, says that the total of all the voltage drops around any closed path (loop), such as *ABEDA* in Fig. 7-16*c*, is *zero*. To understand the meaning of this statement, imagine that you were very tiny and could walk around the circuit, inside the wires and components. Then if you begin walking through the circuit at any point and come back to the point you started from, all the voltage drops you passed through in your journey will total zero, *provided* that you interpret a voltage drop as *positive* when the current in that component was going in the same direction as you, but as a *negative* voltage drop (a voltage "rise") when you and the current are moving in opposite directions.

7-8 ENERGY AND POWER IN A CIRCUIT

Although the circuit designer is perhaps most concerned with the voltage-current-resistance relationships in the circuit, the user is extremely interested in the energy and power (see Chapter 2, Section 2-9) consumption of the circuit. From the practical standpoint, the energy consumption reflects directly in the user's electric bill; while the power (rate of energy consumption) determines the size of wiring and the circuit breaker rating required to service the outlet to which the circuit in question is connected.

Recall that voltage represents either the energy imparted (within a source) to or the energy expended (within any other component) by a unit of charge. Recall also that the current through the component represents the number of such units of charge which pass through that component in a given amount of time. Thus, if we multiply voltage by current, we obtain energy per unit of time, or *power*. For a source of emf, then

$$\text{Power (provided)} = \text{voltage (supplied)} \times \text{current}$$

and for any other[1] device

$$\text{Power (consumed)} = \text{voltage (drop)} \times \text{current}$$

In performing the power calculation, one measures the voltage in volts and the current in amperes; the power then comes out in *watts*, and for large

[1] This is strictly true for *any* other device only in dc circuits; in ac circuits it is true only for resistive devices.

power consumption we often speak of *kilowatts* (thousands of watts). For instance, a typical light bulb used in a 120-volt socket draws 0.625 amp of current. The power it requires is therefore

$$\text{Power} = 120 \text{ v} \times 0.625 \text{ amp} = 75 \text{ watts}$$

Similarly, most electric irons require about 9.2 amp and so their power consumption is

$$\text{Power} = 120 \text{ v} \times 9.2 \text{ amp} = 1100 \text{ watts} = 1.1 \text{ kw}$$

One requirement for any electrical appliance to be approved by the Underwriters' Laboratory is that it be clearly marked with *both* the voltage for which it is designed and the power which it consumes.

To us, as consumers of "electricity" and also, of course, to the vendor of electricity, the most interesting thing is that which we pay for—the *energy* used. Since power is the rate of energy consumption—that is, how much energy is consumed (or supplied) *per unit of time*—we can figure out the total energy consumed by multiplying the power a device consumes by the length of time that the device is in operation. As a formula, we can write

$$\text{Total energy} = \text{power} \times \text{time}$$

Now a watt of power represents a consumption of one joule of energy each second. Therefore, if we multiply watts by the number of seconds in use, we obtain energy in joules. In talking about electrical consumption, however, it is preferable to keep the compound name *watt-second* instead of joule, to remind us that the energy is obtained by multiplying together watts and seconds. In most cases, the watt-second is too small an energy unit for practical applications, so we normally speak of *watt-hours* and *kilowatt-hours* (thousands of watt-hours) of electric energy.

If the electric iron discussed previously were used for 3 hours, the energy consumed would be

$$\begin{aligned}\text{Energy} = 1100 \text{ watts} \times 3 \text{ hours} &= 3300 \text{ watt-hours} \\ &\text{or } 3.3 \text{ kilowatt-hours}\end{aligned}$$

The electric energy consumed in your home is automatically registered by a device called a *kilowatt-hour meter*, installed outside your

Fig. 7-17. Kilowatthour meter. (Courtesy of Los Angeles Department of Water Power.)

home by the electric company. Such a device is shown in Fig. 7-17. It contains a motor, the *speed* of whose rotation depends on the *power* being drawn, and which drives the little indicating dials. The total number of revolutions of the motor is read from the dials, which are hooked up to indicate total revolutions of the motor, just as the odometer section of an automobile's speedometer reads "total miles covered." The numbers on the four dials of the kilowatt-hour meter, reading from left to right, represent, respectively, the number of thousands, hundreds, tens, and units of kilowatt-hours consumed. The electric company keeps a running dated record of meter readings, and the difference between any two readings represents the electric energy consumed between the corresponding dates.

REVIEW QUESTIONS

1. Describe the major differences between a functional block diagram and an electrical schematic diagram.
2. An incandescent lamp is connected in series with a 6-volt battery, a 10-ohm variable resistance, a 0–1 amp ammeter, and a switch (SPST). Draw a labeled schematic diagram of this circuit.
3. What is the effective resistance of a circuit containing three 47-Ω resistors, two 10-Ω resistors, and a 5-Ω resistor, connected in series?
4. A 20-Ω resistor and a 100-Ω resistor are connected in series with a 240-volt supply. How much current flows? Can you tell how big the voltage drop is across each resistor?
5. What is the resistance through an ideal switch when it is closed? When open?
6. Is the wiring in your home one large series circuit? What evidence can you give to justify your answer?
7. Is the resistance of wires, contacts, etc., always inconsequential? How about the internal resistance of the voltage source? Explain your answer.
8. What is a short circuit? Give an example of how one might arise. Why is a "short" dangerous?
9. What is a network? What priniciples are used to analyze a network mathematically?
10. How do we calculate the power supplied from a source? How do we determine the power dissipated in a resistance?
11. Is electrical energy measured in joules or watt-seconds? What is the difference?
12. An atomic power generating plant has a rated capacity of 100 megawatts. How many kilowatt-hours of energy can it deliver in 30 minutes?

8

SOME CURRENT EVENTS

8-1 TWO IMPORTANT EFFECTS OF ELECTRIC CURRENT

An electric current can be made to produce a variety of effects. Two of the earliest discovered, best known, and most widely used effects are discussed in this chapter—the heating effect and the chemical effect. We have encountered both before.

As indicated in the section on resistance (Chapter 6), the electric "friction," due to interaction of the moving electrons with the neutral atoms of the conductor through which they are passing, causes the atoms of the conductor to vibrate more rapidly. The macroscopic effect of rapidly vibrating atoms is *heat*. The conductor actually gets warm or hot. Depending on the material and how hot it gets, one of two additional things may happen: the conductor may melt or it may begin to glow, emitting visible light. Each of the three occurrences—simply giving off heat, melting, and giving off light—is made use of in various devices and appliances.

The chemical effect represents the reverse of what normally occurs in a voltaic cell. There, chemical reactions were induced in order to liberate electrons. Here we are interested in using a supply of electrons to *cause chemical reactions*. In most such reactions, the electric current promotes the breakdown of a compound into its constituent elements. Hence the process is called *electrolysis* (electrical breakup).

8-2 THE FUSE

The most important application of electrically heating a conductor to its melting point is the *fuse*. A fuse is a simple safety device for preventing a fire in the event of a short circuit. The essential element of a fuse is the *fusible link*, a small strip or wire of a metal that *fuses* (melts) easily, hence the name. Fuses are of two general types, *cartridge type* and *screw-in type*, and are illustrated in Fig. 8-1. The fusible link is so designed that when a

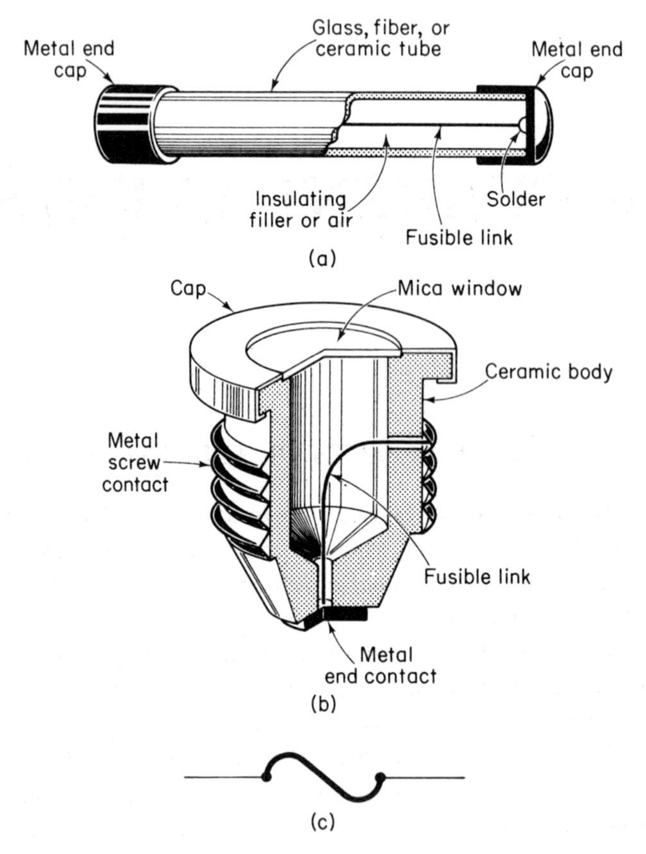

Fig. 8-1. Fuses. (a) Cartridge. (b) Screw-in. (c) Circuit symbol.

steady current of a certain size (called the fusing current) flows in it, the link is heated to its melting point.

A fuse is always connected to a circuit *in series with the power source.* If, then, the current ever reaches (or tries to exceed) the fusing current, the link melts, acting like a switch and opening the circuit.

Note that a fuse is a "one-shot" device—that is, it can only perform its design function once and then is discarded. In applications where current overloads occur frequently, it would be a nuisance to replace fuses constantly, so another device called a circuit breaker, which is reusable, is installed instead of a fuse.

8-3 THE THERMOSTAT

The word *thermostat* means "fixed temperature" and refers to a device that is used to control temperature to a nearly constant value. To understand how such a device works, we must first understand the thermal behavior of what is called a "compound bar" or *bimetallic element*.

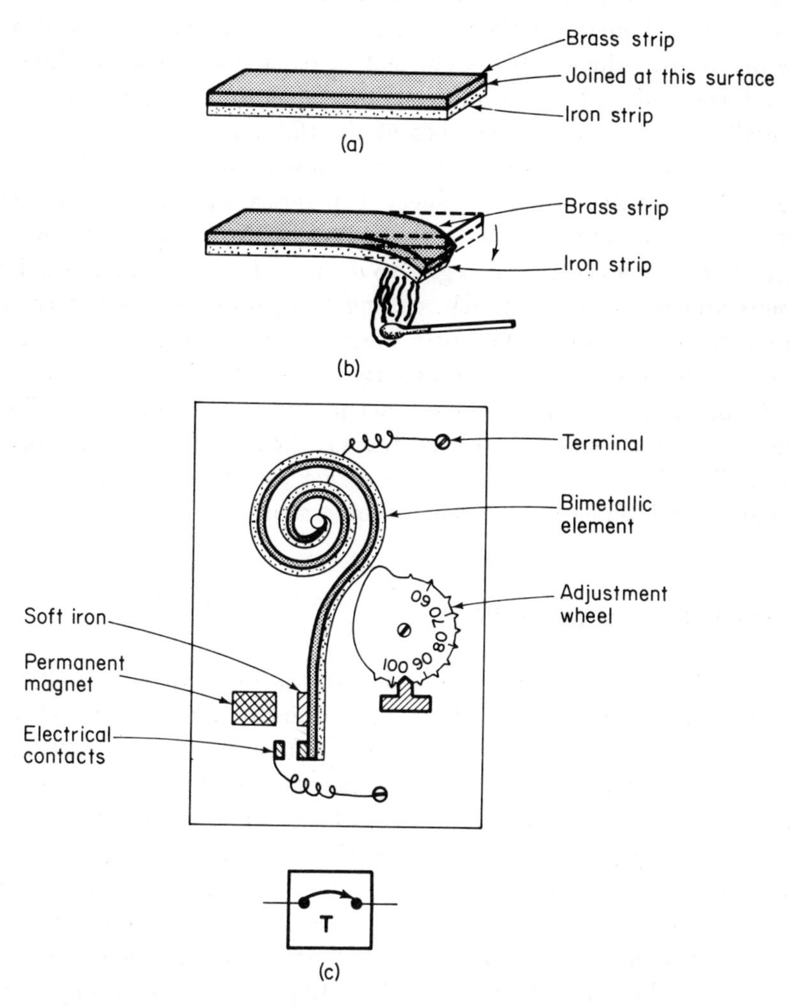

Fig. 8-2. Principle of the thermostat. (a) Bimetallic strip. (b) Effect of heating bimetallic strip. (c) Typical wall thermostat. (d) Thermostat circuit symbol.

A bimetallic strip, shown in Fig. 8-2*a*, consists of two strips of different metals welded, riveted, or bonded together along their lengths. Now, it is a physical property of most materials, and of metals in particular, that they expand somewhat when heated. In other words, their length, width, and thickness all increase as they become hotter. The measure of this property is called the *coefficient of* (linear) *expansion* of the material and represents the average fractional increase of size in any direction per degree of temperature increase. Different materials have different coefficients of expansion, which means that one will change size more than another for the same change in temperature. When a bimetallic strip is heated, as shown in Fig. 8-2*b*, the one metal expands more than the other, and because they are rigidly joined together, the entire assembly bends in the direction of the metal with the smaller coefficient of expansion. Upon cooling to its original temperature, the bimetallic strip returns to its original shape. Such a bimetallic element forms the basis of the thermostat.

A typical home wall-mounted thermostat is depicted functionally in Fig. 8-2*c*. Its terminals are connected in series with the heating device, and it acts as a temperature-sensitive switch to turn the heating device on and off. When the temperature begins to drop too low, the bimetallic element contracts and the contacts touch, closing the circuit to the heater. When the temperature becomes sufficiently warm, the bimetallic element expands, separating the contacts and opening the circuit to the heater. The little piece of soft iron and permanent magnet provide a necessary "snap" action for positively (instead of gradually and erratically) making and breaking contact. The adjustment wheel's cam action sets the nominal position of the element and hence its controlled temperature.

8-4 HEATING APPLIANCES

An almost endless variety of heating appliances is available commercially, all of which are based on a thermostat and a *heating element.* A heating element is a resistance element that is designed to operate hot. It is therefore usually made of an alloy (a mixture of metals) having three special characteristics: *high resistivity*, so that a relatively large resistance can be obtained in a reasonable length of wire; *high melting point*, so that it can be used at high temperature without fusing; and *reasonable structural strength at high temperature*, so that it can withstand forces and vibrations while hot. The most common alloy used for heating elements is called *Nichrome*, an alloy of 60 percent nickel, 12 percent chromium, 2 percent manganese, and 26 percent iron. The resistance wire is supported by, and/or

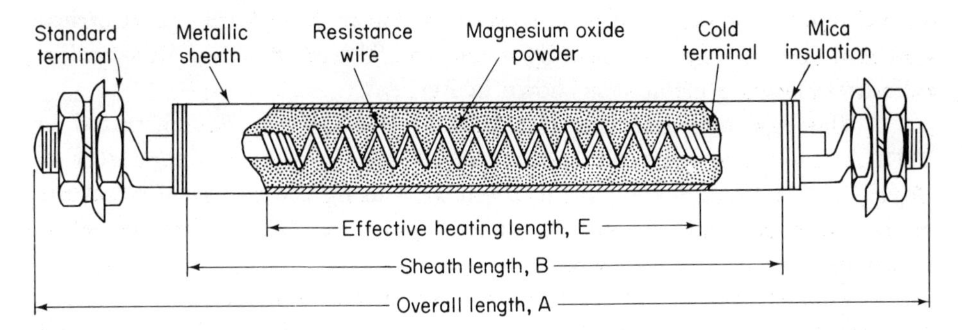

Fig. 8-3. Construction of a typical heating element. (Courtesy of General Electric Co.)

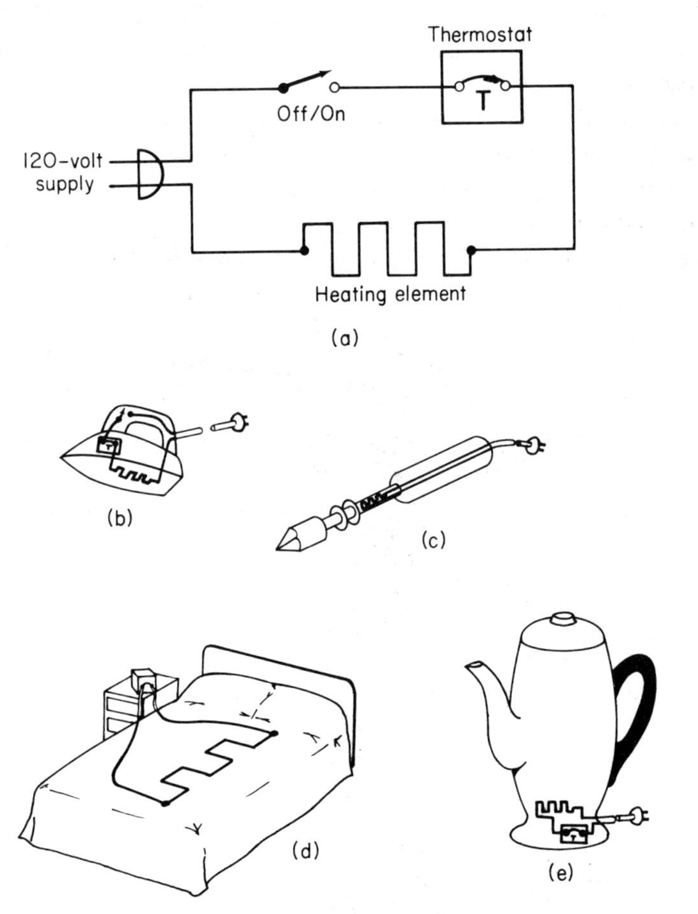

Fig. 8-4. Heating appliances. (a) Typical heater circuit. (b) Electric iron. (c) Soldering iron. (d) Electric blanket. (e) Electric coffee maker.

wound upon, a material that is a good insulator from both the electrical and heat flow standpoints, usually a ceramic. The construction of a typical cylindrical heating element is shown in Fig. 8-3.

The first practical use of a heating element was demonstrated in 1877. At that time Elihu Thomson was assistant professor of chemistry at Boys' Central High School in Philadelphia. During the course of a lecture, he boiled eggs with an electric current, using a coil of German silver resistance wire to heat the water.

Figure 8-4 shows the typical thermostatically controlled heater circuit and its use in several common heating appliances. In many such appliances, particularly where a high temperature is required and the control need not be too precise, the "thermostat effect" is provided simply by the way the device loses heat to its surroundings. In other words, the heating element is designed to reach some balanced temperature without a true thermostat in the circuit.

Electric heating elements are extremely versatile, for they can be fabricated in any desired shape and lend themselves to any installation where the necessary electrical power is available. They can be shaped as straight rods, circular coils, spirals, flat strips, etc. A commercial liquid circulation heater using U-shaped rod elements is shown in Fig. 8-5, and

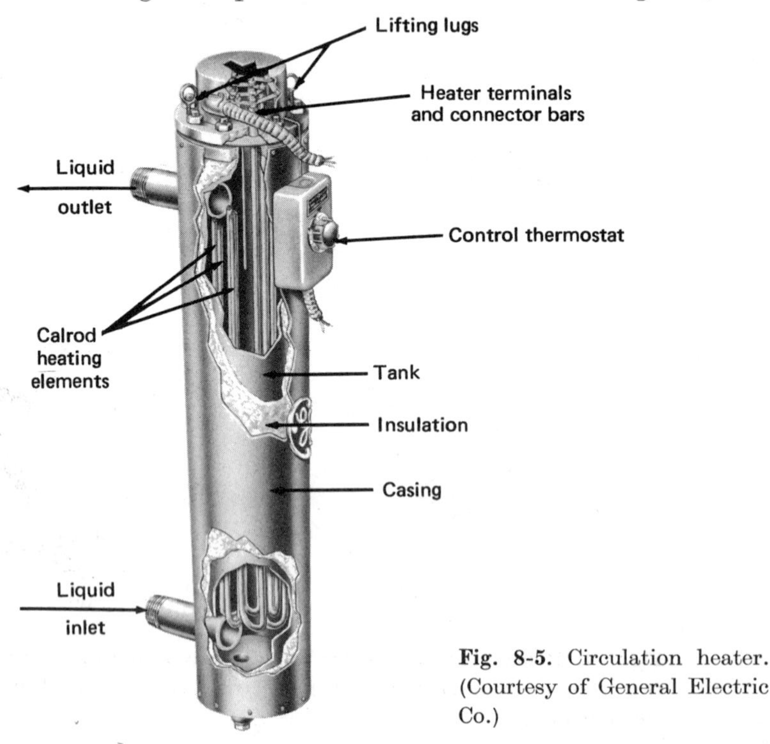

Fig. 8-5. Circulation heater. (Courtesy of General Electric Co.)

Fig. 8-6. Space heater. (Courtesy of General Electric Co.)

a coiled-rod comfort heater, such as might be used in a home or office, is shown in Fig. 8-6.

8-5 THE THERMAL CIRCUIT BREAKER

It was stated at the conclusion of Section 7-6 that electric power sources are usually protected against short circuits by either fuses or circuit breakers. The fuse, a one-shot device, has already been discussed in Section 8-2. We now consider one form of circuit breaker, the alternative type of protective device. A circuit breaker is, basically, an automatic switch that senses the amount of current passing through it and opens when this current exceeds the value for which the breaker is rated. When the cause of the excessive current has been determined and remedied, the circuit breaker may be reset, making the circuit functional once again. Thus the circuit breaker acts like a fuse that is indefinitely reusable. There are both *thermal* and *electromagnetic* types of circuit breakers. We shall discuss the thermal type here.

The principle of the thermal circuit breaker is illustrated functionally in Fig. 8-7. Notice that it is a close relative of the thermostat in that the bending of a heated bimetallic element is used to provide a switching action. It is also much like a manual switch, for a lever or button is used to close a set of contacts mechanically. Current through the breaker flows either through the bimetallic element itself or (in a large breaker) through a heating

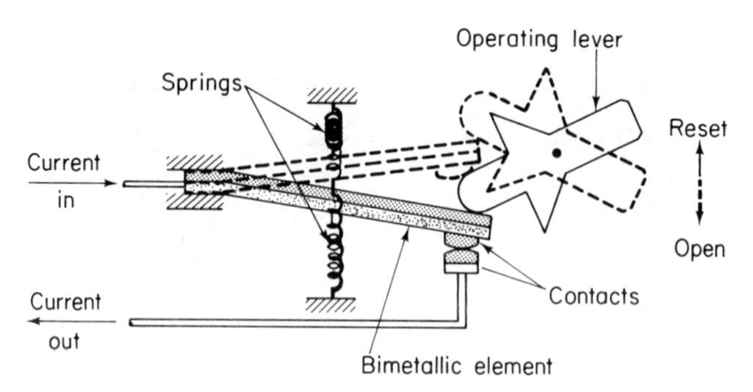

Fig. 8-7. Principle of the thermal type of circuit breaker.

coil element. When this current is excessive, the additional spring force developed by the bending of the bimetallic element opens the contacts and pushes the lever or button to the "open" position. When the element cools off, the spring force it develops diminishes to a value where the lever or button can be manually actuated once again, resetting the breaker.

Some thermal circuit breakers incorporate an additional spring mechanism, which takes the place of the human hand and makes the breaker *self-resetting*.

8-6 LIGHT FROM A BLACKBODY

What usually comes to mind when you think of *light*? To the physicist, "light" is a rather crude term for certain species of a form of energy more properly called *electromagnetic radiation*, which we discuss in depth in the course of this book. Most electromagnetic radiation is invisible, whereas, to the average person, the term light implies "that which enables us to see." For the present, we may consider all electromagnetic radiation to consist of waves moving through space like water waves moving over the ocean. One way to characterize these waves is by the distance between two successive peaks or crests, and this distance is called the *wavelength*. The wavelengths of the various kinds of electromagnetic radiation range from miles or more in size down to the dimensions of atomic particles. Within the whole range of wavelengths, only a very small group can affect the human eye and produce the sensation of vision. These visible wavelengths are between roughly 4×10^{-7} and 7×10^{-7} meter (or about 16 to 27 millionths of an inch). The psychological characteristic that corresponds to wavelength, in the visible wavelengths, is *color*. Every schoolboy knows that ordinary "white light" from the sun is really a mixture of all the colors and that this fact can be demonstrated by breaking up white light with a glass prism or with raindrops (i.e., a rainbow). Wavelengths near 4×10^{-7} meter produce the sensation of violet; wavelengths near 7×10^{-7} meter cause the sensation red; and those in between elicit the intermediate colors of the rainbow. Thus, when we speak of *visible light*, we mean electromagnetic radiation with wavelength(s) between 4 and 7×10^{-7} meter. Wavelengths shorter than 4×10^{-7} meter are called *ultraviolet*, those longer than 7×10^{-7} meter are called *infrared*, and both kinds are, of course, invisible.

It has long been known that if one heats a noncombustible object to a high enough temperature, it glows—that is, emits visible light. Thus, for instance, when the blacksmith heated the piece of iron in his forge, at about 1000 °F it began to glow in a dull red color. As it got still hotter, the glow became brighter and brighter and its color changed from red through yellow to bluish white. This effect was explained quantitatively by the great Max Planck in 1900. He derived the law that predicts how much light at each wavelength is emitted by a *perfect blackbody* (an ideal body that absorbs all light that falls on it) as its temperature is raised. Although

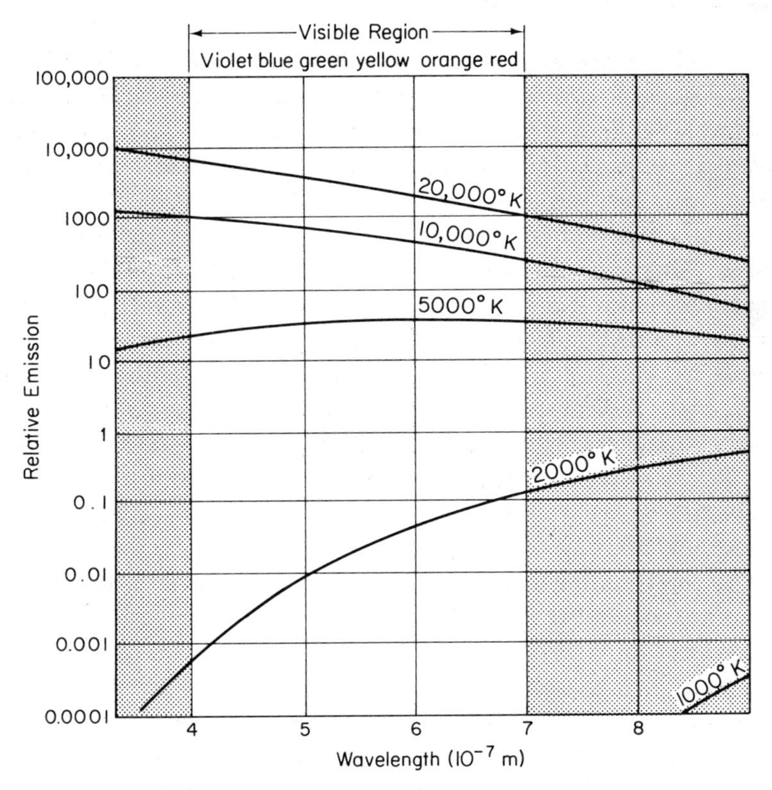

Fig. 8-8. Planck's radiation law for an incandescent blackbody.

no real substance is a perfect blackbody, many substances follow Planck's radiation law fairly closely and are called *graybody* radiators.

Planck's radiation law is illustrated by Fig. 8-8 for several different temperatures. In connection with this law, temperatures are usually specified in degrees Kelvin (°K), which represent "degrees Celsius *above absolute zero*" (1000 °K is about 1300 °F; 5000 °K is approximately 8500 °F; and 10,000 °K is about 17,500 °F). Note that in Fig. 8-8, the vertical scale is not the usual type, increasing by some constant unit, but rather increases by *multiples of 10.* Such a scale is called *logarithmic* and is necessary to show a very rapidly increasing quantity, in this case, the relative amount of energy emitted. We can immediately observe two important facts from this figure. (1) As the temperature of the body is raised, the total quantity of energy radiated at each wavelength increases. (2) As the temperature of the body is raised, the color of the visible energy radiated goes from predominantly red to predominantly blue.

8-7 INCANDESCENT LAMPS

A body heated hot enough to emit visible light, as described in the previous section, is said to be in a state of *incandescence.* Since the resistive effect of an electric current is heat, it was natural to seek some means of obtaining a practical illumination device using electrically produced incandescence—in other words, an *incandescent electric light.*

At any mention of discovery relating to the electric light, the name of Thomas Alva Edison immediately comes to mind. Nevertheless, the first major efforts in this direction were made by the British. The first recorded demonstrations of deliberate, electrically produced incandescence were performed by Sir Humphry Davy. In 1802 Davy showed that strips of platinum and other metals could be heated to incandescence electrically. Dissatisfied with the feeble and short-lived light from his incandescent strips, Davy tried another approach. In 1809 he produced a "brilliant arch-shaped flame" by passing a high current from a 2000-cell battery through two sticks of charcoal separated by an air gap—thus was born the *arc light.* From this point on, both the incandescent filament lamp and the arc lamp went through a parallel development.

Perhaps the first functional incandescent lamp was built in 1820 by another Englishman, Warren de la Rue, renowned for his pioneering work in astrophotography. His lamp, shown in Fig. 8-9*a*, consisted of a coil of platinum wire, sealed into an evacuated glass tube fitted with brass end caps. This same idea was carried over when, twenty years later, de la Rue's

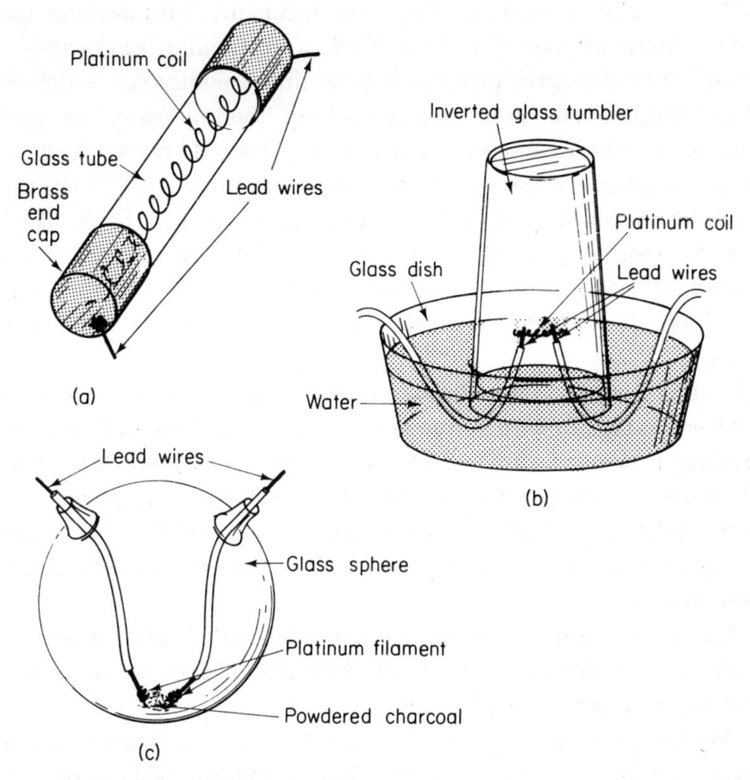

Fig. 8-9. Early incandescent lamps. (a) De la Rue's lamp. (b) Grove's lamp. (c) De Moleyn's lamp.

compatriot, Sir William Grove, illuminated an auditorium with lamps constructed as shown in Fig. 8-9*b*. In the following year, 1841, the first incandescent lamp patent was granted to still another Englishman, Frederick de Moleyns, for the design of Fig. 8-9*c*. Several other Englishmen devised various types of incandescent lamps, the most notable of whom was the physicist Sir Joseph Swan, whose carbon filament design in 1860 anticipated Edison's by 20 years.

French and Russian inventors were also busy in the latter part of the century. A scene at the Paris Opera was electrically illuminated in 1849, while the Czar ordered the St. Petersburg Admiralty dock to be lighted with lamps invented by a Russian physician, Alexandre de Lodyguine, in 1872.

Nevertheless, the lion's share of the show went to Edison. Already famous as the "Wizard of Menlo Park" with scores of inventions behind him, including the phonograph, Edison announced to the world in 1878 that he was going to build an electric light. The business world was so awed with Edison's ability that, at this announcement, illuminating gas stock prices fell drastically on the New York and London exchanges. Edison understood that the main problem was to find a conductor which could be heated to incandescence in a vacuum without "boiling away" (evaporating). It had to be kept in a vacuum because, in air, it would merely burn up when heated hot enough to glow. At last, after hundreds of experiments, he found what he wanted—not some exotic metal but a little filament of carbon obtained by scorching a cotton thread. In 1879 Edison built a bulb with such a filament and it remained lighted continuously for 40 hours. Somewhat later he found that filaments of carbonized Bristol board would last about 200 hours.

Carbon-filament incandescent lamps remained in use until the tungsten-filament lamp, first introduced in 1907, came into widespread use. In 1913 Irving Langmuir discovered that filling the lamp with an inert gas slowed down the evaporation of the tungsten filament, giving longer life and better efficiency. At first pure nitrogen was used for this purpose, but later lamps used nitrogen and argon mixtures. Thus the modern incandescent lamp was born.

The construction of a typical incandescent "bulb" is illustrated in Fig. 8-10. Some idea of the variety of shapes and sizes commercially available can be obtained from Fig. 8-11.

Arc lamps were also developed to the point of practicality, the first such lamp having been patented by Thomas Wright of London. The earliest widely successful arc lamp was the Jablochkov "candle" invented by a

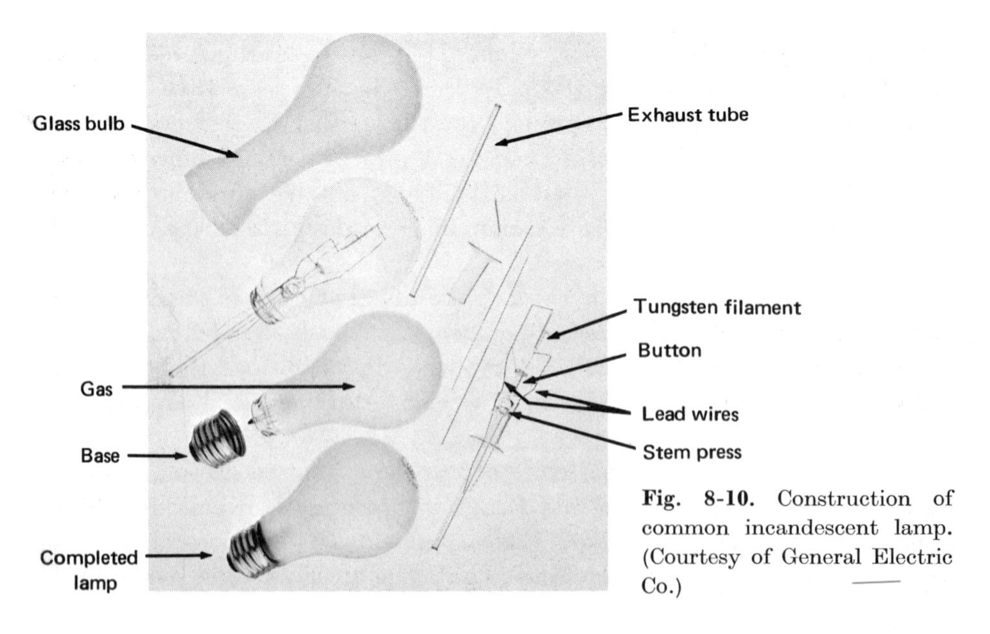

Fig. 8-10. Construction of common incandescent lamp. (Courtesy of General Electric Co.)

Fig. 8-11. Variety of modern electric lamps. (Courtesy of General Electric Co.)

Russian, Paul Jablochkov, in 1876. Modern arc lights are used where a very high illumination intensity is required, such as in searchlights and in movie theater projectors.

Incandescent illumination is not, physically, very efficient. The unit in which the flow of visible energy is measured is called a *lumen.* Recall from the discussion of Section 7-14 that an incandescent body emits radiation at all wavelengths. Yet the average human eye is most sensitive to a wavelength of about 5.55×10^{-7} meter and cannot see wavelengths longer than 7×10^{-7} meter or shorter than 4×10^{-7} meter at all. If one watt of electrical power could be completely converted to visible radiation with a wavelength of 5.55×10^{-7} meter, it would produce 668 lumens. Therefore, in electric lighting, 100 percent efficiency corresponds to 668 *lumens per watt* (abbreviated lpw). The Edison 100-watt lamp had an efficiency of 1.6 lpw, or slightly better than 0.2 percent. A modern 100-watt lamp operates at an efficiency of about 17 lpw, or 2.5 percent while, a modern carbon arc lamp may offer an efficiency as high as 65 lpw, still only about 9.7 percent.

Presently we shall discuss other types of electric lamps that do not rely on incandescence.

8-8 LIGHT FROM IONS

You will recall from Section 4-12 that an ordinarily neutral atom may acquire an electric charge by either gaining or losing one or more electrons. Such a charged atom is called an ion. When a sufficiently large potential difference is created across a column of gas, some electrons are ripped away from the atoms in the region of high positive potential. These electron-deficient atoms are positive ions that tend to drift toward the region of high negative potential (or tend to steal electrons from their nonionized neigh-

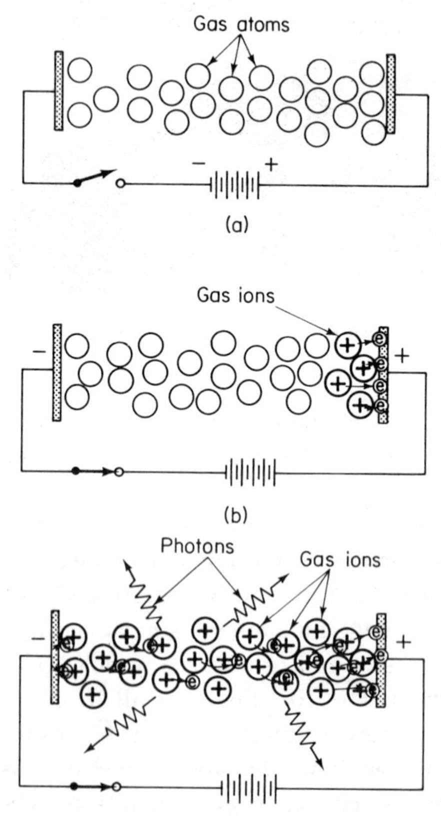

Fig. 8-12. Principle of gas discharge lamp. (a) Gas contained between electrodes. (b) Switch closed, electrons torn away from atoms near positive electrode. (c) Steady electron current set up, photons emitted by excited ions.

bors, which gives the same effect). The net result is a current through the ionized gas, as shown in Fig. 8-12. Some of these very excited ions give up a portion of their energy as photons of radiation, often at visible wavelengths. A number of lamps are based on this principle, initially studied by Heinrich Geissler circa 1850. Since the light is produced by discharging an electric current through an ionized gas, these lamps are categorically called gas discharge lamps.

Several gas discharge lamps are in common use. The *sodium vapor lamp* is used in lighting streets and highways. It has an efficiency of about 50 lpw. Neon-glow bulbs, producing a soft, red-orange glow, are used extensively in night lights and indicator lights. Photographers' "strobeflash" units contain a rare-gas discharge lamp, which is supplied a very brief, high-intensity "spurt" of electrical energy from a capacitor instead of a continuous, low-intensity supply. The most common discharge lamp is a special type known as the *fluorescent lamp*.

8–9 THE FLUORESCENT LAMP

Two distinct energy transformations take place in a fluorescent lamp: (1) electrical energy is transformed into invisible ultraviolet radiation through the gas discharge process previously described; (2) the invisible

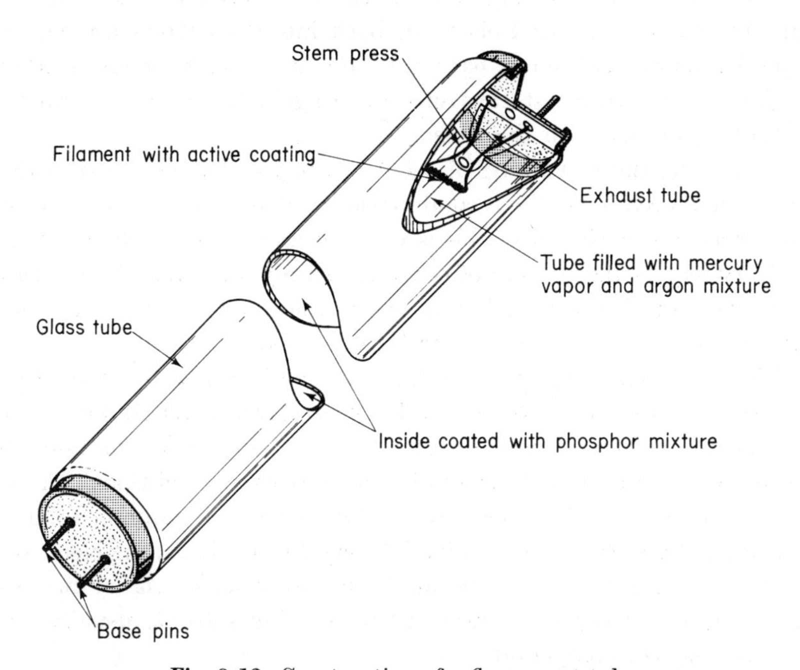

Fig. 8-13. Construction of a fluorescent tube.

ultraviolet radiation is transformed into visible radiation by a *phosphor*. A phosphor is a fluorescent chemical; that is, one which absorbs photons of energy with ultraviolet wavelength and re-emits energy as photons of visible wavelength. The basic construction is shown in Fig. 8-13. In operation, the electron-emissive filaments maintain a current through the mercury vapor (the argon just aids in "starting" the tube) such that the dominant wavelength emitted is 2.537×10^{-7} meter. The coating is a mixture of phosphors chosen to emit a desired color "white" (i.e., pinkish, yellowish, bluish, etc.) when stimulated by this wavelength.

Fluorescent lamps are more efficient than incandescent lamps. A 40-watt fluorescent tube converts about 20 percent of its input wattage to visible light.

8-10 ELECTROLYTIC PROCESSES

It has already been pointed out (Section 8-1) that a process in which electrical energy is used to cause a chemical compound to decompose is called an electrolytic process, or simply electrolysis. Charging a storage battery furnishes a good example of such a process. In using electrical energy from the battery, we caused some metal atoms to ionize by giving up electrons, thereby acquiring a positive charge. They go into solution in the electrolyte, effectively forming a chemical compound of that metal. In charging the battery, we feed electrons back into the battery and the positive metal ions are attracted back into the negative terminal (cathode). There they regain their lost electrons, once again becoming neutral metal atoms that "plate out" on the cathode.

Of course, the recharging of a battery is possible only if the original elements remain available for recombination and are not lost to some other reaction or as an escaping gas. This is why not all cells are rechargeable.

A battery in the process of charging is only one example of a practical application of electrolysis. Before discussing others, let us take a closer look at what is necessary in an electrolytic process.

The first requirement is to have an *ionic compound* (see Section 5-5), which is called the *electrolyte*, to be electrolyzed (broken down electrically), and it must be in such a state that the *ions are free to move*. Most ionic compounds are crystalline solids under normal conditions. The individual ions, like the sodium ions and chloride ions in common table salt, are locked tightly in the latticework of the crystal by the strong electrostatic forces. The ions must therefore be made free to move in one of two ways: either by (1) *dissolving* the compound in a polar solvent, usually water, or by (2) *melting* the compound.

In the case of dissolving, one or both products of electrolysis may, in turn, react chemically with the solvent so that, from the standpoint of the end products of the process, it is not simple electrolysis. This reaction happens because the elements that form ionic compounds are fairly active substances. In point of fact, electrolysis of naturally occurring compounds is actually the only way in which many of these pure elements can be produced.

The other two requirements of an electrolytic process are for an external source of *direct current* and for two conducting rods or plates of metal (or sometimes graphite) to lead the current into and out of the substance being electrolyzed.

The electrolysis of molten sodium chloride is illustrated schematically in Fig. 8-14. When a dc voltage is applied to the graphite electrodes, the

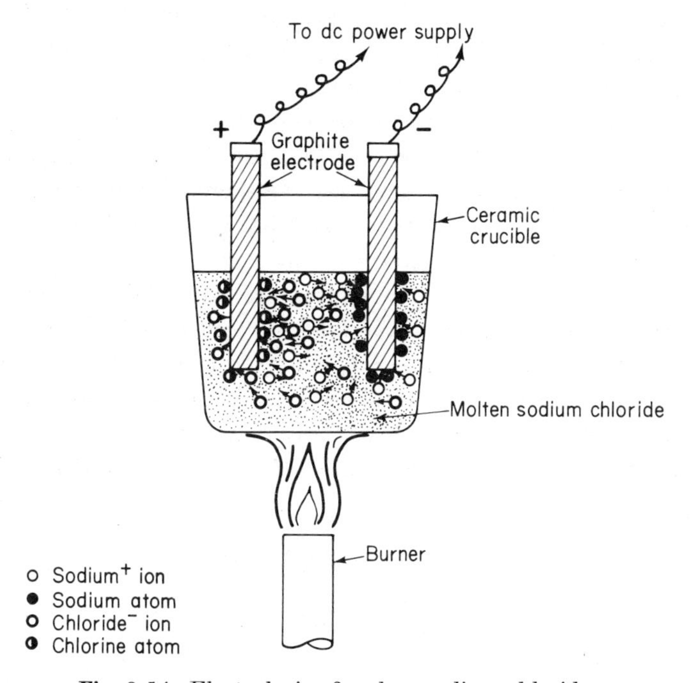

Fig. 8-14. Electrolysis of molten sodium chloride.

positively charged sodium ions (Na^+) are attracted by and drift over to the negative electrode (cathode). There, each arriving Na^+ ion accepts an electron and becomes a neutral sodium atom, according to the reaction

$$Na^+ + \textcircled{e} = Na$$

so sodium metal collects at the cathode. At the same time, the negatively charged chloride ions (Cl^-) are attracted to the positive electrode (anode), where each gives up an electron and becomes a chlorine atom. Subsequently, the chlorine atoms combine in pairs to form chlorine gas molecules (Cl_2). The reactions are

$$Cl^- = Cl + ⓔ$$

$$2Cl = Cl_2$$

Hence chlorine gas is given off at the anode.

Notice that unlike the electrical conduction in wires and resistors, the current in the electrolytic cell is carried by heavy, slow-moving ions rather than by electrons.

Another classic among electrolytic processes is the electrolysis of copper sulfate solution. This process is interesting for two reasons. First, the negative ion in the compound, copper sulfate ($CuSO_4$), is not a simple ionized atom (like the chloride in sodium chloride) but the ionic group (radical), the sulfate ion, $SO_4^=$. Secondly, in this process, the water in the electrolytic solution also participates. As you would expect, the doubly charged (i.e., valence $= +2$) copper ion, Cu^{++}, migrates to the cathode, picks up two electrons, and "plates out" as a metallic copper atom. At the anode, however, the end product is not what you would expect. Although it is true that the $SO_4^=$ ions migrate to the anode, they do not give up their electrons there. Instead, oxygen atoms are liberated by electrolysis of hydroxyl ions from the water, while the hydrogen ions from the electrolyzed water stay in solution with the $SO_4^=$ ions to form sulfuric acid (H_2SO_4). Thus the overall effect that one observes from outside the copper sulfate cell is that copper plates out on the cathode, oxygen gas bubbles out at the anode, and sulfuric acid accumulates in the electrolyte around the anode.

8-11 FARADAY'S ELECTROLYTIC LAWS

The story of electrolysis goes back almost as far as the story of electricity, and many men have contributed to its understanding. Nicholson and Carlisle, in 1800, while repeating Volta's "pile" experiments with various materials, deposited copper from a copper sulfate solution on a copper wire. (Nicholson also first demonstrated the electrolysis of water.) William Wallaston achieved similar electrolytic results with charges from a static machine. But the first to put electrolysis on a firm theoretical footing was Michael Faraday, perhaps the greatest of the scientific giants of that period. We encountered Faraday in Chapters 3 and 4, and the name of this remarkable man will appear again and again in the course of this book.

Born in 1791, one of ten children of a poor Surrey blacksmith, Faraday had no formal education. He was apprenticed to a kindly London bookbinder who permitted him to read in his spare time and to attend scientific lectures. It was at a series of such lectures that Faraday heard the celebrated Sir Humphry Davy. He immediately set about establishing correspondence with Davy, eventually obtaining a position as Davy's laboratory assistant and secretary (and whenever Mrs. Davy had her way, as a servant). As the years passed, Faraday's work surpassed that of Davy, causing the latter to become very embittered. In 1824, when Faraday was elected to the Royal Society, Davy had cast the only vote against him, but the deeply religious Faraday never responded with malice. Faraday declined many of the honors awarded him and this self-made, confirmed bachelor devoted his entire frugal life to the laboratory. He was never the great theoretician, but most of our modern electronic world is a tribute to his empirical scientific ability.

Carrying on Davy's work in electrochemistry, in 1832 Faraday discovered and enunciated the two fundamental laws of electrolyis:

1. The mass of a substance produced at either electrode by electrolysis is directly proportional to the total quantity of electricity (i.e., charge) that has passed through the electrolyte.
2. The mass of substance liberated by a given quantity of electricity is proportional to the atomic weight of that substance divided by its valence (i.e., the number of electrons per atom involved in forming the compound from which it was liberated).

His pronouncement of these laws was the more astonishing because atomic theory was new and the structure of the individual atom was unknown at the time.

The first law says that if a given number of coulombs pass through a copper sulfate solution and yield a pound of copper, then twice as many coulombs will produce 2 pounds of copper.

The second law is easier to deal with if we understand what is meant by a *gram-atom.* A gram-atom of a pure substance is, by definition, a lump of that substance whose mass in grams is numerically equal to its atomic weight. Thus, since sodium's atomic weight is 23, a gram-atom of sodium means 23 grams of it. Copper's atomic weight is about 64, so 64 grams make approximately one gram-atom of copper. Now, in sodium chloride, the sodium ion is singly charged so its valence is said to be $+1$. Let us suppose that we have a quantity of electricity which will plate out 23 grams of sodium from molten salt; then Faraday's second law says that this *same quantity* of electricity will plate out 32 grams of copper from copper sulfate solution. This is so because

$$\underset{(23\ \text{grams})}{1\ \text{gram-atom of sodium}} \div \underset{(1)}{\text{valence of sodium ion}}$$

is equivalent to

$$\underset{(64\ \text{grams})}{1\ \text{gram-atom of copper}} \div \underset{(2)}{\text{valence of copper ion}}$$

It turns out that the quantity of electricity required to deposit exactly one gram-atom of an element, which has a valence of one in its compounds, is 96,500 coulombs. This quantity has been given Faraday's name to honor his work in electrochemistry.

$$1\ \text{faraday} \equiv 96{,}500\ \text{coulombs of charge}$$

8-12 INDUSTRIAL APPLICATIONS OF ELECTROLYSIS

Many important substances are produced commercially by electrolysis. One large class includes *pure active* elements such as sodium, potassium, aluminum, magnesium, chlorine, and fluorine. A second large class includes *compounds formed in secondary reactions* after some active substance required is temporarily liberated by electrolysis. In the latter category are the manufacturing processes for caustic soda, hypochlorites (laundry bleach is a dilute hypochlorite), chlorates, and perchlorates.

The Hall process, discovered by the young American chemist Charles Martin Hall, is a good example. In the early nineteenth century, aluminum was considered a precious metal and was used in jewelry. (Napoleon III had his banquet cutlery and his child's rattle made from aluminum. The top of the Washington monument is also made of aluminum, for this metal was still precious when the monument was constructed.) It was well known that great quantities of aluminum ore (Al_2O_3), such as bauxite, were readily available in the earth's crust. The problem was that aluminum ore would not melt or dissolve easily and, you will recall, the ions must be free to move for electrolysis to take place.

Hall's teacher remarked one day that the man who discovered a way to produce aluminum economically would become wealthy and famous. Taking the teacher at his word, Hall worked in his own laboratory and in

1886, when 23 years old, he made the key discovery that did indeed bring him fame and fortune. He learned that aluminum ore could be readily dissolved in molten *cryolite* (a white mineral fluoride of sodium and aluminum found in Greenland and near Pike's Peak, Colorado). All of the world's aluminum is today produced by electrolysis of aluminum ore dissolved in cryolite, which, because it was discovered independently about the same time by Paul Heroult of France, is now called the Hall-Heroult process.

Other important commercial electrolytic processes are the production of ozone (O_3) from oxygen, and the fixation of nitrogen in air by electrolytic formation of nitric oxide.

REVIEW QUESTIONS

1. What effects of electric current are stressed in this chapter? What devices do they help us to understand?
2. Which electrochemical effect was noted first, the "battery effect" or electrolysis? Why?
3. What might you say to a friend who "fixes" a blown cartridge fuse by wrapping tinfoil around it and putting it back?
4. Why is the thermostat's active element made from *two* metals?
5. Suppose you had to design an electric frying pan. What important components would you need? Describe how you would put them together.
6. Home electrical circuits used to be protected by fuses; now circuit breakers are installed instead. Give several reasons for this change.
7. Unlike appliances, circuit breakers are *not* marked with a "wattage rating." Why not?
8. Give several examples of things you have seen which were (approximately) blackbody radiators.
9. Why is it obvious that the temperature of the filament of a lit 100-watt bulb is greater than 500 °F.?
10. What is the benefit of filling incandescent lamp bulbs with an inert gas?
11. Why do fluorescent lamps have an internal coating of white powder?
12. Knowing from Section 8-11 how sodium and chlorine can be produced, how do you suppose one might produce potassium and bromine in a laboratory?
13. Explain Faraday's laws of electrolysis in your own terms.

9

BETWEEN TWO POLES

9-1 MEET THE MAGNET

As small children, most of us discovered the "magic" of the magnet. By playing with magnets and bits of iron and steel, we learn some empirical facts about magnets. They attract and "stick to" iron or steel but (apparently) nothing else. When two magnets are brought together, depending on which parts of the respective magnets are in closest proximity, they either attract or repel each other. We may even know that they have regions called *poles*. But are these attractive little devices of any value other than as children's toys and curiosities? The answer is an emphatic and resounding Yes, for magnetic poles share the glory about equally with electric charges in producing today's electronic marvels. For this reason, we shall devote a chapter to the study of magnets, magnetism, and the relationship between electricity and magnetism.

9-2 THE MAGNESIAN STONE

The original discoverer of magnetism, like the discoverer of electricity, is unknown. Perhaps some small boy first noticed the strange, attractive properties of the odd piece of rock he had picked up when it was held near iron. In any case, centuries before Christ, the Greeks, knew that the mineral commonly called *lodestone* would attract bits of iron and other pieces of the same material. The name lodestone means "way stone" or "stone that leads the way," from its later use as a natural compass. For many centuries, however, only the attraction for other pieces of lodestone and for iron was noted. It is first mentioned by the natural philosopher, Thales of Miletus, around 600 B.C. Later, Plato has Socrates say to Ion:

> . . . there is a divinity moving in you, like that contained the stone which Euripides calls a *magnet* [author's italics] but which is commonly known as the stone of Heraclea. This stone not only attracts iron rings, but also imparts to them a similar power of attracting other rings; and sometimes you may see a number of pieces of iron suspended from one another so as to form quite a long chain: and all of them derive their power of suspension from the orignal stone. . . .[1]

Writing around 60 B.C., the Roman poet Lucretius says ". . . iron can be drawn by that stone which the Greeks call Magnet by its native name, because it has its origin in the hereditary bounds of the Magnetes [i.e., the inhabitants of Magnesia in Thessaly, now Asia Minor]."[2] Hence all our English *magnet-* and *magneto–* words having to do with magnets or magnetism come from the Greek, *ἡ Μαγνησία λίθος* (*i Magnesia lithos*), meaning "the Magnesian stone," just as the *electr(o)-* words come from the Greek word for amber.

Lodestone is actually a black magnetic iron ore whose proper name is now *Magnetite*, a chemical compound of the type called ferrites, with the composition $Fe\,(Fe_2O_4)$. Although not the most common, magnetite is the richest of iron ores and serves as a major source of iron in some parts of the world. In the United States, strongly magnetic lodestone can be found at Magnet Cove, Arkansas. Figure 9-1 shows a "chain" of iron rings suspended from a piece of lodestone.

The ancients had no explanation for the strange powers of the lodestone. They concocted many odd theories, the most prevalent of which was that the lodestone, like amber, had a "soul" and there were certain substances (iron) that it "liked."

There is no indication in the ancient literature that either the early Greeks or Romans were aware of the directional characteristics of a suspended magnet which make it a compass. As early as A.D. 121, the Chinese made artificial magnets by rubbing iron needles on lodestone. It is only a millennium later, however, when reference is made to the compass.

[1] Plato, *Ion*, p. 533.

[2] Lucretius Carus, *De Rerum Natura*, Book vi, pp. 906 ff.

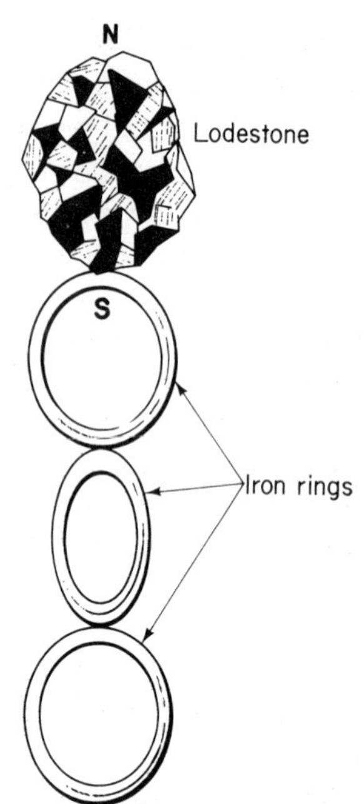

Fig. 9-1. Lodestone is a natural magnet.

9-3 MAGNETIC POLES; THE COMPASS

Every magnet, natural or man-made, has two distinct regions called *poles*. If a magnet, shaped like a bar or rod, is supported in such a manner that it is free to rotate (as by hanging it from a string, or by placing it on a cork floating in water), it will line itself up in a north-south direction as shown in Fig. 9-2. The pole of the magnet on the end that points northward is called a *north-seeking pole,* or simply *north pole* for short. The pole on the other end of the magnet, which points southward, is called a *south-seeking pole,* or *south pole* for short. Thus every magnet has a north and a south pole. The reason that a pivoted magnet lines up in the north-south direction is that our planet, Earth itself, is a giant magnet. Perhaps the most fundamental law of magnetism is that

unlike poles attract each other, and like poles repel each other.

This being the case, it necessarily follows that the geographic North Pole of the earth is in fact a south(-seeking) magnetic pole, since it attracts a north-seeking magnetic pole. Similarly, the South Pole of the earth is a north(-seeking) magnetic pole.

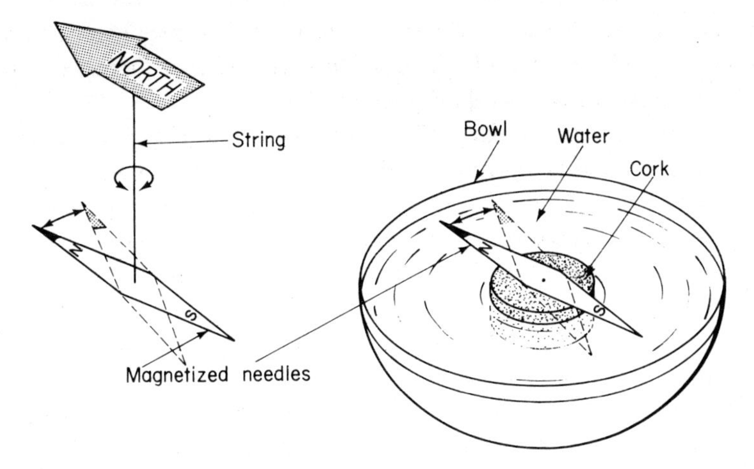

Fig. 9-2. Crude compasses. Freely supported magnets tend to align themselves along the north–south direction.

When a magnet is suspended so that it is free to rotate and thus indicate geographic directions, such a device is called a *compass*. The first mention of a magnetic compass is by the Chinese mathematician and instrument maker Shen Kua early in the eleventh century. However, his writing only describes is use on land. It is not until about the year 1086 (the Chinese writer Chu Yu tells us,) that "foreign" ships on the Canton-Sumatra run were guided by compass. The foreigners (to the Chinese) who had a corner on that trade route at that time were the Moslems; therefore we must

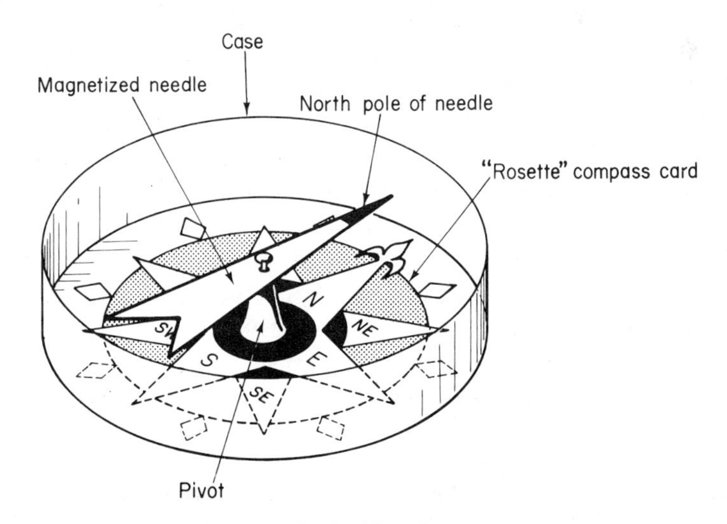

Fig. 9-3. Typical hand compass.

tentatively conclude that they were first to use the compass at sea. Recent evidence seems to indicate that certain Scandinavian peoples may have had the compass by the year 1000. In any case, Arabs and Europeans were all using the compass in the twelfth century.

A more modern hand compass, illustrated in Fig. 9-3, makes use of a pivot-mounted, arrow-shaped needle. The background against which it turns is a card marked with the principal directions, in a pattern called a *rosette.* In use, the compass is held level and the needle is permitted to come to rest, pointing north. The case is then carefully rotated under the needle until the N indicator of the rosette is under the north-pointing needle. One may then determine any other direction from the appropriate pointer of the rosette.

The compass was the first and only practical use of magnetism for several hundred years. Superstitiously, compasses were also used as potent charms because of their "magick" power of attraction.

9-4 THE PILGRIM'S PROGRESS

Even the medieval armies had their "missile engineers"—men whose job it was to design devices for hurling boulders, Greek fire, and other items at the enemy's fortifications. Such a man was Master Peter de Maricourt, better known by his Latin name, Petrus Peregrinus (Peter the Pilgrim). Peregrinus was a Crusader with the army of Louis IX of France, which eventually became the army of Louis' brother Charles. Charles, hungry for power, had made a deal with the papacy whereby he became king of the two Sicilies in return for ridding Italy of German influence. So it happened, in the year 1266, that Charles lay siege to the southern Italian city of Lucera, and Peter Penegrinus was there serving him. The siege was a long, dull affair, lasting over 3 years, and the demands on Peregrinus' time were not heavy. Consequently, the scholarly Peregrinus had time to think about his favorite subject, mechanics. He had been intrigued by the idea of a drive mechanism that required no human effort and wondered if magnetic forces might be used. (This is the first recorded suggestion that magnetic energy might be convertible to mechanical energy.) Such speculation led Peregrinus to study and experiment with the magnet. His results have come down to us from a long and remarkable letter, *Epistola de magnete,* written to his friend, Sigerus de Foncaucourt, in 1269. Some of the facts he discovered were

1. How to determine the north and south poles of a magnet.
2. The law of attraction and repulsion.

3. That there is no such thing as an isolated magnetic pole, since breaking a magnet only results in two or more smaller magnets, each having *both* a north and south pole.

He was also the first to study the "magnetic field" around a magnet, using bits of iron. Finally, Peregrinus devised a pivoted compass with a circular scale, an improvement over the then-universal floating compass without a scale. His compass was the forerunner of those that gave early explorers the confidence to sail far out of sight of land. Peregrinus' only major error was in his explanation of how the compass worked. He believed that the magnetic north poles pointed ". . . by the will of God. . ." to the "pole of the celestial sphere" (the outermost of the concentric spheres of the universe as Ptolemy conceived it).

9-5 THE MAGNETIC FIELD

Just as in the case of electrically charged objects, it was also difficult to explain how magnetic objects could produce "action at a distance"—that is, how one could produce a force of attraction or repulsion upon another without there ever being a contact between the two. As you may recall from Chapter 3, a convenient description of a region of space in which energy is (somehow) stored, and which has the ability to produce forces and do work, is to say that a "field of force" exists in that region. It is therefore convenient to say that a *magnetic field* exists in the region around a magnetic pole, just as we defined an electric field around a charge.

In Section 3-12 we saw how an uncharged body could acquire a charge by electrostatic induction when placed within the electric field of a charged body. It is also true that a piece of soft iron (or of certain other materials) can become a magnet itself by *magnetic induction* when placed within the field of a magnet, even if the two do not touch.

The reality of the magnetic field is most easily demonstrated by the most classic of magnetic experiments, first performed by Peter Peregrinus and since repeated by every schoolboy. This experiment, illustrated in Fig. 9-4*a*, consists of placing a magnet beneath a sheet of paper (or other thin nonmagnetic material) and sprinkling iron filings (Remember meeting them in Section 4-2?) on top of the paper while lightly tapping it. Each iron filing becomes a tiny magnet by induction and lines up with the direction of the magnetic field at the point where it falls. The resulting pattern, shown in Fig. 9-4*b* for a straight bar magnet, represents the direction of the field at each point. Note the clustering of filings near the poles, where the field is strongest.

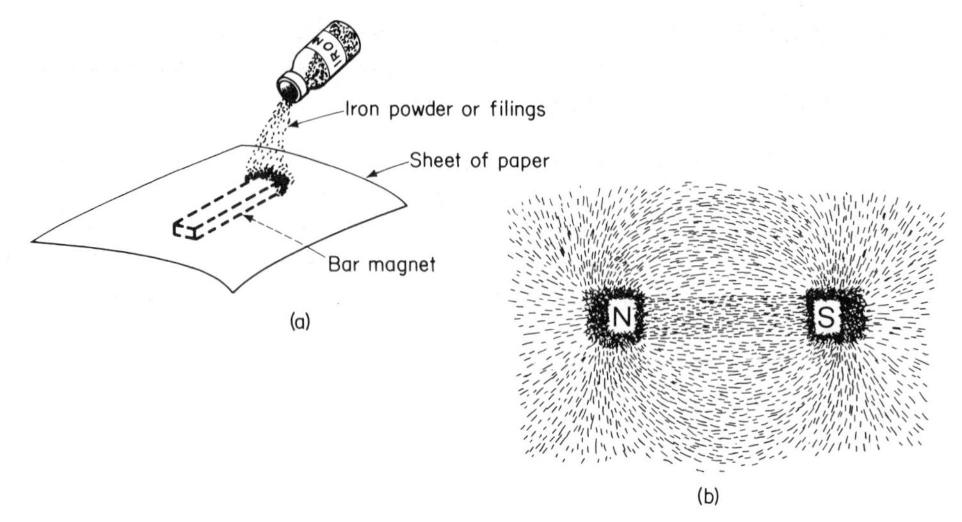

Fig. 9-4. Demonstrating the magnetic field around a bar magnet. (a) The technique. (b) The pattern obtained.

It is convenient to speak of the magnetic field in terms of *magnetic lines of force* that run from pole to pole through the surrounding medium (and back through the magnet) and whose direction at each point along their length shows the direction of the field. The total quantity of lines passing through a region is called the *magnetic flux*. It is usually symbolized by the Greek letter phi (ϕ) and is measured in units of *webers* (just as charge is measured in coulombs), named for the physicist Wilhelm Weber.

The quantity defining the magnetic field at any point is a vector (see Section 2-7), analogous to the electric field intensity for the electric field. This vector quantity is called the *flux density* (or *magnetic induction*) and is represented by the symbol $\mathbf{B}$. Its magnitude tells how densely the lines of force are packed together, that is, how many webers of flux pass through a unit area (perpendicular to the direction of the field). Thus the natural unit[3] of flux density is the weber per square meter, but, to shorten the name, the unit of flux density is called a *tesla* after Nikola Tesla, the famous engineer.

To get some feeling for the size of the unit of flux density, it should be noted that the magnetic field of the earth, at sea level, is about 3×10^{-5} tesla; that near a typical permanent magnet is about one tesla; and some of the strongest man-made magnets (electromagnets) produce flux densities on the order of 100 teslas.

[3] B is sometimes still measured in the older and smaller (CGS) unit, *gauss*. 1 tesla $= 10^4$ gauss.

The direction of the **B** vector is the direction of the field at that point, which, in turn, is defined as the direction in which a hypothetical tiny isolated north pole would tend to move if placed at that point. This definition is only a formal one, and not practical from the experimental standpoint. An isolated north pole is a fiction that we know cannot exist from the experiments of Peregrinus and others, in which a magnet was broken several times and, in every instance, the smaller fragments each had both a north and a south pole, as indicated in Fig. 9-5. (The theoretical argument to be pre-

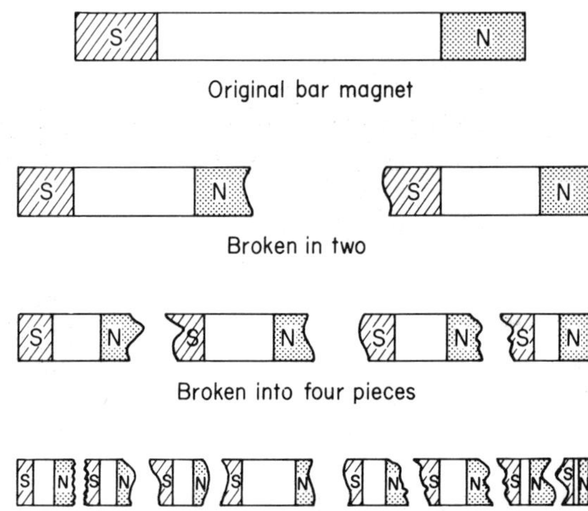

Fig. 9-5. A north or south pole cannot be isolated.

sented later will show that this indeed must be so.) Furthermore, the isolated north pole, even if it could exist, must be so tiny that its own field cannot alter the field into which it is inserted. This is not strictly possible for any finite magnet.

In any case, the **B** vector as defined, at any point in a magnetic field, shows the direction of the field. Our arbitrary selection of a tiny isolated *north* pole as our "test particle," for determining the direction of a magnetic field, then fixes the path we must assume for our lines of force. *External to a magnet, the path of a line of force is from the north pole to the south pole, and within the magnet it is from the south pole to the north pole.*

Several kinds of magnets are shown in Fig. 9-6, as well as the types of fields they produce. The **B** vector is drawn for a few locations in each field. The drawings of the bar and horseshoe magnets only show their fields in the plane of the page, and you must use your imagination to visualize these fields in three-dimentional space. The ring magnet (Fig. 9-6*d*) is shown both in perspective and in cross section to make clear its peculiar doughnut-

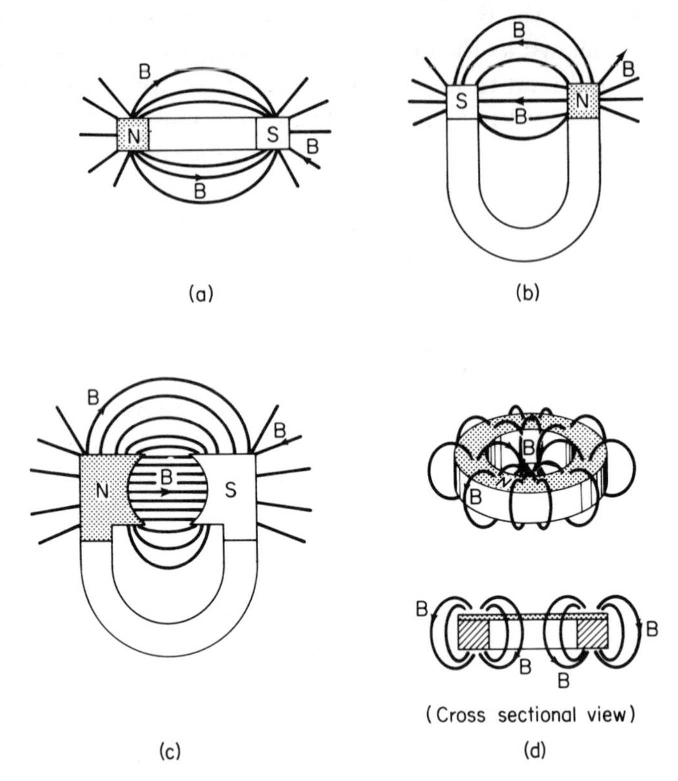

Fig. 9-6. Various types of permanent magnets. (a) Bar magnet. (b) Horseshoe magnet. (c) Field-shaping horseshoe magnet. (d) Ring magnet.

shaped field. The field-shaping horseshoe magnet (Fig. 9-6*c*) has many important commercial applications, such as in small motors and meter movements, because of the uniform field (equally spaced, parallel lines of force) between its poles.

9-6 THE EARTH AS A MAGNET

After Peter Peregrinus, little progress was made in the study of magnetism for over 300 years. The navigator had his compass and thought he understood why it worked. Meanwhile, some interesting myths had grown up around the magnet. One had it that there were islands to the far north which were enormously strong magnets capable of pulling all the nails out of a ship from a distance. Another widely held belief was that the aroma of garlic would destroy magnetism; hence sailors who had just eaten food spiced with garlic were forbidden to breathe on the compass.

A good deal of light was shed on the magnet by the English physician William Gilbert, whom we previously encountered in Chapter 3. Gilbert, educated in medicine at Cambridge, had subsequently traveled widely in Europe. He most probably received a "Doctor of Physic" degree from a continental university before returning home to England to medical practice. In 1576 he was admitted to the College of Physicians, becoming its president 23 years later. During these years, while distinguishing himself as a physician, Gilbert also became known as a savant in cosmology, chemistry, and physics. An avid experimentalist, he repeated the almost-forgotten experiments of Peregrinus and carried them much further. The results of 18 years of work are documented in a book, published in 1600, with the most un-Hollywood title, *De Magnete Magneticisque Corporibus et de Magno Magnete Tellure Physiologia Nova* (Concerning the Lodestone and Magnetic Bodies and Concerning the Great Magnet the Earth a New Theory), usually referred to simply as *De Magnete*. Among other things, Gilbert disproved the garlic myth. His major contribution to magnetism, however, was in his explanation of the behavior of the compass. In experimenting with a terrella (a lodestone ground into a roughly spherical shape), Gilbert noticed that a freely shaped magnet dipped downward when brought near the terrella's poles. He had also noticed that a freely suspended magnet dipped toward the earth as well as aligning itself north-south, and the farther north one went, the more pronounced the dip. He reasoned that the earth itself was a great magnet, a huge terrella in effect, and that the compass merely pointed to its magnetic pole rather than to the sky as Peregrinus believed.

Gilbert believed that the magnetic north pole actually coincided with

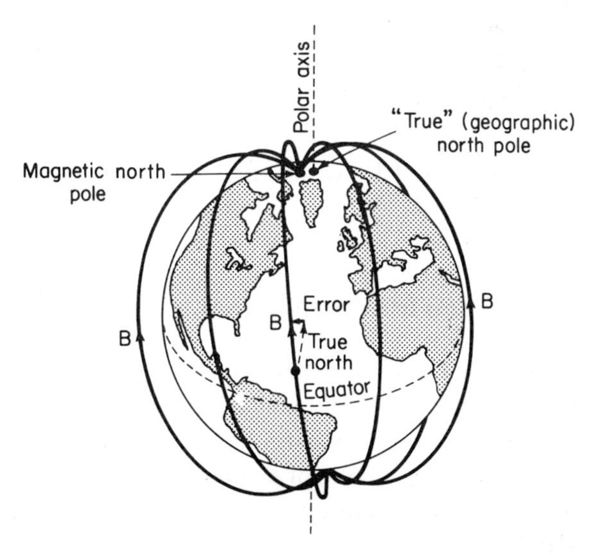

Fig. 9-7. The magnetic field of the earth.

the geographic North Pole. Today we know that the magnetic pole slowly and randomly wanders about and will rarely, if ever, exactly coincide with the geographic (i.e., the true) North Pole. Consequently, as shown in Fig. 9-7, a compass will align itself parallel to the earth's local **B** vector, which is slightly inclined to the direction of true north. The angle of error between the two directions is called the magnetic *declination* or *deviation*. The magnetic north pole is presently about 500 miles from the true North Pole.

There are always places on the surface of the earth at which magnetic north and true north just happen to lie in the same direction. If, on a map, such places are connected by a line, this line is called the *agonic* line. In the United States, the declination is from zero to 20 degrees east and west of the agonic line.

More is said about the earth's field in Section 9–17.

9-7 COULOMB'S LAW

Charles Augustin de Coulomb, the same Coulomb we met in Section 3-6, also discovered the first quantitative law of magnetism. What is most remarkable is that his mathematical law describing the force of attraction (or repulsion, as the case may be) between two magnetic poles is of identically the *same form* as his law describing the force between two electric charges. The law, enunciated in 1785, states that the size of the force depends directly on the strength of each of the two poles and falls off very rapidly (inverse square) with increasing distance between the poles. Examples illustrating these relationships are given in Fig. 9-8. The force is, of course, attractive if the poles are of different kind and repulsive if they are alike. To get some appreciation for the intensity of magnetic forces, you may consider the fact that the force of attraction between two unlike magnetic poles, each of which is confined to a fairly small region and has a strength of 1 weber, when separated by a distance of 1 meter (about 3 feet), is approximately 90 tons!

9-8 ELECTROMAGNETISM, A HAPPY MARRIAGE

For some time it had been suspected that there was some connection between electricity and magnetism. Ben Franklin and other electrical experimenters had noticed that a lightning bolt or discharge from a static machine could reverse the poles of a magnet. Yet it was not until 1819 that the critical discovery was made by an obscure Danish physicist named Hans Christian Oersted.

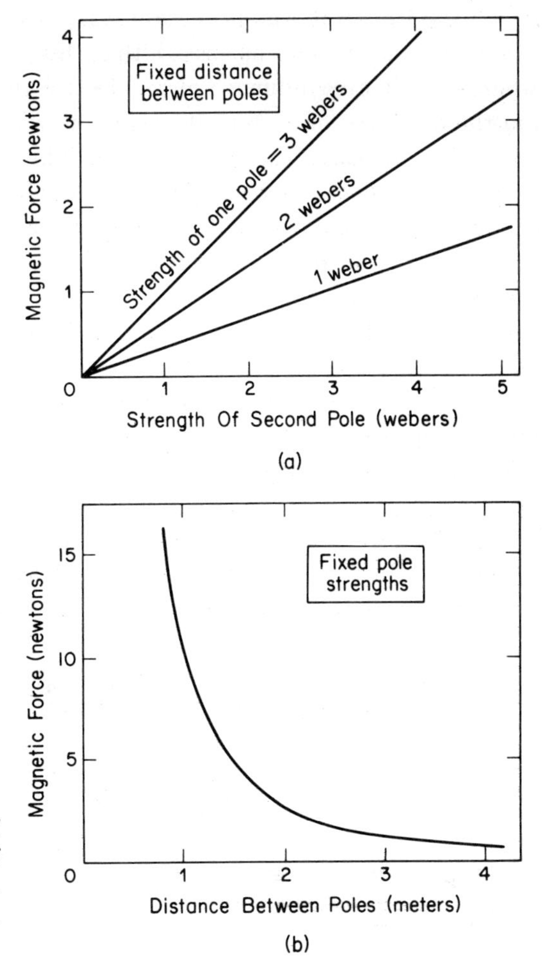

Fig. 9-8. Coulomb's law of magnetism. (a) Variation of force with strength of poles. (b) Variation of force with distance between poles.

Oersted, the son of a poor apothecary, received a piecemeal primary education from an assortment of his father's friends, acquaintances, and business associates. This unusual education proved sufficient for him to gain admittance to the University of Copenhagen. He earned a portion of his expenses as a part-time teacher. Imbued with the metaphysics of his favorite philosopher, Kant, Oersted believed in the fundamental unity of all forms of energy. He therefore believed he could find the relationship between electricity and magnetism. Volta's invention of the battery provided him with the tool he needed, a steady current. It is said that he made the discovery which was to immortalize him during the course of a classroom demonstration. He brought a compass near a wire carriyng a current form a battery. Evidently the needle of the compass was lined up more or less parallel with the wire because when Oersted placed it near the wire, the needle deflected until it was at a right angle to the wire, as shown in Fig. 9-9*a*. Oersted was no less surprised than his students. How very queer! One would expect any influence of an electric current on a compass needle to be in the direction of current flow, just as wind pushes on a sail or river current against a boat; yet the magnetic influence of the electric current acts *perpendicular to the direction of the current.* (Remember that the structure of the atom was unknown in Oersted's day and his electric current was conceived as a flow of positive electric fluid, opposite in direction to the way we know electrons move.)

Oersted had discovered *electromagnetism*, the fundamental principle of which is that a magnetic field is associated with every electric current (i.e., with every *moving* electric charge). The field at any point along a current-carrying wire consists of concentric circular lines of force about the wire. This situation is shown in Fig. 9-9*b*, together with the helpful *left-hand rule* for remembering how the direction of the field is related to the direction of the current: Hold your left hand so that the thumb is parallel to the wire and points in the direction of electron travel; your other fingers then indicate the direction of the paths of the lines of force.

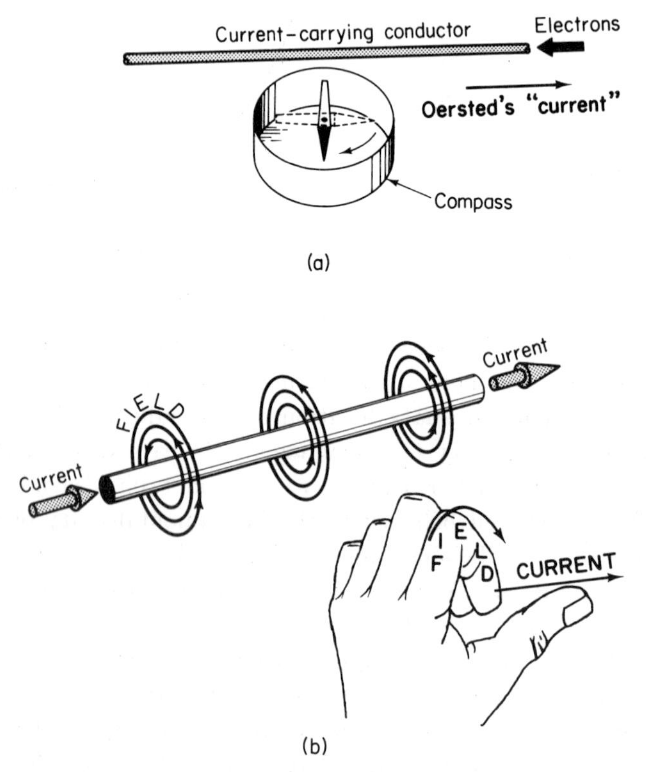

Fig. 9-9. Magnetic field around a current-carrying wire. (a) Oersted's experiment. (b) The left-hand rule.

It is important to understand the significance of Oersted's discovery. Prior to his experiment, it was thought that electricity and magnetism were separate phenomena, and it is true that magnetic poles and electric charges *at rest* do not influence each other. Oersted had shown, however, that a magnetic field is associated with the moving charges in an electric current. It is also true, as we shall see presently, that a moving magnet is capable of producing an electric voltage and current. Hence electricity and magnetism are just two different aspects of the much broader subject, which is the marriage of the two, electromagnetic theory.

9-9 THE FIELD OF A COIL

Now let us see what happens to the magnetic field of a current-carrying wire when that wire is bent into a loop, as in Fig. 9-10*a*. Choosing a direction for the current, and making use of our left-hand rule, we can establish the pattern shown in perspective in Fig. 9-10*a* for two planes passing through the loop at right angles to each other (the vertical plane

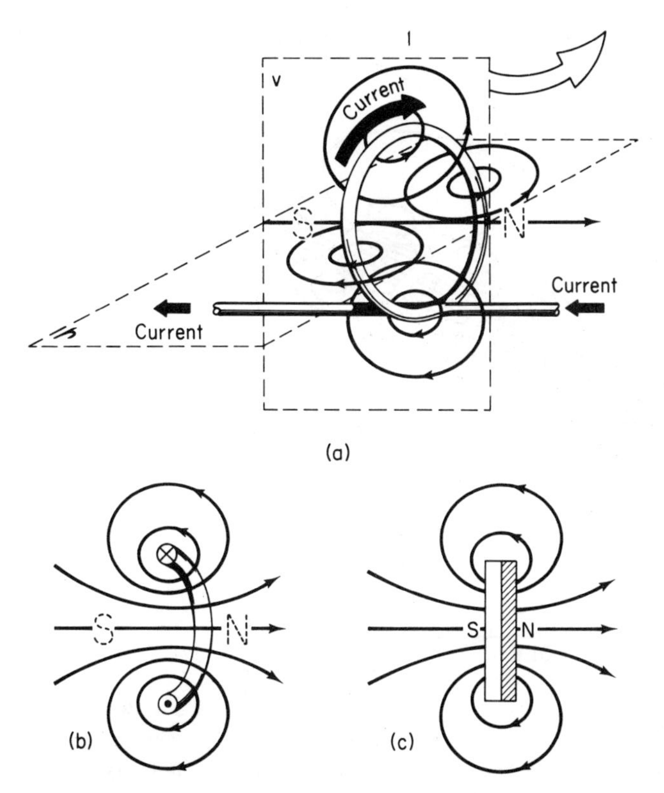

Fig. 9-10. (a) Field of a circular loop. (b) Vertical section view. (c) Field of a short bar magnet.

v and the horizontal plane h). Such perspective drawings are sometimes confusing to the eye, so in Fig. 9-10*b* we show only the vertical cross section, as if the loop had been sliced with a knife along the plane v. In this sectional drawing, we use the very popular conventional symbols, ⊙ and ⊗, representing, respectively, the flow of current out of and into the page. (These symbols are meant to suggest the head of an arrow coming out of the page and the tail of an arrow going into the page.) Notice how the lines of force created by the current at opposite sides of the loop tend to reinforce each other.

Figure 9-10*c* shows the field of a short bar magnet, which, depending on how large it is in the direction perpendicular to the page, might perhaps be more appropriately called a *rod magnet* or a *plate magnet*. Note that its field and that of the current-carrying loop are indentical! If either the bar magnet or the loop were enclosed in a cardboard box, you could not tell which it contained from the way the box influenced a compass or iron filings.

If we wind a coil of many turns, like a spring, instead of just the single loop, the fields from the individual turns of the coil aid each other, as shown in Fig. 9-11, producing an overall field that cannot be distinguished from

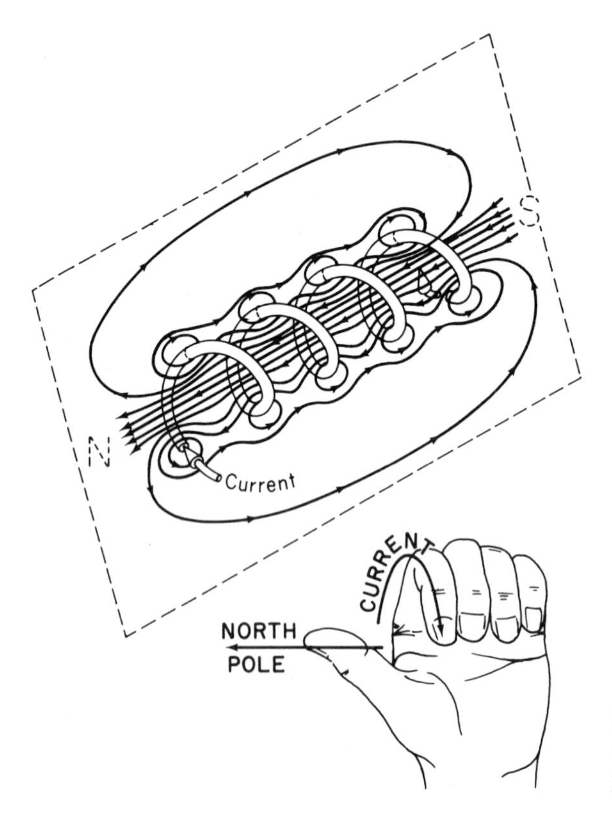

Fig. 9-11. Magnetic field of a coil and the left-hand rule of polarity.

that of a bar magnet. Compare this field with that of the magnet shown in Fig. 9-6*a*.

To help remember the relationship between the current in the coil windings and the polarity of the resulting field, we have another left-hand rule, illustrated in the inset to Fig. 9-11: Hold your left hand with the thumb extended and such that the other fingers encircle the coil in the same direction as the electron current; the thumb then points to the north-pole end of the coil.

A coil wound in this fashion is called a *solenoid*. For any given current, its strength as a magnet is directly proportional to the number of turns in its winding. In other words, a 200-turn solenoid will produce twice as many teslas of flux density as a 100-turn coil, everything else being the same.

9-10 MAGNETISM IN MATTER

At this point, we digress somewhat to consider a question that (the reader may have noticed) we have dodged since the beginning of this chapter: Why are certain substances naturally magnetic? Three obvious targets for this question are lodestones, compass needles, and the entire planet Earth. A moment's reflection will show that they have something in common—iron. The compass needle is almost 100 percent iron; the lodestone, as you may recall, is a rich iron ore; and the earth's core is believed to be rich in iron. "All right, then," you might ask, "why does iron possess magnetism while other materials do not?" This statement is not strictly true.

All natural substances respond weakly to a magnetic field. The kind of strong magnetism we observe in lodestones and lumps of iron was known to the ancients. Today it is called *ferro*magnetism (from the Latin for iron, *ferrum*). Nickel and cobalt were discovered in the eighteenth century, and it was recognized almost immediately that nickel was ferromagnetic. It was not until the years 1845–1846, however, when the great experimentalist, Michael Faraday, thoroughly investigated the magnetic properties of materials, using a strong electromagnet (a solenoid with an iron core). Faraday found that all substances could be classified into one of two categories on the basis of their response to an intense magnetic field. Substances that are attracted are called *paramagnetic*, while those repelled are termed *diamagnetic*. (To remember which is which, it is helpful to note that *p*aramagnetic and *p*ulled both begin with *p*.) Most substances—for instance, glass, copper, wood, and water—are weakly diamagnetic. A few, like air, aluminum, ebonite, liquid oxygen, and all ferromagnetic materials, are paramagnetic. Thus a ferromagnetic material represents merely a special type of paramagnetic material, but one that profoundly influences the field into which it is placed, becoming itself strongly magnetized in the process. The rest of this discussion is devoted to ferromagnetism, for it is of greatest practical importance.

Theories of magnetism have existed for as long as magnets have been around. We need not dwell on ancient "soul" theories and the like. As might be expected, the "fluid" theorists of the eighteenth century, notably Franklin, Poisson, and Coulomb, were ready with one- and two-fluid theories.

The first to propose a theory that fit most of the observed phenomena was Wilhelm Edward Weber, the German physicist, for whom the unit of flux is named. In 1852 Weber theorized that every iron molecule was itself a tiny permanent magnet. Normally their north poles point randomly in all directions. When a piece of iron is placed in a magnetic field, their individual north poles are forced to line up in more or less the same direction, and the entire piece of iron becomes "magnetized." Weber's theory was modified somewhat by the Scottish physicist Sir James Alfred Ewing in 1891. Ewing allowed that the tiny elementary magnets, which are now called *domains*, might be single molecules or they might comprise groups of molecules acting as tiny iron crystals. Soft iron would therefore magnetize more readily, and also lose its magnetism more easily, than a hard steel because of the energy required to rotate the domains against the interference of their neighbors. Figure 9-12 conceptually shows the orientation of the

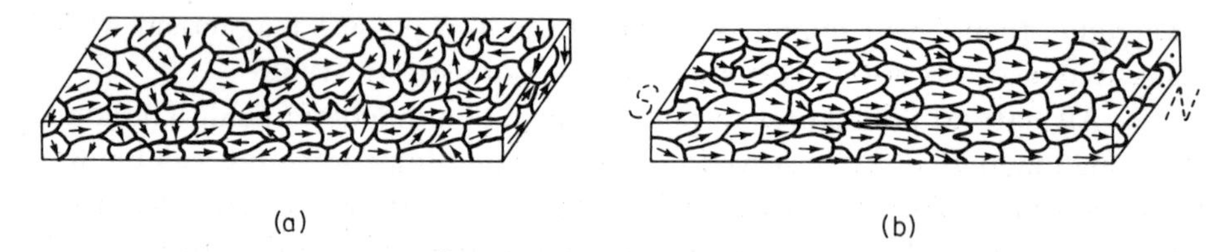

Fig. 9-12. Magnetic domains in an iron bar. (a) Before magnetization. (b) After magnetization.

domains, according to the Ewing theory, in a bar of iron before and after magnetization. There is good evidence to support this theory. If an unmagnetized iron bar is held in a north–south direction and tapped with a hammer, it will become magnetized. The Weber-Ewing theory here is that the hammer blows momentarily jar the individual domains loose from the influence of their neighbors, and in this temporarily freed condition, they tend to line up with the earth's field. Another phenomenon that seems to support this theory is *magnetostriction*. When an unmagnetized bar is placed in the magnetizing field, careful measurements show that its dimensions change slightly, as might be expected if large numbers of domains rearrange themselves.

Still a third phenomenon, which also supports this theory, is the *Curie point*, named for its discoverer, Pierre Curie (who, with his wife, Marie, gained renown for the discovery of radium). The Curie point is a temperature, characteristic of each ferromagnetic material, above which the property of ferromagnetism disappears (1418 °F for iron, 676 °F for nickel, and 2048 °F for cobalt). We can easily imagine the little domains driven out of their orderly lineup when they feel the "hot foot."

The Weber-Ewing theory also makes it clear why we cannot have an isolated north (or south) pole. If we cut a magnet, the domains in each piece remain aligned; hence a pair of poles appears at the separation and each piece has both a north and a south pole.

One unsatisfying point about the Weber-Ewing theory is that it does not explain why the domains themselves are magnetic. Referring back to Figs. 9-10 and 9-11, we know that the field of a bar magnet looks and behaves just like that of a solenoid. This fact did not escape the notice of the celebrated French physicist André Ampère. In 1823 Ampère proposed the theory that *all* magnetism was due to circulating electric currents; therefore there must be tiny electric currents flowing eternally within a bar magnet. This is indeed the modern "electron theory of magnetism." Ampère, of course, knew nothing of the structure of atoms, so he was far ahead of his time in his belief. We know that the electrons in the atom are in motion; and since a moving electric charge is, by definition, an electric current, each moving electron constitutes a tiny direct current. In addition, quantum mechanics tells us that electrons have *spin.* Now the "real" meaning of this statement is, of course, a nonsense question, for we cannot draw a true picture of an electron. If, however, we conceive of the electron as the planet-like object of the Rutherford-Bohr *model,* then the spin can be imagined as the spinning of the electron around its own axis like a top. This spinning charge is also a tiny current with an associated magnetic field. In most atoms, the magnetic fields associated with the electron spin are oriented in all directions, having essentially zero net effect. In the ferromagnetic atoms like iron, however, it is believed that a number of electrons have their spin axes oriented in the same direction; thus the atom is a small solenoid. Groups of these atoms "cooperate" to form a (still submicroscopic) magnetic region, which is the domain of the Weber-Ewing theory.

Do not be misled by the foregoing discussion. Today we know a good deal about how to produce and how to work with magnetic fields, but we are not much closer than Aristotle or Franklin to understanding their fundamental nature. Quantum mechanics seems to be leading us in the direction of understanding. Perhaps you, the reader, may be the one to give the world a better theory of magnetism.

9-11 MAGNETIC FIELD INTENSITY AND PERMEABILITY

Before proceeding to the study of applications of magnetism, we need to delve a little deeper into the magnetization of iron. It will be helpful to define another vector quantity associated with the magnetic field. That quantity is the *magnetic field intensity*, defined at each point in the field as the *force* experienced by a north magnetic *pole of unit strength* when placed there. The magnetic field intensity is always represented by the letter **H**. Since **H** is a force per unit of pole strength, its natural unit of measurement is the newton per weber.[4] However, by some juggling of fundamental units, it can be shown that the unit newton per weber is equivalent to the unit ampere per meter, and this is the more frequently used from for the unit of **H**. Thus, in various texts, or even in different places in the same text, you may see magnetic field intensity measured in either newton per weber or amperes per meter.[5]

The **H** vector, of course, points in the direction in which the unit north pole is pushed by the field, which is, you will recall, the same direction in which the **B** vector points. In fact, the **B** and **H** vectors are closely related to each other and in a very simple way—through a property of the material, in which the magnetic field exists, which is called its *magnetic permeability*.

The permeability, represented by the symbol μ (Greek letter mu), is the measure of how the medium, in which the magnetic field exists, tends to concentrate lines of force—that is, to bunch them together. This situation is illustrated in Fig. 9-13, which shows how a magnetic field in wood, glass, and water is essentially the same as it would be in empty space, but almost all the lines of force are concentrated in a ferromagnetic material like iron (which itself, of course, becomes a magnet by induction). The permeability of empty space is represented by μ_0 (mu sub zero) and has a value of about 1.26×10^{-6} weber per ampere meter. The permeabilities of diamagnetic substances are slightly less, and those of paramagnetic substances (other than the ferromagnetic ones) are slightly greater, than that of empty space, but all their permeabilities are approximately the same. In contrast, the permeabilities of ferromagnetic materials are hundreds to tens of thousands of times greater than the permeability of empty space, because the alignment tendencies of their "atomic magnets" alter the flux greatly.

[4] There is also the older (CGS) unit for **H**, called the oersted. One oersted = 79.58 amperes per meter.

[5] The reader may see this written as ampere *turns* per meter, the word turns referring to the number of turns in a solenoid's winding. The number of turns is, however, a pure number with no units, and so the "turns" serves as a helpful reminder but does not properly belong in the unit of **H**.

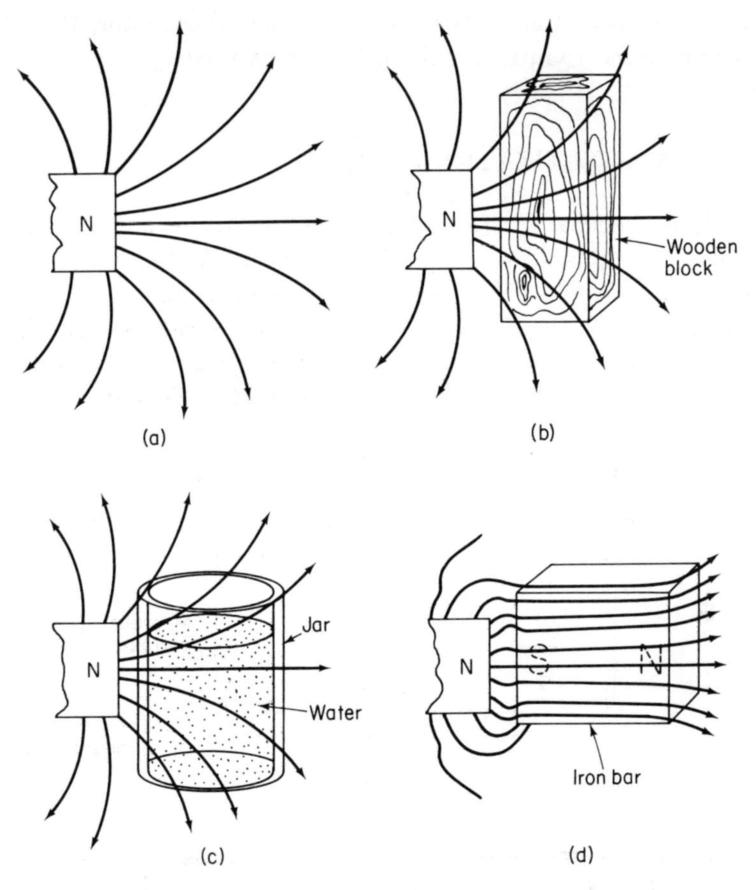

Fig. 9-13. Permeability of various materials. (a) Free space. (b) Wood. (c) Glass and water. (d) Iron.

The relationship between flux density (B), permeability (μ), and magnetic field intensity (H) is

Flux density = permeability × field intensity

Since the permeability multiplies the field intensity, it is apparent that for a given value of the magnetic field intensity, the greater the permeability, the greater the flux density. This is why iron cores are used so extensively in the windings of magnetic devices. Thus we can see how the flux density depends on two factors. One, the capacity of the solenoid to produce a magnetic field, depends only on the geometry (i.e., length of winding, number, shape, and size of turns) of the coil and the current in it, but not on the material in the core. This capacity is measured by H, which is therefore also sometimes called the *magnetizing force*. The other factor, the capacity of the core material to intensify the field, is measured by μ.

9-12 THE MAGNETIZATION CURVE

It is instructive to study the magnetization of a piece of iron. A schematic of the apparatus is given in Fig. 9-14*a*. As the resistance of the variable resistor is adjusted, the current in the solenoid changes, and the magnetic field within the solenoid increases with increasing current.

The relationship we wish to study is that between the flux density and the magnetic field intensity in the core of the solenoid. The resulting

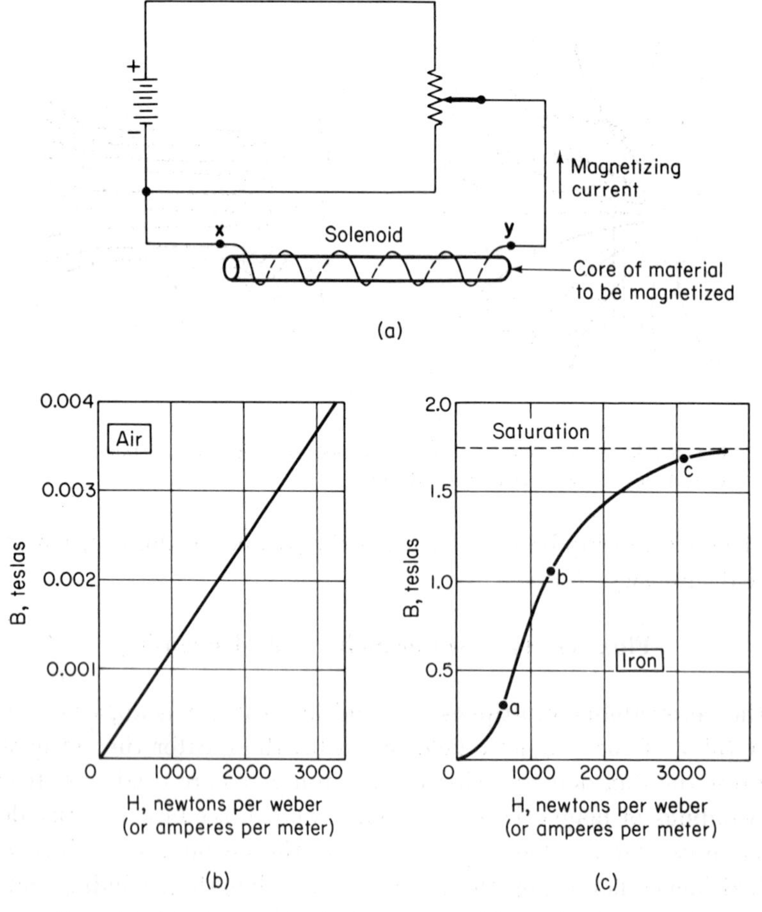

Fig. 9-14. The magnetization curve. (a) Schematic of magnetization circuit. (b) B-H curve for air. (c) B-H curve for iron.

graph of how B varies with H is called the *magnetization curve*, or simply the *B-H curve*, for the core material.

As a standard for comparison, we study first the B-H curve for the core of the empty solenoid—that is for air—given as Fig. 9-14*b*. Air at ordinary temperatures and pressures, is only very weakly paramagnetic, and its molecules do not cooperate enough to affect any magnetic field significantly. Consequently, its permeability is small and nearly constant, and the B-H curve is a straight line, showing that the flux density increases and decreases in direct proportion to the field intensity.

Now observe what happens when the solenoid has an iron core. The corresponding B-H curve is shown as Fig. 9-14*c*. Notice how B increases slowly at first with increasing H (say, from 0 to point a on the curve), as it takes some initial "hard pushing" to get the domains to start to align. B is then almost a straight-line function of H, just as in air, for a short range of H (from point a on the curve to point b). From there on, most of the domains are aligned, and it takes a greater and greater increase in H to cause a few more to line up and make a noticeable change in B, and at this point (c on the curve) we say that the iron is *saturated*. The flux density at saturation can be read from the vertical scale, which (you will note) has a range 500 times as great as that for the "air" curve, even though the H scales are the same.

We learn some very important facts from these two curves. The permeability of air (and of all substances that are *not* ferromagnetic) is small and essentially a constant. The permeability of iron (and of all ferromagnetic substances), on the other hand, is large and changes with the value of H, due to the phenomenon of saturation.

9-13 HYSTERESIS AND PERMANENT MAGNETS

The magnetization curve given in Fig. 9-14*c* is adequate for the purpose intended in the previous section, but it is not complete. We are now going to study the complete magnetization curve for a ferromagnetic material.

Consider the closed curve shown in Fig. 9-15. Note the directions of the arrows on the curve. Note, also, that we are now concerned with negative as well as positive values for both B and H, the negative values representing a *reversal in direction* of those vectors. The section of the curve $\overrightarrow{0abc}$ is the same as Fig. 9-14*c* and represents the saturation of an initially unmagnetized bar, as before. If the current in the solenoid is now gradually reduced, many of the domains remain aligned, and although H returns to zero, B

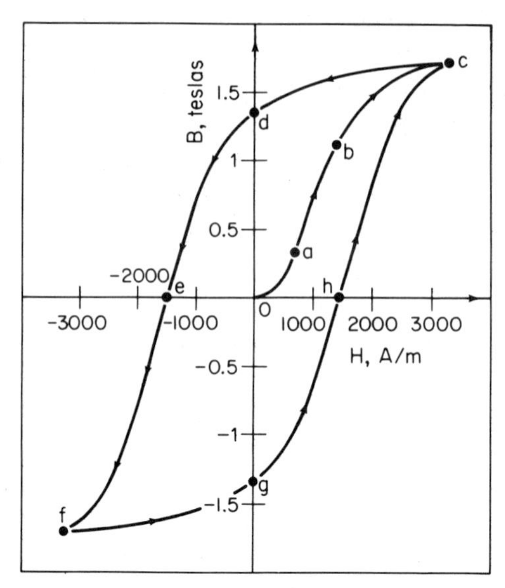

Fig. 9-15. Hysteresis curve.

does *not* return along the original curve but follows the path to point d instead. Thus, with no applied magnetizing force, there is a *residual* or *remanent magnetism* left in the bar.

At this point, we reverse the connections on the solenoid, at point x and y in Fig. 9-14a. Doing so causes the current in the solenoid to flow in the direction opposite to the one it had previously and therefore reverses the direction of the H vector. As the current is increased in the new direction, the domains begin to reverse; and when we reach point e on the curve, we have nullified the remanent magnetism in the bar. The amount of reverse magnetizing force needed to demagnetize the bar is called the *coercive force.* As the reverse current is increased beyond this point, the domains continue to realign themselves until the bar is saturated in the reverse direction, indicated by the point f on the curve. When the reverse magnetizing current is removed, the bar again has residual magnetism, indicated by point g on the curve, which is of the same magnitude as at point d, but with the north pole now on the other end of the bar.

Finally, we reconnect the wires to the solenoid as they were originally. Increasing the magnetizing current once again brings the bar back to a demagnetized condition, at point h on the curve, and beyond that back to saturation at point c.

The complete magnetization curve for a ferromagnetic material is thus a closed loop. The portion of the curve $\overrightarrow{0abc}$ is called the *normal magnetization curve.* The phenomenon of residual magnetism, because of the tendency of the domains, once aligned, to remain so, was called *hysteresis* by Ewing, who first studied it, and the loop $\overrightarrow{cdefghc}$ is called an *hysteresis loop*.

Energy is required to rotate the domains and this energy shows up as heat. The energy dissipated in one complete magnetization cycle (once around the hysteresis loop) is measured by the area of the hysteresis loop. Suppose that we intend to use an iron core in a solenoid as an electromagnet, that is, a device which will act like a magnet or not, as we choose, merely by throwing a switch. In this case, we would like the core to lose all its magnetism immediately when the current is shut off, and we would not want to waste any electric energy in producing heat. These ideal requirements are represented in the left diagram of Fig. 9-16*a*, an hysteresis "loop"

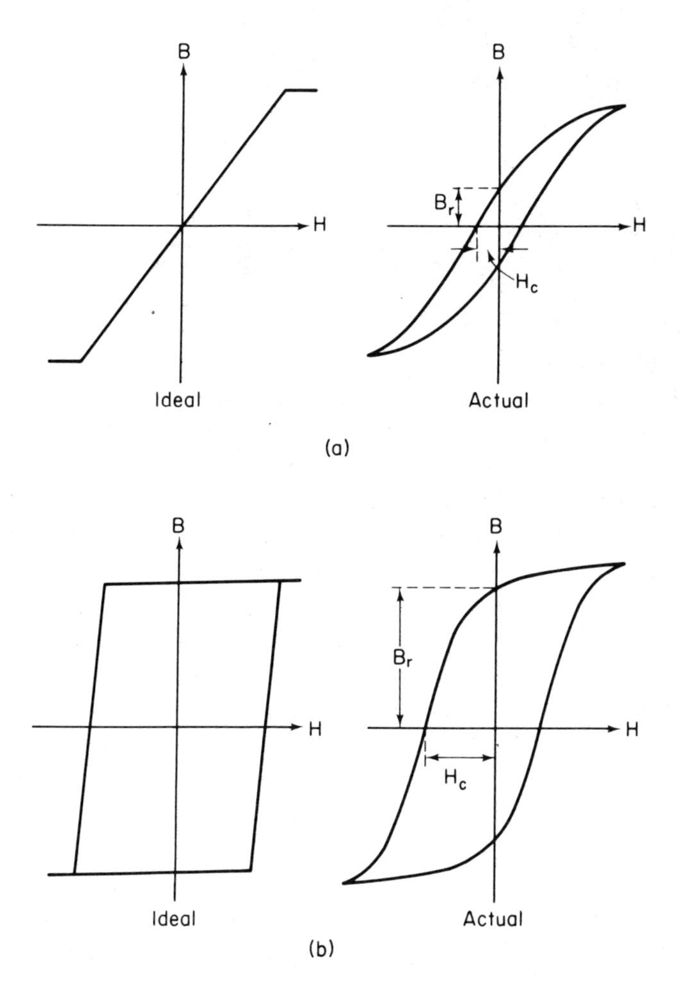

Fig. 9-16. Desirable hysteresis curves. (a) For an electromagnet core. (b) For a permanent magnet.

of zero area with no remanent magnetism. Of course, there is no such ideal material, but very soft and pure iron, represented by the right diagram of Fig. 9-17*a*, has a narrow hysteresis loop with a low remanent magnetism

Fig. 9-17. Commercial ceramic magnets. (Courtesy of the Allen-Bradley Corp)

(B_r) and small coercive force (H_c) for demagnetization; hence it is commonly used for this purpose.

If, on the other hand, we wish to make a permanent magnet, we must look for a ferromagnetic material with a high residual magnetism requiring a high coercive force to demagnetize it. Such a material would have an ideal hysteresis loop that looks like the left diagram of Fig. 9-15*b*, while that of an actual permanent magnet material might look like the right diagram in Fig. 9-16*b*. Compare the magnitudes of B_r and H_c in the two right-hand parts of Fig. 9-16.

Permanent magnets are made from a variety of materials, but all contain at least some ferromagnetic material. One of the most popular permanent magnet materials is Alnico 5 (also called *alcomax*), an alloy of 24 percent cobalt, 14 percent nickel, 8 percent aluminum, and 3 percent copper. Typically it has a coercive force 550 times as great as that for pure iron. Before the discovery of Alnico alloys, hard cobalt steel was used for permanent magnets.

In applications where a metallic permanent magnet is undesirable, as, for instance, where a high electrical resistance is required, a *ceramic magnet* may be used. This comparatively new type of magnet is made from oxides of iron (ferrites) and other metals, rather than the metals themselves. The powdered oxide (e.g., Fe_2CoO_4) is pressed into a cake, sintered at high temperature, and allowed to cool in an intense magnetic field. Some commercial ceramic magnets are shown in Fig. 9-17.

9-14 AIR GAPS AND THE MAGNETIC CIRCUIT CONCEPT

Consider the toroidal (doughnut-shaped) iron-core solenoid shown in Fig. 9-18*a*. When current is flowing in the coil, a magnetic flux is developed

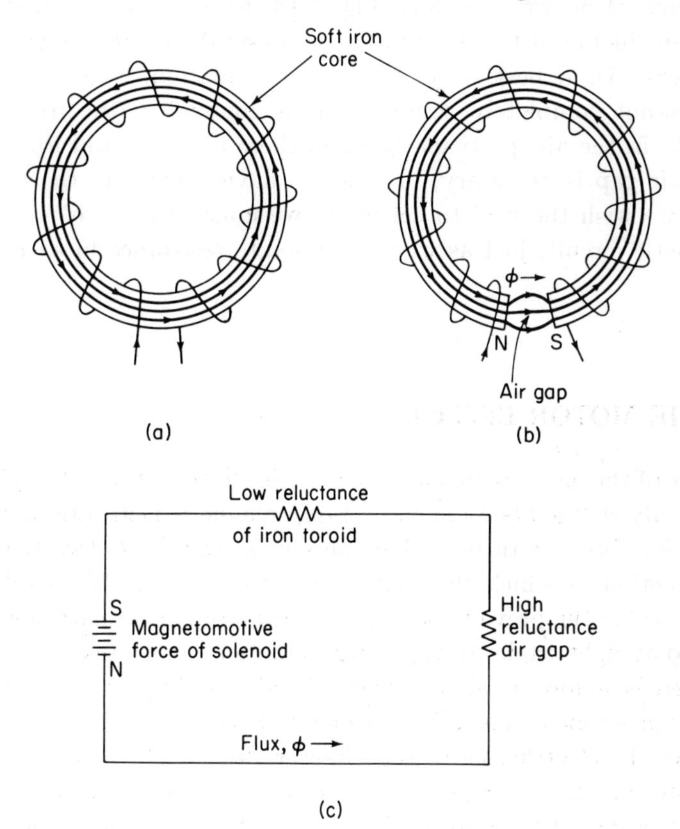

Fig. 9-18. (a) Toroidal solenoid. (b) Toroidal solenoid with air gap. (c) Magnetic "equivalent circuit" diagram for (b).

in the core. The lines of force are unbroken and there are no distinct poles. Now let the core be cut as in Fig. 9-18*b* to produce an air gap. A pair of magnetic poles immediately appears on the opposing faces of the gap, and intense magnetic field appears across the gap. The shorter the gap, the stronger the field (for any given pole strength) because of air's low permeability. Analysis of such devices led to the idea of a *magnetic circuit*, by analogy to the electric circuit. We can think in terms of magnetic flux, ϕ, as analogous to the electric circuit current, and of materials in the flux path having different "magnetic resistances." With this kind of thinking in mind, engineers defined a "Magnetic Ohm's Law" for the magnetic circuit. In this approach, the "current" is the *flux*, the "resistance" to formation of flux is called *reluctance*, and the driving potential is called *magnetomotive force* (mmf) by analogy to electromotive force; and these quantities are related mathematically in the same way as their electric circuit counterparts. In magnetic circuit terms, then, we may draw Fig. 9-18*c* to represent schematically the arrangement in Fig. 9-18*b*. It is to be emphasized that the Magnetic Ohm's Law approach is, at best, only a good approximation. Flux, unlike electric current, is not completely confined to its conductor (the iron core) but "leaks out" in the air, particularly when there is a gap, and iron saturates. Since an air gap is necessary, in order to have a place in the field to put whatever we wish the field to act upon, we must "live with" reluctance in any magnetic circuit, just as we must tolerate resistance in an electric circuit.

9-15 THE MOTOR EFFECT

One of the most useful electromagnetic effects is the *motor effect*. When an electrically charged body moves *across* a magnetic field, the charged body experiences a *force*, or thrust, which acts *perpendicular to both* the field and to the direction in which the charged body is moving. We emphasize the word "across" in the previous sentence, because only that part or component of the field at right angles to the charge's motion exerts any force. (In other words, there is no force at all on a charged body moving parallel to a magnetic field.) This phenomemon is called the motor effect because it is the underlying principle of the electric motor (see Chapter 10).

Note that this force is not the magnetic attraction or repulsion—the charge does not get tugged in the direction of the field—but a sideways push, shown as upward in Fig. 9-19*a* for a moving negative charge. A positive charge would, of course, be pushed in the opposite (downward) direction. The directions involved can be easily remembered by a *right-hand rule.* Extend the first three fingers of the right hand perpendicular to each other, as shown in Fig. 9-19*b*; if the *m*iddle finger points in the direction of the charge's *m*otion, and the *f*orefinger points in the direction of the *f*ield, then the *t*humb shows the direction of *t*hrust (force) on a negative charge.

Although only a single moving electron is shown in Fig. 9-19, the "moving charge" may be current in a wire, in which case the entire wire experiences the sideways thrust. This situation is illustrated by Fig. 9-20.

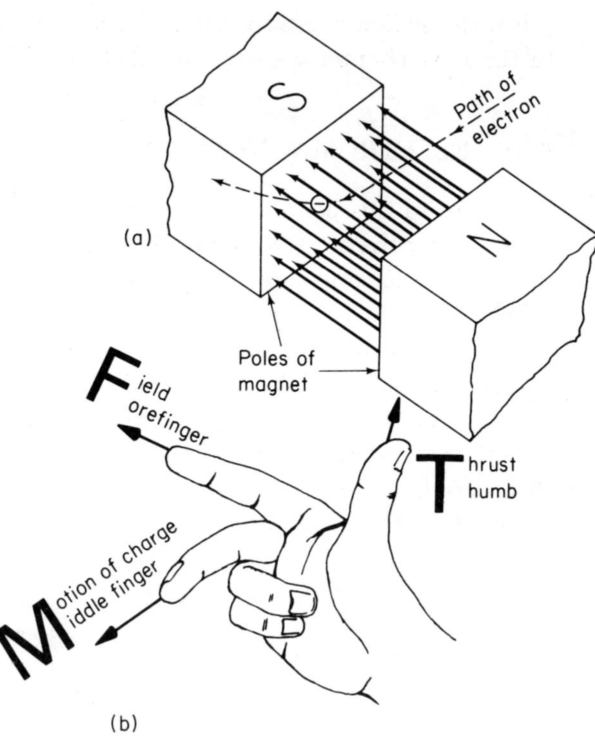

Fig. 9-19. The motor effect. (a) Path of an electron through a magnetic field. (b) Right-hand rule for a negative charge.

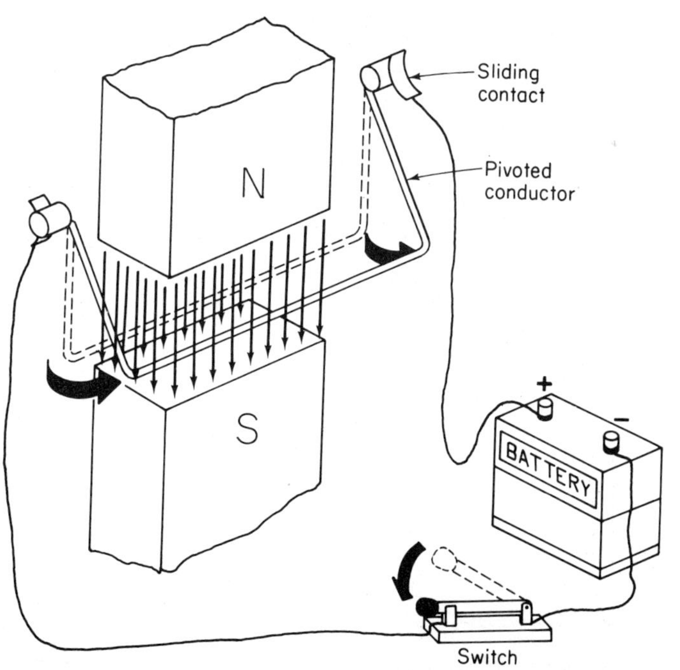

Fig. 9-20. Motor effect upon a current-carrying conductor.

When the switch is closed, the movable section of wire will swing outward. In the next chapter we see a number of applications of this effect.

9-16 ELECTROMAGNETIC INDUCTION; THE DYNAMO EFFECT

Up to this point, we have seen how magnetism can be produced by an electric current. We might suspect, as did Michael Faraday, that we should be able to use magnets to produce electricity. There is indeed such an effect. It is at least as important in our lives as the motor effect discussed in the previous section, for without it our homes would probably still be lit by gas. This effect is the *dynamo effect.*

Recall that the key to producing magnetism from electric charge was that the charge had to be moving (i.e., a current). It is also true that when there is relative motion between a conductor and magnetic field—that is, when either the conductor or magnet moves such that the *conductor cuts across magnetic lines of force*—a voltage is *induced* in the conductor. If the conductor forms a circuit, then the induced voltage, of course, produces a

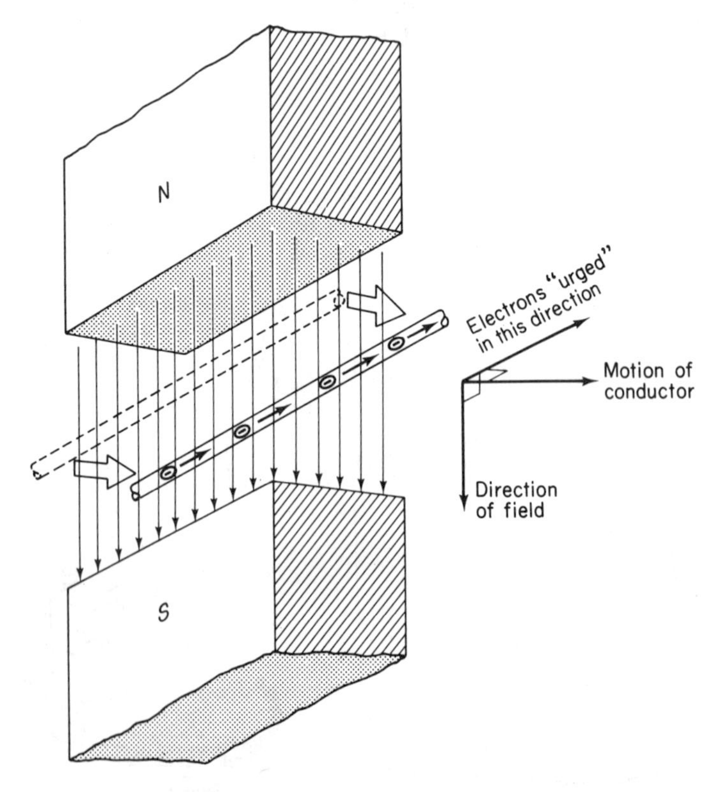

Fig. 9-21. Electromagnetic induction, the dynamo effect.

current. Here, again, the word across is emphasized, for only that part of the field acting at right angles to the conductor is effective in inducing voltage. This effect is shown schematically in Fig. 9-21. The section of the conductor moves across the field between the poles, and the electrons within the conductor are pushed in the direction of the colored arrows.

We can use the same right-hand rule for induced emf that we use for the motor effect if, as in Fig. 9-22, we associate the *m*iddle finger this

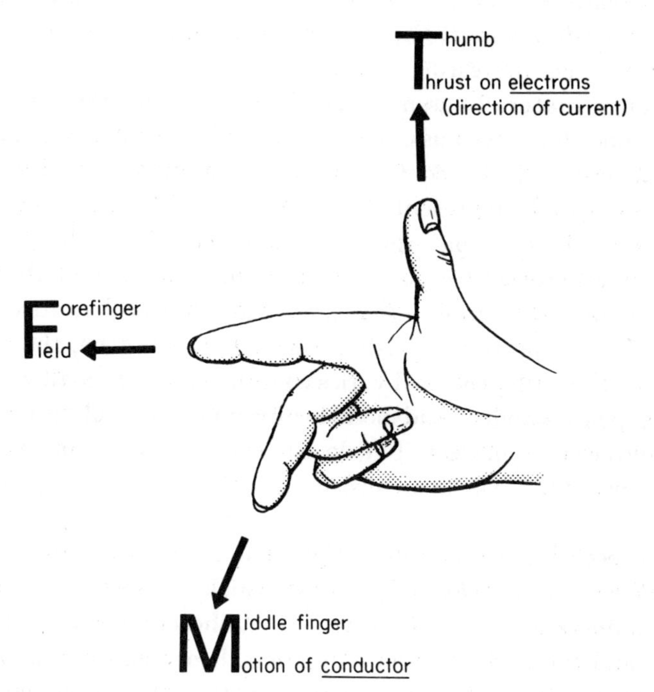

Fig. 9-22. Interpretation of right-hand rule for electromagnetic induction.

time with *m*otion of the conductor relative to the field, the *f*orefinger with the direction of the *f*ield, and the *t*humb with the direction (inside the wire) in which the electrons are *t*hrust (i.e., toward the positive end of the conductor).

The production of the voltage (and current, if a circuit exists) by cutting magnetic lines of force with a conductor is called *electromagnetic induction*. Note that this is the third way we have used the word induction. In Section 3-12, we talked about *electrostatic* induction, which had to do with an electric charge causing another electric charge; in Section 9-5, we saw how a magnet could produce another magnet by *magnetic* induction; here we are concerned with how a magnetic field is used to produce an electric voltage by *electromagnetic* induction, probably the most common use of the word induction.

9-17 GEOMAGNETISM, THE FIELD OF THE EARTH

As yet we have said nothing about the cause of the earth's magnetic field, or *geomagnetism*, as it is usually called. There is a good reason for this lapse: *No one really knows the cause of geomagnetism*! However, there have been, and are, some interesting theories.

One might naturally assume, as Gilbert did, that the earth is simply a giant lodestone. Johann Gauss, the great German scientist and mathematician, even determined the configuration and properties of the lodestone that would adequately represent the earth's field. There are several major difficulties with this simple approach, however. First, the temperature increases very rapidly as we go down into the ground, and there is good evidence that the earth's iron-nickel core is much hotter than its Curie point, even at the terrible pressures existing there. If this is true, the core is not ferromagnetic. Secondly, the field varies continuously in a small way (remember how the poles wander) and periodically does a complete reversal (the magnetic poles change places), if geologists can believe the magnetic record in rocks. Neither of these facts is consistent with a "giant permanent magnet" theory.

Until recently, the accepted explanation came from the new field of *hydromagnetics* (or *magnetohydrodynamics*), which deals with electromagnetic effects in conducting fluids. The earth's metallic outer core is believed to be a liquid, and the hydromagnetic dynamo theory had it that the earth's rotation caused patterned vortex motions within this outer core, thereby resulting in internal electric currents from which the geomagnetic field arose. Several difficulties also exist with this theory.

The most recent theory, due to Dauvillier,[6] contends that the outer core of the earth is virtually a *superconductor*—that is, its resistance is so low that a current, once set up in it, will continue to flow for millions of years. Circulating currents, the theory goes on, are set up by electromagnetic induction as shown in Fig. 9-23, when the sun occasionally emits a detached jet of protons (i.e., a current of positive charges) which passes in the vicinity of the earth's orbit. The field produced by this proton current, which is ostensibly on the order of 10^{11} amp and occurs once in every hundred thousand to million years, induces the circulating currents in the outer core.

[6] Alexandre Dauvillier, *Concerning Reversals and Origin of the Geomagnetic Field*, Observatoire Pic-du-Midi, Bagneres de Bigorre; 26 September 1966.

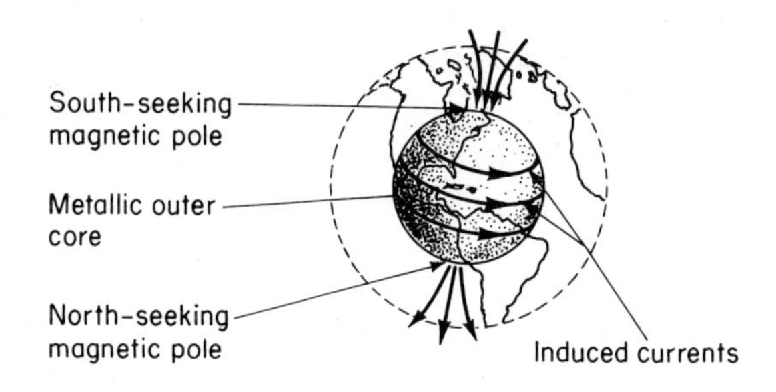

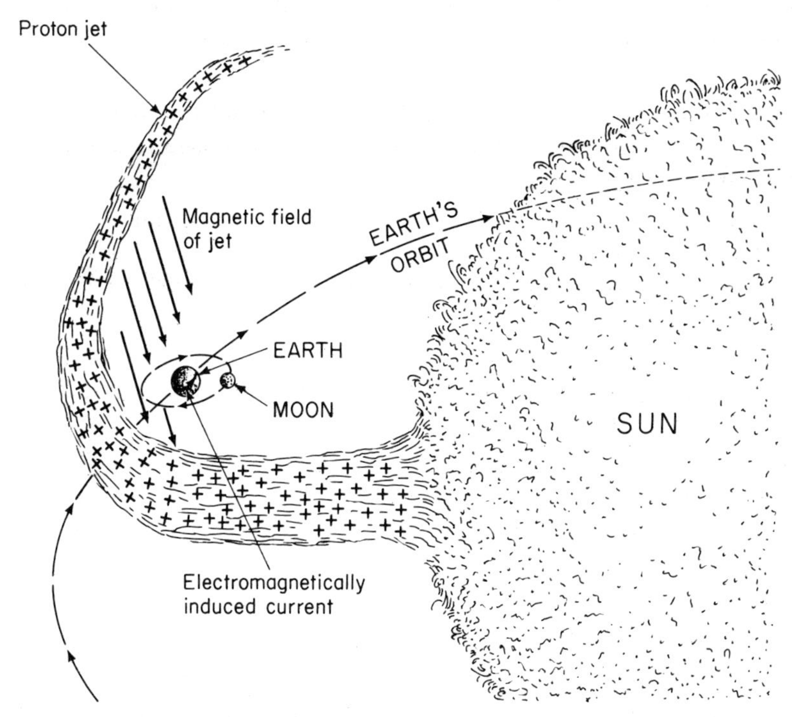

Fig. 9-23. The solar proton jet as the cause of geomagnetism (orbits not to scale).

The solenoid effect of these currents is to maintain the geomagnetic field, as per the inset to Fig. 9-23. Although this theory explains most of the phenomena of the geomagnetic field, including the possibility of sudden reversals, the last word has not yet been said on the subject.

REVIEW QUESTIONS

1. How does lodestone get its name?
2. If Santa Claus were to suspend a bar magnet in his sled, what position would the magnet assume when he arrived home? Why?
3. If the ancients had been correct, how might the Spanish Armada have been sabotaged with loaves of garlic bread?
4. Draw the magnetic field for an L-shaped bar magnet.
5. Can a **B** vector point toward a north pole? Why?
6. What does it mean when the value of **B** for one magnet is twice that of another?
7. What was the most significant contribution made by Gilbert in *De Magnete*?
8. Explain why motion plays such an important role in electromagnetism.
9. How must our left- and right-hand rules be modified, if we talk about a positive current like Benjamin Franklin did, or about the stream of protons in Fig. 9-23?
10. If your bedroom were completely lined with iron and the coilsprings in your mattress were solenoids, would your mattress be attracted to a wall or to the floor? Why?
11. Given a sufficiently powerful coil, which of the following can be permanently magnetized: a table knife, a drinking glass, a cobalt steel cylinder, a scissors, an aluminum frying pan, a piece of copper wire, a nickle, a tree branch? Explain why in each case.
12. If you wrap a screwdriver with a coil of wire and connect the ends to a battery, the screwdriver becomes a magnet. Why? When the battery is disconnected, the screwdriver is still a magnet. Why?
13. How hot would you have to get the screwdriver in problem 12 (assuming it's pure iron) to make it lose its magnetism? Is this a good idea? Is there a better way to demagnetize it?
14. Using the Weber-Ewing theory, explain how you would make a bar magnet with two south poles.
15. A solenoid wrapped on a cardboard spool produces a flux density of one tesla within the spool. If a slug of material with a permeability 200 times that of air is inserted into the spool, how is the field intensity changed? How is the flux density changed?
16. Using magnetic circuit terminology, explain why an air gap between poles should be as small as practically possible.

10

CHILDREN OF THE MARRIAGE

10-1 AFTER THE HONEYMOON

In Chapter 9 we saw how Oersted performed the marriage ceremony for electricity and magnetism, leading to the unified science of electromagnetism. What followed was truly a honeymoon of discovery—discovery of an enormous diversity of applications for electromagnetism. Some of the people involved, like Ampère and Faraday, we have already met; others, like Arago, Henry, Morse, and Lawrence, we shall meet presently. It is significant that all their discoveries and inventions are based on only three electromagnetic effects, the same three we discovered in the last chapter: magnetic force (attraction and repulsion), the motor effect, and the dynamo effect. In this and the following chapter, we see the results of these discoveries.

10-2 THE SOLENOID: MANY FORMS AND MANY USES

The simplest and one of the most useful electromagnetic devices is the solenoid. The word *solenoid* comes from the Greek and literally means "like a pipe or conduit." To the physicist, and as we saw in the last chapter, any electrically energized coil of wire may be considered a solenoid. In modern engineering terminology, however, a solenoid is usually understood to be wound on a ferromagnetic (usually soft iron) core and, in particular, on a *movable* ferromagnetic core, while any other electromagnetic winding is simply a *coil*. Here we try not to be too restrictive about the terminology and use both words, coil and solenoid, freely when an electromagnet application is intended. For certain other uses, to be discussed presently, the same type of solenoid may also be referred to as an *inductor* or *choke*.

Whatever the name used in any given instance, certain drawing conventions and schematic symbols are always used in conjunction with solenoids. The drawing convention is illustrated in Fig. 10-1 and the schematic symbols are given in Fig. 10-2.

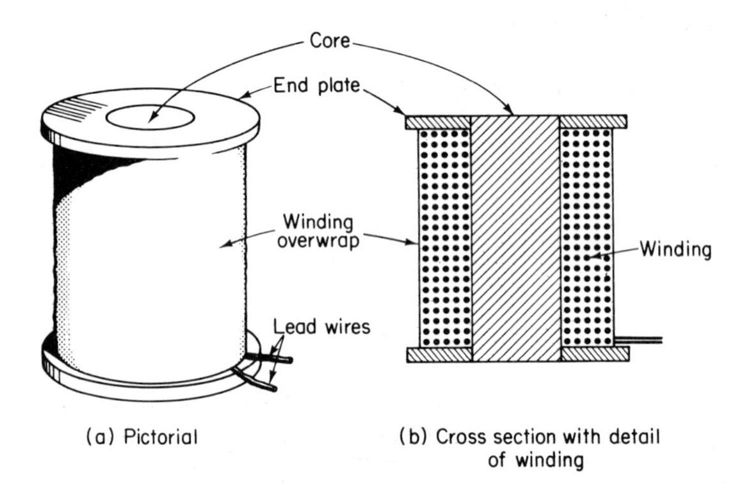

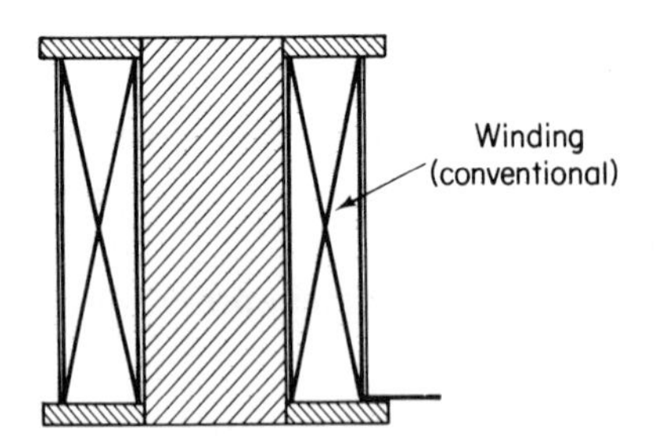

Fig. 10-1. Pictorial and sectional drawing of a coil. (Note convention for sectional view.)

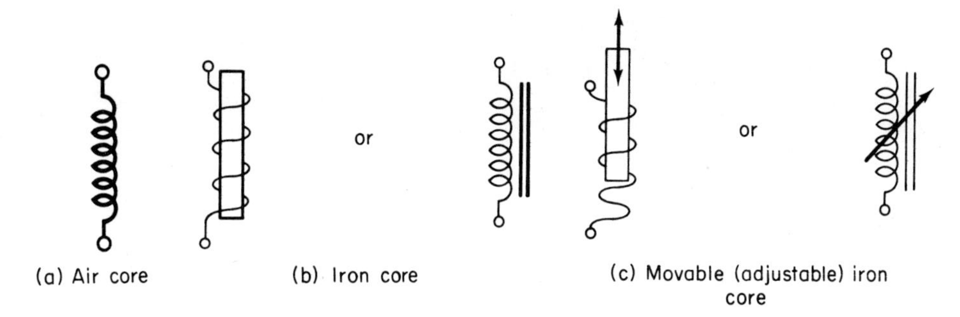

Fig. 10-2. Schematic circuit symbols for solenoids (or *coils* or *inductors*).

10-3 THE ELECTROMAGNET

One of the most obvious direct uses of electromagnetic force is for lifting masses of iron and steel. Perhaps the first electromagnet was made by the French physicist, Dominique François Jean Arago.

The eldest of four brothers, all of whom had interesting careers, Arago started out on an Army career, only to be appointed secretary of the Paris Observatory because of his scientific excellence. He was a personal friend of such greats as Biot, Laplace, and Lalande. His exploits while on a surveying assignment with Biot in France and Spain read like an adventure novel. He was active in politics, as well as in mathematics and physics, entering the Chamber of Deputies in 1830. When Arago heard of Oersted's experiments, he first verified and then extended them. He showed that a current-carrying coil attracted iron filings like a bar magnet. He magnetized steel needles by placing them inside the coil. He was also first to show a relation between the *aurora borealis* (Northern Lights) and variations in the earth's magnetic field.

Arago did not propose to use his electromagnet as a lifting device. Credit for the first lifting magnet must go to the English physicist, William Sturgeon. Sturgeon's story reads like Horatio Alger fiction. Starting as a shoemaker's apprentice, he educated himself while in the Army and became a respected physicist after returning to civilian life. Either accidentally or otherwise (we do not know which), Sturgeon wrapped his coil around an iron core, and he discovered that this core tended to concentrate the magnetic force. His device, as shown in Fig. 10-3, was constructed in 1825. Eighteen turns of wire were wound on a soft iron horseshoe 5 in. long. Electric power was supplied from a *crown of cups* battery (only two cells of which are shown in the illustration). When the circuit was completed, current flowed with

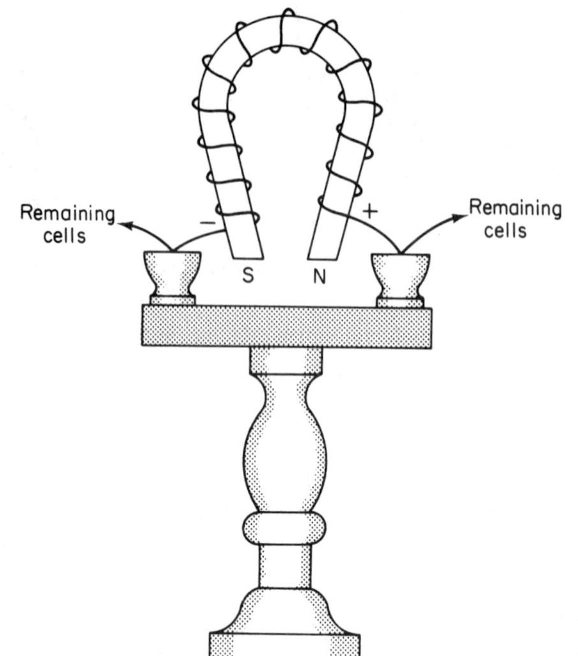

Fig. 10-3. Sturgeon's electromagnet.

the indicated polarity. This little electromagnet, which weighed less than half a pound, could lift 9 lb.

A modern lifting electromagnet is shown in Fig. 10-4. Such magnets are used for moving large quantities of scrap iron and steel and may lift

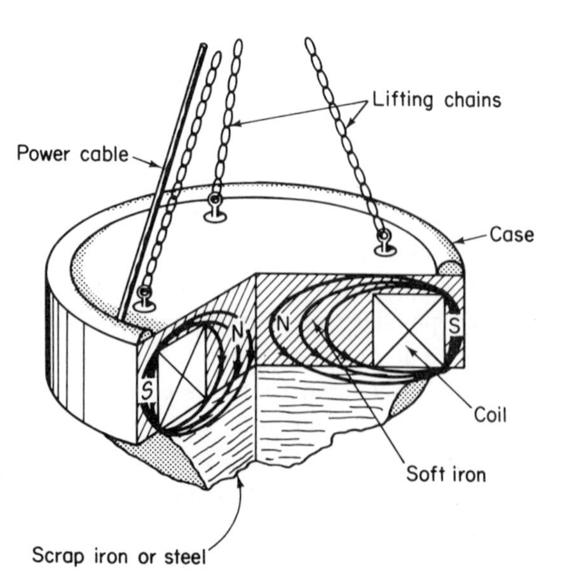

Fig. 10-4. Modern commercial lifting magnet with load (sectioned view showing construction).

a few tons of metal at one time. Tiny electromagnets of this type are used by doctors to remove steel chips from delicate tissue, such as the eye or the brain.

10-4 THE SOLENOID ACTUATOR

One important industrial use of the solenoid is as an *actuator*, an electrical device for imparting a simple pull or push to a mechanical device. A solenoid actuator is basically an electromagnet in which the iron core is movable and is held away from being centered within the coil by a spring. One end of the movable core, which is called an *armature*, is shaped appropriately for its pushing or pulling function. Figure 10-5a shows a pull-type solenoid actuator. When the current is off, the spring holds the armature in the position shown. When the current in the coil is turned on, the armature becomes a magnet by induction, with its poles opposite to the apparent

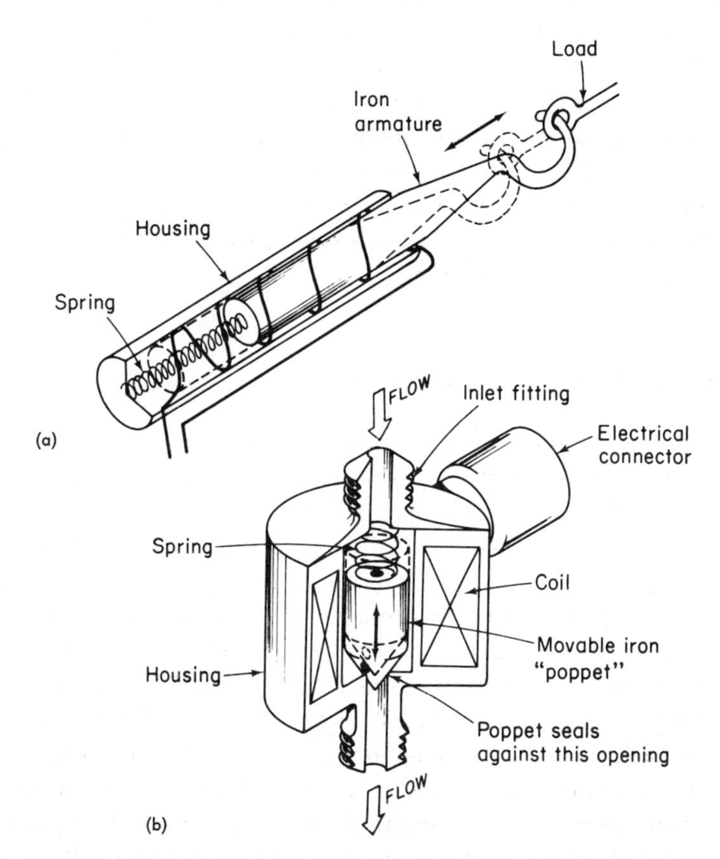

Fig. 10-5. (a) Solenoid actuator. (b) Solenoid valve.

poles of the coil. As a result, the armature is attracted deeper into the coil, indicated by the position shown in phantom, compressing the spring and pulling the load. When the current is turned off again, the demagnetized armature is pushed back to its original position as the spring returns to its uncompressed shape.

Solenoid actuators are in widespread commercial use, wherever a short stroke and modest force are involved (otherwise *hydraulic* or *pneumatic* actuators are used instead). Two familiar examples are the starter mechanism in your automobile (when you turn the key, you energize a solenoid actuator, which, in turn, closes contacts on the starter motor itself), and the rudder, elevator, and aileron controls on a radio-controlled model airplane.

Another application of solenoid actuation is the *solenoid valve,* a typical example of which is shown in Fig. 10-5*b*. Construction is quite similar to that of the solenoid actuator. Here the movable iron armature is called a *poppet.* In the unenergized condition, the spring holds the poppet in the position shown, its conical end blocking off the exit port of the valve. When the coil is energized, the poppet is attracted up into the coil, as shown in phantom, compressing the spring and allowing the fluid to flow out the exit port. When the coil is de-energized again, the spring returns the poppet to the shut off position. Electrically operated valves of this type are used extensively in industry. Homely applications are in automatic clotheswashers and dishwashers, where solenoid valves are used to control the admission of water at specific points in their cycles.

10-5 RELAY, BUZZER, AND BELL

In a slightly different vein, the solenoid actuation principle can be put to use in making an electromagnetically operated switch. Such a device is called a *relay.* The general principle is illustrated in Fig. 10-6. The coil is wrapped around a soft iron core called the *pole piece.* A movable armature is pivoted so that it will be attracted by the pole piece when there is current in the coil, and a spring holds the armature away from the pole piece. The relay is fitted with one or more sets of contacts. In each set, there is a movable contact that is bent by the action of the armature. (In some cases, the movable contact is physically mounted right on the armature). The other contact or contacts in each set will be so positioned that the motion of the armature will either break contact, make contact, or both. A pair of the contacts that are closed when the relay is de-energized and that open on being energized are called "normally closed," abbreviated *NC*. Those contacts that are open

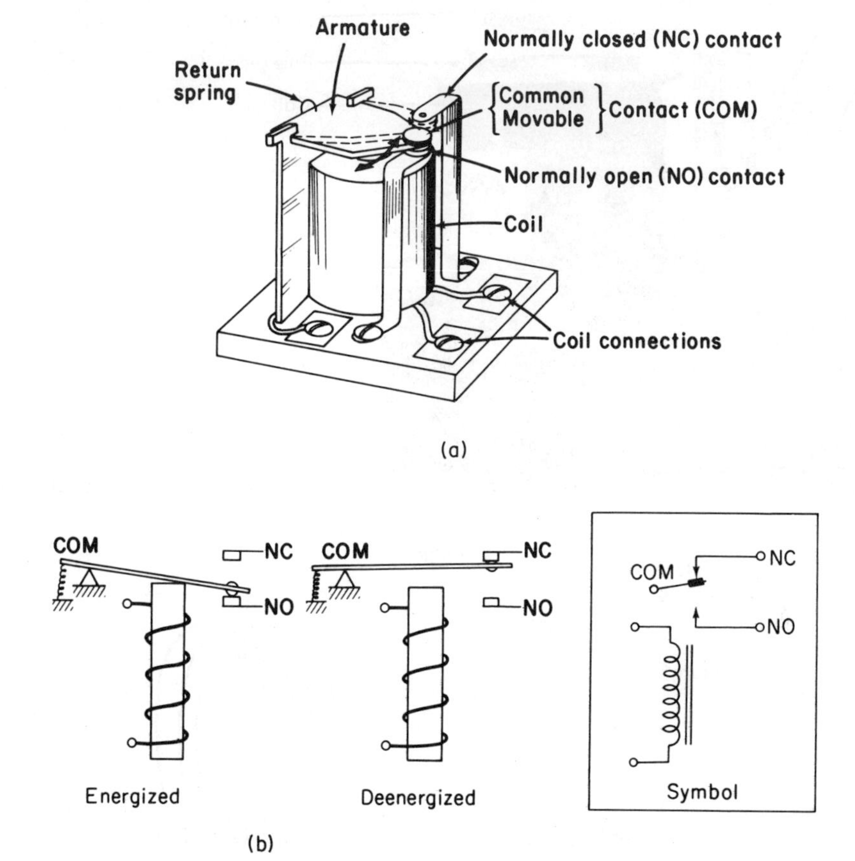

Fig. 10-6. The relay. (a) Construction. (b) Operation.

in the de-energized condition and that close on being energized are called "normally open," abbreviated *NO*. When both types of contact exist in a set, the movable contact is "common" to both and so is usually marked *COM* or simply *C*.

The action of the relay is shown in Fig. 10-6*b* and its proper circuit symbol appears in the inset. A typical commercial relay is shown in Fig. 10-7. Relays may have only one set, or a great many sets, of contacts. Two sets and four sets are most common.

The major use of a relay is to operate a high-power device with a low-power signal. The coil can, for instance, be designed to energize at less than one ampere. The contacts, meanwhile, can be made very heavy so that they can handle hundreds of amperes.

Relays are used extensively in remote control and automatic control applications.

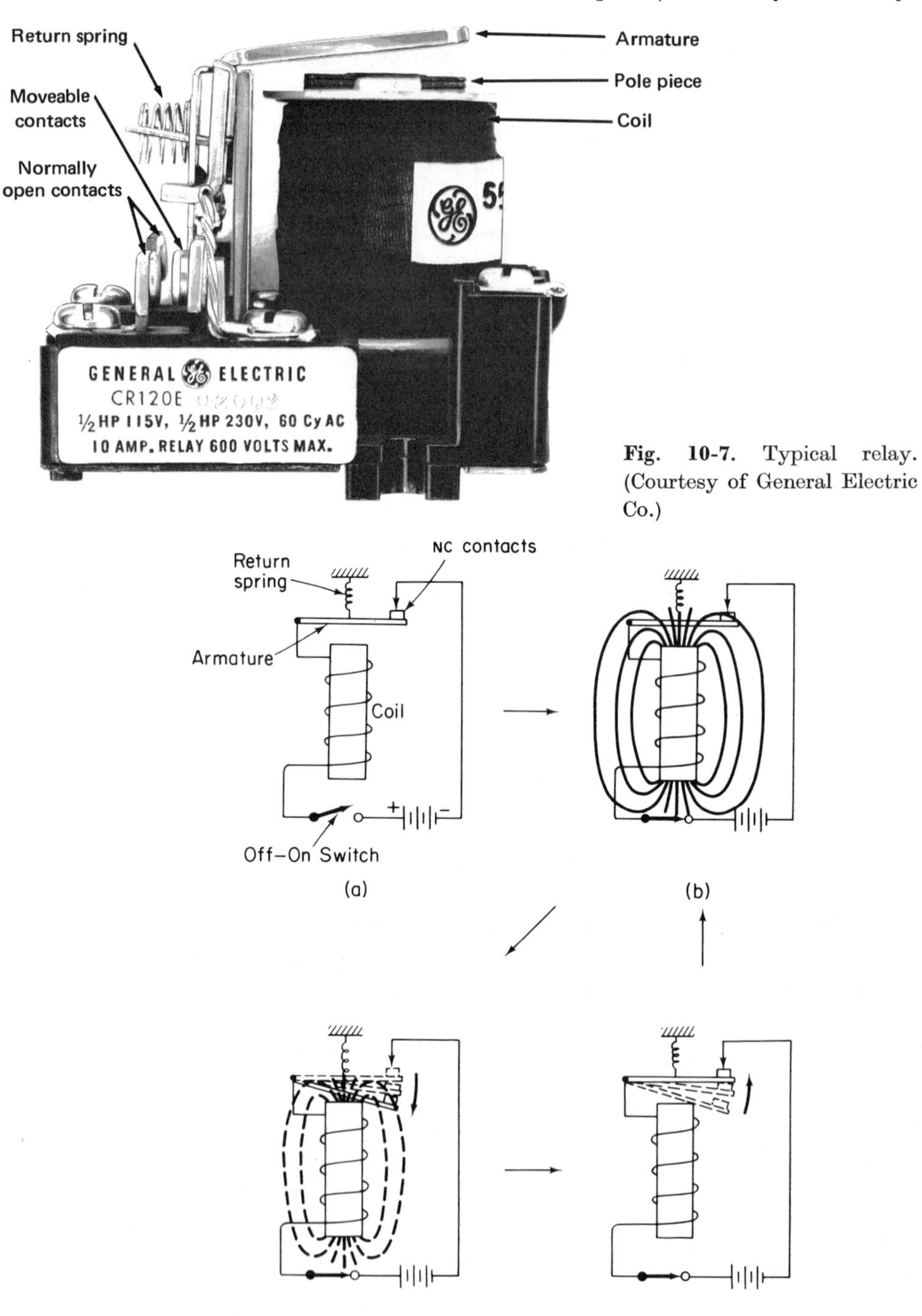

Fig. 10-7. Typical relay. (Courtesy of General Electric Co.)

Fig. 10-8. Principle of the buzzer.

If we now wire the relay such that its NC contacts are in the power supply circuit to its own coil, we obtain a new type of device—a *buzzer.* Figure 10-8 shows the buzzer principle. The de-energized circuit is shown in Fig. 10-8*a*. When the switch is closed, as in Fig. 10-8*b*, current flows in the coil, causing the armature to be attracted. This step opens the NC contacts, as in Fig. 10-8*c*, which de-energizes the coil and releases the armature. The spring pulls the armature back, as represented in Fig. 10-8*d*, closing the contacts once more. This action permits the coil to energize again, and the entire process repeats over and over, very rapidly. The result is a buzzing sound, hence the name *buzzer*.

The electric bell, as used for home doorbells, fire alarms, and burglar alarms, is nothing more than a buzzer to which a gong and striker are added. The striker is merely an extension of the armature with a knob on the striking end. Bells may use two coils with two sets of contacts and two springs, like the one shown in Fig. 10-9, or they may simply have one of each.

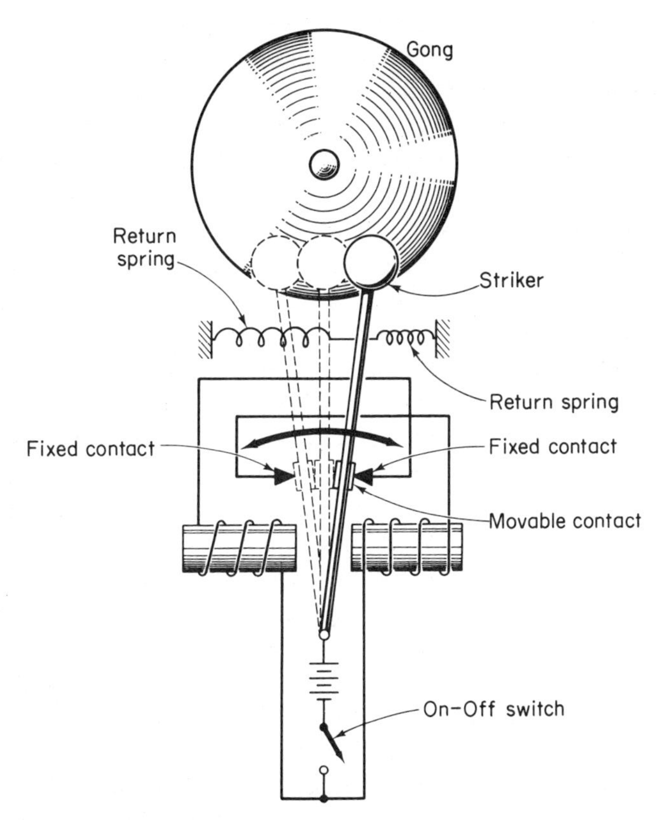

Fig. 10-9. The electric doorbell (or alarm bell).

10-6 THE TELEGRAPH

The word *telegraph* almost automatically conjures up the name Morse as the inventor of both the instrument and the code for messages sent thereby. In point of fact, however, the idea of a telegraph is very old, and several models were shown feasible (if not practical) long before Morse. Furthermore, the real hero of the story, who deserves at least as much credit as Morse if not more, is the comparatively unlauded American physicist, Joseph Henry.

The telegraph, in the modern sense, is, of course, a device for communication over long distances by signals, that is, by a *code* representing letters, numerals, and punctuation (as opposed to voice transmission via telephone). The first recorded suggestion of telegraphic device was made by Giovanni della Porta in 1558, a rather fanciful device involving "sympathetic magnetisation" of steels by the same lodestone. The first functioning telegraph was demonstrated by George Le Sage in 1774, his design being based on that of an anonymous Scottish writer in 1758; it used the charge from a static machine to cause a pith ball to strike a bell. Various other electrostatic and electrolytic ideas were tried. Volta's discovery of the electrochemical cell provided the necessary low-voltage dc source that would make the telegraph practical. The idea of using an electromagnet for telegraphy occurred to André Ampère in 1820, but the real precursor of our modern telegraph must be credited to Henry.

Joseph Henry's career, like that of many other "greats," began unimpressively. In 1810, at the age of 13, he completed a rural education and was apprenticed to a watchmaker. He apparently had little interest in study until, we are told in a somewhat apocryphal story, he chanced across a book entitled *Lectures on Experimental Philosophy* (i.e., a science book), which fired his interest and ambition. Returning to school at the Albany Academy, teaching part-time at primary schools to help pay his way, Henry took what we would call a "premed" course today, intending to become a physician. An interim surveying job changed his direction to engineering, however, and he returned to teach mathematics and science at the Albany Academy. There began the career of discovery and invention that distinguished Henry in the field of physics, particularly electromagnetism. Among Henry's great administrative contributions were the plans for the present charter and operation of the Smithsonian Institute (he became its secretary in 1846) and the establishment of the U.S. Weather Service.

Henry built a telegraph in 1831, using an electromagnet to momentarily attract a bar magnet, which was suspended horizontally like a compass needle, and cause it to strike a bell. He actually demonstrated that this

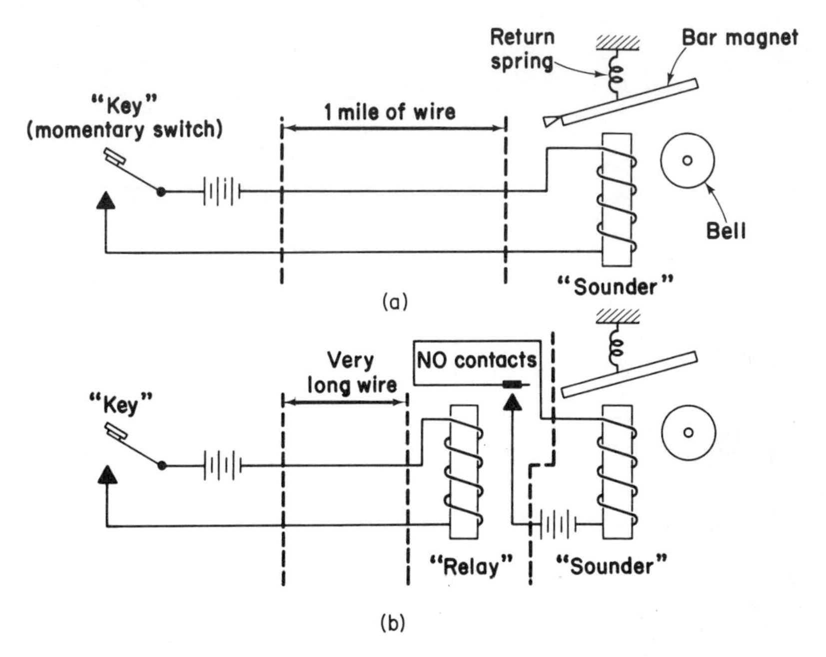

Fig. 10-10. Henry's telegraph. (a) Original version. (b) With relay.

device would work with a mile of wire between the switch and the sounder. The device is shown schematically in Fig. 10-10*a*. As the wires get longer, however, we know that the resistance goes up and thus the current available from the battery becomes too feeble to move the sounder. Henry overcame this difficulty, too, by inventing the relay in 1835. He then used the weak current from the remote switch and battery to operate the relay at the receiving end, which, in turn, operated the sounder from a second battery located there, as shown in Fig. 10-10*b*. Henry, the idealist, did not patent these ideas, for he believed that they belonged to the world. He thus set the stage for Morse.

Samuel F. B. Morse was an artist, not a scientist. What is more important, he was an opportunist and promoter. This is not to detract from his accomplishment, for without his foresight and promotion, the commercial telegraph would not have been realized. He met Henry by accident, and using Henry's freely given ideas, planned and patented a telegraph. His real ability was demonstrated when he persuaded the Congress, which, as usual, was dragging its collective feet, to appropriate 30 thousand dollars to build a 40-mile stretch of telegraph from Washington, D. C. to Baltimore, Maryland. Henry would probably never have tackled such a venture. The first message transmitted over this link was the now-famous "What hath God wrought?" in Morse's original code. The Morse key and sounder, shown

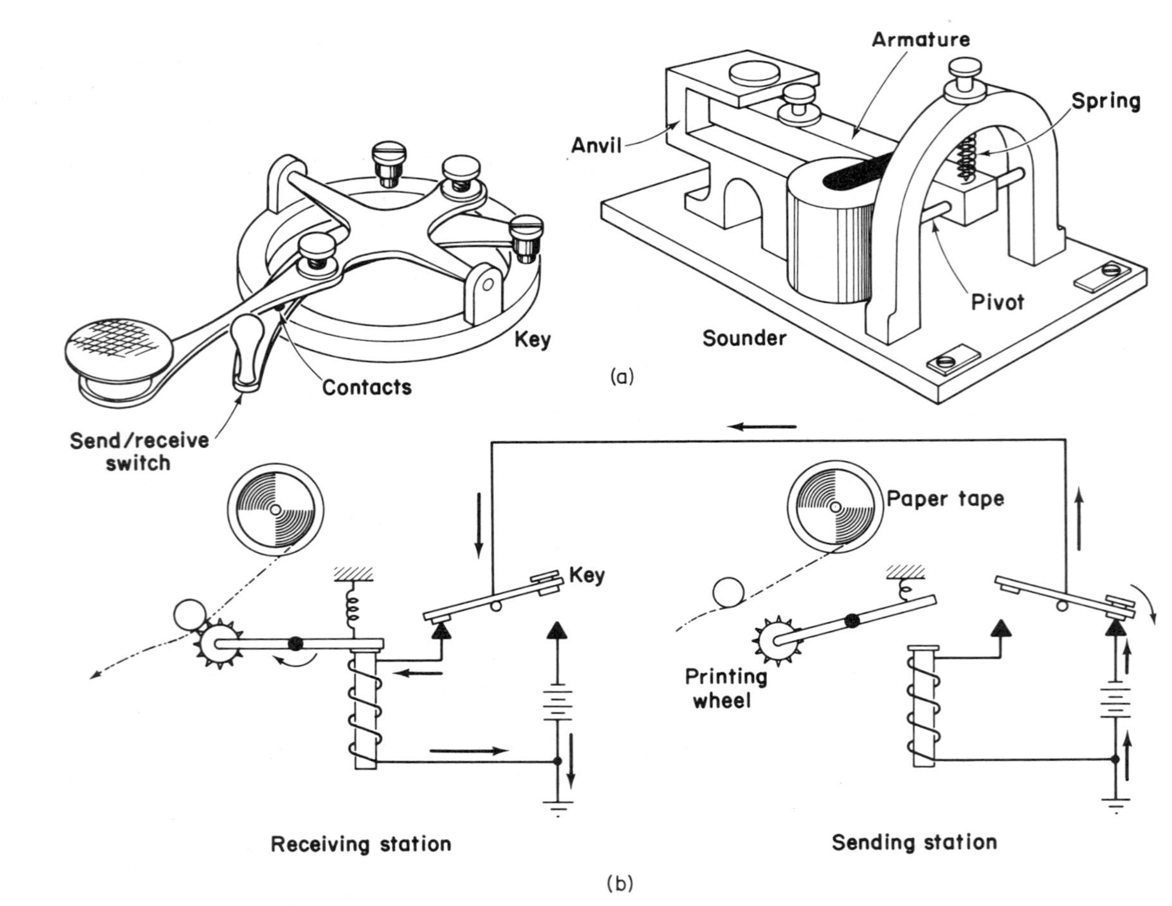

Fig. 10-11. (a) Early Morse telegraph equipment. (b) One type of telegraph circuit.

in Fig. 10-11*a*, are now museum pieces, but a slightly modified form of Morse's dot-dash code is used today throughout the world. Figure 10-11*b* shows schematically the operation of a recording telegraph. Modern telegraphy is completely automatic, using a combination of wire and radio linkages, and many messages are sent over the same line simultaneously via a technique called *multiplexing*. The basic principle remains the same, however.

The development of telegraphy equipment paved the way for the invention of the telephone, to be discussed later.

10–7 THE GALVANOMETER

The quantitative laws of electricity could not have been discovered had there not been devices to measure electrical quantities. Most of today's electrical measuring instruments are adaptations, in one form or other, of a single basic type of device—a device for *measuring very small current*, called a *galvanometer*. (Electric currents were frequently called *galvanic* currents, after Luigi Galvani.)

The earliest galvanometer of record was made by Schweigger in 1820 and consisted simply of a compass needle placed inside a coil of wire, both parallel to the earth's magnetic field. Such a device is depicted in Fig. 10-12.

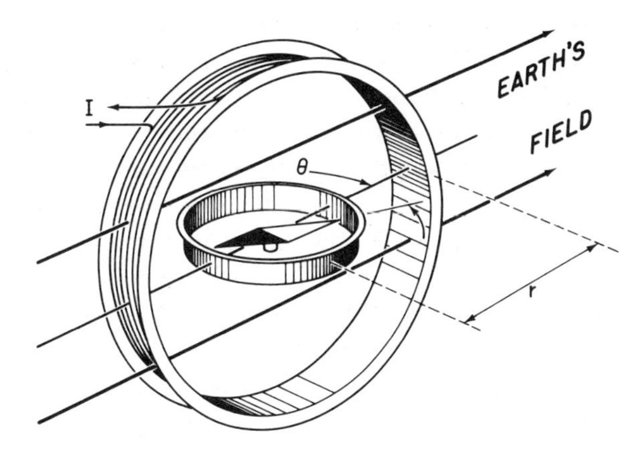

Fig. 10-12. The tangent galvanometer. (Courtesy of Weston Instruments Inc.)

The angle, θ, through which the needle is deflected depends on the current I, the distance r, the number of turns in the coil, and the horizontal intensity of the earth's magnetic field at that location. Thus, for a given set of hardware and at a given location, the angle through which the needle turns depends only on the current. This device is called a *tangent* galvanometer because of the mathematical relationship between the angle and the current.

You can see that the tangent galvanometer is not a practical instrument. Among other things, a serious shortcoming is its dependence on the earth's field. This deficiency has been corrected by using permanent magnets. One particularly simple permanent magnet movement is the *polarized iron vane movement*, shown in Fig. 10-13. When the current I enters and exits

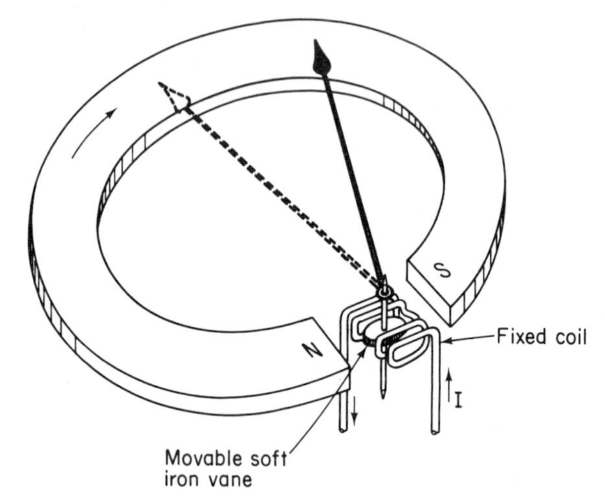

Fig. 10-13. Polarized iron vane movement. (Courtesy of Weston Instruments, Inc.)

the fixed coil as shown, the little vane of soft iron becomes magnetized. Its poles are so oriented with respect to the large permanent magnet that attraction and repulsion cause the vane to rotate, as indicated by the pointer movement. Had the current been introduced into the coil in the opposite direction, the vane (and pointer) would also have moved in the opposite direction. In other words, the motion of the vane depends on the polarity of the connections to the coil, hence the name "polarized" iron vane movement. This type of movement is not very sensitive and is used today more as an indicator than as a measuring instrument. It can be produced in large quantities at low cost. If your automobile has a "charge-discharge" gage (rather than the so-called "idiot light"), it is a crude galvanometer using the polarized iron vane principle.

A more sensitive and useful development in vane-type galvanometers is the *moving iron vane mechanism* depicted in Fig. 10-14. This movement is based on the principle that two similar iron bars placed together within a current-carrying coil will be similarly magnetized. Since their corresponding poles oppose each other, a repulsive force is developed between the two bars. In use as a galvanometer movement, one vane is held fixed, while the other, to which a pointer is attached, is free to rotate on a shaft. Other elements of this design include: jewel bearings to minimize friction; a fine

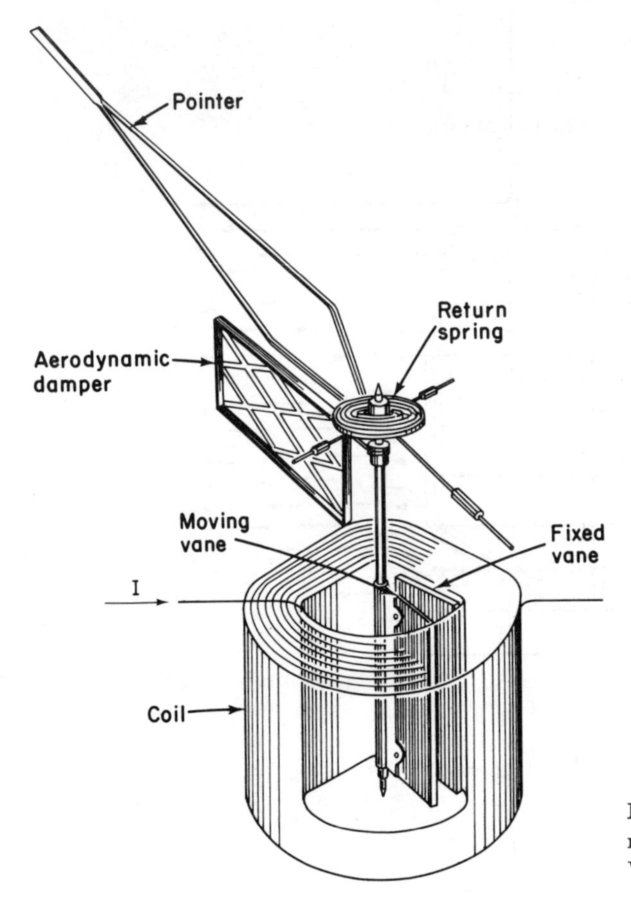

Fig. 10-14. Moving iron vane mechanism. (Courtesy of Weston Instruments, Inc.)

spiral return spring; and an aluminum damper vane that rotates in a tight-fitting cylinder and quickly slows the pointer's movement by air drag forces. When current flows in the coil, in either direction, the pivoted vane separates from the fixed vane like the opening of a book. The fact that the direction of the current in the coil is immaterial makes this movement especially useful as an alternating current instrument, but it is used for dc as well. Although its sensitivity can be considerably greater than that of the polarized vane movement, the moving vane is still not sensitive enough for many applications. Another drawback is that its scale is nonlinear—that is, increasing the current in equal steps does *not* result in the pointer's deflection increasing by equal angles.

Electrical-measuring instruments really came into their own with the invention of the *moving coil* galvanometer, which was introduced by Arsène d'Arsonval in 1882 and is usually referred to as a *d'Arsonval move-*

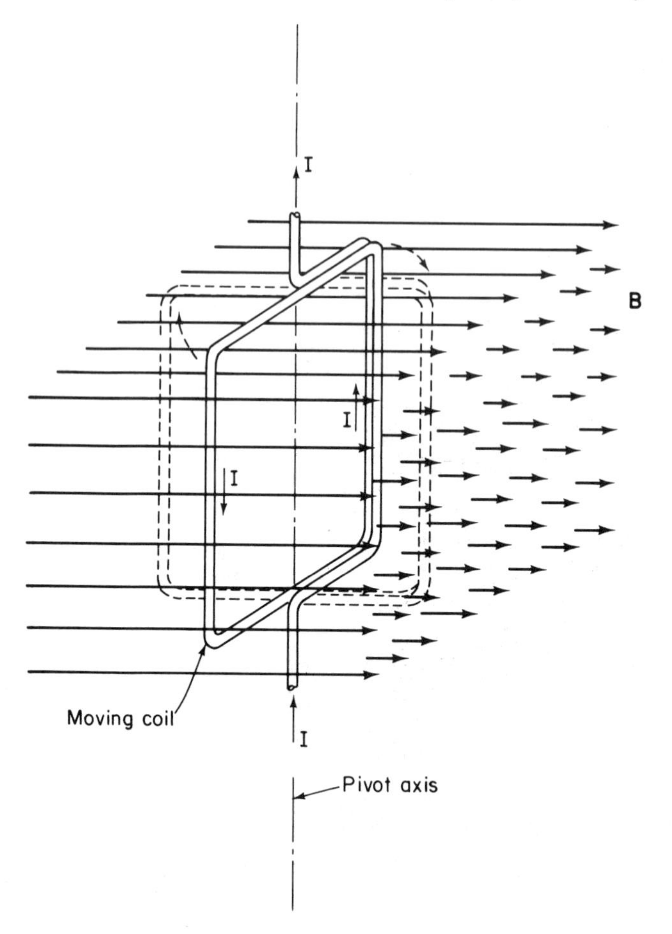

Fig. 10-15. Principle of the d'Arsonval movement.

ment. Its principle, illustrated in Fig. 10-15, is simply the motor effect. A pivoted coil is suspended in the field of a permanent magnet. When current flows in the coil in the direction indicated, the force developed on the wires causes the coil to rotate as shown. A pointer attached to the coil moves over a suitably marked scale. (The reader should convince himself of the direction of rotation by using the right-hand rule.) D'Arsonval's galvanometer, like the one in Fig. 10-16, was a reflecting type—that is, it used a beam of light instead of a mechanical pointer—the direction of the reflected beam being governed by the angle through which the mirror has been turned. It was constructed with a rectangular coil of fine wire suspended between the poles of a vertical horseshoe magnet by taut connecting wires made as thin as possible. A soft iron core within the coil intensified the field.

In 1888 Edward Weston devised a ruggedized and more portable version of the d'Arsonval movement by using a more compact, stable mag-

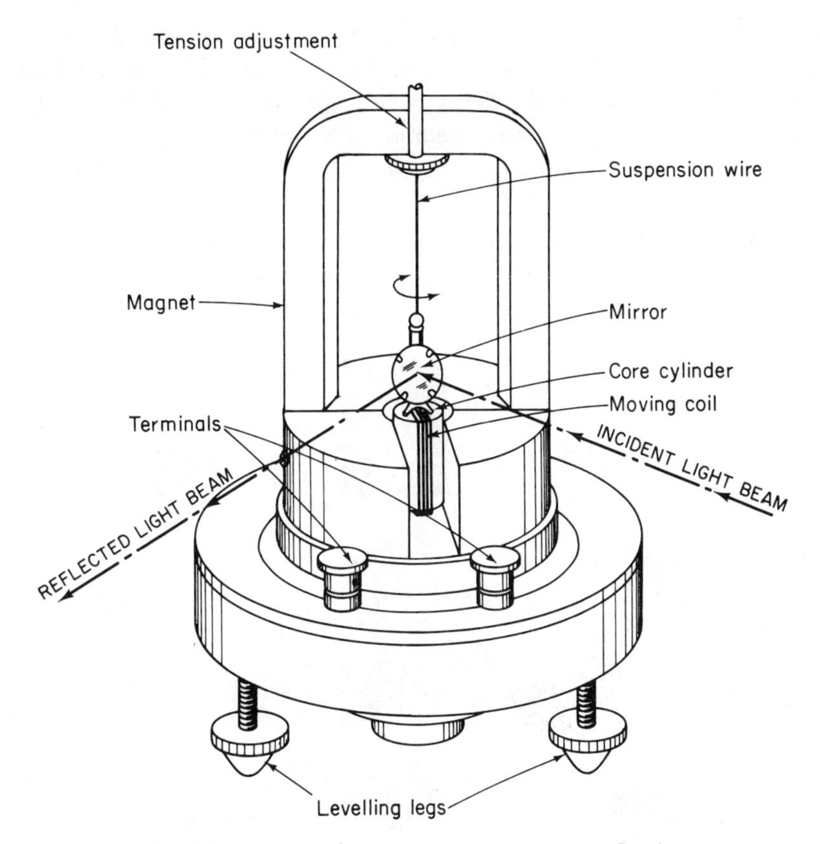

Fig. 10-16. D'Arsonval moving coil galvanometer, reflecting type.

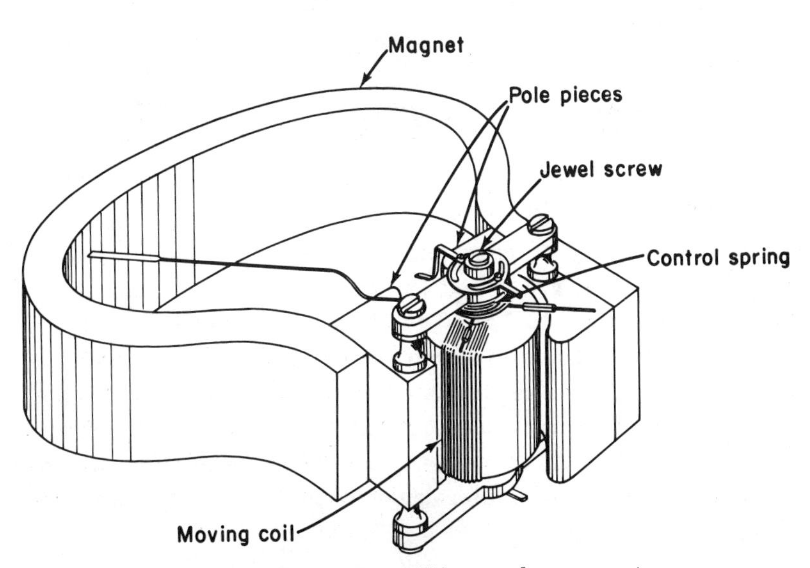

Fig. 10-17. Weston-type d'Arsonval movement.

netic system and by replacing the taut wires with sapphire pivot bearings and spiral bronze current-carrying control springs. In this configuration, shown in Fig. 10-17, the d'Arsonval movement (the name "Weston movement" was never adopted) has become the accepted galvanometer for most measurements and is the "heart" of about 90 percent of all indicating instruments.

Today d'Arsonval movements are made in an almost endless variety of shapes and sizes, those in Fig. 10-18 being representative. They can be

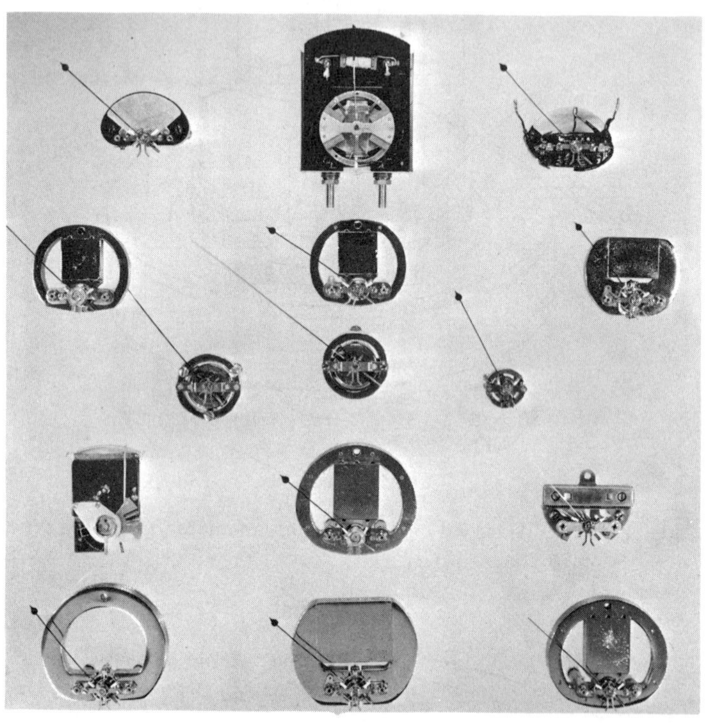

Fig. 10-18. Variety of d'Arsonval-type movements. (Courtesy of Simpson Electric Co.)

designed so that the current required to deflect the pointer over the full range of its scale, which is called the *full-scale current*, is anything from 5 microamperes (i.e., 5×10^{-6} amp) to about one ampere. Both the Weston-jewelled bearing suspension and a modern version of the taut-wire suspension are used, as indicated in Fig. 10-19. Since the direction of the pointer movement depends on the direction of current in the coil, the d'Arsonval movement is a polarized or dc movement, but it can be adapted to alternating current use by incorporation of a device called a rectifier (discussed later in the text).

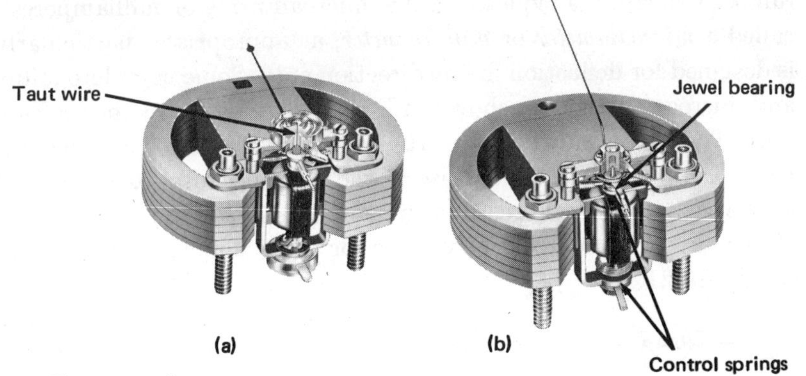

Fig. 10-19. Two modern d'Arsonval movement suspension techniques. (a) Spring and bearing. (b) Taut band. (Courtesy of Simpson Electric Co.)

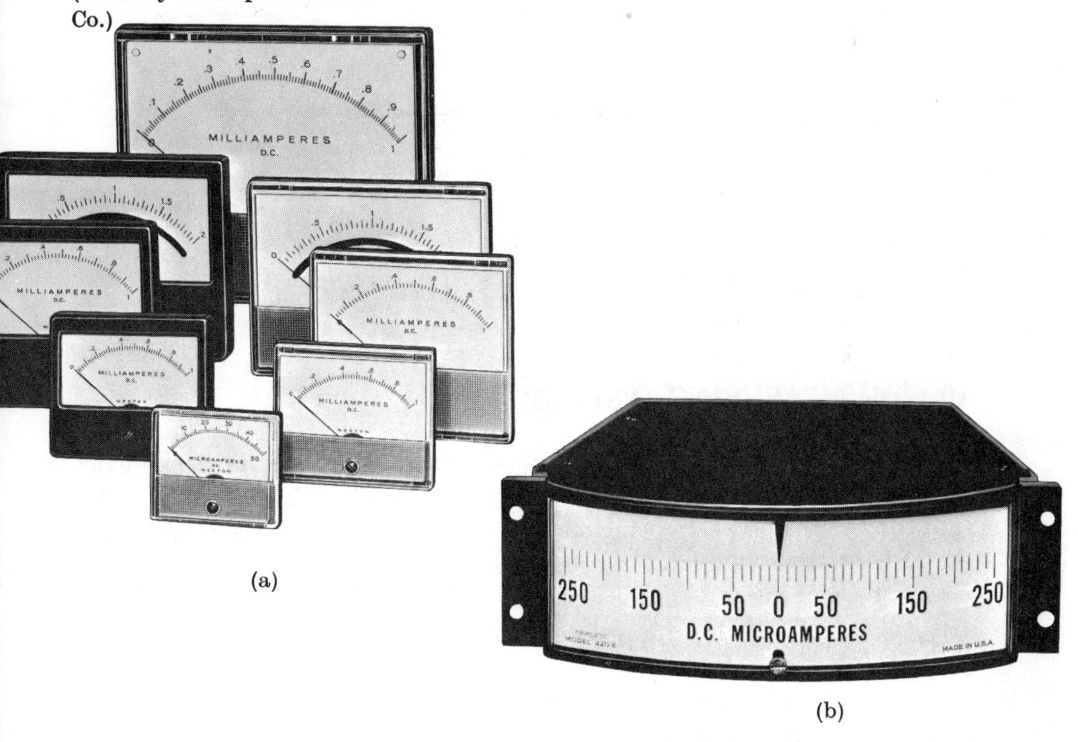

Fig. 10-20. Panel meters: (a) various milliammeters and microammeter. (courtesy of Weston Instruments Inc.); (b) 250-0-250 μA galvanometer. (courtesy of the Triplett Electrical Instrument Co.).

Since the galvanometer is a current-measuring device, or *meter*, whose full scale current is typically a few microamperes or milliamperes, it is also called a *microammeter* or *milliammeter*, as appropriate, particularly when it is designed for deflection in one direction only. Some complete milliammeters and microammeters are shown in Fig. 10-20*a*. Note how the scales are graduated. The meter shown in Fig. 10-20*b* is properly called a "true" galvanometer in the present restricted use of the term because "zero" is in the center of the scale and it will read dc currents (up to 250 μa) flowing through it in either direction. Schematic symbols for various meter movements are given in Fig. 10-21.

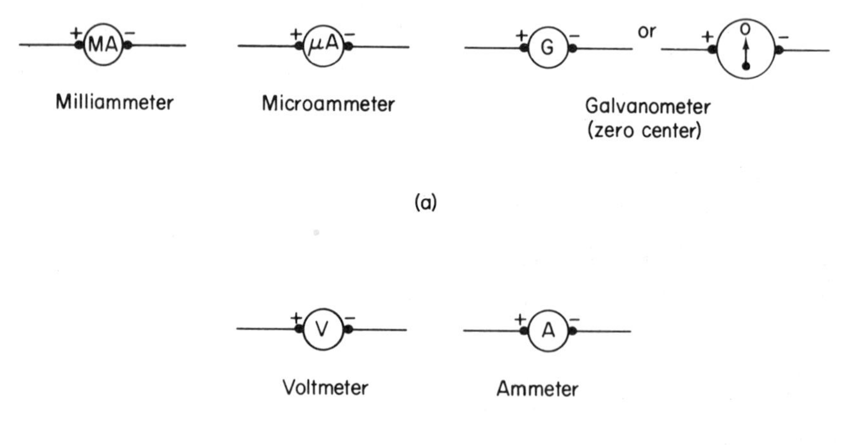

Fig. 10-21. dc meters. (a) Basic movements. (b) Composite instruments.

10–8 dc MEASURING INSTRUMENTS

Many useful measuring instruments can be built around the milliammeter or microammeter type of galvanometer. In all of these, the basic idea is to convert the property being measured into a current that is proportional (or, at least, directly related) to that property. This current is then read out on a meter and the scale of the meter is usually marked in units of the property instead of milliamperes or microamperes. The device that converts the property into an electric voltage is called a *transducer*. Some circuitry is usually required between the transducer and the meter itself, to convert the tranducer's voltage output into a current that the meter can read. Often amplification is involved, for many transducers put out a very small voltage and have a relatively high internal resistance. This intermediate circuitry is called the *signal conditioning* circuitry.

Our primary interest at the moment is in measuring electric circuit variables themselves—current, voltage, and resistance. For such measurements, a transducer as such is not required because we are already dealing with electrical quantities. We now discuss the required signal conditioning circuits for each type of measurement.

Consider first the measurement of current. A current-measuring meter, as we already know, is called an *ammeter* (an euphonious way of spelling *amp*-meter), since the unit of current is the ampere. You will recall that an ammeter is connected in series with the rest of the circuit whose current we are trying to measure. Now the basic d'Arsonval movement has a full-scale current that may be as little as a few microamperes or as much as several milliamperes. As long as the current to be measured is less than the full-scale current, the ammeter is merely the basic meter movement itself. But what if we wish to measure a current of, say, one ampere? Clearly, such a current would burn out the coil of a milliammeter. The answer is to *bypass* most of the current around the meter movement, such that only a small (but definite) fraction of the total current passes through the movement. This step is accomplished with a very low resistance precision resistor called a *shunt*, which is connected in parallel with the meter movement. The parallel combination then forms the ammeter, which is connected in series with the circuit under test. Suppose, for instance, that we have a milliammeter whose full-scale current is one milliampere (i.e., 10^{-3} amp) and whose internal resistance is 100 ohms, with which we wish to measure currents up to one ampere. This means that the current we wish to measure is one thousand times as great as the capability of the meter movement. We must therefore connect a shunt across the meter, as shown in Fig. 10-22, whose resistance is $\frac{1}{999}$th that of the movement itself. In this case, the shunt must have a resistance of about 0.1001 ohm. Then one thousandth of the current will go through the movement, while the remaining 999 thousandths will go through the shunt. In a meter that is to be used permanently as an ammeter, the shunt is

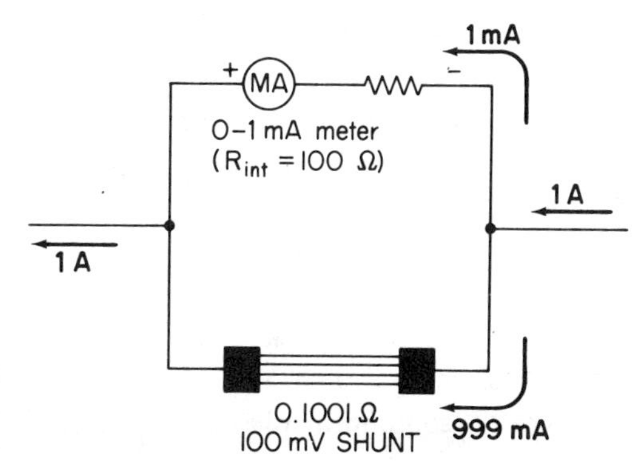

Fig. 10-22. Using a shunt to extend the range of a 0–1 milliammeter to 0–1 ampere.

usually installed inside the meter housing. In other cases, when the milliameter or microammeter is to be used only temporarily as an ammeter (or it is otherwise convenient to be able to change its range), shunts like those shown in Fig. 10-23 are made and sold separately for external connection. These shunts are rated in terms of the number of millivolts voltage drop across them at their maximum design current. Thus the shunts in Fig. 10-23 are 50-millivolt shunts with design maximum currents of 100 and 2000 amp.

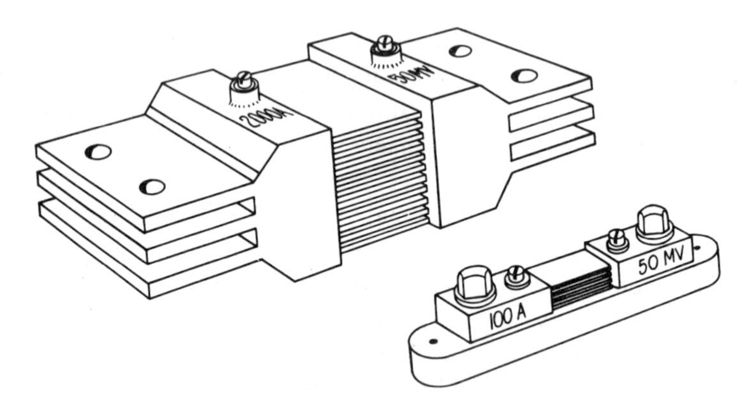

Fig. 10-23. External shunts. (Courtesy of Weston Instruments Inc.)

Next, we turn our attention to the measurement of emf (i.e., voltage). The instrument is called a voltmeter. In a voltmeter, the signal conditioning circuitry consists simply of a high resistance in series with the milliammeter or microammeter. If, for instance, it were desired to measure voltage up to 500 volts with a 0 to 500-μa movement, the total series resistance (including that of the meter itself) must be such as to produce full-scale deflection when maximum design voltage is applied. In other words, we need a resistance such that a current of 500 μa (5×10^{-4} amp) flows when 500 volts dc is applied. With Ohm's law, we can compute the required resistance, one megohm (10^6 Ω). If the internal resistance of the meter movement itself is 200 ohms, then we need an additional series resistance of 999.8 kΩ. The circuit is shown in Fig. 10-24*a*. The series resistor in a voltmeter is called a *multiplier*, just as the parallel resistance in the ammeter is called a shunt. Often it is convenient to have a voltmeter capable of measuring several ranges. In that case, there will be a selector switch to choose the multiplier resistor appropriate to the desired voltage range. Figure 10-24*b* shows a multirange voltmeter circuit, using a 0 to 50-μa movement. Note that the ranges differ from each other by some multiple of 10. This is usually done for convenience,

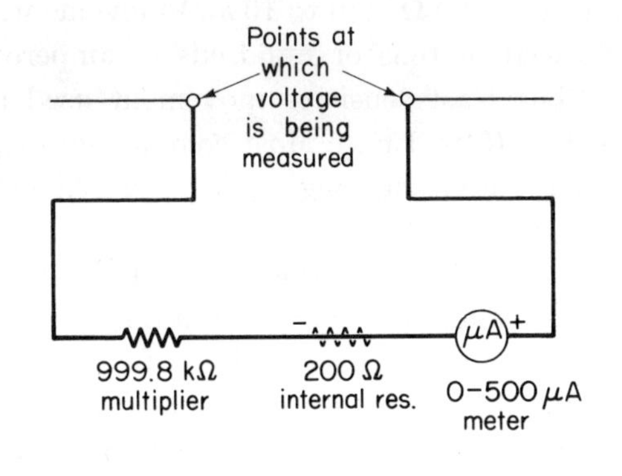

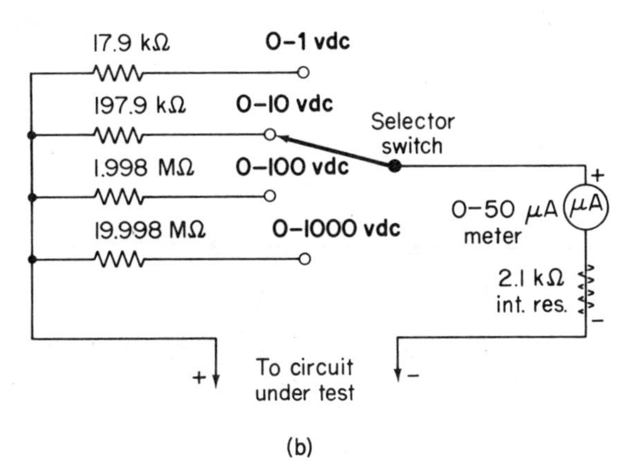

Fig. 10-24. (a) 0–500 dc voltmeter circuit. (b) Multirange dc voltmeter circuit.

for it is then necessary to print only one scale (numbered here, for instance, from 0 to 100) on the meter face, and the user can easily multiply or divide mentally by 10 or 100.

From the preceding examples, it may have occurred to the reader that any meter movement, together with a suitably chosen multiplier, can be used to make a voltmeter of any desired range. To a large extent, this is true. One factor does have to be kept in mind in designing a voltmeter, however, and that is circuit loading. Remember that the ideal voltmeter draws no current at all from the circuit it is measuring, and a real voltmeter should draw as little as possible; otherwise its very presence will significantly change the voltage it is trying to measure. This tendency to load the circuit is expressed as the ohms-per-volt rating of the voltmeter. It is inversely related to the full-scale current of the meter movement used, that is, the higher the ohms-per-volt (Ω/v) rating, the lower the full-scale current and hence the lower the loading on the test circuit. Thus panel meters in industrial power supplies may use as low as a 100 Ω/v (0 to 10 ma) movement, for an extra 10 ma would not be noticed in tens or hundreds of amperes of current output from the supply. The least sensitive movement used in a test instrument, however, is 1000 Ω/v (0 to 1 ma). Most commercial instruments designed for general test and troubleshooting work use a 20,000 Ω/v (0 to 50 μa) movement.

For measuring high voltages, there is another type of device, *not* based on the d'Arsonval movement, called an *electrostatic voltmeter* or *electrometer*, which draws essentially zero current. Its principle of operation is like that of the gold leaf electroscope (discussed in Chapter 3)—the electrostatic forces between charged foils. Here, however, instead of repulsion between like charges, the force of attraction between unlike charges is used to cause one set of plates to rotate toward another fixed set. A modern electrostatic voltmeter movement is illustrated in Fig. 10-25. This is the

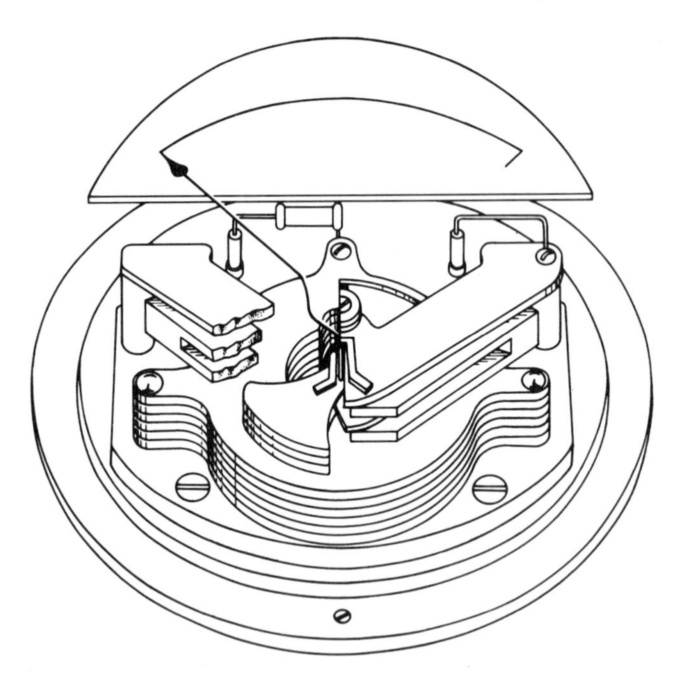

Fig. 10-25. Electrostatic voltmeter movement. (Courtesy of Weston Instruments Inc.)

only movement activated by voltage directly, rather than by a current produced from the voltage.

Finally, we consider the measurement of resistance. This measurement is accomplished in several ways, depending on the accuracy required. In any case, the current to drive a d'Arsonval meter movement cannot be had from a resistance alone, and so the signal conditioning circuitry of a resistance measuring device must necessarily include a voltage source. The simplest and quickest resistance-measuring instrument is called an *ohmmeter*. Figure 10-26*a* shows the circuit for a "series" ohmmeter, which is nothing

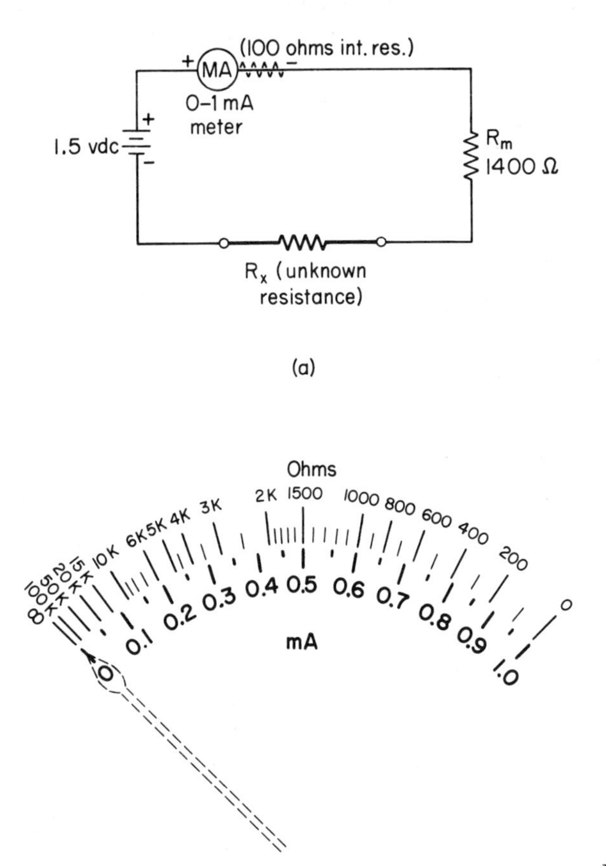

Fig. 10-26. Series ohmmeter. (a) Circuit diagram. (b) Meter scale.

more than a battery and a meter movement, together with a series resistor, R_m, of such a size that if points a and b are connected, the meter will deflect full scale. The unknown resistance is connected between points a and b, and its value is read from the meter's scale which is calibrated directly in ohms. The meter's ohms scale is shown in Fig. 10-26*b*, together with the actual current values for comparison. Note that the ohms scale runs "backward" from the current scale and that it is *not* a linear (i.e., equal spaces for equal number of ohms) scale. Note, also, that the left end of the scale is marked infinity (∞), but any resistance greater than about 100 times the center scale value is practically infinite for this ohmmeter range. This type of ohmmeter is not suitable for measuring low resistance, but does offer the advantage of simplicity in certain applications. The most commonly used ohmmeter circuit, and one that can be designed with selectable ranges to measure resistances from 1 ohm to several megohms, is given in Fig. 10-27*a*. The "zero adjuster," R_z, is used to set the meter reading to

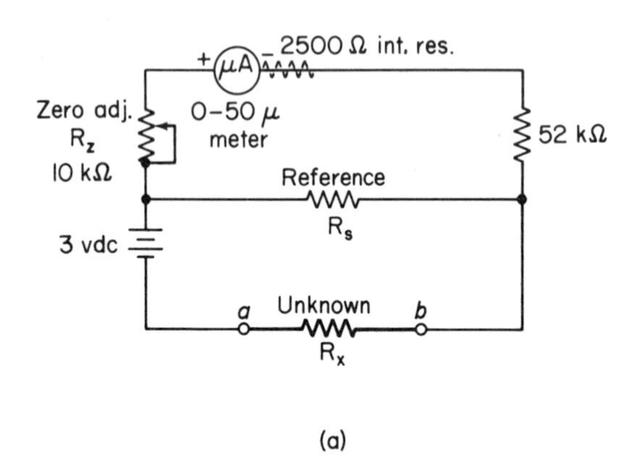

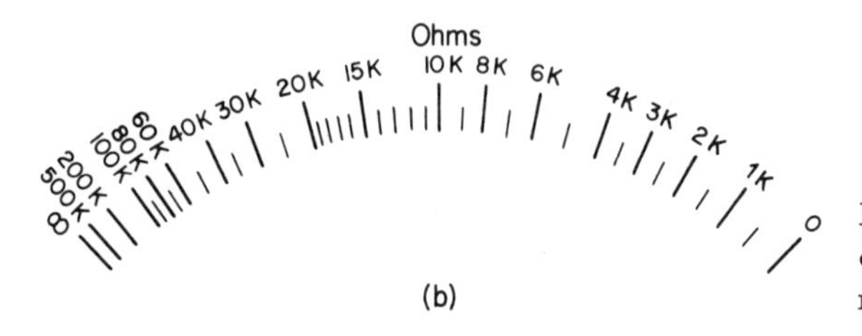

Fig. 10-27. (a) Most common ohmmeter circuit. (b) Scale of meter for $R_s = 12{,}000\ \Omega$.

exactly full scale while points *a* and *b* are momentarily short-circuited. This adjustment compensates for slight changes in battery voltage and drift in resistance values. The unknown resistance, connected across points *a* and *b*, determines the voltage appearing across the reference resistor, R_s, and therefore the current through the meter. The range of the meter is determined by the value of R_s, once the battery has been selected. In a multiple-range ohmmeter, several different values of R_s are selectable with a switch. The ohmmeter scale, for $R_s = 12\ \text{k}\Omega$, is shown in Fig. 10-27*b*. Note that here, once again, the ohms scale increases from right to left.

A more accurate measurement of resistance than that obtained with an ohmmeter can be attained by using a *Wheatstone bridge*, whose circuit is given in Fig. 10-28. We have already encountered this network in Section 7-7, although its purpose was not discussed there. In this device, the value of the unknown resistance, which is connected between points *c* and *d*, is

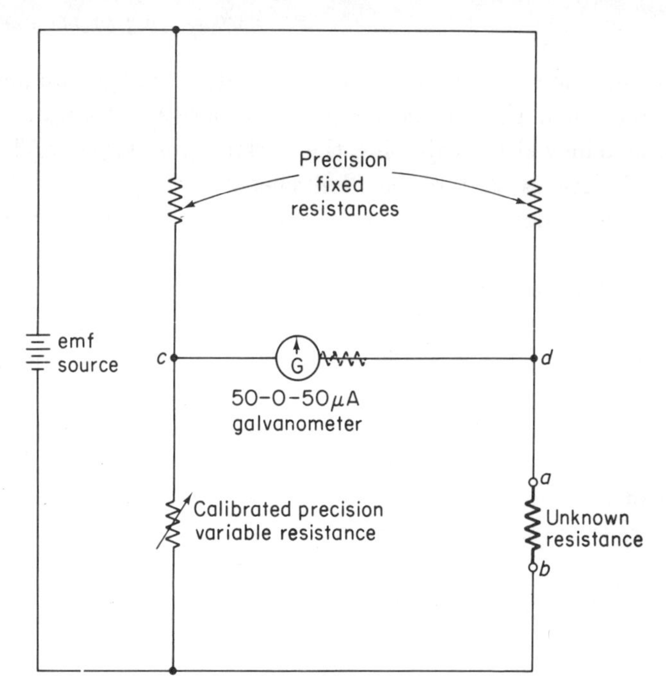

Fig. 10-28. Wheatstone bridge for resistance measurement.

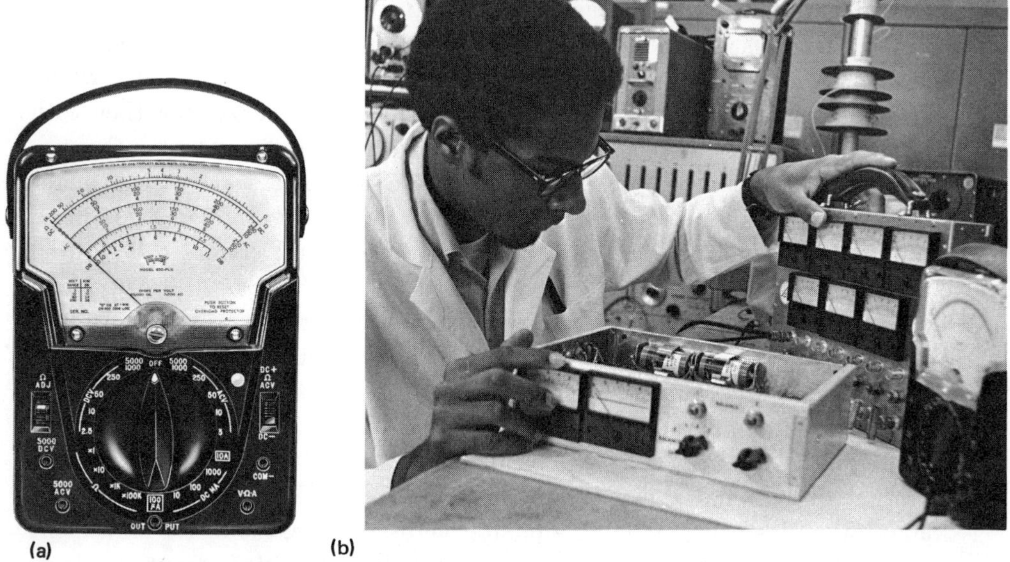

Fig. 10-29. (a) A commercial portable volt-ohm-milliammeter. (Courtesy of the Triplett Electrical Instrument Co.) (b) Volt-ohm milliameter being used to test electronic equipment. (Courtesy of the Aerospace Corp.)

not read from the scale of the meter. Instead, a true galvanometer is used to determine when there is *no current flow* between points *c* and *d*. This condition is achieved by adjusting the variable resistance, and in this condition the bridge is said to be "balanced." The value of the unknown resistance is then determined from the reading of a scale on the variable resistance.

The laboratory technician and repairman must often measure all three major circuit parameters: voltage, current, and resistance. It is therefore convenient to have a single instrument capable of doing just that. A multipurpose measuring device of this type is called a volt-ohm-milliammeter, or VOM for short. Each of the major meter manufacturers carries one or more VOM's in his product line. One example is shown in Fig. 10-29*a*. The type of measurement and range are selected with the central selector switch. Note that the VOM also measures ac voltage. Observe the variety of scales and information on the meter face. Figure 10-29*b* shows such a VOM in use in an electronics laboratory.

10-9 OTHER dc INSTRUMENTS

In the preceding section we discussed instruments for measuring dc circuit parameters. Here we touch briefly on two examples of electrical instruments for measuring nonelectric parameters.

The first is the *thermoelectric thermometer*. This is a temperature-measuring device that consists of a thermocouple (see Section 5-10), which acts as the transducer from heat energy to electric energy, together with what is essentially a voltmeter, to read the voltage produced by the thermocouple. The circuit is illustrated in Fig. 10-30*a*. The scale of the meter is, of course, calibrated directly in degrees Celsius or degrees Fahrenheit, rather than in voltage units. Such electrical thermometers are widely used in monitoring temperatures in industrial processes.

A similar device is the basic *photometer* circuit, shown in Fig. 10-30*b*, for the measurement of light intensity. Here the transducer is a photocell (see Section 5-10), which develops a dc voltage proportional to the intensity of the light illuminating it. A voltmeter-type circuit is used to measure this voltage. The scale calibration depends on the use of the photometer. In one version, it becomes the photographer's hand-held lightmeter. In another adaptation, instead of moving a pointer across a scale, the meter movement is used to drive a gear train that opens and closes the iris in front of a lens. This step automatically controls the amount of light admitted, and we have the automatic exposure control system used on modern cameras.

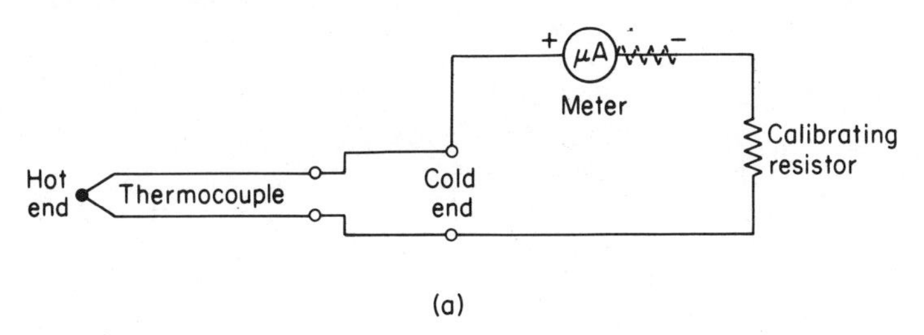

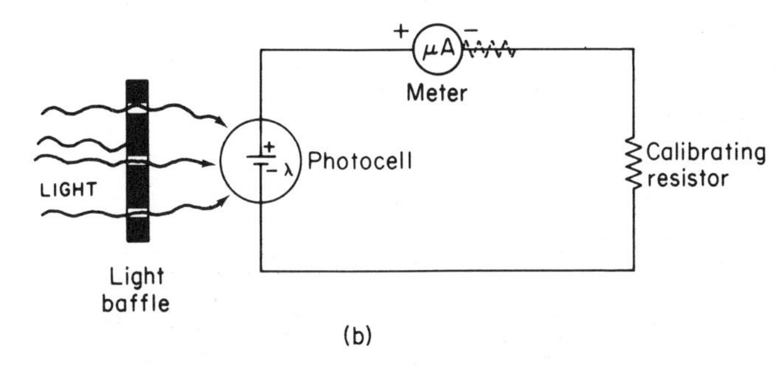

Fig. 10-30. Two dc instruments. (a) Electric thermometer. (b) Lightmeter.

10-10 THE MOTOR

An electric motor is a device for converting electric energy into mechanical energy, specifically into continuous rotary motion. Perhaps the granddaddy of all electric motors was invented by our old friend, Michael Faraday, in 1821. Today it is one of the most prevalent electric devices. If asked, we all could give literally dozens of uses for the electric motor.

The meter movements we have already discussed are essentially simple electric motors, but with one limitation. They can only rotate through a limited angle (about 90°). The motor, on the other hand, must be capable of turning continuously. We must therefore provide a different way of getting current into, and out of, the rotating coil. This process is accomplished through the use of sliding contacts. The first practical motor was devised by Henry.

Consider the device shown in Fig. 10-31. It consists of a single-turn "coil" mounted so that it can rotate between the poles of a permanent magnet. The leads from the coil are brought out and connected to two halves of a split ring, which is called a *commutator*. A dc source of emf is connected to the coil by means of sliding contacts, called *brushes*, which press against the two segments of the commutator. The whole assembly that is free to rotate

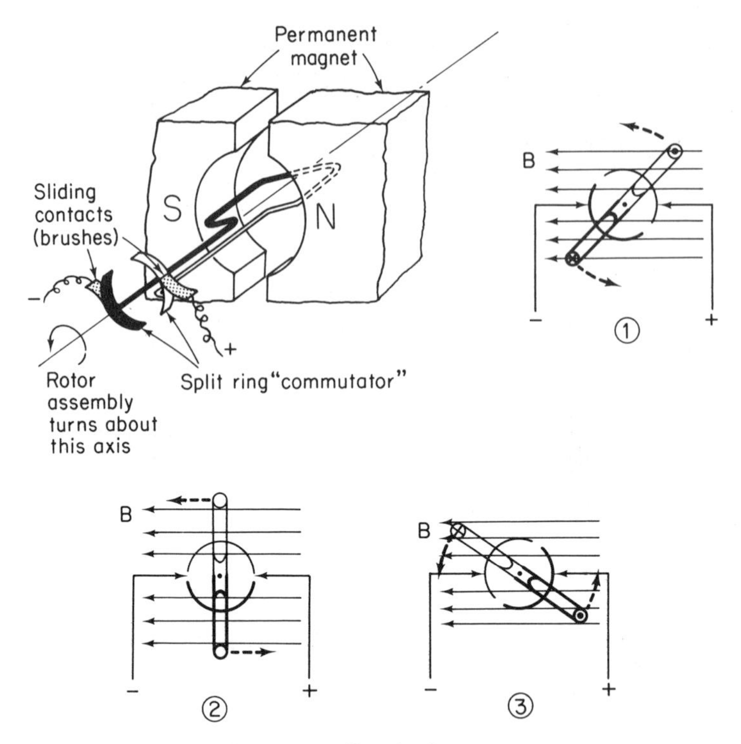

Fig. 10-31. Simple dc motor.

is called the *armature assembly*, or simply the *rotor*. The magnetic portion of the motor that does not rotate is called the *stator*. Some structure is required to support the rotor and the stator, and this structure is called the *frame*. The entire motor assembly is usually enclosed in a protective case.

We now study the directions of currents and forces in our simple motor of Fig. 10-31 as emf is applied. Once again we use the "coming and going arrows" symbolism, $\odot$ and $\otimes$, to represent electron flow out of and into the plane of the page, respectively. Also, to make it easier to keep track of the position of the rotor, we have darkened half the coil and the segment of the commutator to which it is connected.

In the sketch of the assembly, the position of the rotor is position 1 of Fig. 10-31. With the emf applied and the direction of the field B as shown, the current is into the page in the darkened half of the coil and out of the page in the other half. Applying the right-hand rule, we determine that the motor effect produces a net "push" on the two halves of coil in the directions shown by the dashed darkened arrows. This causes the rotor to turn counterclockwise as we view it. As the rotor reaches position 2, the brushes momentarily break contact with the commutator. However, the momentum of the rotor carries it past this position to position 3. Here the current is out of the page in the darkened half and into the page in the nondarkened half, which produces forces such as to maintain rotation in the counterclockwise direction. Thus the major function of the commutator is to reverse the direction of the applied emf each time the rotor coil reverses its position. In other words, the commutator and brush arrangement ensures that the negative side of the coil is always the one moving downward on the left and the positive side is always that moving upward on the right, which, in conjunction with the fixed direction of the field, causes uniform rotation in one direction.

The two important parameters of a motor are its *torque* and *speed.* The torque is the amount of "twist" it produces and is specified by the amount of force it exerts at a given distance from the hub of the shaft. In American engineering units, the torque is measured in foot-pounds (ft-lb), or inch-ounces (in.-oz) for a very small motor. The "speed" of the motor refers to its rotation rate and is generally specified as the number of revolutions per minute (rpm) through which the shaft turns. The mechanical output power of a motor depends on both the torque and speed and is usually specified in horsepower (hp). The relationship connecting the three quantities is given by Fig. 10-32.

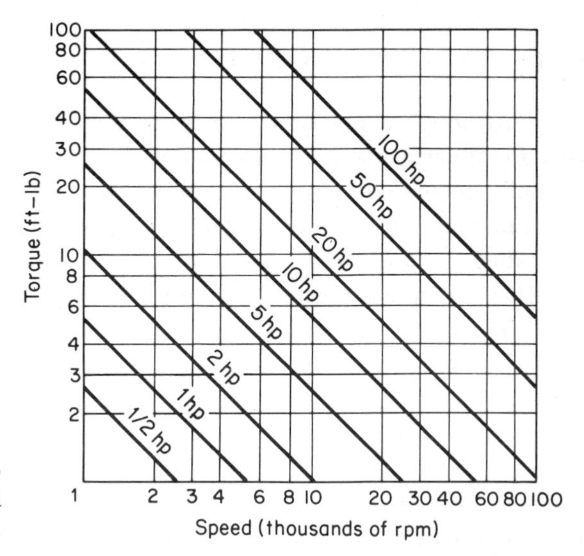

Fig. 10-32. Relationship between speed, torque, and horsepower in motor ratings.

Ideally the rotor winding has no resistance. Since it is made from a real metal, copper, it actually does have a small resistance, but it is not this resistance that determines the rotor current. The effective resistance to current flow comes from a *counter emf.* Recall the dynamo effect from the previous chapter. In the motor, we have a moving conductor (the rotor coil) cutting across a magnetic field—just the right conditions for electromagnetic induction. We would therefore expect an emf to be induced in the rotor's coil. This does indeed happen, and it turns out that the polarity of this emf is opposite to that of the applied emf, hence the name counter emf. It therefore opposes or "bucks" the driving emf and limits the current. Thus a motor with no load (i.e., only friction from air drag and bearings) will run very fast in order that its rotor cut lines of force rapidly enough to induce a sufficiently large counter emf to oppose that applied. When a load is applied to the shaft, the rotor slows down, resulting in a smaller counter emf that permits a greater current flow in the rotor, which, in turn, yields the required increase in magnetic torque. If the shaft is loaded down to the point where the rotor will not turn at all, the current is limited only by the rotor's resistance (i.e., there is no counter emf). This current may be very high, and if the rotor is not designed to withstand it, it may burn out the rotor coil. For this reason, many motor circuits are protected by circuit breakers, in case the motor should be inadvertently stalled or jammed. Note, also, following this same line of reasoning, that the starting current of a motor is always considerably higher than its running current, because there is no counter emf until the rotor begins to move.

10-11 TYPES OF MOTORS

Although a permanent magnet may be used for the stator, as was used for illustrative purposes in our simple motor of Fig. 10-31, this use generally occurs only in very small dc motors. Most often, the stator is an electromagnet with one pair or several pairs of poles. This stator winding is also called the *field winding*. A typical dc machine is shown schematically in Fig. 10-33. Power is normally supplied to the field coil from the same source as to the rotor.

The manner in which the field coil is connected, relative to the rotor, serves as a basis for classifying dc motors. When the field coils are connected in series with the armature winding, as in Fig. 10-34*a*, the motor is called a *series* motor. In this type, the field coils usually consist of comparatively few turns of heavy wire, for they will carry the same heavy current as the rotor. Alternatively, the field winding may be connected to the power supply separately, in parallel with the armature as in Fig. 10-34*b*. This type is called a *shunt* motor and the field will have a larger number of turns of finer wire. Finally, one can obtain a little of the "best of both worlds" by using both a series and a shunt field winding, which makes the device a *compound* motor. This device is shown schematically in Fig. 10-34*c*. When drawing a circuit

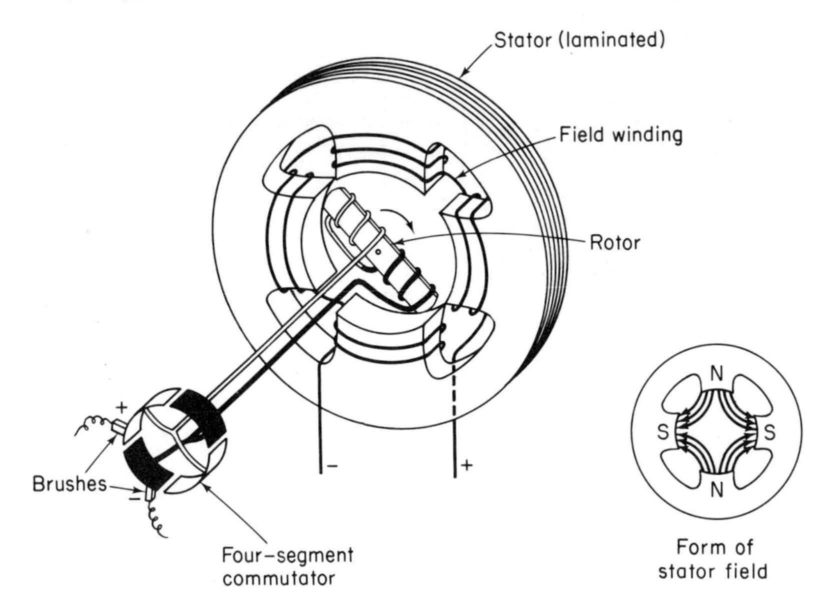

Fig. 10-33. Typical dc motor (schematic).

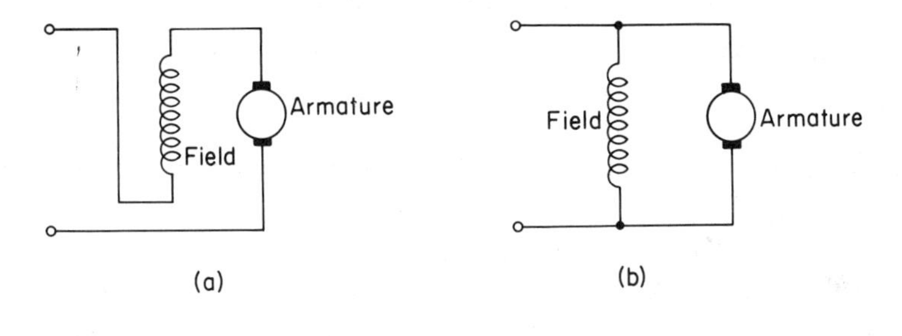

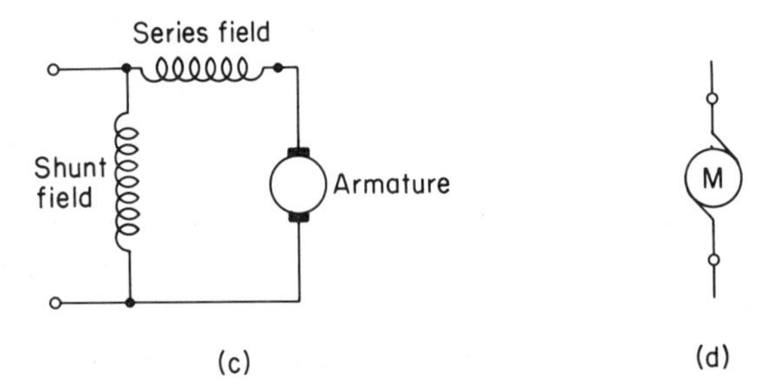

Fig. 10-34. Types of dc motors: (a) series, (b) shunt, (c) compound, (d) conventional simplified circuit symbol.

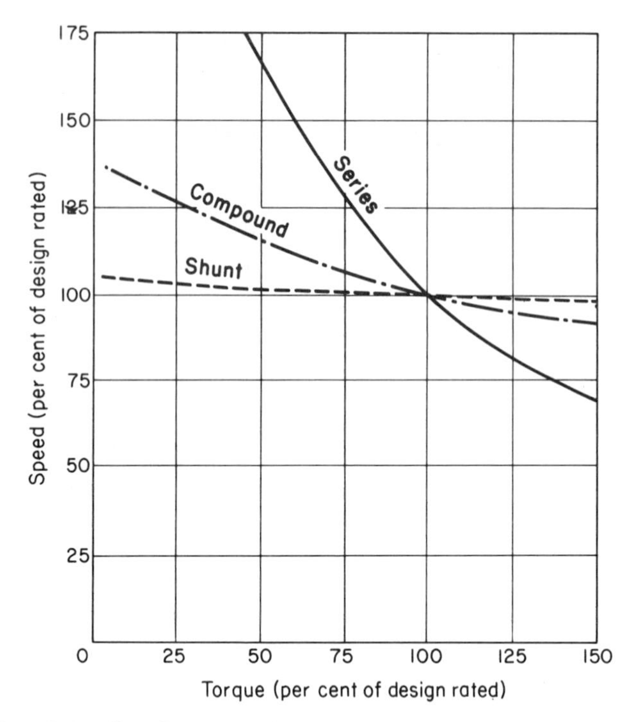

Fig. 10-35. Speed-torque characteristics of various types of dc motors.

containing a dc motor, where the type of motor is not important, the simplified symbol of Fig. 10-34*d* is often used.

Which type to use in a given application depends on the requirements. Figure 10-35 shows the speed-versus-torque characteristic curve for each type. Note that the shunt motor operates at almost constant speed irrespective of the torque it is required to deliver. Moreover, its operating speed can easily be adjusted by varying the voltage applied to its field coils. Thus the shunt motor is particularly useful in a controlled-speed device. The series motor, on the other hand, is more of a constant-power device whose speed drops to compensate for higher torque loads. Its most significant feature is its high starting torque, which makes it useful for quickly spinning up heavy loads. The compound motor has some of the best features of both, the drop of speed with load depending on the relative number of turns in the series and shunt fields.

The shunt motor, unlike the others, can also be operated with an ac supply. Consequently, a commutated series motor is often called a *universal* motor. Figure 10-36 shows a modern universal motor.

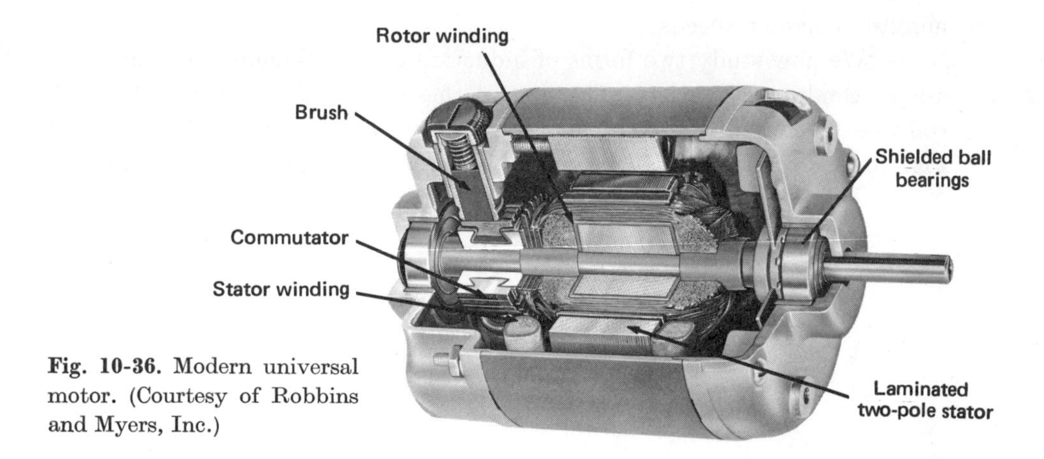

Fig. 10-36. Modern universal motor. (Courtesy of Robbins and Myers, Inc.)

10-12 MORE ABOUT ELECTROMAGNETIC INDUCTION

We already know that for a voltage to be induced in a conductor, the conductor must be subjected to a changing magnetic field. This step can be brought about in two fundamentally different ways: (1) the conductor may be moved about so that it cuts across lines of force in a field of constant flux; or (2) the flux crossing a stationary conductor may be varied, as by varying the current in the winding producing the flux. The direction of the induced emf is given by *Lenz's law* and is named for Heinrich Lenz, a Russian physicist contemporary to Faraday and Henry who also experimented with electromagnetic induction. Lenz's law states that the diection of an induced emf is such that the effect of any current it produces tends to *oppose* the change that induces the emf. If this were not so, and the induced voltage tended to aid the inducing change instead, induced currents of any desired magnitude could be obtained with no additional work put into the process, and this does not happen. In other words, Lenz's law is another way to state that energy is always conserved.

Faraday studied the quantitative relationships involved and gave us the following law:

$$\text{Induced voltage} = (\text{number of effective flux links}) \times (\text{flux change rate})$$

Thus, if a conductor of 100 turns is immersed in a field changing at 0.25 webers per second, the induced voltage is $100 \times 0.25 = 25$ volts. Faraday's law shows us why the counter emf in our permanent magnet motor is greater at higher speed; the turns in the rotor winding are cutting more lines per minute at higher speeds.

We now study two forms of induction of the "changing-flux-through-a-fixed-conductor" variety. The first of these involves emf induced in one coil due to changing flux in another nearby. The principle, illustrated in Fig. 10-37, is called *mutual induction*, and was discovered almost simultane-

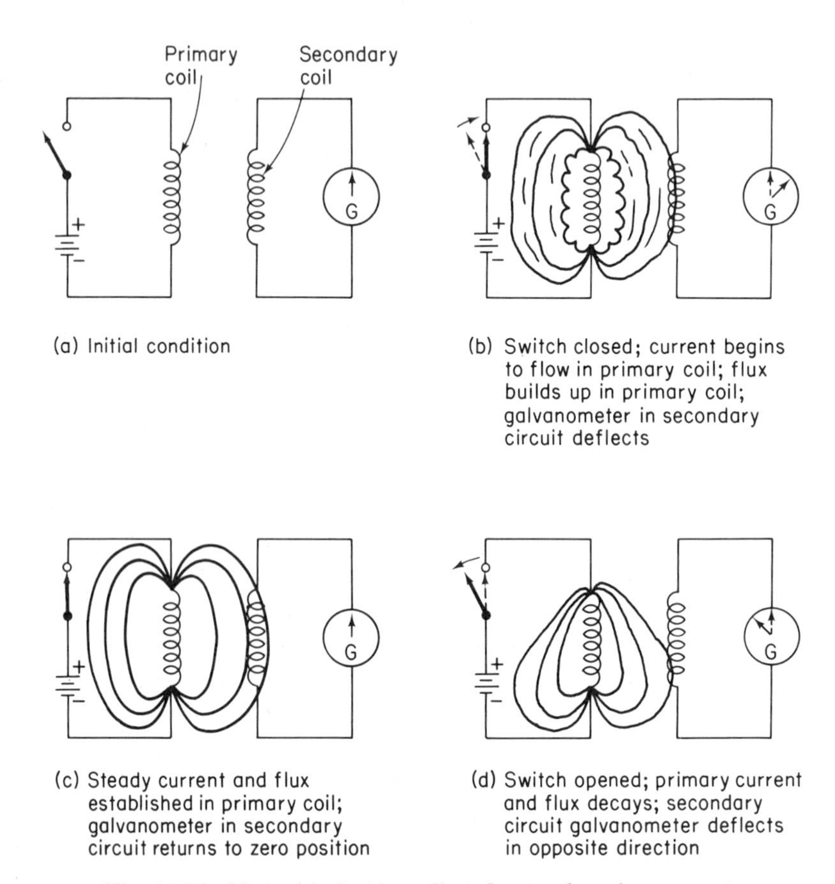

Fig. 10-37. Mutual induction effect due to changing current.

ously by both Faraday and Henry (although the former published his discovery first). A changing magnetic field is set up about one coil, which is called the *primary*, usually by varying the current in the winding. If another coil, called the *secondary*, is placed in the varying magnetic field of the primary, a voltage is induced in the secondary. The polarity of the secondary voltage can be determined from Lenz's law. Its magnitude is determined by the rate at which the primary circuit's current is changing and by the geometry and saturation condition of the cores of the two coils. The latter are essentially properties of the windings and are "lumped" together into a quantity called the *mutual inductance* of the two-coil assembly, which is measured in units called *henrys*. (The mutual inductance is defined in such a way that, when multiplied by the rate at which the primary current is changing, it gives the secondary emf in volts.) We observe therefore, in Fig. 10-37, that whenever the primary current is steady at any value (including zero), the secondary circuit galvanometer does not deflect. However, when the primary circuit switch is opened or closed, the primary current changes suddenly and the secondary galvanometer does deflect. Thus the secondary winding can be connected to some other device and the induced emf used to produce its own current and do useful work. Two extremely important types of devices depend on mutual induction, the *transformer* and the *induction coil*.

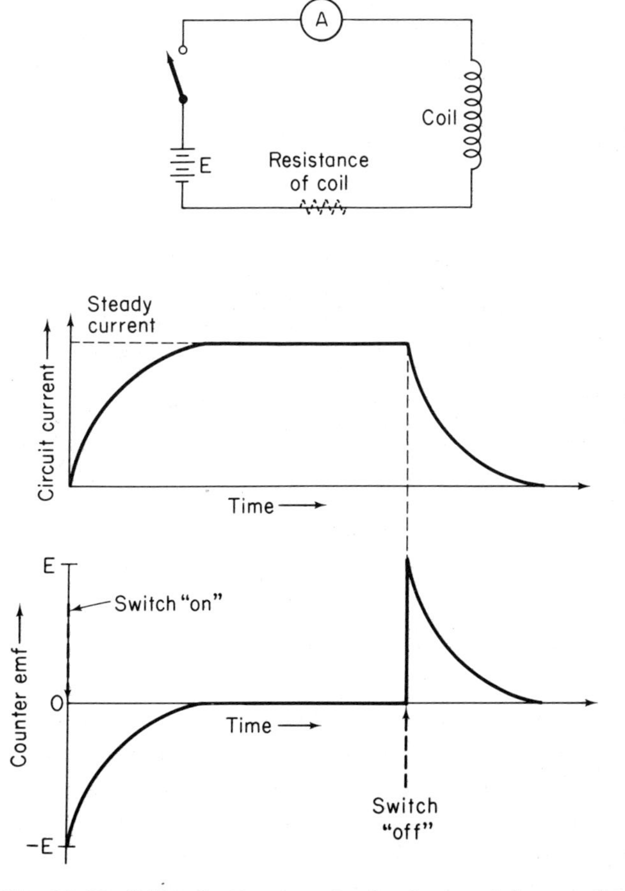

Fig. 10-38. Self-induction in a dc circuit containing a coil.

The other form of induction we consider here is evident whenever we connect a coil to a dc source, as in the upper illustration of Fig. 10-38. When the switch in the circuit is closed, instead of rising instantaneously to the level dictated by Ohm's law, the current rises gradually; and when the switch is opened, the current decays gradually to zero, instead of dropping immediately. These effects are shown graphically by the central illustration of Fig. 10-38. The explanation for this phenomenon lies in the fact that the electric current is propagated at finite speed (fast though it may be) and hence the magnetic field begins to build up around the first turn (relative to the direction of current) of the coil before the second, and around the second turn before the third, etc. As the total field of the coil builds up, therefore, flux from prior turns cuts across subsequent turns, as indicated in Fig. 10-39, inducing a counter emf in the coil. When the current is switched

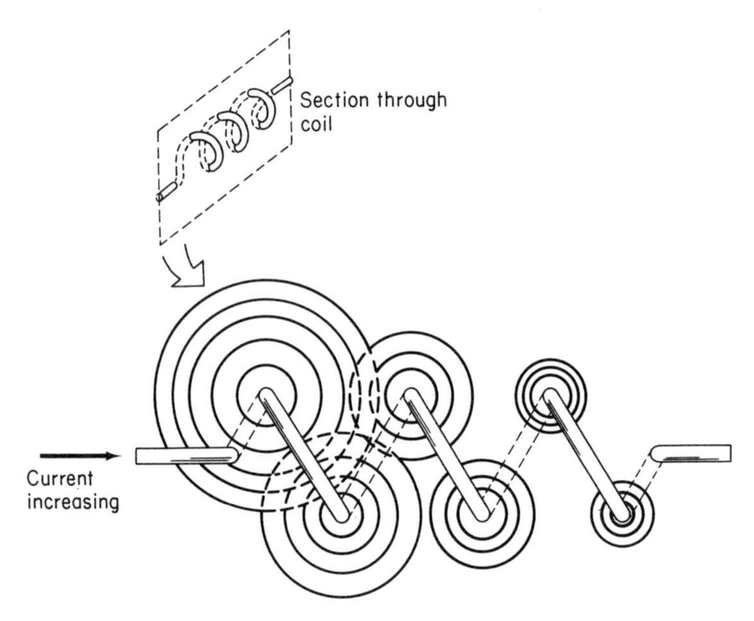

Fig. 10-39. Self-induction due to the changing current in a coil.

off, the field collapses and the same process occurs in reverse. With a dc voltage applied, this counter emf is, of course, temporary and decays to zero eventually. The lower graph in Fig. 10-38 shows the amplitude and direction of this counter emf, on the same time scale as the current in the graph above it, for comparison. Since the induced emf is, in this case, in the same coil with the current that causes it, this type of induction is called *self-induction.*

It was discovered by Henry in 1830, and independently by Faraday four years later. The induced emf depends on the rate of change of current in the coil and on the geometric and core properties of the coil. As with mutual induction, the latter are included in a quantity called the *self-inductance*, or simply *inductance*, of the coil, which is also measured in henrys and is symbolized by the letter L. (The self-inductance is so defined that when it is multiplied by the rate at which the current through the coil is changing, the result is the counter emf.)

10–13 THE INDUCTION COIL

It is often necessary to produce a high voltage when the only source available is a low-voltage dc source, such as a battery. One way of accomplishing this is with an induction coil, a device illustrated schematically in Fig. 10-40.

In the previous section we learned that a voltage will appear in one coil when a changing current is produced in another nearby coil, due to mutual induction. Moreover, the voltages in the two coils are related to the

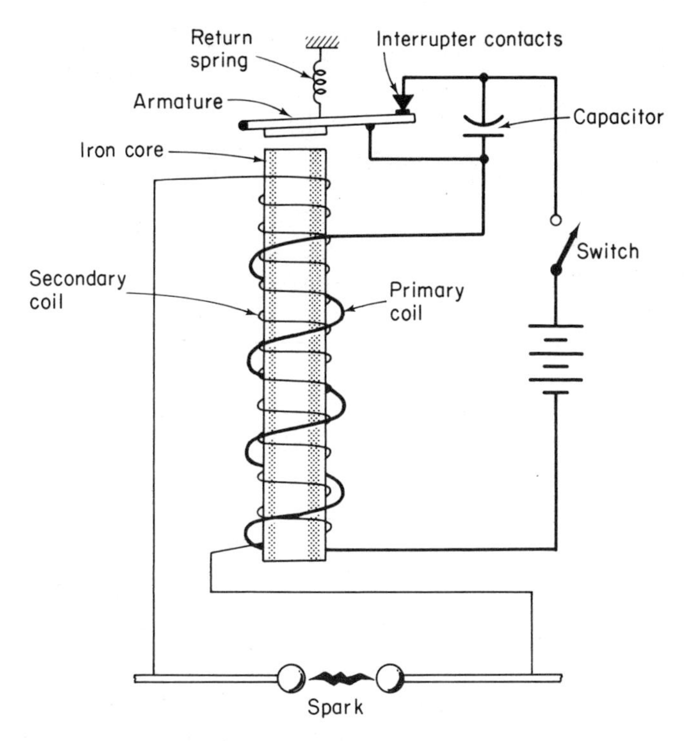

Fig. 10-40. Induction coil.

number of flux linkages involved. Thus, if both windings have the same number of turns, the applied and induced voltages will be equal; if the secondary has twice as many turns as the primary, the secondary voltage will be twice as great as that in the primary and, so on. In the induction coil of Fig. 10-40, the primary winding comprises comparatively few turns of fairly heavy wire,, while the secondary winding consists of many thousands of turns of relatively fine wire. Both coils are wound on the same soft iron core, either a bundle of wires or thin plates. The primary is connected through an interrupter, nothing more than a buzzer circuit, which ensures a constantly changing (make-and-break) current in the primary circuit. The capacitor connected across the interrupter contacts prevents burnout of those contacts by arcing. The changing flux due to the primary current cuts across the many turns of the secondary winding, inducing a very high voltage—high enough to produce a spark across a gap of one inch or more.

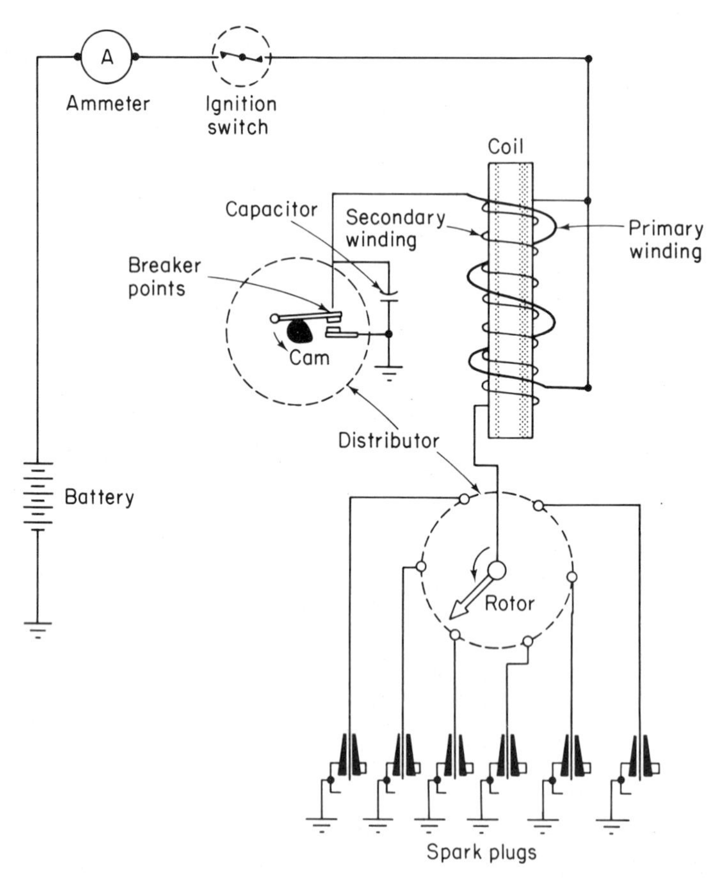

Fig. 10-41. Automobile ignition system (typical, six-cylinder engine).

At one time induction coils were in widespread use in radio transmitters, and, in fact, the induction coil was instrumental in the invention of radio, as we shall see. Today, however, induction coils as such are primarily laboratory curiosities. They do have some commercial use, the most significant of which is in the automobile's ignition system.

10-14 FEEDING YOUR SPARK PLUGS

The heart of the ignition system in your automobile is an induction coil circuit. Here, however, the "interrupter" function is performed mechanically rather than magnetically, the points being opened and closed by the action of a cam driven from the engine's crankshaft. The output of the secondary is fed to the proper spark plugs in the correct sequence by a rotary switch, also driven from the crankshaft. Both the interrupter points and the rotary switch are contained in a single housing called the distributor. The circuit for a typical 6-cylinder engine is shown in Fig. 10-41. The circuits for 4- and 8-cylinder engines are similar. It is essential that the spark be delivered to each cylinder at just the right instant, and various automatic mechanisms are now coupled into the distributor to advance and retard the spark when required.

10-15 THE GENERATOR

In discussing the electric motor, you may recall, we observed a counter emf induced by its rotation. Could we not use such a rotational induction to produce a *source* of emf? As a matter of fact, the major source of all electric power is a rotational electromagnetic device called a *generator* or sometimes a *dynamo*.

In principle, a generator is nothing more than a motor used in a "backward" manner: instead of putting electricity in and obtaining mechanical rotation, the shaft of a generator is mechanically rotated by some other device and electric energy is taken out.

Michael Faraday must again be credited with the invention. Having discovered that electricity could be produced from magnetism (induction), he wanted a continuous source instead of "spurts." On the basis of the results of an earlier experiment of Arago's, Faraday set up a copper disk that could be rotated between the poles of a permanent magnet. He borrowed the most powerful magnet then available, which belonged to the Royal Society. Faraday added iron armatures to concentrate the flux in a small gap, only slightly wider than the copper disk. Sliding contacts (which

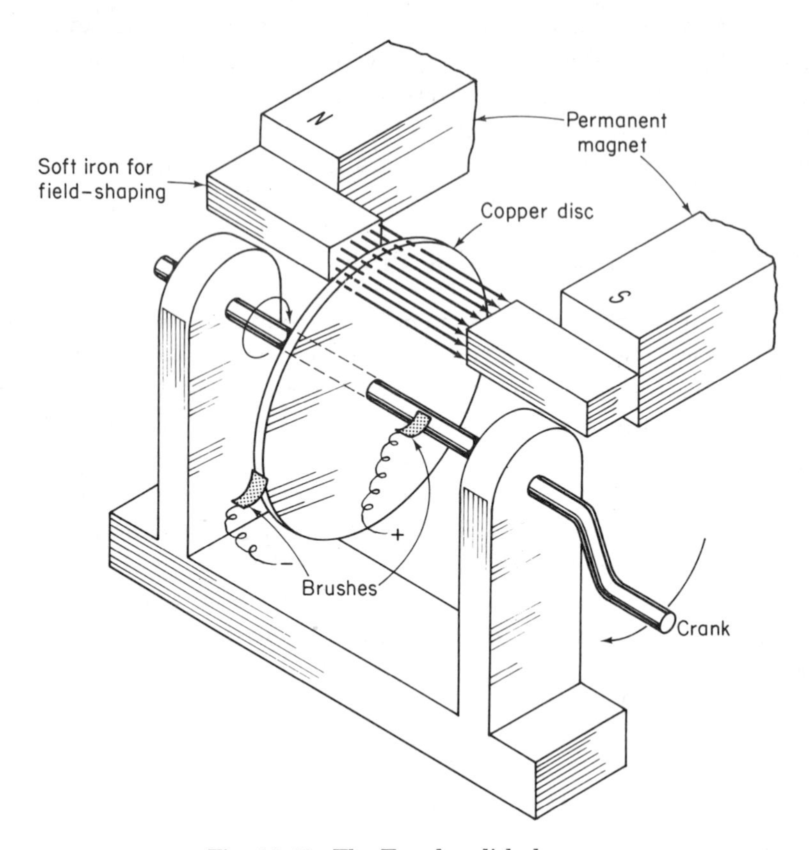

Fig. 10-42. The Faraday disk dynamo.

we would call brushes) pressed against the periphery of the disk and the shaft on which it turned. The essentials of the apparatus, which is called Faraday's disk dynamo, are depicted in Fig. 10-42. When the disk was rotated via the crank, a feeble dc voltage appeared between the contacts.

This, the invention of the generator by Faraday in 1831, was probably the greatest single electrical discovery in history. Man was set free from dependence on the costly, small-scale battery for his source of electric power. All he required was a means, such as a water wheel, gasoline engine, or steam turbine, to turn the crank of his generator, and he had a constant and essentially inexhaustible supply of electricity.

The disk dynamo was not a practical source, and it remained for others to refine it. In 1832 Hippolyte Pixii constructed a permanent magnet generator of the type shown in Fig. 10-43, which uses a wound armature instead of a disk. The commutator is necessary to maintain a unidirectional (dc) voltage output. Here, again, one half of the single-turn armature "coil" is darkened to make the action in illustrations 1 through 3 clear. In position

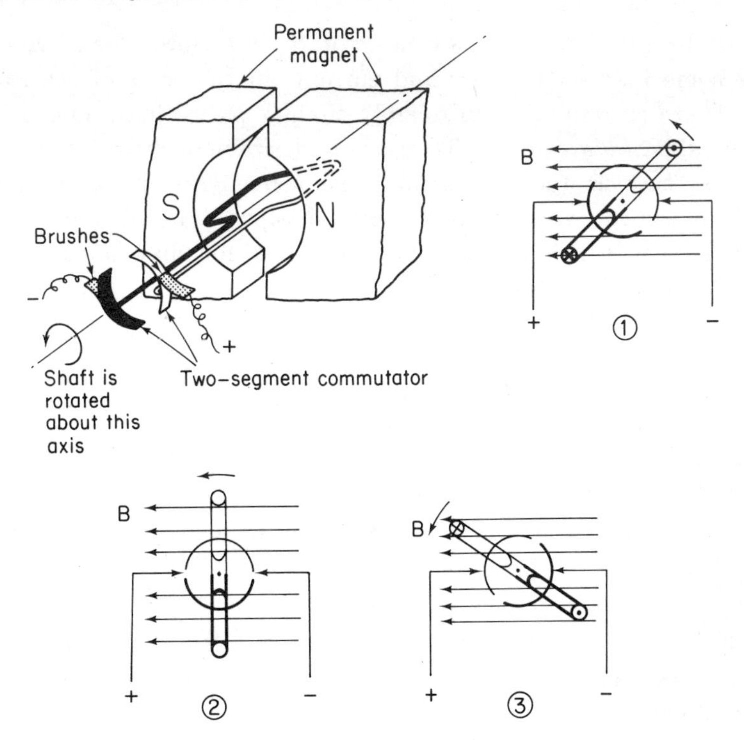

Fig. 10-43. Simple permanent-magnet dc generator ("magneto").

1, the darkened half of the coil is moving down through the field; therefore electron flow is into the page and the darkened commutator segment is positively charged. As the coil turns through position 2, the output circuit is momentarily broken. At position 3, the darkened half of the coil is now moving upward through the field, resulting in electron flow out of the page and a negative charge on the darkened commutator segment. Note that the charge on the brushes is always of the same polarity (except for the momentary break between commutator segments), for this is a dc generator. Compare Fig. 10-43 with Fig. 10-31.

Permanent magnet dc generators are often called *magnetos* and are limited to small sizes. They find wide application in small gasoline engines (lawn mowers, marine outboards, minibikes, etc.) and field telephones. Larger dc generators (discussed later), have electromagnetic stators rather than permanent magnets.

10-16 WHAT ARE EDDY CURRENTS?

Whenever an emf is induced in a conductor that forms a closed path, a current is, of course, also produced. An iron core used in an electromagnetic device is also (unfortunately) a conductor. So in a motor, for instance, there will be some induced voltages and currents in the cores of the rotor and stator. These currents flow in roughly circular paths, like eddies in a pool, and are called *eddy currents*. They are undesirable for two reason: (1) they are losses; that is, they represent consumed energy for which no useful work is obtained; and (2) the energy dissipated in eddy current losses appears as heat, which, if not properly designed for, may damage or at least shorten the life of the device.

REVIEW QUESTIONS

1. Among the list of accessories for a certain camera is a "solenoid shutter release." How do you suppose this operates?
2. From your knowledge of the relay and solenoid, describe the construction of an electromagnetic circuit breaker.
3. I would like to install some device that, at the touch of a single switch, would turn three circuits "on" and four circuits "off." What would you suggest?
4. What elements are necessary for a simple electric telegraph? What was the crucial invention?
5. What makes the d'Arsonval movement superior to other types of meter movements?
6. You have a 0 to 50-ma milliammeter whose internal resistance is one ohm. You would like to use this meter as the basis for a 0 to 10-amp ammeter. What other major component(s) do you need?
7. When measuring resistance, what are the advantages and disadvantages of an ohmmeter? What alternative do you have?
8. Why should the ohms-per-volt rating of a "universal" voltmeter be as high as possible?
9. Why do the different ranges of a VOM usually differ from each other by factors of 10?
10. Name and describe the function of the major parts of a dc motor.
11. Would you use a series or a shunt motor to drive a phonograph turntable? Why?
12. What is the difference between mutual induction and mutual inductance?
13. Explain the behavior of an induction coil in terms of Faraday's and Lenz's laws.
14. How would Fig. 10-41 have to be changed for a 4-cylinder engine? For an 8-cylinder engine?

15. Use Lenz's law to verify the polarity of the generator voltage in Fig. 10-42.
16. What are the major power losses in a generator?
17. How does an increase in the electric power drawn from a generator reflect itself in the mechanism driving the generator?

11

THE CHANGING CURRENTS

11-1 THE EXCEPTIONAL CHILD

Of all the offspring of the marriage between electricity and magnetism, one outshines and surpasses all the rest. That one is the *alternating current.* Without the means to produce and control alternating currents, electronic technology as we know it today could not exist.

For a good many years, the only available form of electric energy was the direct current. Static machines, batteries, photocells, thermocouples, and electric eels all produce dc voltages, you will recall. Yet, although there are still many household and industrial uses for dc, well over 90 percent of all the generating stations in the United States produce only ac. *The major advantage of ac is that its voltage can be easily increased or reduced to suit the application.*

There are still applications in which dc has the advantage: charging batteries; electroplating and other electrolytic processes; and in controlled speed heavy-motor applications like electric trains and trolleys, elevators, and rolling mills. Even in these cases, however, the original sources of power today are ac sources, with some intermediate device used to convert the ac to dc (This process is called *rectification*).

In this chapter we study the fundamentals of alternating voltages and alternating current circuits.

11-2 WHAT IS AN ALTERNATING VOLTAGE?

When we use the words *alternates* or *alternating* in connection with a voltage (or a current), we generally mean the same thing as when we say that a child alternates between being good and being naughty, or someone's mood alternates between being happy and being sad. The idea we are trying to convey is that the thing which is alternating reverses its ways regularly. With regard to voltage and current, it is the *polarity* of the voltage, and the *direction* of the corresponding current, which reverses in a regular pattern.

The difference between ac and dc can be illustrated by the following idealized experiment. Suppose that we have two zero-center dc voltmeters—that is, dc voltmeters which can read either positive or negative voltages—and suppose further that these voltmeters will respond to very rapid changes in the voltages they are measuring (ordinary d'Arsonval movements cannot do this). Now let one voltmeter be connected to a dc source,

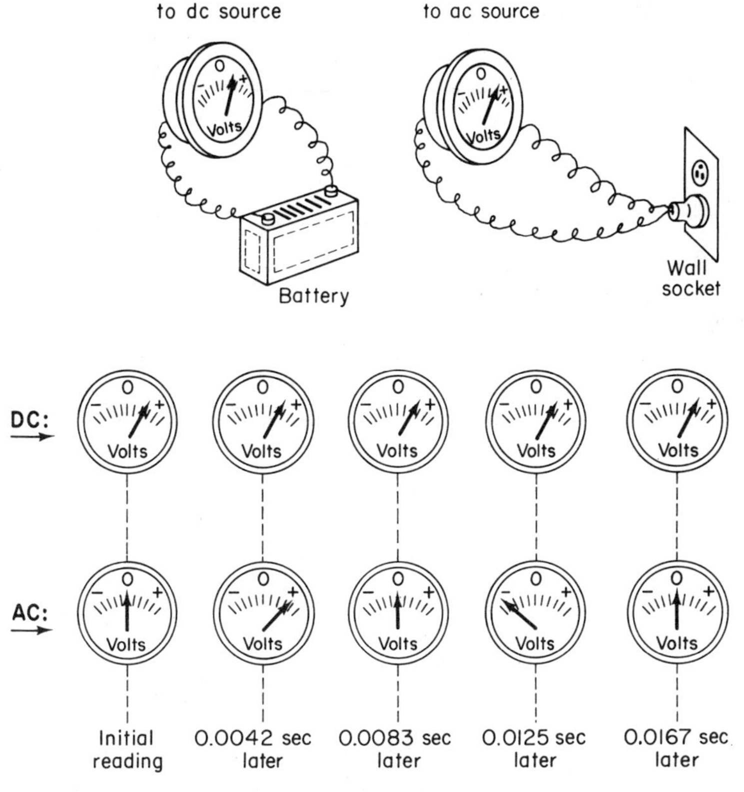

Fig. 11-1. Comparison of direct and alternating voltages.

such as a battery, and the other voltmeter be connected to an ac source, such as the ordinary wall socket, as shown in the upper part of Fig. 11-1. If we take a series of snapshot photographs of the voltmeters, we obtain a sequence something like that shown in the lower part of Fig. 11-1. The dc voltage we see is unchanging, as shown by the fixed position of the pointer in the upper group of readings. The ac voltage meanwhile changes from zero to some maximum positive value, back down through zero to some maximum negative value, back again through zero to the maximum positive value, etc., as in the lower group of readings, and repeats this pattern over and over again.

Now, any voltage or current that changes with time from positive to negative and back again, *in a regular pattern*, can legally be called an alternating voltage or current. If we draw a graph of the instantaneous value of the voltage versus time the graph will show a "wiggle" back and forth across the horizontal axis in some pattern. Figure 11-2 shows alternating voltages of

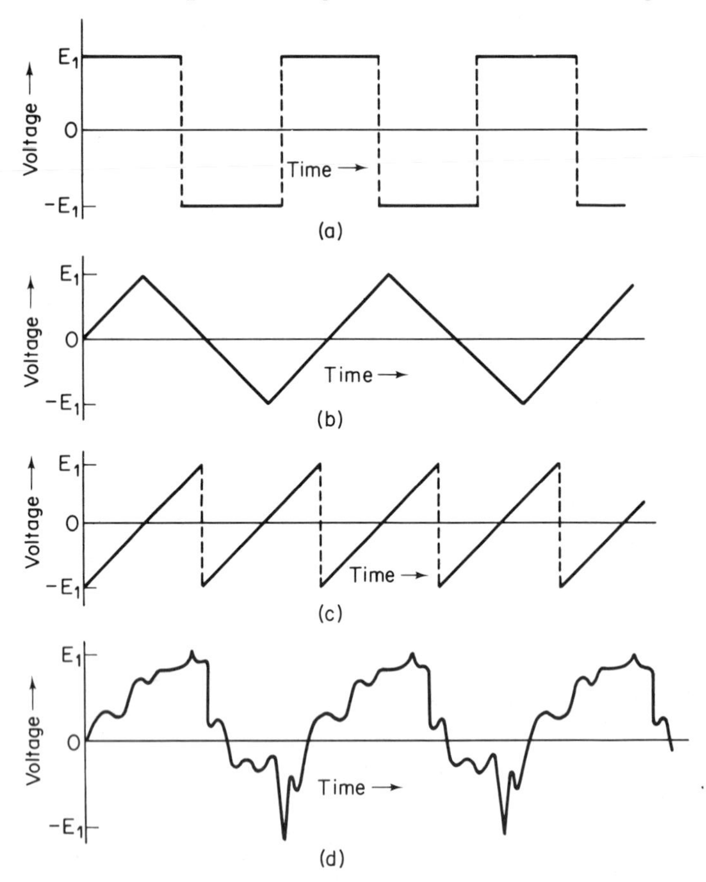

Fig. 11-2. Several alternating voltage waveforms. (a) Square. (b) Triangular. (c) Ramp. (d) Irregular.

several differant *waveforms*. The one represented in Fig. 11-2*a* is called a *square* waveform and consists of a voltage that is constant at some positive value for a short period of time, then suddenly changes to an equal constant negative voltage for the same period of time, repeating this pattern over and over.

The second waveform (Fig. 11-2*b*) is termed *triangular*. It represents a voltage that increases at a steady rate to a maximum positive value, abruptly decreases at an equally steady rate to a maximum negative value, abruptly increases again, and so on.

The pattern presented in Fig. 11-2*c* is called a *ramp* waveform. Here the voltage starts at some maximum negative value and increases steadily to some maximum positive value, much as in the triangular waveform, but then jumps suddenly back to its maximum negative value, only to begin its steady rise again.

The lowermost waveform (Fig. 11-2*d*) has no particular name; it is one of an endless number of possibilities that may be called "irregular." Notice, however, that its irregular form does repeat time after time, as any persistent alternating voltage must.

From the foregoing discussion, it should be clear that a voltage or current can follow any one of an infinite number of time-varying patterns and still be called alternating. Yet the term ac as ordinarily used, in household current for instance, is much more restrictive in meaning. It refers to only one type of waveform, and that type is the subject of most of this chapter.

11–3 WHAT'S YOUR ANGLE?

Before we can properly discuss ac voltages and how they are produced, it is necessary to know something about angles and how they are measured. (Those readers who are already familiar with angles and angle measurement are invited to skip the rest of this section.) We will merely require a passing acquaintance and will not belabor the subject.

An angle is formed whenever two straight lines or edges intersect each other, the lines being called the *sides* of the angle. As one side is held fixed and the other is rotated away from it, using the intersection as the pivot point of rotation, the angle is said to increase.

The most common way of measuring angles is in *degrees of arc*, called simply *degrees* when there is no chance of confusion with degrees of temperature. It has been agreed that the angle formed by rotating one side through a complete circle shall contain of 360 *degrees* of arc. Therefore an angle formed by rotation through half a circle contains half of 360, or 180 degrees. Similarly, a quarter circle contains 90 degrees, and so forth. To save writing the word degrees each time, we use the symbol °, just as for degrees of temperature. Angles of several different sizes are illustrated in Fig. 11-3. Notice particularly the angles 0° and 360° as well as the angles 45° and 405°. Without the different line contrast and the dashed curves to show how they are formed, we could not distinguish an angle of 0° from one of 360°, nor one of 45° from one of 405°. This merely illustrates the fact that angles repeat their positions in space every time they are increased by 360°; hence the properties of a 45° angle are identical to those of a 405° angle (as well as angles of 765°, 1125°, etc.).

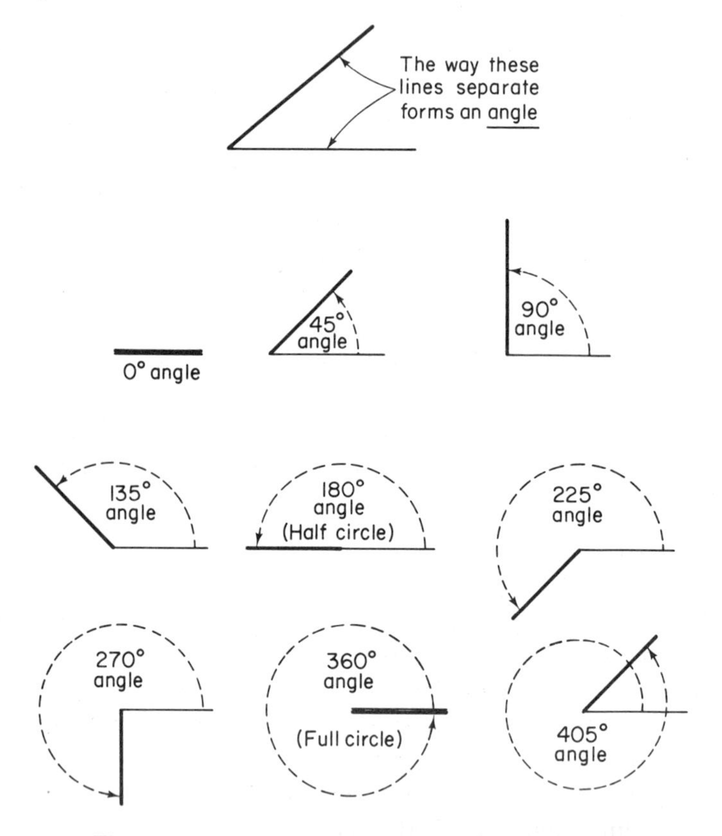

Fig. 11-3. Angles and their measurement in degrees.

11-4 A GOOD SINE

We are now in a position to discuss the common alternating voltage and how it is generated. Hereafter we use the universal terminology *ac voltage* (and *ac current*) instead of writing out alternating voltage (or alternating current) in full.

Consider the simple single-turn generator shown at the top of Fig. 11-4. This generator is just like the simple dc generator in Fig. 10-43 with one important difference: instead of being connected to the segments of a commutator, the ends of our one-turn rotor "coil" are each connected to a continuous ring. These rings are called *slip rings* and electrical energy is extracted from them via the brushes. Note that each brush is thus always connected with the same end of the rotor coil. We are now going to rotate the coil in the direction indicated and observe the reading of the voltmeter at several rotor positions. Again, for ease in following, we have darkened one side of the coil and its corresponding slip ring.

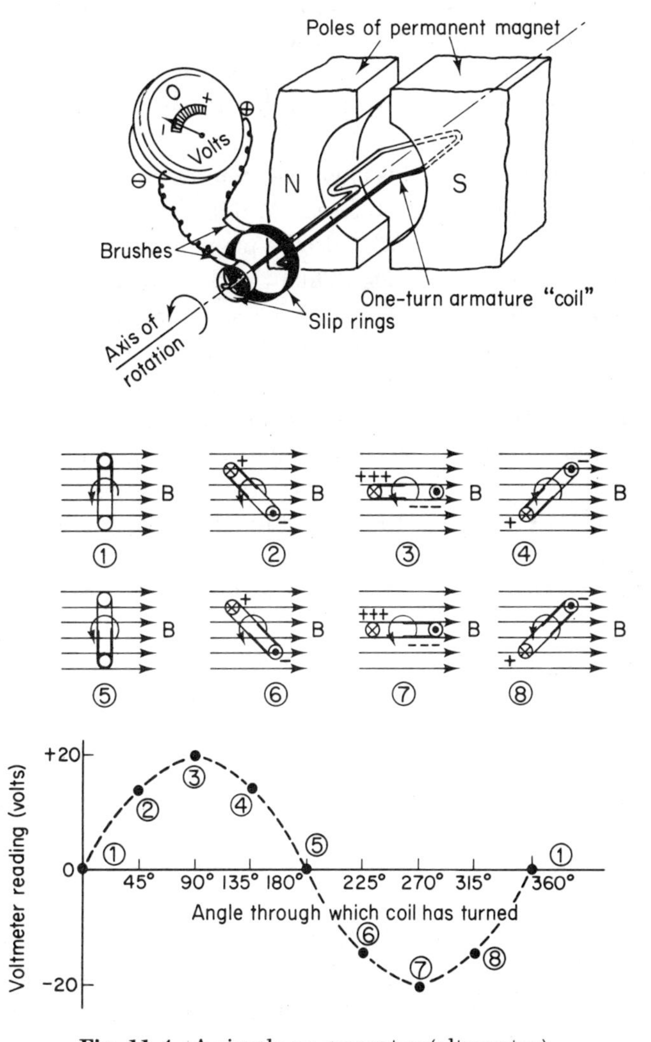

Fig. 11-4. A simple ac generator (alternator).

The group of illustrations in the center of Fig. 11-4 show, in a schematic manner, a cross section of the armature coil in several positions, relative to the fixed magnetic field, **B**, provided by the permanent magnet. The ⊙ and ⊗ symbolism shows the direction in which electrons tend to move, and the curved arrow ↶ indicates the rotation of the armature.

The graph at the bottom of Fig. 11-4 is a plot of the voltmeter reading against the angle through which the rotor coil has turned, for each position shown in the center group.

If we assume that the rotor starts in the position shown as 1, it has not yet turned at all — the angle is 0°. In this position, the legs of the coil are moving parallel to, rather than cutting, the magnetic field. We already know that an emf is induced only when a conductor cuts across lines of force; therefore in this position there is no induced voltage and the voltmeter reads zero. Thus, on the graph, we put a dot at zero volts, corresponding to the angle of 0°.

As the rotor turns (in this example, counterclockwise), the legs of the coil begin to cut across lines of force. Application of our right-hand rule shows us that electrons are pushed into the plane of the figure in the darkened leg and out of it in the other leg. Thus the darkened slip ring, from which electrons are trying to run away, acquires a positive charge, while the other ring acquires a negative charge. By the time the rotor has turned through 45°, as in position 2, the voltmeter shows a considerable positive voltage, let us say 14 volts. So we next put a dot on our graph at 45° and 14 volts.

When the coil has reached position 3, its legs are cutting straight across the magnetic field. Therefore it is cutting the greatest number of lines per second and is producing the maximum positive voltage at the rings, which is about 20 volts (because of the value we assumed at the 45° position). Corresponding to this condition, we put a dot on the graph at 90° and 20 volts.

At the 135° point, in position 4, the coil is oriented with respect to the field, and hence cutting lines at the same rate, as in position 2. Therefore the voltage has dropped back to 14 volts at this point and we place a dot on the graph at 135° and 14 volts, corresponding to position 4.

In position 5, after 180° (half a circle) of rotation, the coil is again momentarily moving parallel to the field, cutting no lines at all and having no induced voltage. We note the 180°, zero-volts condition by a dot on the graph.

In position 6, having rotated through 225°, the coil is in the same relationship to the field as in position 2, except that now the undarkened segment is moving down on the left and is positively charged, while the darkened segment is moving upward on the right and is negatively charged. In other words, the coil in position 6 has an induced voltage of the same magnitude as in position 2, but of opposite polarity. Thus, on the graph for position 6, we plot −14 volts at 225°.

Similarly, positions 7 and 8 at 270° and 315° correspond to positions 3 and 4, respectively, but with the opposite polarity. We therefore plot points on the graph at 270°, −20 volts, and at 315°, −14 volts, to represent conditions 7 and 8.

Finally, after turning through a complete circle of 360°, the coil is back in its original position 1 and the voltage is, of course, again zero. We put a dot on the graph at 360° and zero volts.

From this point on, the pattern repeats over and over and we have nothing to gain from plotting additional points. Let us look instead at the graph we have obtained. If we connect the plotted points together with a

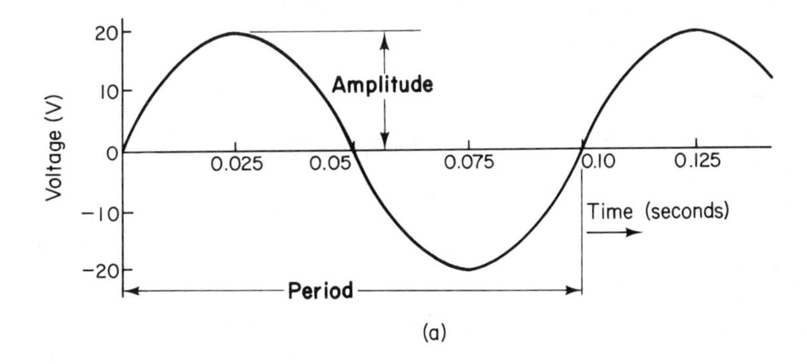

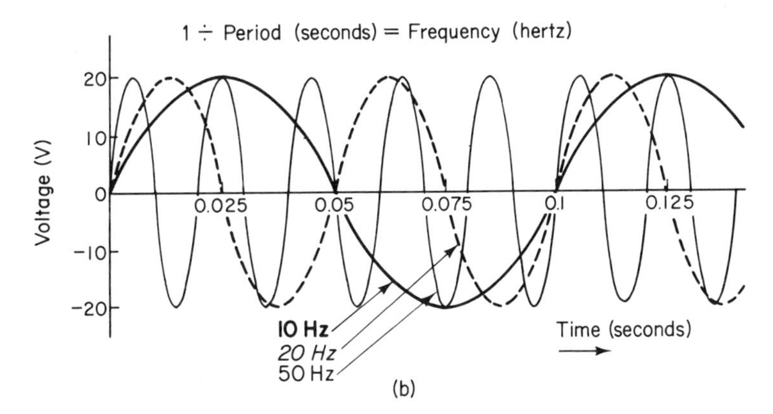

Fig. 11-5. (a) ac waveform. (b) ac voltages of the same amplitude (20 volts) but different frequencies (10, 20, and 50 Hz).

smooth curve, shown as a dotted line in Fig. 11-4, we obtain one of the most famous and important graphs in all of mathematics, the *sine* curve. This S-type curve is also called a *sinusoid.*

Since the coil in Fig. 11-4 is rotating at a constant speed, the total angle through which it has turned at any instant is in direct proportion to the elapsed time. Instead of plotting the voltage against the angle, we could just as well have plotted the voltage against *time*. Suppose, for instance, that the coil was rotating at a speed of 600 rpm, that is, at 10 rps. That means the coil would pass through an angle of 360°, or one revolution in one-tenth (0.1) of a second; through 180°, a half revolution, in one-twentieth (0.05) of a second, and so on. When we replot the voltage against time, we obtain Fig. 11-5*a*, which shows that the voltage does indeed vary *sinusoidally* with time.

It is this waveform, specifically, that is meant when we talk about an ac voltage (or current), as opposed to those shown in Fig. 11-2. The voltage appearing at the wall sockets in your is home sinusoidal.

11-5 TALKING ABOUT A SINUS CONDITION

There is some terminology with which you will have to become familiar in order to discuss these sinusoidal ac voltages and currents. In general, a sinusoidal waveform has two distinguishing characteristics. The first is the maximum value it attains, in either the positive or negative direction, and this characteristic called the *amplitude*. The other distinguishing characteristic of the sinusoidal waveform is the time required for the pattern to be completed once. One complete variation of amplitude, from zero to the maximum positive value, back through zero to the maximum negative value, and back to zero again, is called *one cycle* of the sine. The time required to complete one cycle is called the *period* of the sine. This nomencalture is illustrated in Fig. 11-5*a*. Here our example is a sinusoidal voltage with an amplitude of 20 volts and a period of 0.10 second. By giving the amplitude and period of a sine, we have said all that there is to say about it.

Normally, however, the period is *not* specified but rather a related quantity that is, in many ways, more convenient. That quantity is the *rate* at which the pattern repeats, and is called the *frequency* of the sinusoid. Numerically it is given as the number of cycles that occur in one second. Since the period represents the time required for one cycle, the frequency is related very simply to the period.

$$\textbf{Frequency} = \mathbf{1} \div \textbf{(period)}$$

The unit of frequency is the *hertz* (abbreviated Hz) and is named in honor of the "father of radio," Heinrich Hertz (and has nothing to do with auto rentals). One hertz represents the completion of one cycle in one second. Thus our example sinusoid of Fig. 11-5*a* with a period of one-tenth of a second to complete one cycle will complete ten such cycles in one second. Its frequency is therefore 10 Hz. High frequencies are expressed in kilohertz (kHz), megahertz (MHz), and gigahertz (GHz), representing, respectively, thousands, millions, and billions of hertz. A one-kilohertz voltage, for instance, completes one thousand cycles in a second, while a 15-GHz signal undergoes 15 billion cycles in one second.

In older literature, written prior to the time that the term hertz was adopted, you will see frequencies expressed in *cycles per second*, abbreviated cps or sometimes (~). The symbol ~ was frequently stamped on appliances. For high frequencies, however, the words "per second" were almost always omitted. One read about frequencies in kilocycles (kc) and megacycles (Mc), instead of the more correct kilocycles per second (kcps) and megacycles per second (Mcps). It was, in fact, to avoid this confusion that the term hertz was adopted. Thus our household electric frequency is often called 60-cycle, but should properly be called 60 hertz (or, if you must, 60 cycles per second).

Figure 11-5*b* shows three different ac voltages on the same graph. They are all of the same amplitude, 20 volts, but of three different frequencies: 10 Hz, 20 Hz, and 50 Hz.

11–6 ac GOES THROUGH A PHASE

Whenever we make a comparison between two ac voltages, or an ac voltage and a current, *of the same frequency*, a third factor becomes important in addition to amplitude and frequency. That factor is the way in which the two sinusoids are time-sequenced with respect to each other, and is called the *phase* relationship. If both start from zero at the same instant and pass through their respective cycles simultaneously, they are said to be *in phase* with each other, as in Fig. 11-6*a*. If they start from zero at different instants, and hence pass through corrresponding parts of their cycles at different times, they are said to be *out of phase*, as in Fig. 11-6*b* through *e*. The degree to which the two sinusoids are out of phase could be stated as a time difference. It is more common, however, to express this phase difference as an equivalent angle called the phase angle. (Remember that we have already seen, in Section 11-4, how an ac voltage varies sinusoidally with both time and angle.) The standard procedure is arbitrarily to select one of the two sinusoids as the refer-

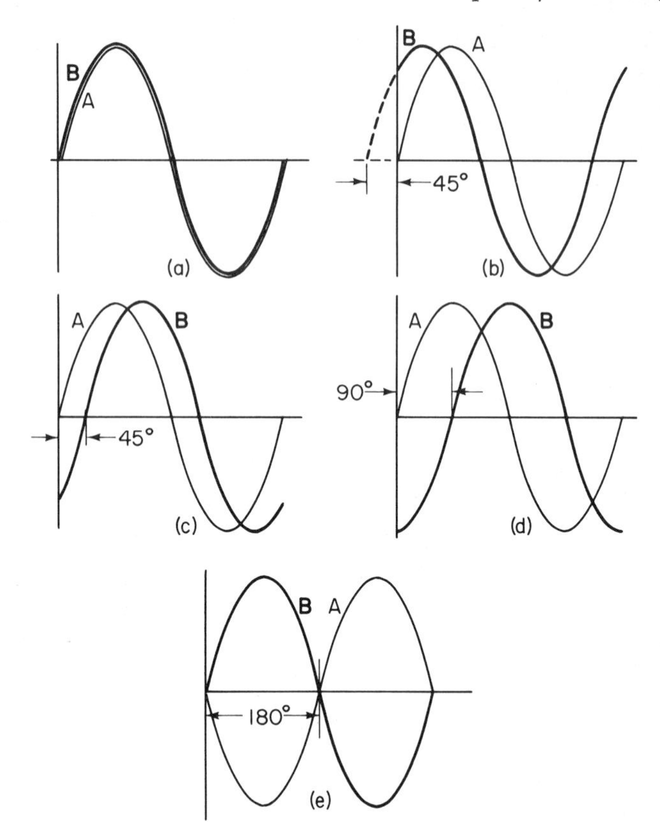

Fig. 11-6. Phase relationships between two ac voltages of the same frequency. (a) A lags B by 45°. (c) A leads B by 45°. (d) A leads B (or B lags A) by 90°. (e) A and B are 180° out of phase.

ence and measure the other's phase angle with respect to it. In Fig. 11-6*b* through *d*, we have decided to take sinusoid A as the reference.

In Fig. 11-6*b*, sinusoid B has already climbed to the amplitude corresponding to a 45° angle when reference sinusoid A is just starting from zero. We therefore say that *B leads A*, or *A lags B, by* 45°.

In Fig. 11-6*c*, the situation is reversed from 11-6*b*. Here A has reached its 45° value when B is just starting from zero, so we say thay *B lags A*, or *A leads B*, by 45°.

In Fig. 11-6*d*, the situation is like that in Fig. 11-6*c*, except that the phase angle is now increased to 90°. Here we say that A leads B, or B lags A, by 90°.

Figure 11-6*e* represents a phase angle of 180°, which is the greatest distinguishable phase angle. (Why ?) Here we say merely that the two sinusoids are 180° out of phase and it is immaterial which one is considered the reference. In this condition, if their amplitudes are the same, the two sinusoids have equal and opposite effects at every instant.

The amplitudes of all the sinusoids in Fig. 11-6 were shown equal. This was done for convenience, and need not be the case in an actual situation. We can talk about the phase relationship of any two sinusoids having the *same frequency*. Phase relations play an important part in understanding the behavior of ac circuits.

11–7 ON THE AVERAGE

We hear a good deal about averages these days. Economists tell us about families of average income. Meteorologists talk about the average rainfall in an area. Batters have batting averages and pitchers have earned run averages. Most of all, we are deluged with reports about what the average man has, thinks, and does.

All these uses of the word have certain things in common. First, *average* means some single specification that more or less typically describes a large group of similar, but somewhat different, things or occurrences. Second, if it is a number, that number may not actually occur in any of the individual instances that the average is supposed to represent. For instance, a family can only have a whole number of children (1 or 2 or 3 or 4, etc.), but the statistician may tell us that the "average" number of children per family is 2.3 in a certain area.

In dealing with ac voltages and currents, it is also convenient to talk about averages. We already know that the ac voltage (or current) has a different value every instant, changing sinusoidally with time. We would like to assign a single number that represents some kind of "average" value for the whole cycle.

We could, of course, specify the amplitude of the sinusoid. Sometimes ac voltages and currents are indeed specified in this way, and the amplitude is called the *peak* value. In electronics, it is frequently important to know the peak value of an ac voltage or current. We know, however, that the sinusoid is at its peak value for only an instant, so the peak value is not representative of most of the cycle—we could never call it an "average."

We might think of measuring the instantaneous value at several instants in a cycle, adding up all of these instantaneous values and dividing the total by the number of measurements (just the way you'd figure a batting average). Indeed, if we do this for a large number of points in the cycle, and treat just the actual magnitudes of the instantaneous values (overlooking the fact that some are positive and some negative), we obtain the *true mathematical average*. This turns out to be *63.7 percent of the peak value.*

There is, however, an even more useful kind of average: the *effective* value. It represents the level of a dc voltage or current that will produce the same amount of work as the ac voltage or current in question. This effective value is *70.7 percent of the peak value*, and is more usually called the *root-mean-square* value, or rms value for short. The name derives from the mathematical way it is calculated, which need not concern us here. Thus an ac voltage whose rms value is 110 volts is capable of producing the same amount of brightness from a given light bulb as 110 volts dc. **Whenever the value given for an ac voltage or current is not specifically identified as peak or average, it is understood to be the rms value.**

Figure 11-7 shows one cycle of our typical household "120 v 60 ~"

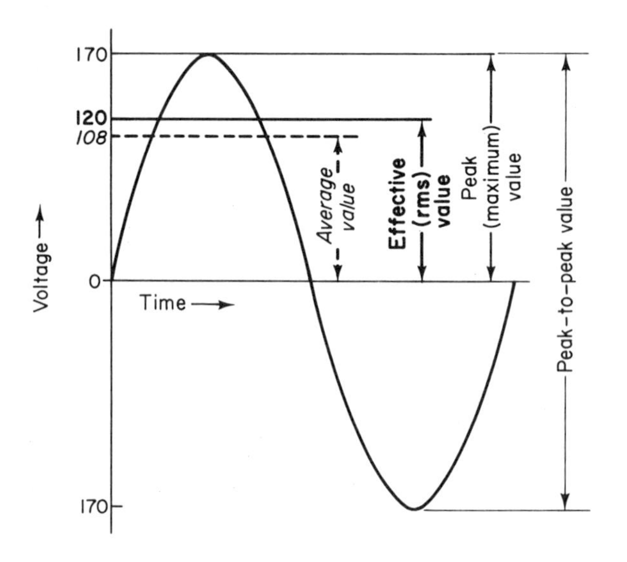

Peak, or maximum, value = amplitude
Effective, or rms, value = 70.7% of amplitude
Average, or mean, value = 63.7% of amplitude
Peak-to-peak value = 2 x amplitude

Fig. 11-7. Peak, effective, and average values of a sinusoidal waveform.

ac voltage. Note that the 120 volt figure is the rms value, the peak value being about 170 volts. The true average voltage is 63.7 percent of that peak value or 108 volts.

While we are on the subject, there is one other value that is sometimes given for a sinusoid, the *double amplitude* or *peak-to-peak* value. This represents the total voltage or current swing, from maximum positive to maximum negative, and is just two times the amplitude. Hence our 120-volt rms sinusoid of Fig. 11-7, with its amplitude of 170 volts, can be said to have a peak-to-peak value of 340 volts (2×170).

11-8 ALTERNATING CURRENT

When we complete the circuit to a dc source of emf, we know that a direct current flows. Similarly, when we complete a circuit to an ac source of emf, an ac current flows. The word "flow" is less appropriate here than for dc, since the electrons never get very far from where they started. Recall from Section 6-3 that in a typical household circuit, the electrons in the wire are moving at a speed of about 10^{-4} feet per second. In an alternating current, electrons move in one direction for half a cycle; therefore, in a 60-Hz ac current, the maximum travel time in one direction is about $\frac{1}{120}$th of a second. Consequently, the maximum distance covered by an electron is only about a millionth of a foot, but it goes back and forth over this distance 120 times in a second.

The energy of this sinusoidal current can be made to do work just like a steady (dc) current. The nature of ac however, requires that we take two other factors besides resistance into consideration when attempting to understand an ac circuit. Those two factors are *inductance* and *capacitance.* Their effects are described in the following sections.

11-9 WHEN ac MEETS RESISTANCE

The simplest type of ac circuit is one that contains only the source and a pure resistance. Such a circuit is shown schematically in Fig. 11-8*a*, and might represent, say, an electric iron plugged into the wall (120 volts, 60 Hz) socket. Ohm's law still applies and enables us to calculate the current at any instant if we know the voltage applied to the resistance at that instant. Because of this proportional relationship, the current follows the voltage: when the voltage is positive maximum, the current is maximum in one direction at that instant; when the voltage is zero, there is no current at that moment; and when the voltage is negative maximum, the current is maximum in the other direction. In other words, in a resistive circuit the current is also sinusoidal and is *in phase* with the applied voltage. This situation is illustrated in the graph of Fig. 11-8*b*, where the voltage and current are plotted to the same time scale so we can see that they are in phase (the vertical scale is not important here).

In actuality, there is no such thing as a purely resistive circuit. At low frequencies (say less than a million hertz), however, capacitive and inductive effects are inconsequentially small in many circuits, which may be treated as purely resistive for practical purposes.

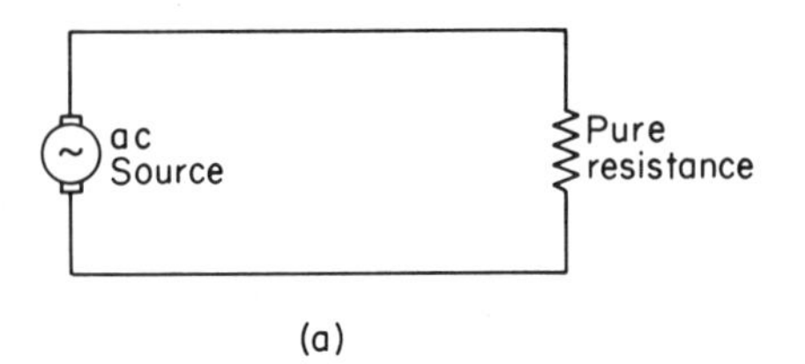

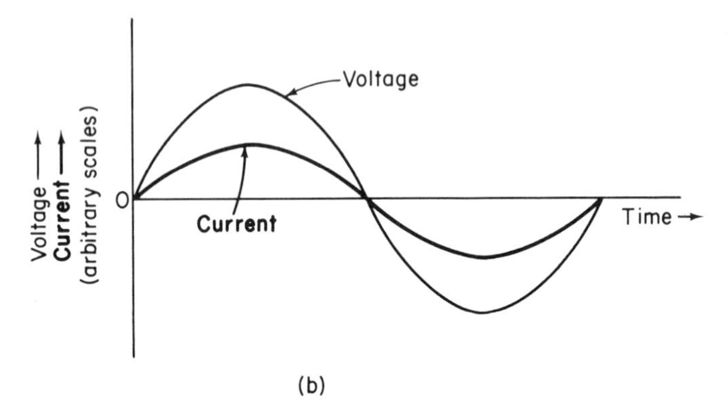

Fig. 11-8. (a) Pure resistive circuit. (b) Voltage and current phase relationship in resistive circuit.

11-10 INDUCTANCE IN AN ac CIRCUIT

In Section 10-12, we learned that the phenomenon of inductance manifests itself in a dc circuit only during a brief period of time immediately following a sudden change in the circuit (such as opening or closing a switch). The reason, you will recall, is that the induced counter emf depends on the rate at which the current through the inductance is changing. Once the current reaches a steady value (is no longer changing), there is no counter emf. A glance at Fig. 10-38 should refresh your memory.

When an inductance experiences an alternating current, however, there is always a counter emf present, because the ac current is continually changing. This counter emf opposes the applied emf and limits the current, in much the same way as resistance does. Since it has this tendency to oppose an alternating current, the inductance is said to produce a quantity of *inductive reactance* in the circuit. Inductive reactance is measured in ohms, just like resistance, and is always represented by the symbol X_L.

It is important to note that inductive reactance results from the interaction between the inductor and the circuit. Its value depends on both the inductance of the coil and the frequency of the applied ac voltage. The resistance of a resistor, on the other hand (at least at the frequencies we consider here), is a property of the resistor alone. The variation of inductive reactance with the applied frequency is shown in Fig. 11-9 for coils with

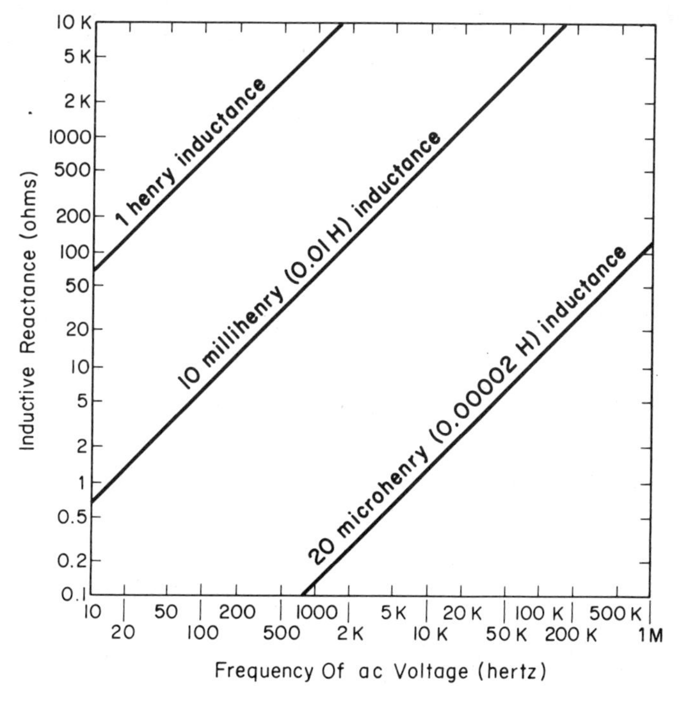

Fig. 11-9. The reactance of an inductor increases with its inductance and with frequency.

three different inductances. Note the uneven (logarithmic) scales on this graph. We see that inductive reactance increases with both increasing inductance and increasing frequency.

It is also important to look at the phase relationship between the current and voltage in an inductive circuit. Let us pretend, for a moment, that we have a "pure" inductance, that is, a coil with no resistance at all. We connect this coil to an ac source, as shown at the left of Fig. 11-10*a*, and observe what happens to the voltage across the coil as it goes through its sinusoidal cycle. When the current is changing most rapidly, the induced voltage is maximum. When the current is changing least, as at its maximum and minimum points, the induced voltage is zero as indicated in the graph at the right of Fig. 11-10*a*. As a result, the current waveform is out of phase with the voltage drop across the coil and is lagging 90° behind it. Furthermore, since there is no other circuit element, the voltage drop across the coil and the applied emf are one and the same voltage. We therefore say

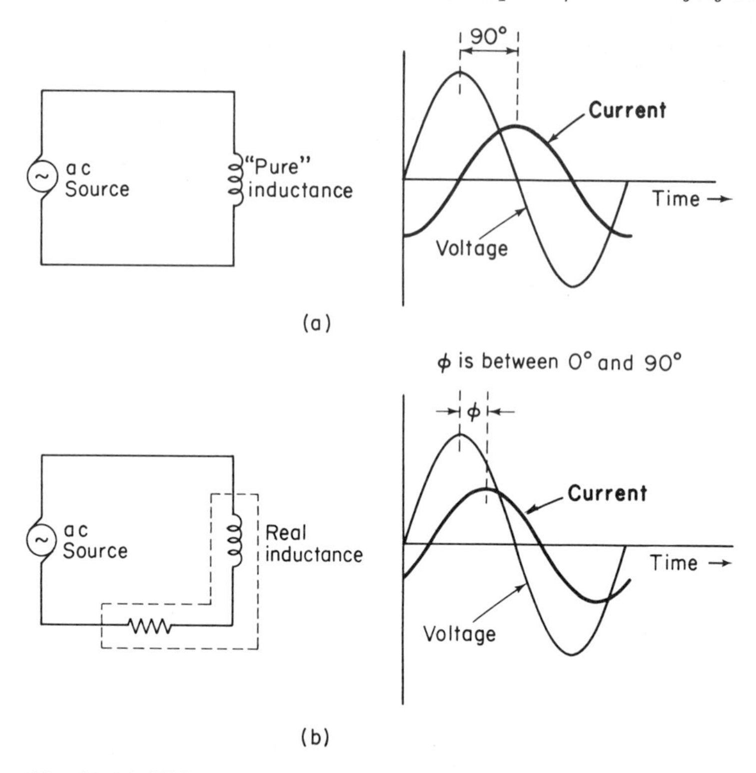

Fig. 11-10. Voltage–current phase relationship in an inductive circuit. (a) "Pure" inductive. (b) Inductive and resistive.

that, in a purely inductive ac circuit, the *current lags the voltage by* 90°.

Under ordinary circumstances, there is no such thing as a *pure* inductance, for the wire from which the coil is wound always has some resistance. Yet, in many such circuits, the resistance is so much smaller than the inductive reactance that the circuit may be treated as purely inductive to a fair degree of approximation.

Now let us see what happens if we do put a resistance in series with the inductance. This resistance may be just the resistance of the coil itself, as in the left part of Fig. 11-10*b*, or it may represent an actual resistor connected to the coil, or even the total resistance of both. The voltage drop across the inductance is still, of course, 90° out of phase with the current, but the voltage drop across the resistive part is, as we saw earlier, in phase with the current. The net result is that the *current still lags* the applied voltage by some angle, which we call ϕ (phi) and which is greater than 0° but less than 90°, as represented by the graph in Fig. 11-10*b*. The size of ϕ depends on the relative sizes of the resistance and the inductive reactance. If the reactance is much larger (say 10 times greater), ϕ is almost 90°; if the reactance is much smaller (say, $\frac{1}{10}$ th or less of the resistance), ϕ is nearly 0°; if they are equal, ϕ is 45°.

So far we have said nothing about the combined effect of resistance and inductive reactance in opposing current. You might think that they are merely additive, since they are both measured in ohms; for instance, a 100-ohm resistance and a 100-ohm inductive reactance should oppose the current like a single 200-ohm resistance. This is *not correct*, because of the phase difference between a resistive and an inductive voltage drop. To be sure, their combined opposition is greater than that of either one alone, but it is *not* as great as their simple total. The "lumped" effect of a resistance and a reactance, in opposing an ac current, is called *impedance*. We will say more about impedance presently.

11–11 THE CAPACITIVE CIRCUIT

Since leaving Section 3-13, we have badly neglected the capacitor, not because of any personal prejudice against capacitors but because capacitors do not play a major role in dc circuits. A quick look back at Fig. 3-14 will refresh your memory as to why this is so. The capacitor, that great-grandchild of the Leyden jar, is in essence two sheets of metal separated by a sheet of insulator. To a dc voltage, this looks like an open circuit, just like a switch in the "off" position. Consequently, a persistent dc current cannot be maintained through a capacitor; hence the minor part they play in dc circuits.

When capacitors are used in dc circuits, it is because of the way they respond to voltage transients (i.e., sudden, momentary changes in applied emf). Remember that the capacitor has the "capacity" (hence its name) to store a charge. Consider the circuit shown in Fig. 11-11, with the switch in position 1 and the capacitor uncharged. There is no current in the circuit, as indicated by the graph below the circuit diagram. Now let the switch be moved to position 2, connecting the capacitor and resistor to the battery through the galvanometer. Immediately the positive potential makes itself felt at one plate of the capacitor, and the negative potential at the other. As a result, electrons begin to rush from the battery's negative terminal into the one plate of the capacitor, and simultaneously, electrons are drawn off the capacitor's other plate into the positive terminal of the battery. This initial "rush" is a current, which is limited only by the circuit resistance, Thus the current read by the galvanometer, at the instant the switch is moved to position 2, is what you would expect from Ohm's law if the capacitor were simply a wire. (We have assigned values to the battery voltage and resistance

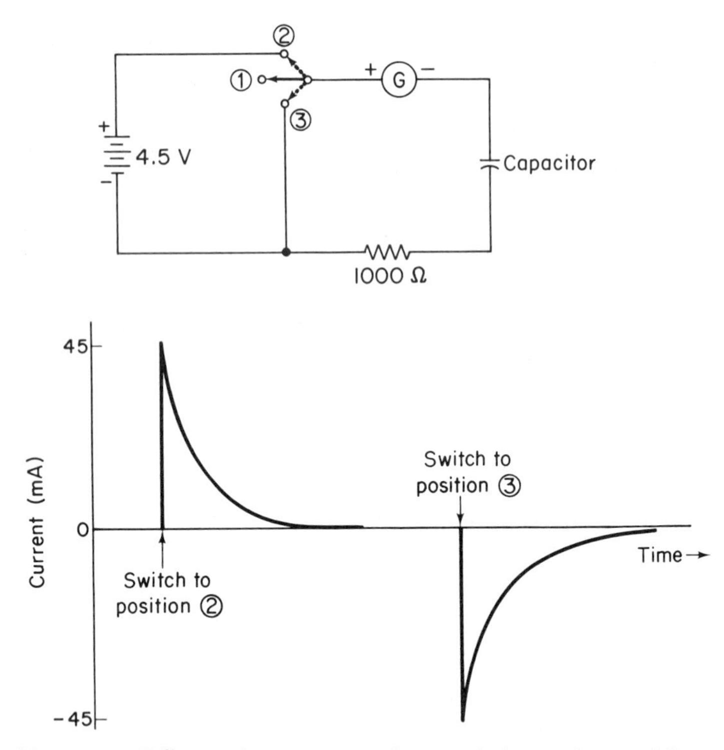

Fig. 11-11. DC transient currents in a resistive and capacitive circuit.

in Fig. 11-11, so that we could show a numerical value for the initial current in the graph.) Immediately, however, the electrons find themselves stopped by the insulating dielectric—they cannot get through to replace those drawn off the positive plate. Consequently, a charge rapidly builds up on each plate, which shows up as a counter emf, and it becomes progressively more difficult to "squeeze" a few more electrons into the strongly repelling negative plate or to pull a few more away from the strongly attracting positive plate. Thus the current quickly drops to zero, as you would expect in an open circuit. In this time, the capacitor has charged to the battery voltage.

Next we rotate the switch from position 2 to position 3, disconnecting from the battery and, instead, connecting the capacitor (through the galvanometer) directly across the resistor. In the first instant, the charged capacitor acts just as if it were the battery, since it does contain a charge and presents the same voltage to the resistor. A current again flows, this time in the reverse direction, and is again equal in magnitude to the Ohm's law value, as shown by the galvanometer. Unlike a battery, the capacitor cannot maintain its voltage as the charge bleeds off its plates; hence the current again drops quickly to zero, leaving the capacitor uncharged once more.

This transient response of the capacitor, while of limited application in dc circuitry, is just what the doctor ordered for ac circuits, where the voltage is continually changing. Indeed, capacitors are used as much as resistors in ac circuits. Since the applied voltage is continually changing, the capacitor always presents an opposing emf to an ac circuit, which limits the current like resistance and inductive reactance. We therefore say that the capacitor produces a quantity of *capacitive reactance* in the circuit, which, like resistance and inductive reactance is also measured in ohms. Like inductive reactance, capacitive reactance (symbol X_c) depends on both the nature of the capacitor (i.e., its capacitance) and the frequency of the ac voltage. But whereas inductive reactance increases with increasing inductance and increasing frequency, capacitive reactance *decreases* with increasing capacitance and increasing frequency. This fact is illustrated in Fig. 11-12,

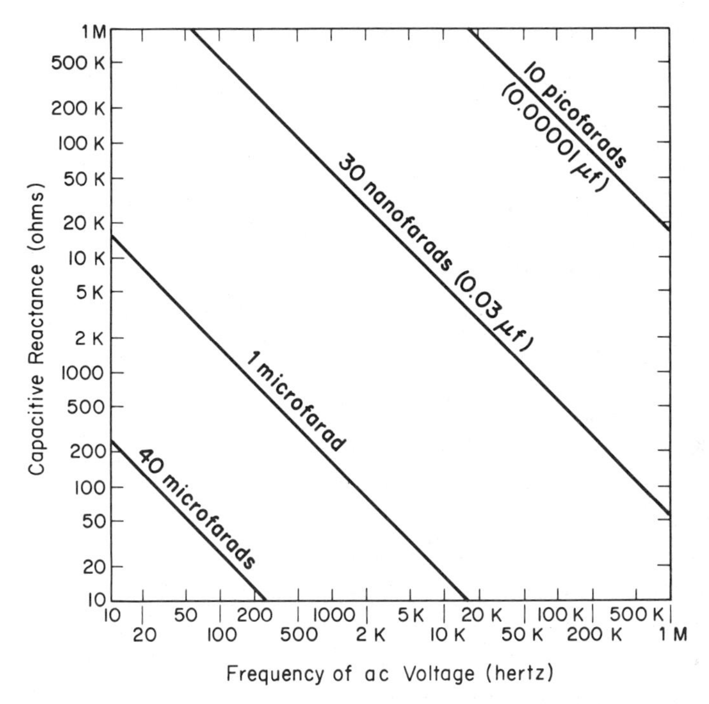

Fig. 11-12. The reactance of a capacitor decreases with its capacitance and with frequency.

which shows how the reactance of capacitors of several sizes changes with frequency. Compare this figure with Fig. 11-9.

The voltage-current phase relationship in a capacitive circuit is also significant. Figure 11-13*a* shows a pure capacitance corrected across an ac source, and we may follow the capacitor voltage and current through a cycle

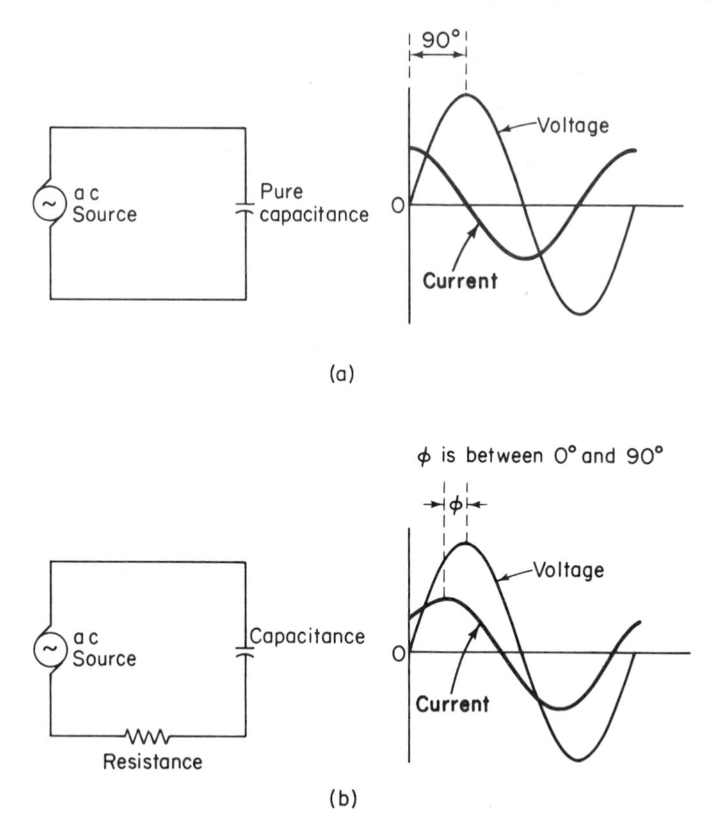

Fig. 11-13. Voltage–current phase relationship in a capacitive circuit. (a) Pure capacitive. (b) Capitive and resistive.

in the accompanying graph. When the voltage has the value zero, it is changing at its maximum rate, caused by the maximum flow of electrons into one plate, and out the other, of the capacitor—in other words, the maximum current. When the current is zero, the capacitor is charged to its maximum value, in one direction or the other; hence the voltage across the plates has either the positive or negative maximum value. The current is thus 90° out of phase with the voltage, as with an inductance, but in this case the current *leads* the voltage instead of lagging. Since the voltage across the capacitor and applied voltage must be one and the same we say that in a pure capacitive circuit *the current leads the voltage by* 90°. Capacitors, in general, can be treated as a "pure" capacitances, unless we have reason to believe that a significant number of electrons can leak through the dielectric.

If we connect a resistance in series with the capacitance, as in Fig. 11-13*b*, the situation is analogous to the resistive-inductive circuit of Fig. 11-10*b*. The capacitor's voltage drop is 90° out of phase with the circuit current, while the resistor's drop is in phase with it. The overall drop, which is the same as the applied voltage, is therefore out of phase with the current by some angle ϕ, which is between 0 and 90°. The value of ϕ depends on the relative sizes of the resistance and capacitive reactance, but the current leads the voltage in any case.

Again the combined opposing effect of resistance and capacitive reactance is a net "impedance," which is *not* simply the total number of ohms but something greater than either the resistance or reactance and less than the total. The reason here, too, is the difference in phase effect between a resistance and a capacitive reactance.

11-12 THE GENERAL ac CIRCUIT

In the preceding sections we learned that an ac circuit can contain any combination of three types of current-opposing quantities: resistance, inductive reactance, and capacitive reactance. Their combined effect is called the *impedance* of the circuit, usually symbolized by the letter Z. The way in which resistance and the reactances are combined to arrive at a net

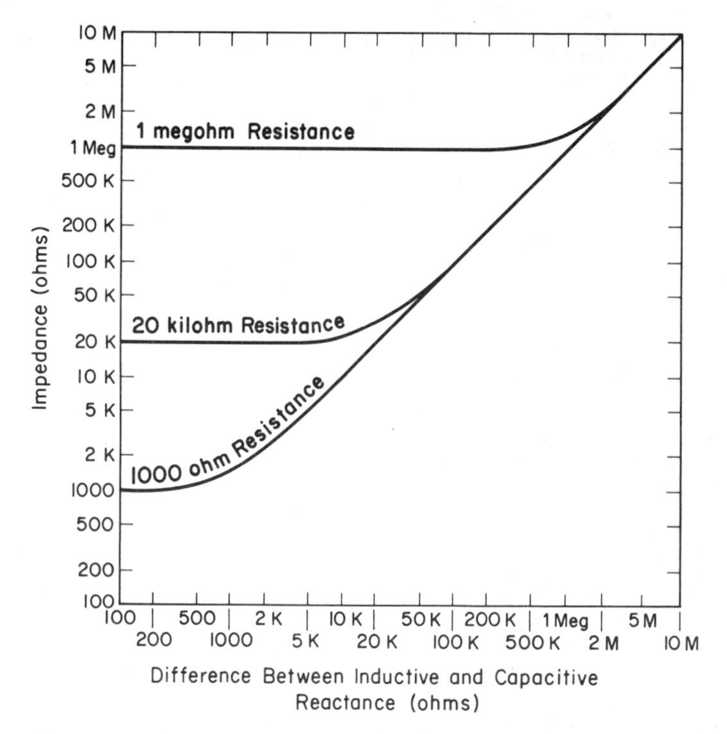

Fig. 11-14. Combining resistance and net reactance into impedance.

impedance results from the way they act in a circuit. The capacitive and inductive reactance effects both act 90° out of phase with the resistance effect, and 180° out of phase with each other. In other words, inductive and capacitive reactance directly oppose and tend to cancel each other. The *net reactance* of an ac circuit is thus the *difference* between its overall inductive reactance and its overall capacitive reactance, and will itself be effectively either inductive or capacitive, depending on which of its two constituents is the larger. This net reactance, which acts 90° out of phase with resistance, combines with resistance to make up the net impedance. The mathematical way of combining them is called root-sum-square, or rss for short, and need not concern us. Figure 11-14 illustrates the results of the computation for several resistances and a range of net reactances.

Ac circuits may, of course, be of the series, parallel, or more complex types, just like dc circuits. The ac circuits involve one additional consideration not present in their dc counterparts, the phase relationships.

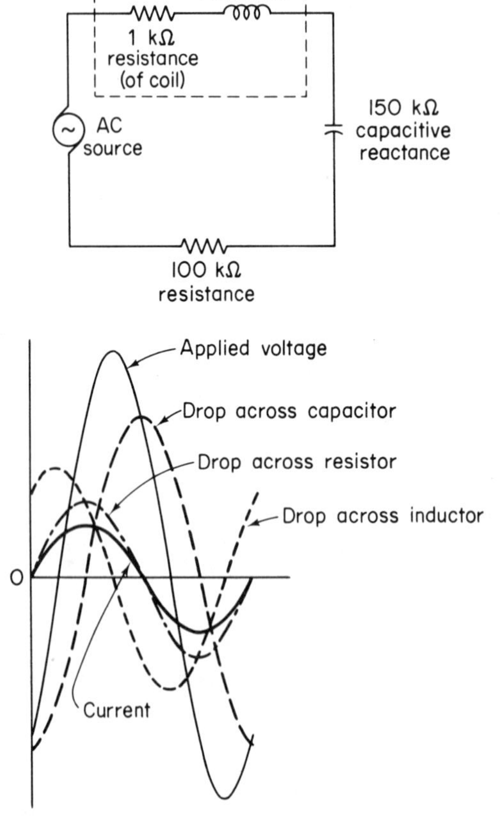

Fig. 11-15. Phase relationship in a series ac circuit.

Figure 11-15 shows a series ac circuit containing resistance and both types of reactance. For sake of a concrete example, we assume here that the capacitive reactance is greater. The phase relationships are shown in the graph. Since the current is everywhere the same in a series circuit, we use it as our timing reference. Looking then at the voltage drops, we see how the drop across the resistor is in phase with the current; the voltage across the capacitor lags behind it by 90°; the voltage across the inductor leads the current by something less than 90° (due to the coil resistance); and the applied voltage, or overall voltage drop, lags slightly because the net circuit reactance is capacitive.

A parallel circuit containing resistance, inductance, and capacitance is illustrated in Fig. 11-16. Here, for definiteness, the inductive reactance

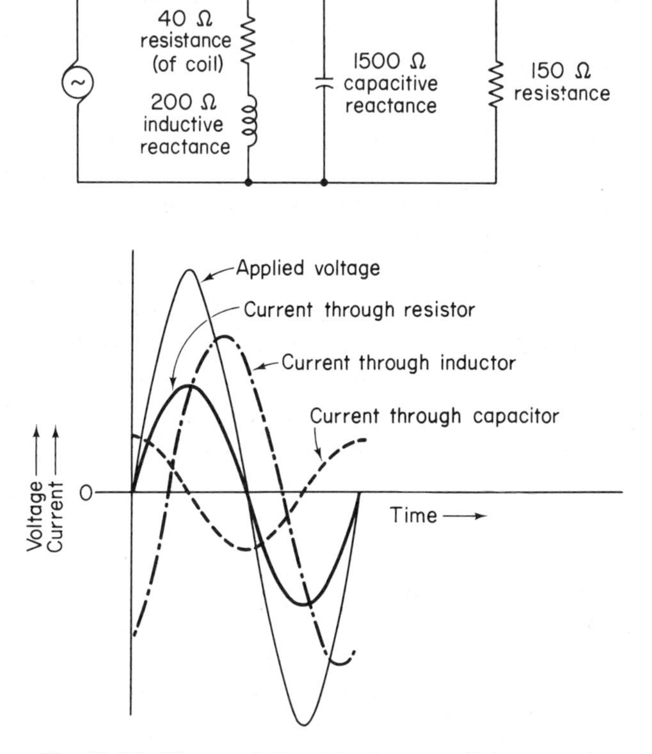

Fig. 11-16. Phase relationships in a parallel ac circuit.

was again chosen to be the smaller. In the parallel circuit, it is the voltage that is the same across each device and the currents through them are different. In the phase graph, we therefore take the voltage as our reference this time and look at the phase relationships among the currents. The resistor's current is, of course, in phase with the voltage; the capacitor's current leads the voltage by 90° while the inductor's resistance causes its current to lag by a little less than 90°. Since the inductance has the smaller reactance, it passes more current than the capacitor, hence the overall circuit behaves inductively. Thus, in a series circuit, the larger reactance determines whether the circuit acts inductively or capacitively, while in the parallel circuit it is the branch with the smaller reactance that governs.

11-13 RESONANCE

Almost everyone has heard about how a sturdy bridge can be knocked down by soldiers marching across it with a certain rhythm, or about how an operatic soprano can shatter a particular type of drinking glass by singing a certain note. Both examples demonstrate how seemingly small vibrations can be made to produce effects that are out of all proportion to their apparent strength. These are two instances of the phenomenon called *resonance*.

The mechanism of resonance is quite simple. Everything that can be made to vibrate has one or more "favorite" frequencies at which it "prefers" to vibrate because of its size, shape, mass, "springiness," etc. These preferred frequencies are called the *natural* frequencies of the vibrating members. Now, a body can be *forced* to vibrate at any desired frequency, given the right conditions and enough energy, but when it is driven at just the right frequency—its natural frequency—then resonance occurs. The motion of the driving mechanism is actually amplified by the vibrating body's "desire" to move at that frequency. If there is nothing to *damp* (i.e., hinder) the extent of the vibration, the amplitude tries to become infinitely large, and of course, something breaks.

By extension, we may think of a "vibration" of sorts occurring in the ac circuit, since the potential varies sinusoidally between some maximum positive and negative value. Whenever an ac circuit contains both inductance and capacitance (and they all contain some of both, although frequently one or both may be negligibly small), it turns out that the circuit does have a natural frequency. Recall that inductive reactance increases with frequency, while capacitive reactance decreases with frequency. Consequently, for any given inductance and capacitance, there is one certain frequency at which their reactances are just exactly equal. This frequency is called the *resonant frequency* for that particular inductance and capacitance. Figure 11-17 illustrates this point for the particular case of a 240-microhenry (μH) inductance and a 105-picofarad (pF) capacitance, whose resonant frequency is about one million hertz or one megahertz. Now let us see what happens if we use these elements in a circuit at their resonant frequency.

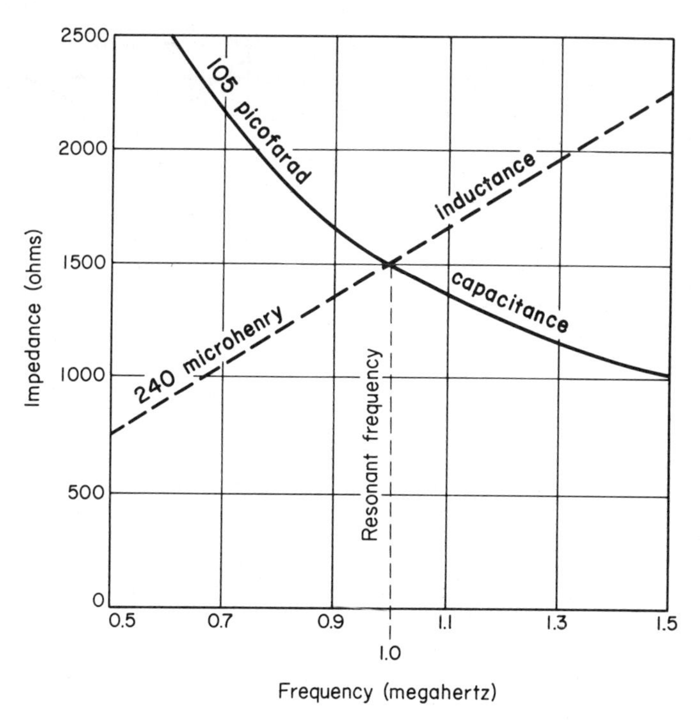

Fig. 11-17. The resonant frequency.

In the circuit of Fig. 11-18*a*, the inductance and capacitance have been connected to form a series circuit. The voltage drops across the inductance and the capacitance are, as we know, 180° out of phase with each other. At the resonant frequency, they are of exactly equal size as well; hence they exactly cancel each other out. The only remaining impedance is the resistance of the coil (in a "pure" inductance it would be zero). This situation is shown in the graph to the right of the circuit diagram, where, at resonance the impedance drops from thousands of ohms to a mere 10 ohms. The characteristic of a series-resonant circuit is, therefore that the *impedance is minimum* and the *current is maximum.*

The same circuit elements have been connected in parallel in Fig. 11-18*b*. In the parallel circuit, remember, it is the currents that are out of phase with each other. At resonance, the reactances are equal and so are the currents through them. If there were no resistance in the circuit, the inductor's current and the capacitor's current would be exactly equal and 180° out of phase, making the net overall current zero. This is equivalent to infinite impedance! The presence of some resistance results in a large but finite, effective impedance and a minimum but not zero, current. Therefore the characteristic of a parallel-resonant circuit is *maximum impedance* and *minimum current.*

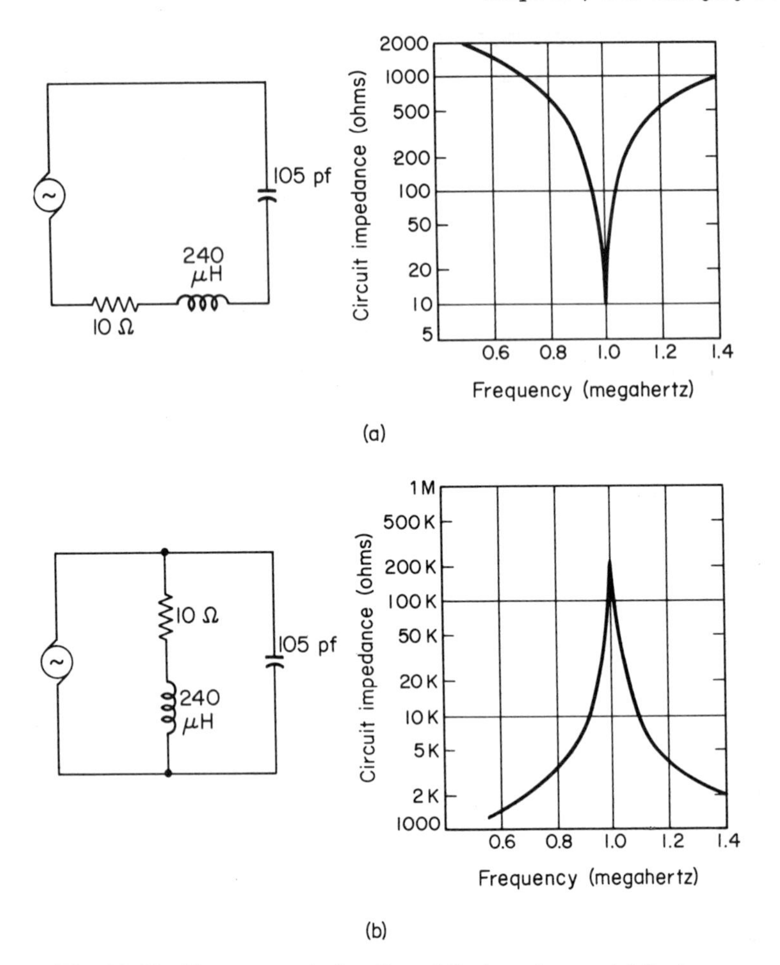

Fig. 11-18. The resonant circuit and its impedance. (a) Series resonant. (b) Parallel resonant.

In these resonant circuits, the resistance does the damping. A useful quantity associated with resonant circuits is the *quality factor*, Q. The symbol Q is defined as the reactance of the coil at resonance (which, of course, is the same as the capacitor's reactance) divided by the resistance of the circuit (which is usually just the resistance of the coil). This figure gives a quick measure of the effectiveness of the circuit. Q values of hundreds and thousands are easily attainable. In a series-resonant circuit, the voltage drops across the reactances are Q times the applied voltage (!); while in a parallel-resonant circuit, the impedance is Q times either reactance.

Resonant circuits are essential in radio broadcasting and reception, and anywhere else that it is desired to select or suppress one out of a group of frequencies contained in an ac source.

11-14 POWER

The question of power consumption arises for an ac circuit. Among other reasons, we must be able to determine how much to pay the electric company each month. We may still use the method discussed in Section 7-8 for dc to calculate the instantaneous power level, multiplying instantaneous voltage by instantaneous current (with due regard for positive and negative quantities). If this is done over one complete voltage cycle and the results plotted, then for various kinds of circuits we obtain graphs like those in Fig. 11-19. Note that these graphs, although cyclical, do not necessarily vary sinusoidally about the zero line and that they complete two "cycles" for every one voltage cycle. The shaded portions, which represent "negative" power, mean that the device is returning power to the circuit instead of drawing power from the circuit.

In a purely resistive circuit, the situation is similar to that in a dc circuit. Energy is dissipated in the resistance as heat, and the resistance doesn't care in which direction the current is moving at any given instant. As a result, the power cycle looks like the upper waveform in Fig. 11-19. This resistive power cycle is always positive, indicating that power is always being drawn from the circuit. The average power for the cycle is one-half of the maximum value.

In a pure inductive or capacitive reactance, the situation is quite different. During parts of the voltage cycle, when the magnetic field is building up around the inductor in either direction or when the capacitor is being charged in either direction, power is being drawn from the circuit. This energy is stored in the magnetic field of the inductor or the electrostatic field of the capacitor. During the remaining parts of the cycle, the magnetic field of the inductor collapses or the capacitor discharges, returning the stored power to the circuit. In these cases, we have a *reactive power* cycle, which is sinusoidal about the zero line, with twice the voltage cycle frequency. The average power consumed for the cycle is zero. The capacitive cycle is, of course, 180° out of phase with the inductive cycle, as shown in the central graphs of Fig. 11-19.

The power cycle in any real circuit lies somewhere between the extremes of pure resistive and pure reactive, with more power being drawn from the circuit than is returned to it. The lower graph of Fig. 11-19 is for a circuit that is predominantly resistive but slightly capacitive. A slightly inductive power cycle would be similar, except that the negative parts of the cycle would precede the positive parts.

The measure of the relative reactivity of an ac circuit is its *power factor*, abbreviated pf and defined as the net resistance divided by the total

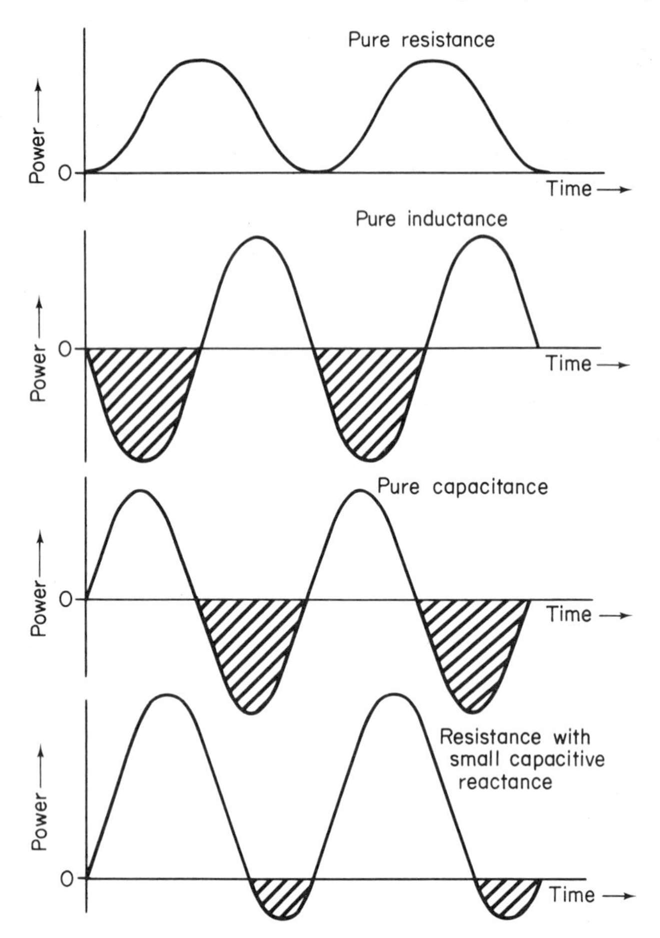

Fig. 11-19. Power waveform, for one voltage cycle, in various ac circuits.

impedance of the circuit in question. It is often given as a percentage. Thus, in a pure resistive circuit, the impedance consists of resistance only and the pf is unity (or 100 percent); while in a purely reactive circuit, the impedance comprises only reactance with no resistance and the pf is zero (or 0 percent). For an ac circuit, the true power consumption in watts (for which you receive an electric bill) is computed by multiplying the rms voltage by the rms current, which gives the *apparent power* in volt-amperes (va), and multiplying this in turn by the pf.

Reactive power, which, as we have seen, averages out to zero, costs nothing on the electric bill. It is, however, a nuisance in most cases, for it puts an instantaneous load on the equipment for part of cycle, just as if it were a resistive power demand. For this reason, large industrial consumers of power from ac sources will deliberately introduce the "opposite" kind of reactance into a circuit (e.g., put a capacitor into an inductive circuit) to reduce the level of reactive power and bring the power factor nearer to 100 percent. This is also the reason why ac generators are always rated in terms of the total apparent power, in kilovolt-amperes (kva), which they can produce. For loads, on the other hand, it is important to know the two individual components of the apparent power—the reactive power in kilovolt amperes-reactive (kvar) and the true power in kilowatts (kw).

REVIEW QUESTIONS

1. What is the major advantage of alternating current over direct current?
2. In alternating voltage and alternating current, what "alternates"?
3. What is the waveform called which is associated with ordinary ac voltages and currents?
4. I have an old automotive magneto (a permanent magnet dc generator). What internal change must I make to obtain ac from it?
5. What is the frequency of an ac voltage that has a period of $\frac{1}{50}$th of a second?
6. The radio announcer says that station KPOL broadcasts at a frequency of 1540 kilocycles. Strictly speaking, why is he incorrect? What should he say?
7. What is the greatest distinguishable angle by which two sinusoids of the same frequency can be out of phase? Why?
8. The amplitude of an ac voltage is 100 volts. What is its average value? Its rms value? Its peak value? Its peak-to-peak value?
9. What is most significant about a purely resistive ac circuit? A purely inductive one? A purely capacitive one?
10. In an ac circuit, what parameter is equivalent to total resistance in a dc circuit?
11. Why were we not too concerned with dc circuits containing a capacitor?
12. A circuit contains a net resistance of 1000 ohms and also a net reactance of 1000 ohms. Using Fig. 11-14, estimate the total impedance of the circuit. Why is it not 2000 ohms?
13. What is the major characteristic of resonant circuits? When are they useful?
14. A 50-volt series resonant circuit contains an inductor whose resistance is 10 ohms and whose inductive reactance is 10,000 ohms. What is the reactance of the capacitor in the circuit? What is the Q of the circuit? What is the voltage drop across the capacitor?

15. What makes power in an ac circuit different from power in a dc circuit?
16. Which quantity, used in computing the true power in an ac circuit, takes the difference discussed in problem 15 into account?
17. What is the true power consumption of a 120-volt circuit in which the total current is 5 amperes and the power factor is 60 percent?

12

PUTTING ac TO WORK

12-1 FROM THEORY TO HARDWARE

With ac theory in back of us, we now take a look at some of the ac counterparts of dc devices we have already met, as well as a device that has no dc equivalent, the transformer. A quick trip is taken through the world of power generation and distribution to show where the money for the monthly electric bill goes.

12-2 ac MOTORS

Alternating-current motors are used much more extensively than dc motors. One major reason for this fact is that only ac utility service is available in most areas. There are other reasons, however. Ac motors are more trouble-free, for the type most widely used requires no brushes. Elimination of commutators and brushes not only cuts down maintenance but also eliminates a possible explosion hazard in areas where the air carries combustible gases or dust, for the commutator and brush arrangement is prone to sparking.

In Section 10-11 we learned that the universal motor, a series-wound motor with brushes and commutator, could be used with ac as well as dc.

Indeed, it frequently is used when high starting torque is required, or when the user wants the ac/dc flexibility.

Most ac motors, however, are of a type called *induction motors*. They are, in general, simpler and hence cheaper to build than equivalent dc machines. They have no commutator, slip rings, or brushes and there is no electrical connection to the rotor. Only the stator winding is connected to the ac source, and, then, as their name implies, induction produces the currents in the rotor. A common and particularly simple form of rotor for this type of motor is the *squirrel-cage rotor*, shown in Fig. 12-1*a* and so named

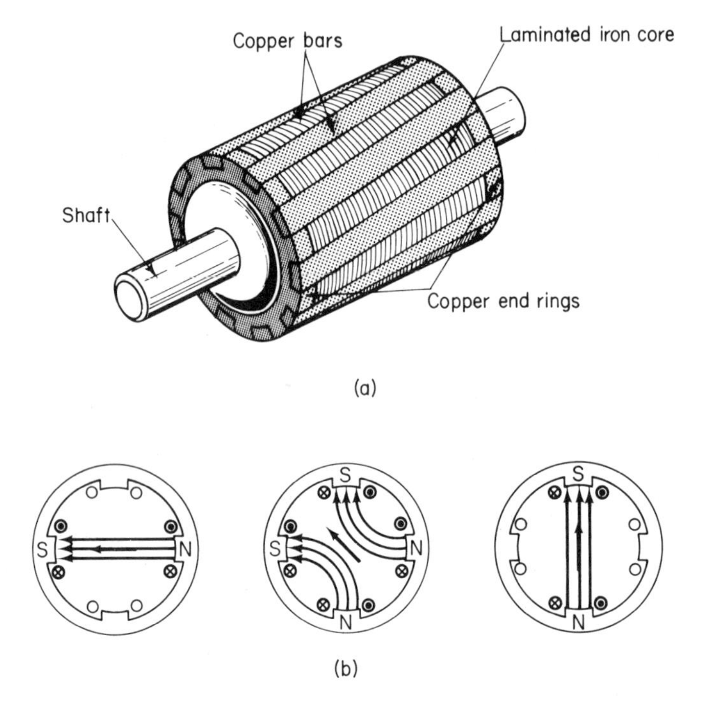

Fig. 12-1. Induction motor principles. (a) Squirrel cage rotor. (b) Rotating field for a four-pole stator.

because of its resemblance to a treadmill-type squirrel cage. The induction motor is based on a rotating magnetic field. This is achieved by using multiple stator field windings (poles), each pair of which is excited by an ac voltage of the same amplitude and frequency as, but phase-displaced from, the voltage supplying the neighboring pair. Figure 12-1*b* shows how the magnetic field rotates in a four-pole induction motor, where the voltages to the two pairs of poles are 90° out of phase with each other. When the rotor is placed in the stator's rotating field, the induced currents set up their own fields, which react with the stator's field and push the rotor around. The speed at which the stator's field rotates is called the *synchronous speed* of the motor and is determined by the number of poles and the voltage frequency. The actual speed at which the rotor turns *must always be less* than synchronous speed, in order for the rotor windings to cut lines of force, to induce rotor currents, to produce the reacting rotor field. The difference between the two is called the *slip* of the motor, usually given as a percentage of synchronous speed.

One other important class of ac motors, although not nearly as widely used as induction motors or universal motors, is the *synchronous* motor. This device has a number of stator poles, like the induction motor, but has a wound rotor that is supplied from a dc source. It requires some external means (e.g., a separate little induction rotor winding) to get it started. Once it gets up to speed, the synchronous motor runs like an ac generator in reverse. In effect, the poles of the rotor are attracted by and electromagnetically "locked" to the rotating magnetic field, and the motor turns at synchronous speed. If the motor has the same number of pairs of poles as the supply generator, it will turn at exactly the same speed; if twice as many pairs of poles, it will turn at half the generator's speed; if it has half the generator's number of pairs of poles, it will turn at twice the generator's speed. Synchronous motors are used where constant speed is essential, such as for timing devices and high-quality phonograph turntables and tape recorders. They only handle a limited range of loading torques; once forced out of synchronism, they stall and must be restarted.

12-3 ac GENERATORS

Alternating-current generators, more usually called *alternators*, are very similar in principle to dc generators. In fact, we have already seen (Section 11-4) how a dc generator becomes an alternator merely by replacing the former's commutator with a pair of slip rings. In either case, we must have (1) a constant magnetic field provided by either a permanent magnet (in very small generators), or more usually by a dc field winding; (2) an armature winding in which the dc or ac voltage is to be generated; and (3) rotational motion between the two, to obtain the line-cutting action necessary to induce the voltage.

In dc machines, it is essential that the armature windings be on the rotor, in order to make use of the commutator to change the ac voltage in the armature itself to a dc output voltage at the terminals. In ac machines, it doesn't matter which winding rotates, just so long as there is relative rotation between the armature and field windings. Consequently, alternators have been built both ways, and this distinction serves as a convenient way to classify them. Only small, low-voltage alternators are of the *field stator* type, with the armature winding on the rotor and power drawn off through slip rings. Most commercial alternators, particulary large ones, are of the *field rotor* type, with the armature winding on the stator. Here only the low-power field coil is rotated (with the dc excitation fed in through slip rings), while the high-voltage generated power is drawn off through rigid connections to the fixed armature windings. This process eliminates sparking and simplifies the problems of insulation.

Since a dc source is required for the field winding, a separate little dc generator, called the *exciter*, is usually mounted on and driven from the same shaft as the large commercial alternator. The mechanical power to turn the shaft is usually derived from a turbine, either the "water wheel" variety or a steam turbine.

12-4 THE TRANSFORMER

In the very beginning of the last chapter, we remarked that a notable feature of ac, and the one that makes it so much more advantageous than dc, is the ease with which its voltage can be increased or reduced to suit the application. We now discuss the family of devices that are used to transform the level of ac voltage and are therefore called *transformers*.

The transformer is simply a mutual induction device, comprising two coils so mounted that flux produced by one will pass through the other. Many, if not most, are wound on a soft iron core, especially in low-frequency (less than 20 kHz) applications. The principle of operation is similar to that of the induction coil described in Section 10-13, but since it is an ac device, no interrupter is needed. This is illustrated in the schematic diagram of Fig. 12-2*a*. The voltage appearing across the secondary winding is in the same ratio to the primary voltage as the ratio of the number of turns in their respective windings. In other words, if the secondary has two or three times as many turns as the primary, the secondary voltage will be twice or three times the primary voltage. Actually, this is strictly true only when no current flows in the secondary, and both primary and secondary voltage drop somewhat when there is a current in the secondary circuit, due to counter emfs, but it is a good approximation.

Under no-load conditions—that is, with no current in the secondary—the impedance due to the counter emf in the primary circuit is such that there is almost no current in the primary circuit. When a load is connected to the secondary, a current flows which reduces the counter emf in the pri-

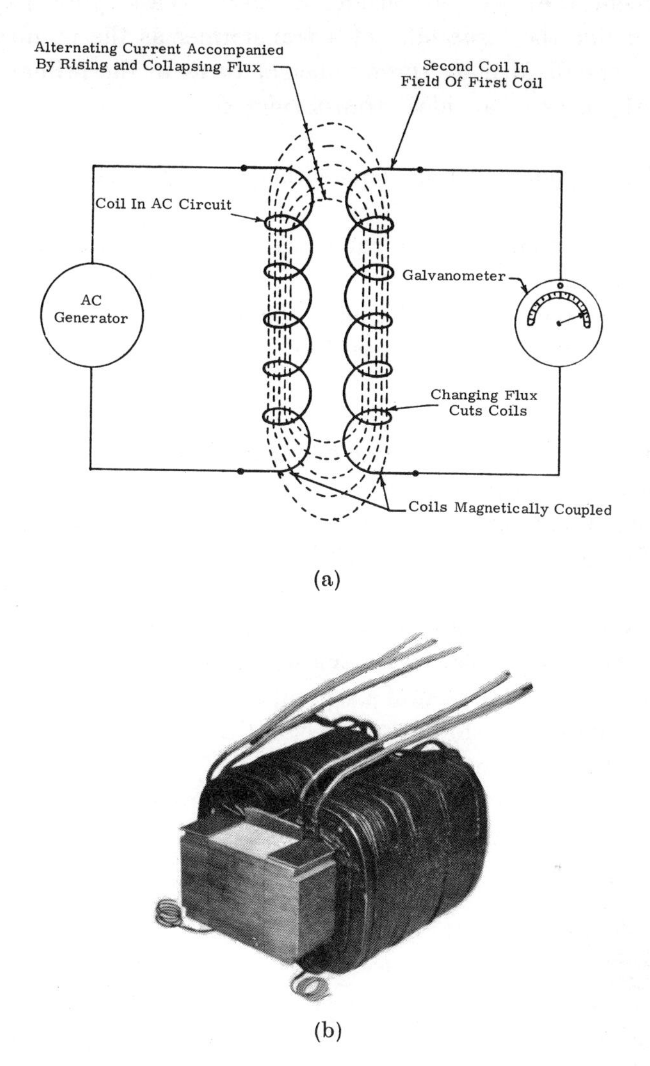

Fig. 12-2. (a) Transformer principle. (b) Typical core-type transformer.

mary circuit. The primary current then increases until the counter emf once again balances the applied voltage. In an ideal transformer, with no losses, the power going into the primary is always exactly equal to the power drawn out of the secondary. A real transformer, like the typical core-type transformer shown in Fig. 12-2*b*, always has some small losses and requires slightly more power into the primary than is gotten from the secondary. Yet, since well-designed commercial transformers operate at efficiencies of

95 to 99 percent, they may be considered "ideal" to a very good approximation. If we define the *turns ratio* of a transformer as the number of turns in the secondary divided by the number of turns in the primary, we may summarize the laws of an ideal transformer as

1. The secondary voltage equals the primary voltage *times* the turns ratio
2. The secondary current equals the primary current *divided by* the turns ratio.
3. The secondary power equals the primary power.
4. The secondary voltage is 180° out of phase with the primary voltage.

Transformers used to increase voltage are called *step-up* transformers, and those used to reduce voltage are called *step-down* transformers. Step-up and step-down transformers are used extensively in power transmission. Step-down transformers are frequently used for safety reasons in conjunction with low-voltage devices, such as doorbells and toy trains. High-current devices, such as arc welders, use heavy step-down transformers. Other uses of transformers are: positively isolating a device from the supply source; matching the impedance of a load to that of a source; and providing adjustable voltage level control. Many transformers, especially those used in power supplies, have more than one secondary winding.

Several different types of transformers are illustrated in Fig. 12-3 and 12-4. Larger transformers tend to be more efficient than smaller ones.

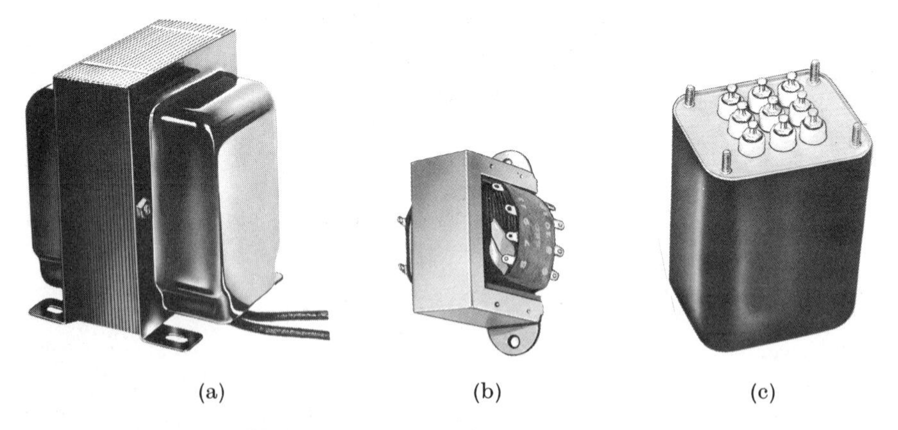

Fig. 12-3. Various small transformers. (a) Power transformer. (b) Multiple tap filament transformer. (c) Audio transformer. (Courtesy of Essex Wire Corp.)

Fig. 12-4. Large commercial transformers. (Courtesy of Los Angeles Dept. of Water and Power.)

The major power losses in a transformer are: power dissipated by the resistance of the windings; flux that "leaks out" and does not couple the two windings (particularly in air-core transformers); and eddy currents in the core (of an iron-core transformer).

12-5 POLYPHASE ac

It is only fair to warn you that we intend to use the same word, at various places in the text, to mean two slightly different things. That word is *phase*, so be careful of its usage each time you come across it.

It is frequently convenient to generate two or more separate voltages in the same alternator. These voltages are of the same amplitude and frequency but are out of phase with each other. Sources comprising two or more such separate out-of-phase voltages are called *polyphase* sources, in contrast to the single-voltage sources we have studied so far, which are called *single-phase* sources. The confusion alluded to in the preceding paragraph may arise because each individual leg of a polyphase circuit is called a *phase.*

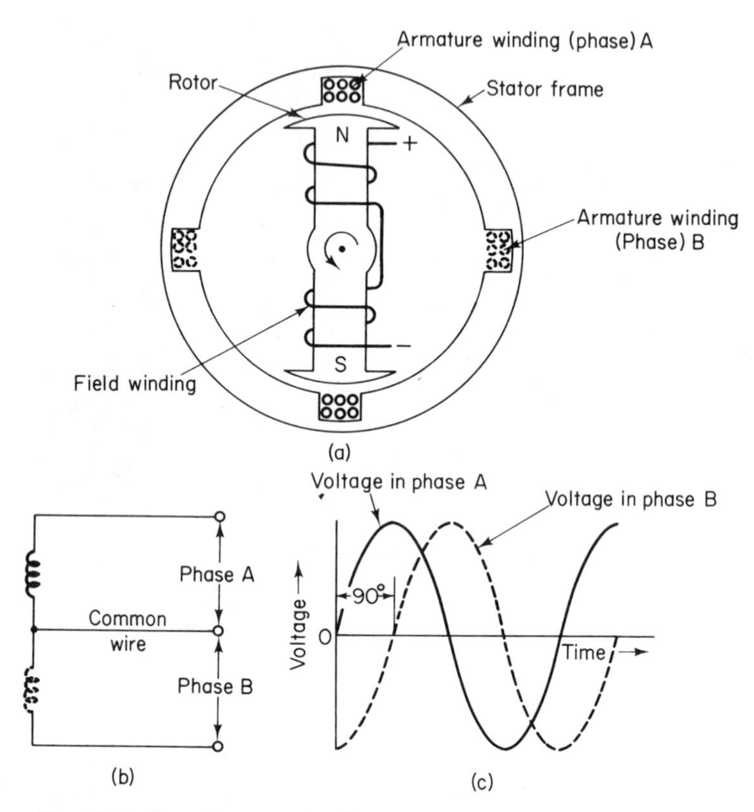

Fig. 12-5. Two-phase ac. (a) Structure of alternator. (b) Connection of armature windings. (c) Phase relationship of voltages generated.

A two-phase alternator is shown diagramatically in Fig. 12-5*a*. The two armature windings are wound, you will note, on the stator, physically 90° apart in space. The field magnet is the rotor, which is excited by dc supplied through slip rings (not shown in the figure). The two armature coils may be wired to separate terminals and used as individual ac supplies, or they may be connected as in Fig. 12-5*b*, with a *common wire* brought out to form a three-wire, two-phase supply. Voltages in the two "phases" are 90 electrical degrees out of phase, as seen in Fig. 12-5*c*. When these voltages are used to supply a two-phase device, such as a motor, there is no instant of time when the applied emf is zero, which results in much smoother operation.

By far more widely used than either a single-phase or a two-phase ac system is the three-phase system, due largely to the pioneering efforts of the famous Nicola Tesla. The three-phase alternator is depicted in Fig. 12-6*a*. Three independent armature "phases" are wound on the stator, 120°

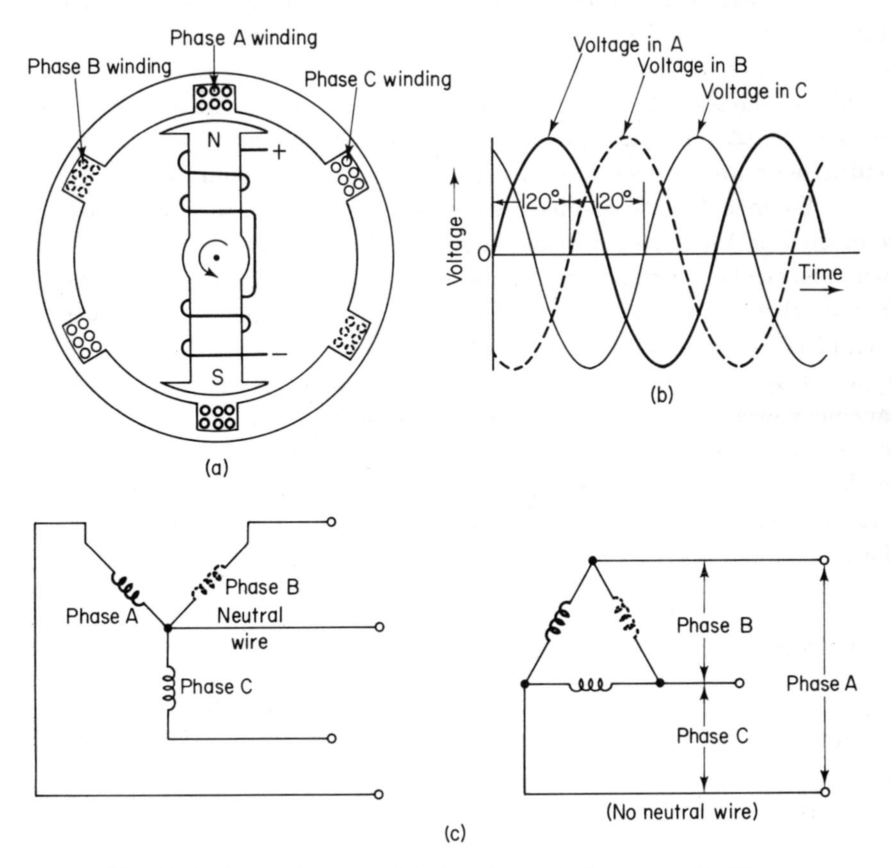

Fig. 12-6. Three-phase ac. (a) Structure of alternator. (b) Phase relationships of voltages generated. (c) Y-connection and Δ-connection for armature windings.

apart in space. Consequently, as the field-magnet rotor turns, it generates equal voltages in these three windings, which are 120 electrical degrees out of phase with each other, as shown in Fig. 12-6*b*.

There are two ways in which the three individual phases may be connected to form a three-phase source. These two modes of connection are illustrated in Fig. 12-6*c*. In the first, one end of each coil is connected to a common point forming (schematically) a Y-shaped configuration, and leads from the other ends of the windings are brought out to the output terminals. A *neutral wire* may or may not be brought out from the common point. This configuration, which then forms either a three-wire or four-wire (if a neutral wire is used), three-phase source, is called *wye-connected*[1] (Y-connected).

Alternatively, the three individual phases may be connected in the second, "triangular," configuration with the leads brought out from the connection points. A three-phase system connected in this manner is said to be *delta-connected*[2] (Δ-connected), because of its resemblance to the Greek letter delta (Δ). The Δ-connected configuration has no place to connect a neutral wire and is therefore always a three-wire configuration.

Y-connected, three-phase sources may be used to supply either Δ-connected or Y-connected loads; similarly, for Δ-connected sources. In any case, practical operation demands that the *loads be balanced* on the phases—that is, that the same amount of true and reactive power be drawn from each phase. Users always distribute loads carefully to keep the phase loading approximately balanced. In a balanced three-phase system, the total instantaneous power is constant. This fact makes for particularly smooth operation of a three-phase motor, giving constant shaft power output and having less tendency to vibration. Considerations of this nature make three-phase industrial systems more economical, and otherwise more advantageous to operate than a single-phase or two-phase system.

12-6 ELECTRICITY FOR THE MILLIONS

Now that we know something about what household ac electricity can do, it is worth while to consider briefly where it comes from and how it gets to us. It is important to realize that electric energy is a *commodity*—an item bought and sold just like beans, bricks, and buns. It is different from these items, however, in that it must be produced as it is used. Large quantities of electric energy cannot be packaged and stored on a shelf, nor can they be produced off-season and sold in peak demand periods. Yet, when you turn on your electric stove or washing machine, you expect it to operate *now*, not in a few minutes or next week. Hence the commercial supply must be capable of supplying a peak load that is many times greater than its average usage. Thus we begin to see some of the problems of electric power *generation* and *distribution*.

[1] Also called "tee-connected" (T-connected) and "star-connected."
[2] Also called "pi-connected (π-connected).

12-7 TURNING THE GENERATORS

We have already seen that the generator (dc as well as ac) requires some external source of mechanical power to turn its shaft. This mechanical device that drives the generator is called the *prime mover*. All commercial

Fig. 12-7. Haynes steam power plant control center. (Courtesy of Los Angeles Dept. of Water and Power.)

Fig. 12-8. The 450,000-kw San Onofre Nuclear Generating Station in Southern California. (Courtesy of Southern Calif. Edison Co.)

Fig. 12-9. The boiling water reactor (BWR), cutaway drawing. (Courtesy of the General Electric Co.)

Fig. 12-10. Nuclear fuel pellets of uranium dioxide. (Courtesy of GeneralElectric Co.)

generating stations use one of three types of prime mover: a *steam turbine*, an *hydraulic turbine*, or an *internal combustion engine*.

Steam plants are the most common types of generating stations. In simplest terms, steam from a boiler is expanded through a turbine wheel (a very efficient many-bladed fan), causing the turbine wheel to turn at a controlled speed of 1800 or 3600 rpm. The alternator is driven from the shaft of the turbine. The steam boilers are of two general types: conventional water-tube boilers and nuclear reactors. The conventional boilers may be heated with coal, fuel oil, or natural gas, depending on which is most available in the steam plant's locale. The complex set of controls required to operate a steam plant is indicated in Fig. 12-7, which shows the central control room at Haynes Steam Power Plant. This plant has a total capacity of 1.58 million kilowatts.

As rapidly as possible, conventionally fired steam plants are being supplanted by nuclear reactor steam plants. The latter provide low-cost, simple, flexible, and *safe* power generation. Public opinion to the contrary, a nuclear steam plant is as safe, or more so, than a conventionally fueled

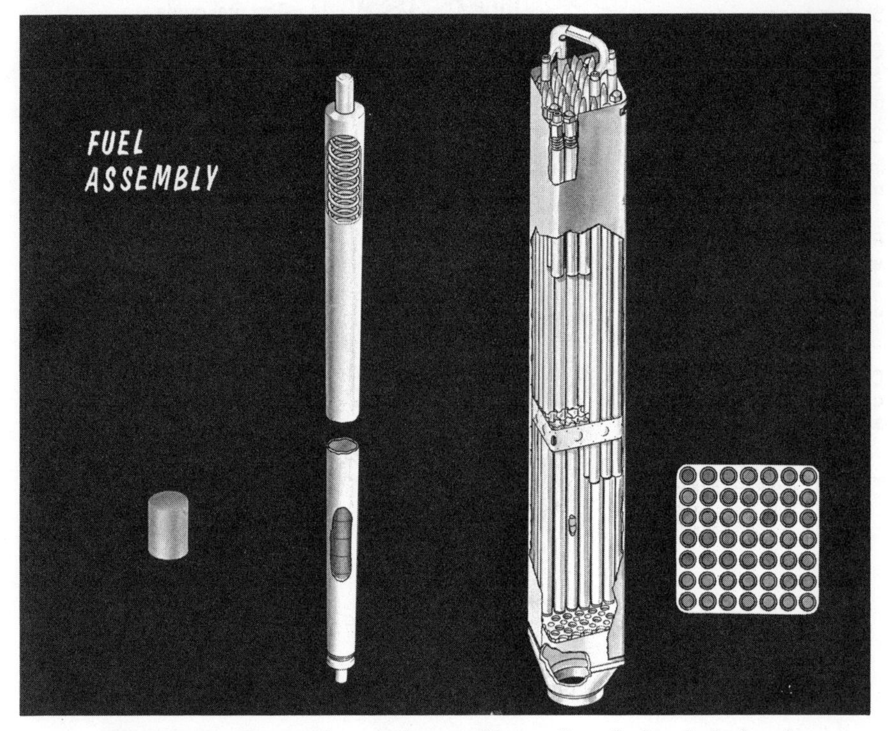

Fig. 12-11. The nuclear fuel assembly: pellet, fuel rod, fuel rod assembly. (Courtesy of General Electric Co.)

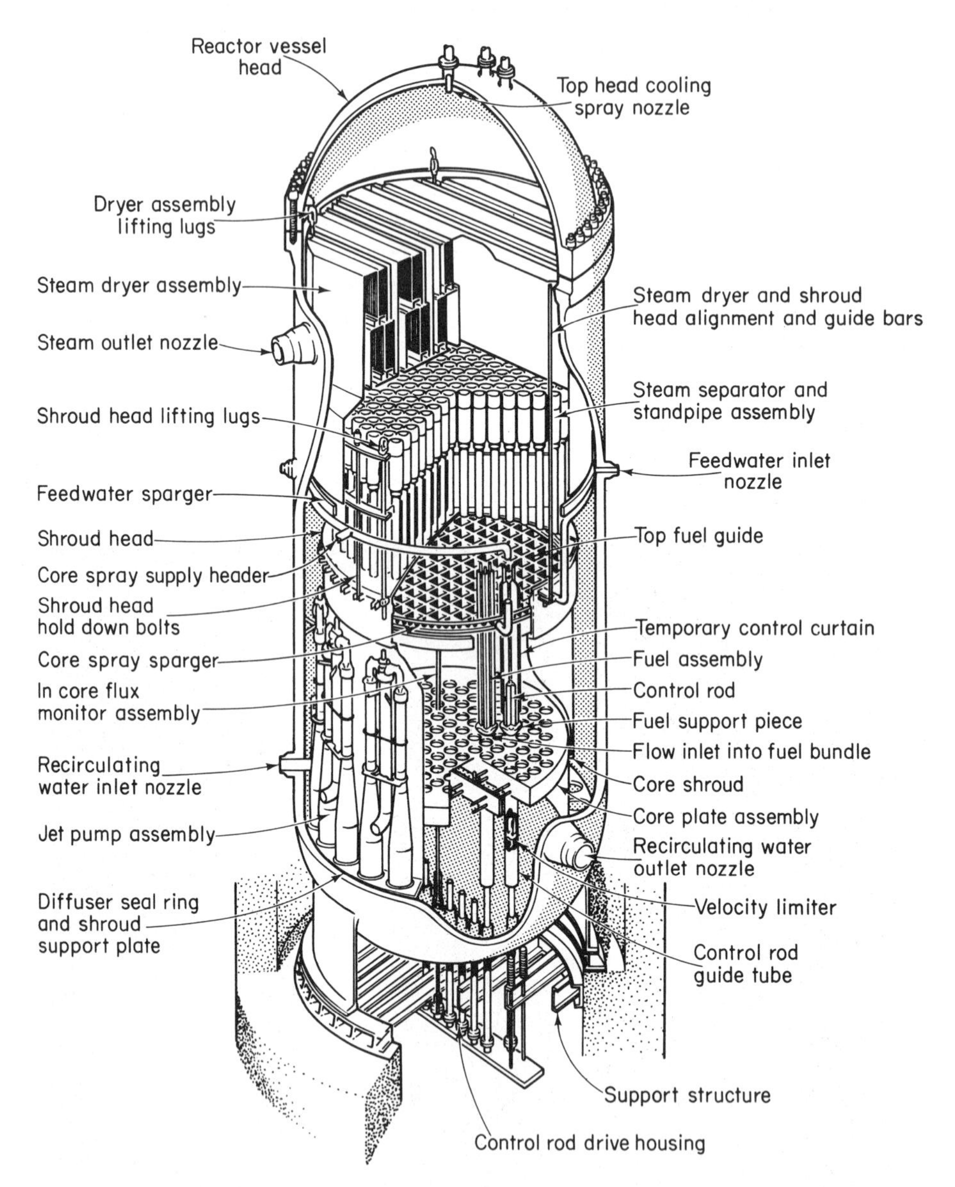

Fig. 12-12. Cutway drawing of reactor. (Courtesy of General Electric Co.)

steam plant. (A nuclear reactor is not designed to, and *cannot*, explode like an atomic bomb!) A major nuclear plant is shown in Fig. 12-8. Nuclear components are housed in the huge steel containment sphere near the center of the picture. The 450,000-kw output of this plant will supply a city of well over half a million people.

One of the latest types of nuclear reactor for the generation of power is the *Boiling Water Reactor* (BWR), a cutaway view of which is given in Fig. 12-9. The reactor itself is the dark cylinder with the dome-shaped cap in the center of the illustration (note the size of the man in the picture). The fuel for the reactor is uranium dioxide, produced in pellet form, as shown in Fig. 12-10. The oxide is used, instead of pure uranium metal, because it is more stable structurally at the high temperatures it must endure; combining the uranium atoms with oxygen atoms does not weaken their activity. The pellets are assembled in stacks inside a *fuel rod* in which they are held together by spring pressure; bundles of 49 of these rods are then mounted together in a *fuel rod assembly*, as illustrated in Fig. 12-11. A cutaway drawing of the reactor is shown in Fig. 12-12, with just a few of these fuel rod assemblies installed. The BWR steam generation cycle is illustrated schematically in Fig. 12-13. Water is fed into the reactor and circulated around the "core" of fuel rod assemblies, where it is heated and converted to steam at 545 °F and 1000 pounds per square inch pressure. Water droplets carried along by the steam are removed in the separator

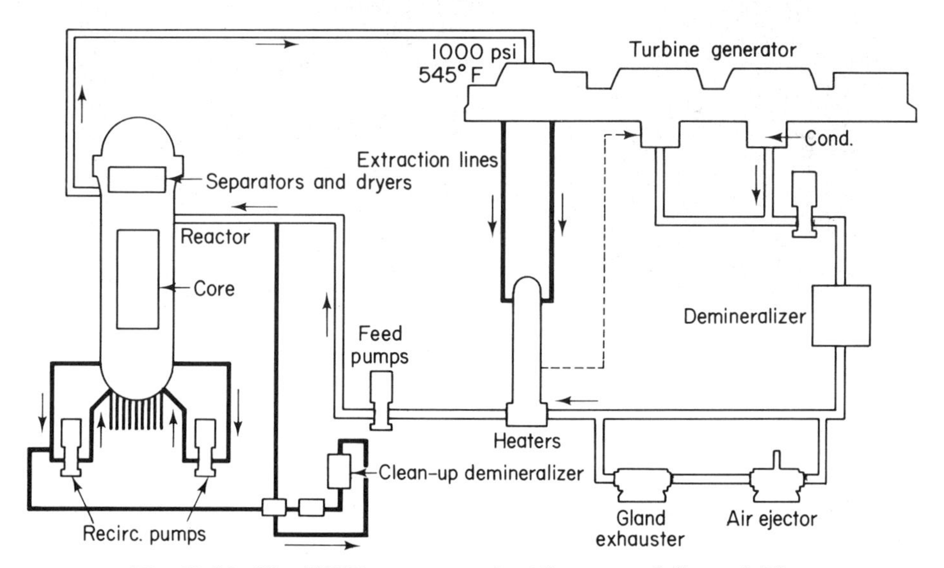

Fig. 12-13. The BWR reactor cycle. (Courtesy of General Electric Co.)

and drier. The dry steam is then fed into the turbine (which drives the alternator). Upon leaving the turbine, with most of its energy gone, the steam is condensed to water again and is pumped back to the reactor. Additional equipment is provided to remove unwanted air and minerals from the water.

Another major type of power plant is the *hydroelectric* station, which uses the energy of rushing water to turn the alternator. One of the most famous, the 1.34-million-kilowatt Hoover Dam Power Station, is shown in Fig. 12-14. In hydroelectric stations, the falling water is directed into either

Fig. 12-14. Hoover Dam Power Station. (Courtesy of Los Angeles Dept. of Water and Power.)

an impulse turbine or a reaction turbine, both of which are modern descendants of the old-fashioned water wheel.

Finally, some small alternators are driven by gasoline or diesel engines. Generally these alternators are privately owned and used to supply only one facility, such as a mountain resort, a hospital's emergency system, and many of today's automobiles.

12-8 HIGH-VOLTAGE DISTRIBUTION

The distribution of ac power always involves high voltages. (The word *tension* is often used for voltage in this connection; you will recall hearing of *high tension* cables.) Before proceeding with our study, it is worth while to consider why high-voltage transmission is more economical than low-voltage transmission. If a bit of arithmetic really disturbs you, then just take the author's word for it and skip over the next few paragraphs; otherwise let us consider a simplified problem in power transmission.

Suppose that there is a little mountain resort which requires 600 kw of power at 240 volts and that there is a small 240-volt generating station 25 miles away. We wish to connect the resort to the generating station with a transmission line consisting of two #1 copper conductors. Each of these 25-mile-long conductors has a resistance of about 16 ohms, or 33 ohms total. Now let us see what happens if we transmit power at 20 times our desired voltage, compared to what happens when we transmit at 500 times the desired voltage. The two alternatives are represented in Fig. 12-15. In both cases, we have a purely resistive 240-volt, 600-kw load, shown on the right. The current in the load and secondary of the receiving transformer is 2500 amp, because

$$(240 \text{ volts}) \times (2500 \text{ amp}) = 600{,}000 \text{ watts} = 600 \text{ kw}$$

In the upper figure, the step-down turns ratio is 20. The current in the primary of the receiving transformer, and, of course, throughout the transmission line, is therefore $\frac{1}{20}$th of the secondary current

$$\text{Transmission line current} = (2500 \text{ amp}) \div 20 = 125 \text{ amp}$$

the voltage drop across the two conductors of the transmission line is, by Ohm's law,

$$\text{Line drop} = (125 \text{ amp}) \times (33\ \Omega) = 4125 \text{ volts}$$

Hence the voltage at the boost transformer's secondary must be 4125 volts higher than that at the receiving transformer's primary, or $4800 + 4125 = 8925$ volts. The voltage into the boost transformer (from the alternator) is 240 volts. Hence its turns ratio must be

$$\text{Boost transformer turns ratio} = 8925 \div 240 = 37.2$$

This means that the current from the alternator into the primary of the boost transformer is

$$\text{Alternator current} = 125 \text{ amp} \times 37.2 = 4650 \text{ amp}$$

Finally, computing the power out of the alternator,

$$\begin{aligned}\text{Alternator power} &= (240 \text{ volts}) \times (4650 \text{ amp}) \\ &= 1{,}116{,}000 \text{ watts} \\ &= 1116 \text{ kw}\end{aligned}$$

Note that we have had to generate 1116 kw to get 600 kw at the load, the difference being lost in the transmission line because we have assumed no other losses.

Now let us see what happens when we transmit at 500 times the load voltage, as in the lower part of Fig. 12-15. In this case, the turns ratio

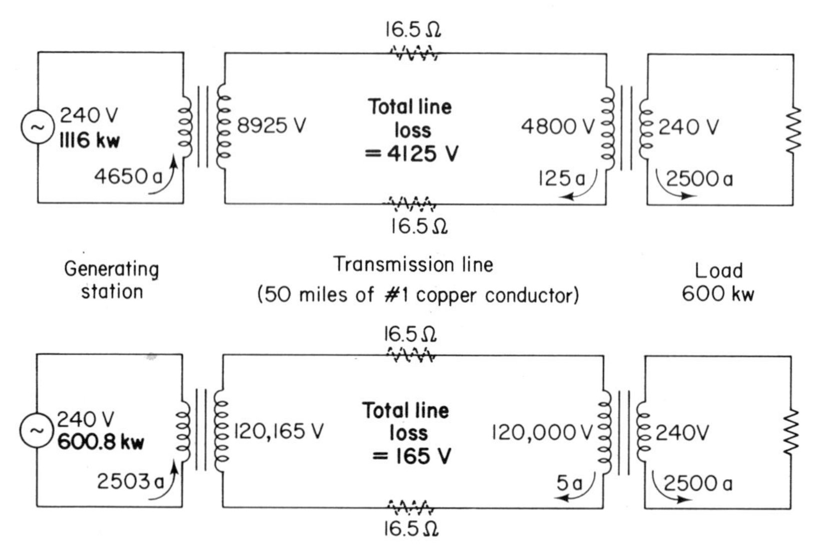

Fig. 12-15. The economics of high-voltage power transmission.

of the receiving transformer must be 500, and the primary (transmission line) current is

$$\text{Transmission line current} = (2500 \text{ amp}) \div 500 = 5 \text{ amp}$$

which gives a voltage drop across the transmission line of

$$\text{Line drop} = (5 \text{ amp}) \times (33\ \Omega) = 165 \text{ volts}$$

Here the voltage out of the boost transformer's secondary only needs to be 165 volts higher than that into the receiving transformer's primary, or 120,000 + 165 = 120,165 volts. The boost transformer's turns ratio is therefore

$$\text{Boost transformer turns ratio} = 120{,}165 \div 240 = 500.7$$

and the current out of the alternator into its primary is

$$\text{Alternator current} = 5 \text{ amp} \times 500.7 = 2503 \text{ amp}$$

In this case, the power required from the alternator is

$$\text{Alternator power} = (240 \text{ volts}) \times (2503 \text{ amp}) = 600.8 \text{ kw}$$

so that we have only had to generate 600.8 kw to obtain 600 kw at the load.

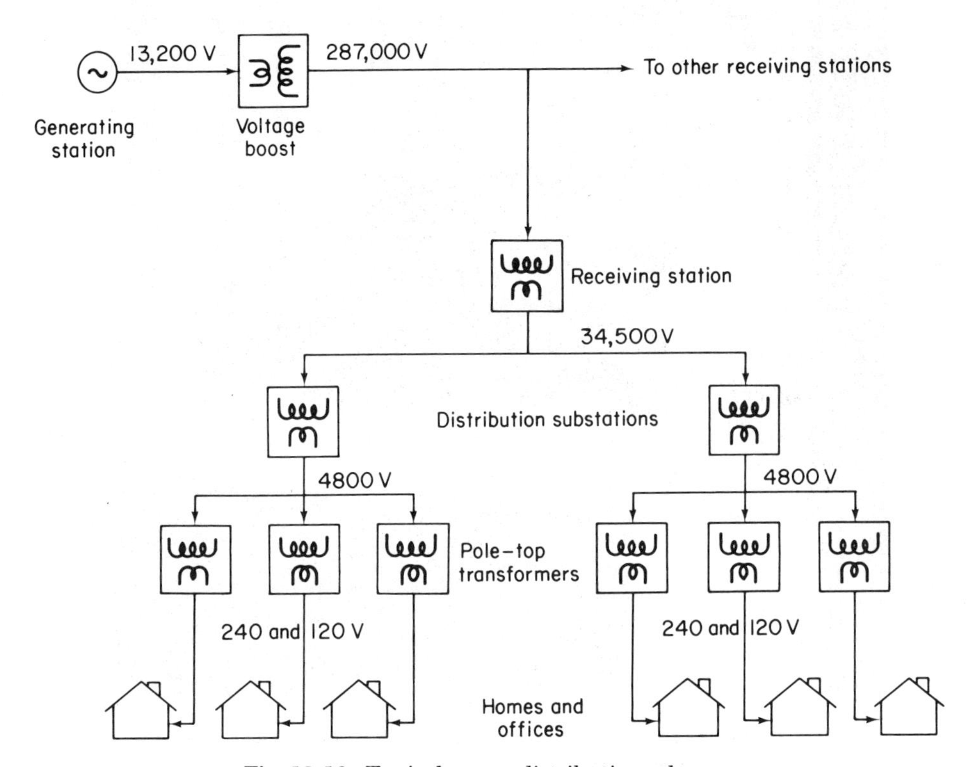

Fig. 12-16. Typical power distribution scheme.

In other words, by increasing the transmission voltage from 4800 volts to 120,000 volts, we have cut the power lost in the lines from 516 kw to only 0.8 kw.

From this grossly oversimplified example, it is evident that higher voltages mean greater savings in line losses, and you may wonder why transmission voltages are not as high as they can possibly be made. Why not a billion volts, for instance? The answer is that there are other practical problems at extremely high voltage, notably the tendency to arc. Present technology places a practical limit on high-voltage transmission at 200,000 to 300,000 volts.

A block diagram for a segment of a typical commercial power distribution system is shown in Fig. 12-16. Power is generated by the alternator at the upper right of the figure. Whether driven by steam turbines or by water turbines, like those shown in Fig. 12-17, the alternators used are almost universally large, self-excited, three-phase machines. In the United States, 13,200 volts has become the most common generating voltage and

Fig. 12-17. The Hoover Dam alternators. (Courtesy of Los Angeles Dept. of Water and Power.)

frequency is rigidly controlled to 60 Hz. In Europe and other parts of the world, 25-Hz and 50-Hz frequencies are more common.

Unless the power is all used nearby, the alternator's output voltage is immediately boosted to a higher level for transmission. The 287 kilovolts used in the illustration is typical for a long transmission distance. The booster transformer's output is then fed, via the high-tension transmission lines mounted on poles and towers, to a number of receiving stations, one of which is shown in the illustration.

At the receiving station, the transmission voltage is stepped down to 34.4 kilovolts. Figure 12-18 shows a typical receiving station. From here

Fig. 12-18. A typical receiving station. (Courtesy of Los Angeles Dept. of Water and Power.)

it is sent to a number of distribution substations over subtransmission lines. The latter may be installed either underground, as in Fig.12-19, or on poles.

At the distribution substation, the voltage is once again stepped down, this time to 4800 volts. Here, also, the phases are split—that is, a given group of customers is fed from only one phase of the three-phase supply.

Fig. 12-19. Installing an underground transmission line. (Courtesy of Los Angeles Dept. of Water and Power.)

There are also special-purpose substations, when required by the customer, for such things as frequency conversion (e.g., from 60 Hz to 400 Hz) or large-scale rectification (conversion to dc).

Finally, power from the distribution substation is delivered to a transformer near the home or office. This transformer, either mounted on the power pole, as in Fig. 12-20, or located in a concrete vault if utilities are underground, does the final step-down to 120 volts, or 240 and 120 volts. Conductors from the secondary taps of this transformer are then connected directly to the electric meter box from which the household circuits are fed.

After reading the preceding discussion, perhaps the reader will feel he has been taking the distribution of electric power a bit too much for granted. It is the installation, operation, and maintenance of this giant complex of equipment that necessitates the existence of the large electric utility company.

Fig. 12-20. Residental area pole-top transformer drops 4800-volt supply to 240- and 120-volt household levels. (Courtesy of Los Angeles Dept. of Water and Power.)

12-9 GRANDCHILDREN: THE ATOM SMASHERS

We have seen how ac was the child of the electromagnetic marriage. The use of ac in turn, in conjunction with magnetism, made possible many novel devices that may be considered "second generation" or grandchildren. At this point, we shall discuss only one small, but spectacular, group of grandchildren—the atom smashers.

In order to study the structure of the atom, the atomic scientist must be able to "take the atom apart" and "see" into it. Of course, he cannot do this in the conventional sense of seeing, but he can infer their structure by shattering the atoms to "pieces" with high-speed "bullets" and observing the effects of the "pieces" as they react with other nearby matter. The high-speed "bullets" in question are really subatomic particles, like electrons, protons, and alpha-particles (helium atom nuclei), which have been accelerated to very high speed and therefore have a large (kinetic) energy. It turns out that higher and higher energy particles are required as we take apart atoms, then nuclei, and, finally, particles themselves. It is no surprise, therefore, that a good deal of the effort in atomic and nuclear research has gone into devising bigger and better particle accelerators.

Here it is well to spend a moment discussing the unit in which particle energies are measured. By now we are all used to joules, watt-seconds, and kilowatt-hours. The atomic scientist, however, finds a different energy unit more useful—the *electron volt*, abbreviated *ev*. Recall from Section 3-7 that an emf actually represents work done by, or energy imparted to, a unit of charge; thus one volt represents one joule of work done per coulomb, or one joule of energy imparted per coulomb. In atomic physics, the charges on all particles are some exact multiple of the electron's charge, so it is natural to take the charge on one electron as the unit of charge rather than the coulomb. Now, if we put an electron into an electric field so that it is accelerated through a potential difference of 100 volts, we say that it has acquired an energy of 100 electron volts, or simply 100 ev. The electron volt is, of course, a very tiny bit of energy; in terms of our more familiar units, 1 ev $= 1.60 \times 10^{-19}$ joule or watt-second $= 4.45 \times 10^{-26}$ kwh. The energies of high-speed particles are measured in thousands of ev (kev), millions of ev (Mev), and billions of ev (Gev[3]).

The principle of accelerating particles to high energies is therefore simplicity itself; put the particle into a high-voltage electric field. Figure 12-21*a* shows a proton (charge $= +1$) placed between plates that are connected to a battery. The proton is repelled by the positive plate and attracted by the negative plate. If the battery represents a 500-volt dc source, then the proton acquires an energy of 500 ev by the time it reaches the negative plate, its charge being equal (but of opposite sign) to one electron charge.

The simplest usable particle accelerator is thus a tube along which the particle is continuously "urged" by a steady electric field. We have already encountered such a device in Section 3-15, the Van de Graaff generator. There is a practical limit, however, to how high a voltage such a device can produce. To obtain still higher energies, it is necessary to have a long, evacuated tube with many sets of electrodes spaced along it. The particle then receives an additional "push" as it passes each accelerating electrode. Such a device is called a *linear accelerator* and several have been built. The trouble with linear accelerators is that they become very long indeed—miles long—to obtain the kind of particle energies desired. Consequently, they are also very expensive, costing over 100 million dollars. There was an urgent need for a relatively compact device to accelerate heavy

[3] Formerly "Bev."

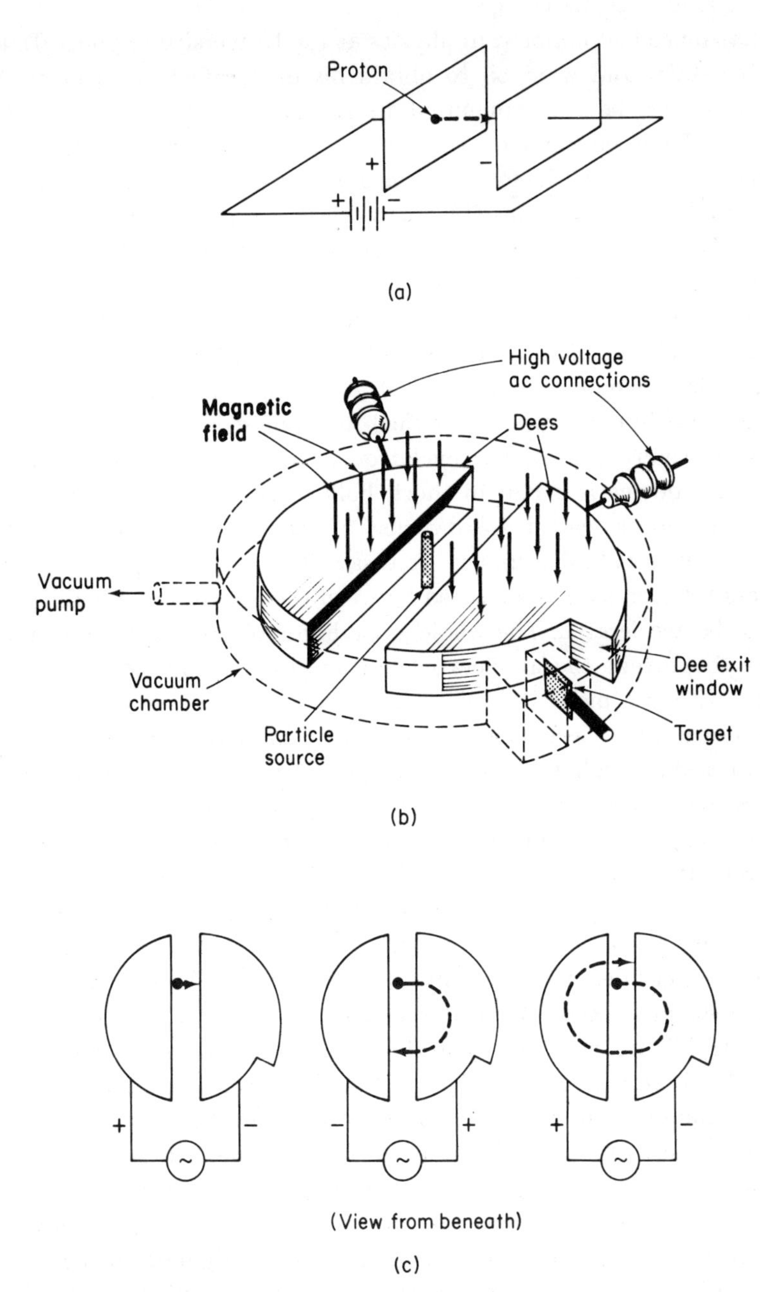

Fig. 12-21. The cyclotron. (a) Principle. (b) Construction. (c) Acceleration of a proton.

particles to high energies. The problem was first solved by the American physicist, Ernest O. Lawrence.

Lawrence had majored in physics at the University of South Dakota. his native state, and went on to obtain his doctorate from Yale in 1925. Two years later, he was appointed to the faculty of the University of California. There, associated with the radiation laboratory, he became engrossed in the search for particle-accelerating techniques. Lawrence believed that it might be feasible to accelerate a particle with a series of small "nudges" instead of building up enormous potentials for one mighty kick. This, he reasoned, could be accomplished by forcing the charged particle to travel in circles, using a magnetic field, and giving it a small "nudge" on each lap. The result is a device called the *cyclotron*, a model of which Lawrence first demonstrated in 1930. Working with M. S. Livingston, Lawrence quickly developed a practical full-scale cyclotron.

The construction of the cyclotron is indicated in Fig. 12-21*b*. The heart of this device is a pair of short, hollow half-cylinders several feet in diameter, which are called *dees* because of their resemblance to the letter *D*. (If a circular ladies' compact were sawed in half, the pieces would be small dees.) An ion source is located between the dees to provide the particle that is to be accelerated. The whole assembly is placed inside an evacuated chamber to prevent the accelerated particle from being slowed down by collisions with gas molecules. This chamber is, in turn, mounted between the poles of a large electromagnet, such that the magnetic field is perpendicular to, and through, the plane of the dees. A high-voltage, high-frequency ac source is connected to the dees through insulators.

The operation of the cyclotron is illustrated in Fig. 12-21*c*. Here you are looking at the cyclotron of Fig. 12-21*b* from below. The magnetic field is coming out of the page at you. Suppose that a proton is injected between the dees when the left dee is positive. The proton is accelerated across the gap into the right dee, where the magnetic field forces it to travel in a semicircle at constant speed. When it returns to the edge of the right dee, as in the central figure, the polarity of the applied voltage has changed so that the left dee is now negative. The proton is then accelerated back across the gap, gaining additional velocity, until it enters the left dee, where its path is again bent into a semicircle by the magnetic field. This time, however, the semicircle is slightly larger because of the proton's higher velocity. When it reaches the edge of the left dee once more, as in the right-hand figure, the polarity of the ac voltage has changed again, the right dee being positive, and the proton rushes back across the gap, gaining still more speed. This process repeats over and over again, and each time the proton gains a little speed and travels in a little wider circle. Hence the proton travels in a widening spiral path until it finally exits, at very high velocity, through a thin "window" provided in one of the dees and impinges on the desired target. The time taken by the proton to travel over any of the semicircular paths is always the same, regardless of the size of the semicircle, because the increased size of each successive semicircle just makes up for the increase in speed from the previous gap-crossing. In this way, for instance, a proton may be made to pass through a potential of only 15,000 volts a thousand times and thus acquire a total energy of 15 Mev.

The cyclotron is a very satisfactory particle accelerator, except at very high energies, where the particle velocities begin to approach the velocity of light. Under those extreme circumstances, Einstein's theory of relativity shows that the mass of the particle begins to increase significantly. The effect is that the time required to travel the semicircular paths inside the dees is not quite constant; thus a constant-frequency ac source cannot be used. Instead, modifications of the cyclotron, called the *synchrocyclotron* and *synchrotron*, are used for extremely high-energy particles. In these devices, the applied voltage frequency, or both the frequency and the intensity of the magnetic field, are variable to compensate for the change in the particle's mass.

REVIEW QUESTIONS

1. What feature of ac motors makes them more maintenance-free than dc motors?
2. Why can't the rotor of an induction motor turn at synchronous speed, while that of a synchronous motor can?
3. What is the major disadvantage of a synchronous motor?
4. Discuss the difference between a field stator alternator and a field rotor alternator. Why didn't we make this distinction with dc machines?
5. The secondary winding of a transformer has 20 times as many turns as the primary winding. Is this a step-down or a step-up transformer? If the secondary current is 1 ampere with 400 volts across the winding, what are the corresponding voltage and current in the primary (assuming the transformer to be ideal)?
6. What advantages does three-phase ac offer from the standpoint of the power company? Of the user?
7. Name the types of prime movers used to turn commercial alternators. Under what circumstances is each type used?
8. Briefly describe the nuclear power generating station. What is its main advantage over a hydroelectric power station?

9. Couldn't the power company save a lot a money by eliminating all that fancy high-voltage equipment, together with all the people who maintain it, and then generating and transmitting power to us at the 240- and/or 120-volt level we use?
10. Describe a typical power generation/distribution system. How many times is the voltage level transformed in such a system?
11. What is the significant advantage of Lawrence's cyclotron over linear accelerator devices?
12. A cyclotron is not a very practical device for accelerating electrons. Can you explain why?

13

CONTROLLERS OF ELECTRON TRAFFIC

13-1 "ELECTRONICS" AT LAST

In this chapter we finally begin to study what the purist would call the "proper subject matter" of electronics, which is, roughly speaking, all the applications of vacuum tubes and transistors. It may seem peculiar that we have spent twelve chapters, over half of this text, in getting to the beginning. Yet, the thorough familiarity with (again, what our friend, the purist, would call) the "fundamentals of electricity and magnetism," acquired in those twelve chapters, is vital to an understanding of both the tubes and transistors themselves and their applications.

Both vacuum tubes and transistors perform a similar type of function. They regulate the flow of one current, using the electric field produced by another independent current or voltage. For this reason, they are called *active* circuit elements, as opposed to resistors, capacitors, and inductors, which are *passive* circuit elements. They control the electron "traffic" in a circuit, much as the speed control systems along certain superhighways control the movement of automobile traffic and as air traffic controllers regulate the takeoff and landing rates at airports.

This chapter is devoted to the specific devices themselves and certain basic circuits in which they are used. In later chapters, we see how these basic circuits are used as building blocks to produce the electronic marvels around us.

13-2 GETTING AN ELECTRON TO BREAK AWAY FROM THE CROWD

Electrons are a funny breed. Like some people, they move around with comparative ease among their own kind, but become quite shy and reticent when urged into an environment strange to them.

We have seen that many of the electrons associated with the atoms inside a metal conductor, at normal temperatures, are moving about quite freely in random fashion, frequently bumping into each other. These conduction electrons, which wander from the valence shell of one atom to that of another, behave much like the milling crowd in Times Square on New Year's Eve. So long as they remain in the bar of metal, the bar as a whole is electrically neutral. Even though individual atoms momentarily become charged ions when they lose an electron, they quickly "steal" one from a neighboring atom and the total positive and negative charge within the bar stays in balance all the while.

The story is quite different if we attempt to get an electron completely out of the bar. Consider a conduction electron at the "surface"[1] of a bar as in Fig. 13-1*a*, trying to escape. Its environment differs from that of its brothers located deeper in the bar. They are surrounded by positive nuclei that exert a more or less uniform pull from all directions. The surface electron, on the other hand, has no nuclei pulling from above it; hence the electrostatic forces tend to hold it in the bar. Furthermore, if the electron did get away, its negative charge would create an opposing positive charge in the bar, by electrostatic induction, which would tend to pull it back. The combination of electrostatic and certain other (quantum mechanical) effects results in what is called a *potential barrier* at the surface of the conductor. The difficulty in overcoming the potential barrier is expressed as the amount of work that must be done on an electron to wrench it out of the conductor and away "to infinity" (i.e., permanently removed from the conductor). This quantity of work is a characteristic of the conductor, different for each material. It is called the *work function* of the material and is usually measured in electron volts (ev). Work functions for the various metals fall between 1 and 7 ev with the majority clustering around 4 to 4.5 ev. Thus electrons in a conductor

[1] This "surface" is not well-defined when we talk in terms of atomic dimensions.

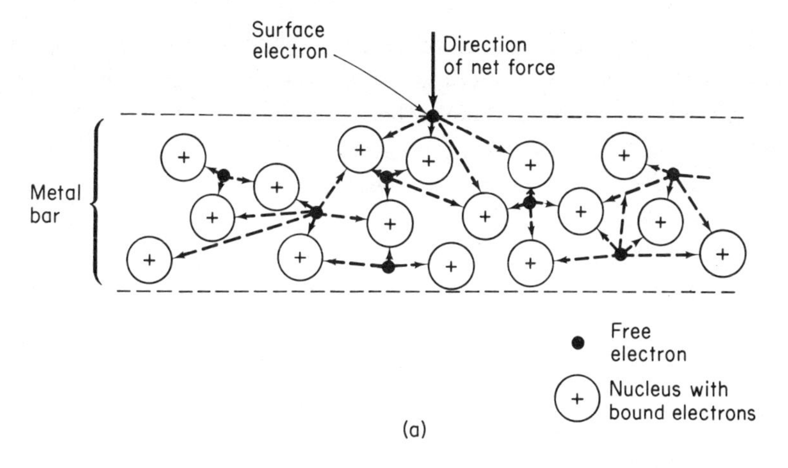

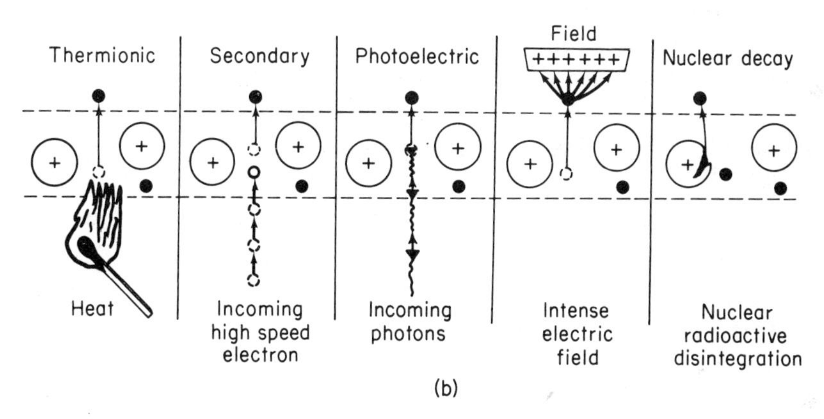

Fig. 13-1. Electron emission. (a) Electrostatic opposition. (b) Modes of emission.

that acquire energy equal to, or greater than, the work function are capable of escaping from the surface of the conductor. This phenomenon is called *electron emission.*

There are a number of ways by which electrons may acquire the energy needed to escape. All of these ways are made use of in various devices. These modes of emission are illustrated schematically in Fig. 13-1*b*.

By far the most widely used means of achieving electron emission is *thermionic emission*, illustrated by the first picture of Fig. 13-1*b*. Thermionic emission occurs when the emitting material is *heated* until the electrons have enough energy to overcome the potential barrier. This electron emission mode is used in most vacuum tubes, and we say a good deal more about it presently.

Another important mode of emission is *secondary emission*, shown second in Fig. 13-1*b*. Loosely speaking, this is the "billiard ball" effect. A high-energy electron from some other source literally "knocks" one or more electrons out of the emitting material. This effect is particularly useful when conditions are such that each incoming electron produces more than one secondary electron.

We have already encountered *photoelectric emission*, the third mode illustrated in Fig. 13-1*b*, in Section 5-10. The effect is similar to secondary emission except that the "knocking out" is done here by photons of radiation having very definite energies, instead of electrons.

The fourth principal emission mode is *high-intensity field emission*, or *field emission* for short. The mechanism here is the intense attractive force due to a strong positive charge brought near the emitting material. The electron is figuratively "sucked out," as if by a vacuum cleaner.

The *nuclear decay* mode is shown for completeness but is really in a different category from the others. Certain naturally radioactive elements decay by emitting an electron from their *nucleus* (not a valence electron). An electron emitted in nuclear decay is more usually called a beta-minus (β^-) particle, and the process is called beta decay. Unlike the other emission processes, this one is not under man's control. Beta emitters keep emitting electrons, regardless of what the experimenter does to them, until all the atoms have decayed.

13-3 THE VACUUM TUBE

A vacuum tube, as the name implies, is a tube of glass or metal from which the air has been pumped and which contains two or more electrodes. The electrodes, usually metallic, are insulated from each other and from the tube itself if it is made of metal. Lead wires are brought out through the wall of the tube to contact pins for connection to the circuit. At least one of the electrodes in the tube must act as an emitter of electrons and this electrode is called the *cathode*. With few exceptions, most vacuum tubes also contain a positively charged electrode thct absorbs the electrons emitted by the cathode. This electrode, although it can properly be called an *anode*, is more commonly called the *plate*. Other electrodes may be provided, depending on the purpose of the tube.

13-4 FLEMING'S VALVE

It was our own Tom Edison who discovered thermionic emission and, in a sense, invented the vacuum tube. Of course, if you want to be a stickler for accuracy, the first vacuum tubes were built decades earlier by

Faraday, Geissler, and Crookes, but theirs did not involve thermionic emission. Edison's discovery came in 1883 while he was trying to perfect his incandescent lamp. These lamps, he noticed, alway became blackened on the inside and Edison resolved to find a remedy. In one series of experiments, he placed a flat piece of metal inside the lamp near the filament and connected it externally through a galvanometer to one of the filament connections at the base. To his surprise, the galvanometer indicated a current flow, even though there was an *open* circuit between the metal plate and filament inside the tube. The current only flowed, however, when the plate was connected to the positive side of the filament, as shown in Fig. 13-2. Since

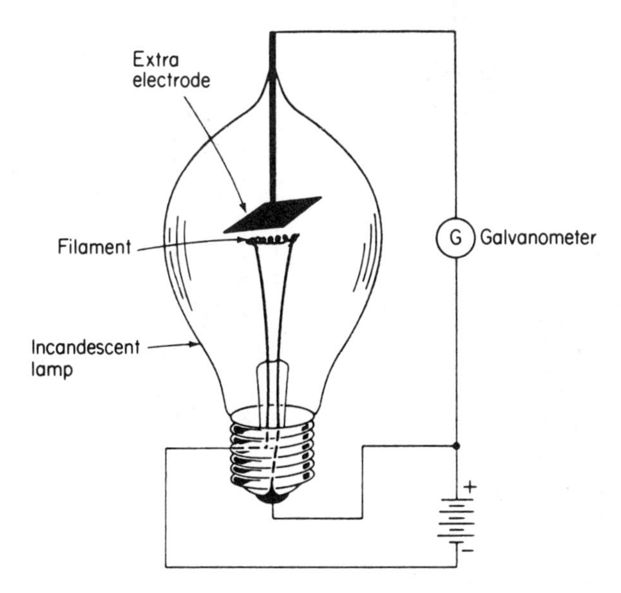

Fig. 13-2. The Edison effect.

the electron had not been discovered yet, Edison was at a loss to explain this phenomenon, which has since been named the *Edison effect* in his honor. Moreover, he could see no practical use for it, but he patented it anyway, just in case it might be usable.

The first true vacuum tube, as we know it, was made by a British engineer, Sir John Ambrose Fleming. A graduate of University College, Fleming entered Cambridge in 1877 and worked under Maxwell. In 1885 he was appointed professor of electrical engineering at University College. He became acquanited with the Edison effect while a consultant to the Edison and Swan Lighting Co. in London, and acquired some of Edison's lamps containing the extra electrode. Then, in the 1890s, Fleming was associated. with Marconi and the newborn wireless (i.e., radio). What the world needed then, among other things, was a good detector of radio waves to replace "crystals and cat whiskers." Fleming believed that the Edison effect could be put to use in this regard.

In 1904 he built a lamp similar to Edison's "extra electrode lamp" except that he formed the plate into a cylinder and mounted it so that it surrounded the filament, as shown in Fig. 13-3*a*. Fleming found that the

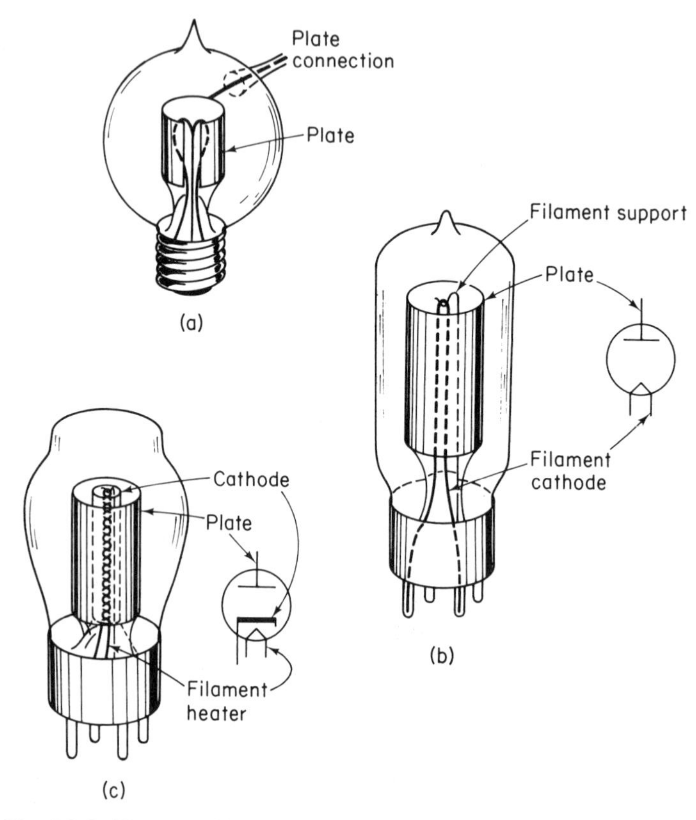

Fig. 13-3. Vacuum diodes. (a) The Fleming "valve." (b) Filament cathode type, with circuit symbol. (c) Indirectly heated cathode type, with circuit symbol.

passage of electricity through the open space in the tube was due to electrons (discovered seven years earlier by Thomson) coming from the glowing filament. The tube will only conduct when a potential is maintained between the cathode and plate, and the plate must be positive with respect to the cathode. Because of this action, in turning current on and off according to whether the plate is positive or not, Fleming called his device an *electric valve*. In the United States, these devices have come to be called *tubes* instead, but in England they are still *valves*.

13-5 KINDS OF CATHODES

Two different types of cathode construction were used in the evloution of the vacuum tube. One type is just like that used in the original Fleming valve, where the heated filament itself *is* the cathode. This construction is called a *directly heated cathode*. The filament is therefore a part of two different circuits at the same time: the heating circuit, which consists simply of the filament and a dc source to supply it; and the plate-to-cathode circuit, which is the working circuit for the tube. The second type of construction uses a specially treated cylinder of metal for the cathode with a separate filament inside it for the heater. This construction is called an *indirectly heated cathode*. The advantage of this construction lies in the ability to keep the heater circuit independent of the cathode circuit. For instance, ac may be used for the heater supply without introducing an extraneous signal into the plate-to-cathode circuit. In modern tubes, the directly heated cathode has all but disappeared. Figure 13-3*b* and *c* shows tubes having both types of cathode construction.

The selection of a cathode material depends on a number of factors. The material must be able to withstand, without melting, a temperature somewhat in excess of that corresponding to electron energies equal to the work function. It must also have adequate structural properties, particularly in the directly heated cathode, where it must be capable of being drawn into filament wire. The most durable is tungsten, which can withstand overloads and other abuse without permanent damage. However, the work function of tungsten is high, 4.5 ev, which means that high temperature is required for emission. The addition of approximately 2 percent of thorium to the tungsten, as a surface layer, results in more than a thousandfold improvement in emission characteristics (work function reduced to 2.6 ev) over pure tungsten. A filament of this type is called *thoriated tungsten*. While more susceptible to overload damage than tungsten alone, the thoriated tungsten filament can be operated at a much lower temperature, saving on filament power. Only directly heated types employ tungsten or thoriated tungsten cathodes. A still more efficient emitter is made by depositing a coating of barium oxide (usually mixed with strontium oxide) on a nickel or molybdenum alloy. This oxide-coated emitter has a work function of only 1.1 ev and is used in all indirectly heated cathodes and some directly heated types. It is much more fragile and short lived than its tungsten brothers, but consumes considerably less heater power.

Figure 13-4 shows the emission characteristics of all three types of emitters in terms of how the electron current emitted per square centimeter

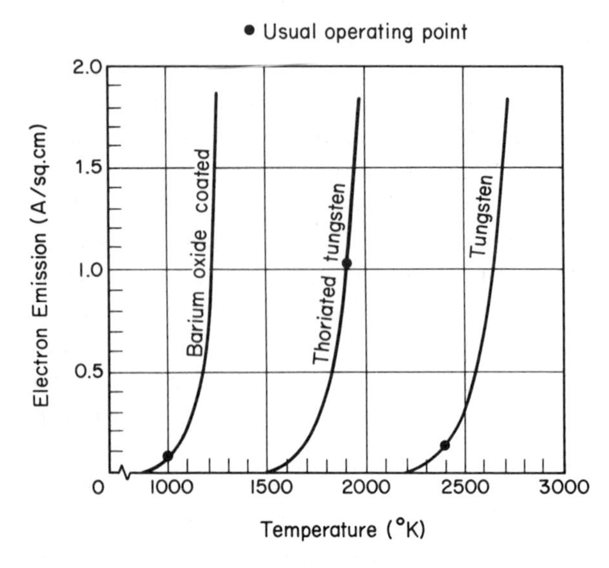

Fig. 13-4. Thermionic emission characteristics of several materials.

of surface varies with the temperature of the emitter. The dots on the curves show the normal operating points: 2400 °K (about 3860 °F) for tungsten, 1900 °K (about 2960 °F) for thoriated tungsten, and 1000 °K (approximately 1340 °F) for the oxide-coated emitter.

13-6 THE VACUUM DIODE

The modern counterpart of Fleming's valve, and the simplest type of vacuum tube, is the *diode*. It takes its name from the fact that it is a two-electrode tube, *di-* being the Greek prefix for "two" and *-ode* from electr*ode*. (In an indirectly heated type, the heater filament is not counted as an electrode, since its only function is to provide heat to the cathode.) The tubes shown in Fig. 13-3*b* and *c* are diodes, one directly heated and the other indirectly heated. Notice the difference in schematic symbols.

When current flows in the filament of a tube, the filament becomes hot. When the temperature of the cathode reaches the emission point, electrons are boiled off into space. Their departure, however, leaves the cathode positively charged, which tends to hold the freed electrons in a "cloud" around the cathode. This cloud of electrons is called the *space charge*. Some of the electrons are pulled back into the cathode while new ones are being boiled off. Soon a balanced condition is reached, where the number of electrons being emitted is equal to the number being drawn back in from the space charge.

If a dc potential is now applied across the tube, with the plate connected to the positive side of the source and the cathode to the negative, some space charge electrons will be attracted to the plate. This action permits more to boil off the cathode while electrons from the negative terminal of the source flow into the cathode to replace those emitted. Thus a steady dc current is established in the plate circuit.

If, on the other hand, the applied potential is connected so as to make the plate negative with respect to the cathode, the plate will repel electrons and no electrons pass from cathode to plate. In other words, there is no current in the plate circuit.

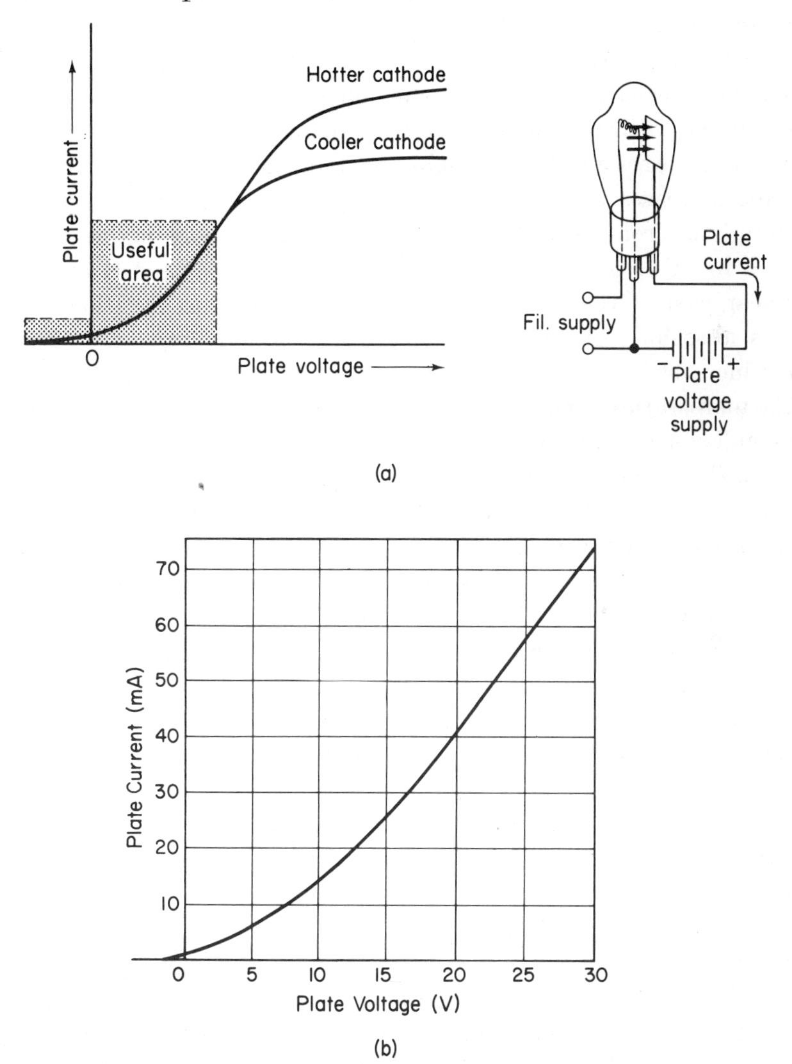

Fig. 13-5. Diode plate characteristics. (a) Complete characteristic curves showing saturation. (b) Useful characteristic for a typical diode.

Probably the most useful piece of data for a control-type vacuum tube (as opposed to a display tube or certain other specialized tubes) is a graph called the *plate characteristic*. This is simply a graph of how the current through the tube, called *plate current*, depends on the voltage between the cathode and plate, the latter being called *plate voltage*. Other operating conditions, such as the filament supply voltage, are considered fixed at some value for which the tube was designed, Figure 13-5*a* shows the form of the plate characteristic for a diode. Note that the plate current is *not* zero when the plate voltage is zero, because some of the space charge electrons manage to wander over to the plate even when no positive potential is applied. It takes a small negative voltage to shut off plate current completely. As the plate voltage is increased, we enter an operating regime where the plate characteristic is essentially a straight line, indicating that the current through the tube increases in proportion to the plate voltage, This portion of the characteristic, from zero current through the linear part, is the most useful, and it is in this region that the diode is normally operated. To the right of the linear part of the characteristic, we encounter the situation where the plate is absorbing electrons as fast as the cathode can emit them. Any further increase in plate voltage is useless, so long as the cathode is operating at that temperature, and the plate current levels out at its maximum value. This condition is called *voltage saturation*. If we supply more power to the filament, the cathode gets hotter and permits a greater maximum current. The net effect is that the linear part of the characteristic extends farther, to a higher plate current, before voltage saturation occurs, as shown by the upper characteristic of Fig. 13-5*a*. The actual plate characteristic of a typical diode is given in Fig. 13-5*b*.

The primary application of a diode is in changing an ac waveform into a pulsating dc waveform. In general, this process is called *rectification*, and in regard to radio signals specifically, it is called *detection*. To understand how a diode rectifies, remember that it allows current to pass from cathode to plate only when the plate is positive with respect to the cathode. In this condition, the diode is said to be *forward-biased*. When the plate is negative with respect to the cathode, the diode is said to be *reverse-biased* and no current flows. This condition is illustrated by the two schematics in Fig. 13-6, which show a diode connected across an ac source at two different instants: one where it is forward-biased and one where it is reverse-biased. To see what happens over a few complete cycles, consider the lower illustration in Fig. 13-6. The ac waveform, below the plate characteristic curve, represents the applied ac voltage. We can trace its excursion, from maximum positive to maximum negative on the plate voltage scale. Projecting upward to the plate characteristic itself, however, we see that the plate current is zero over essentially the whole negative half-cycle and varies

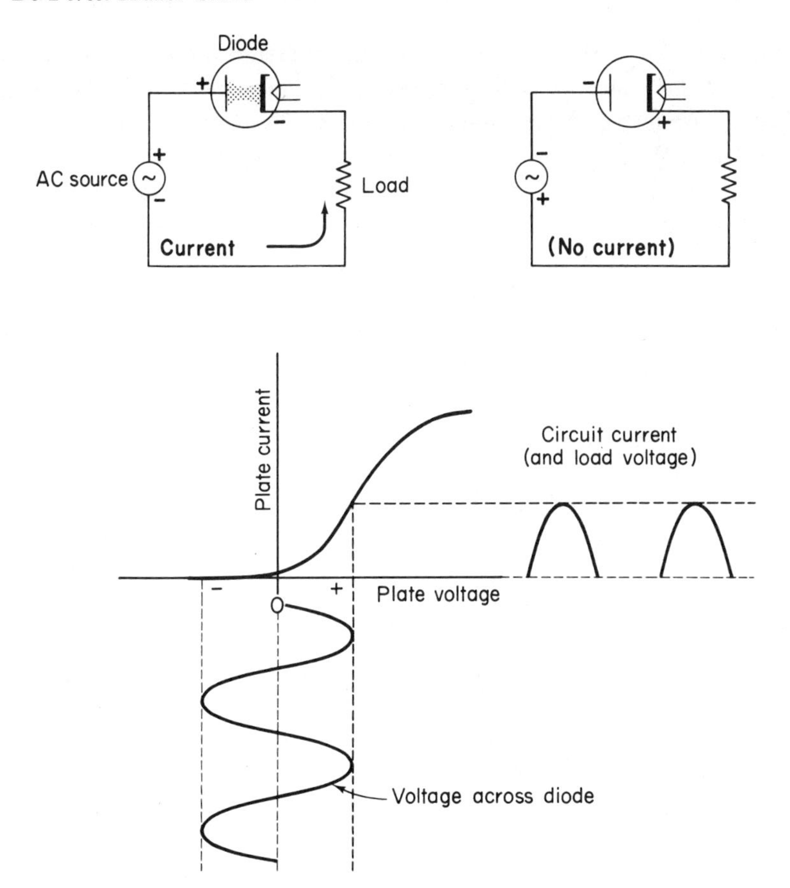

Fig. 13-6. Rectification of ac with a diode.

approximately sinusoidally over the positive half-cycle. Hence the circuit current pulsates; that is, it follows the pattern on-off-on-off, etc., but whenever it is "on" it always flows in one direction. In other words, an ac plate voltage results in a pulsating dc plate current. Since the voltage appearing across the load resistance depends, according to Ohm's law, on the plate current, this voltage is a pulsating dc voltage. Thus the diode has rectified an ac input voltage (the ac source) to a dc output voltage (the voltage across the load). Later we see how this pulsating dc output voltage can be "smoothed out" to a steady dc voltage.

13-7 AND DE FOREST MAKES THREE

Electronics finally came of age in 1906, as a result of a most significant invention by a young man named Lee De Forest. Born in Council Bluffs, Iowa, in 1873, De Forest came from a family of clergy. Despite the fact that both his father and maternal grandfather were ministers, young Lee had a bent for science and majored in that subject at Yale. He served in the Spanish-American War, then returned to Yale, where he obtained his Ph.D. in 1899. Working as an inventor in the electrical industry, De Forest became deeply interested in wireless transmission. In attempting to improve the Fleming valve as a detector, he made the invention that immortalized him. De Forest added a third electrode to the diode, making it a *triode*. He called it an *audion*. The third electrode is a wire formed into a *grid*, which is inserted between the cathode and the plate. De Forest's audion is shown in Fig. 13-7. The addition of the grid makes it possible to use feeble voltages

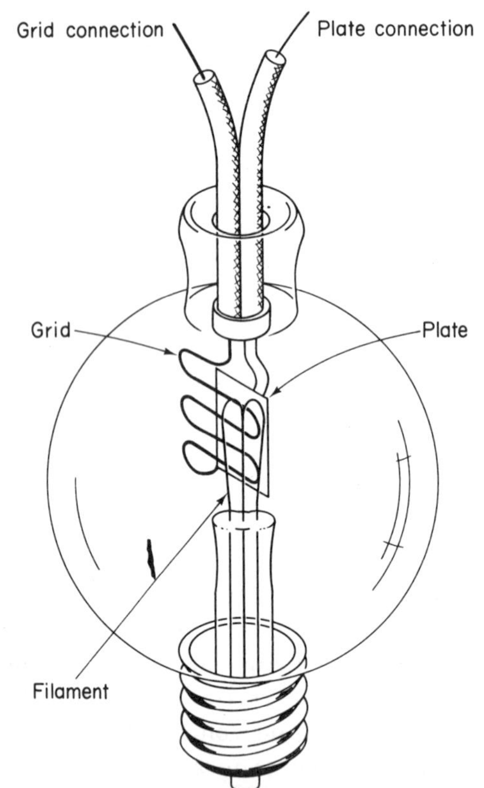

Fig. 13-7. The De Forest "audion" tube.

to control much greater voltages, that is, to *amplify* weak signals. Since the electron was as yet undiscovered, De Forest did not immediately understand and appreciate the implications of his invention. At one time, he and his associates were charged with fraud for promoting the "worthless" device. Eventually De Forest, never a shrewd businessman, sold all rights to the audion to the American Telephone and Telegraph Co. for a total of $390,000. It touched off a $90-billion industry.

When the radio industry developed, with the audion as its heart, De Forest went into the broadcasting business. In 1910 he transmitted the magnificent voice of Enrico Caruso. Not noted for his modesty, De Forest calls himself the "Father of Radio" in his autobiography. Among the other major contributions of his 88-year life span was the pioneering development of talking motion pictures.

13-8 THE TRIODE

Today's three-electrode tube is called a *triode* (*tri*, the Greed prefix for "three"). The construction of a triode and its circuit symbol are shown in Fig. 13-8*a*. The third element is still called the *grid* (in tubes with more than three electrodes, it is called the *control grid*), and it is made as a loosely

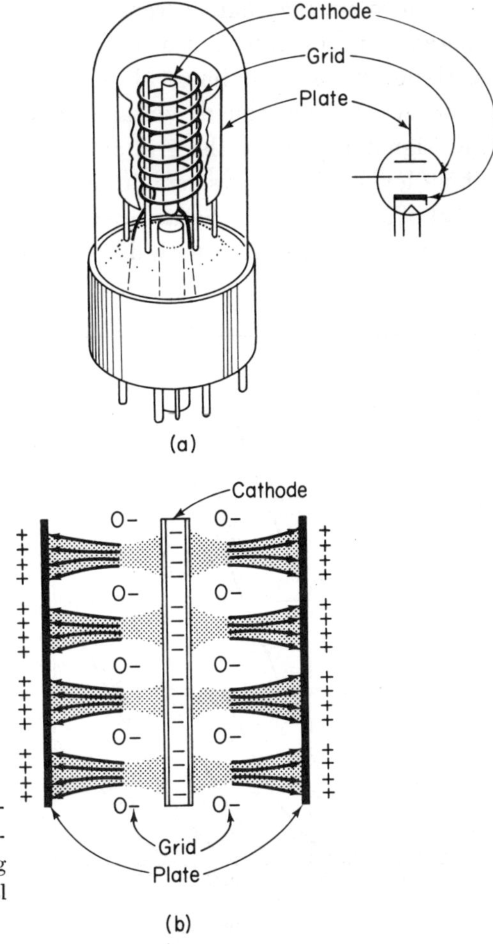

Fig. 13-8. The triode. (a) Construction and schematic symbol. (b) Cross section showing electron flow under normal operating conditions.

wound spiral of wire that closely surrounds the cathode, between it and the plate.

With no voltage applied to its grid, the triode behaves essentially like a diode. The grid blocks off so little area that electrons pass through it like water through a sieve. When a voltage is applied to the grid, however, the situation is different. A negative grid voltage (with respect to the cathode) pushes space charge electrons back into the cathode, cutting down the flow to the plate. The more negative the grid, the more the flow of electrons is throttled down. The effect is something like closing a venetian blind to shut off light, as indicated in the cross-sectional drawing of Fig. 13-8*b*. If the grid is made negative enough, its repulsive force will prevent any electrons at all from going to the plate. This no-current condition is called the *cutoff* point. When a positive voltage is applied to the grid, it aids the plate in pulling electrons away from the cathode; that is, the tube current is increased. However, this also means that current will flow in the grid circuit. The resulting operation is generally not a desirable one, for it distorts the input waveshape. Hence the grid is almost always operated negative with respect to the cathode. A fixed negative dc voltage, called the *grid bias*, is applied to the grid. The signal applied to the grid adds to bias, causing it to fluctuate about the bias level, but always remaining negative.

Instead of a single plate-characteristic curve, the triode has an infinite number, since it can be operated with different grid biases. Its plate charac-

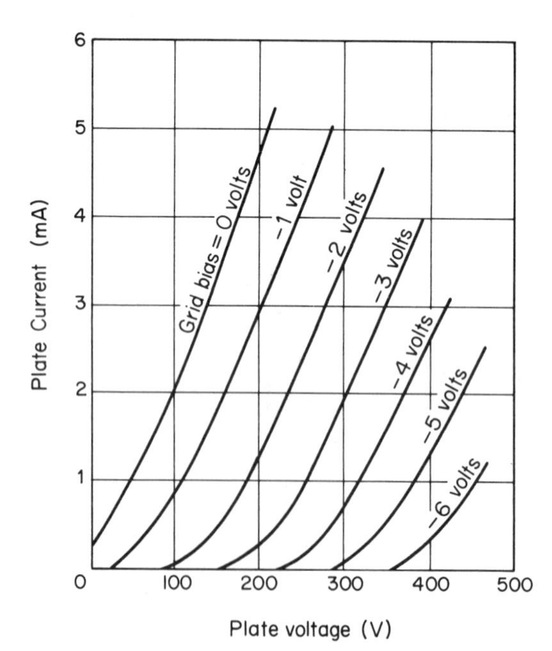

Fig. 13-9. Typical triode plate characteristics.

teristics are therefore usually presented as a *family* of curves, drawn for several selected values of grid bias, as in Fig. 13-9. These, of course, are the useful portions of the curves, which, if drawn in their entirety, would each look like the typical diode characteristic. However, we are not interested in the saturation regions of the curves. The important thing about the triode characteristics is that, over their linear parts, they are more or less parallel to each other and also more or less equally spaced for equal differences in grid bias. These traits give the triode its most marvelous capability, the ability to use a small voltage to produce a magnified version of itself by suitably controlling a sizable dc current. This ability is called *amplification* (more properly, *voltage amplification*)

To understand how a triode amplifies, consider the little circuit shown at the lower left of Fig. 13-10. This is one of several ways a triode may be connected to perform the amplifying task. The plate characteristics for this

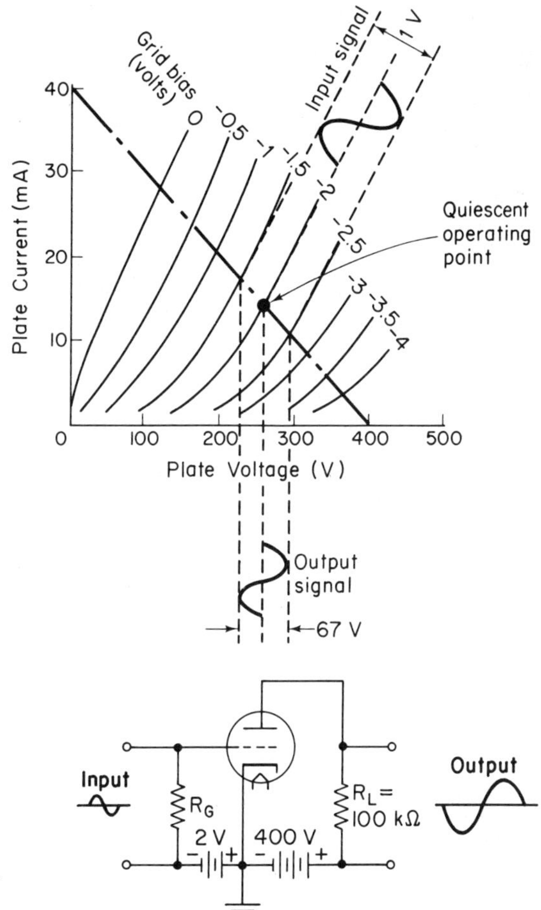

Fig. 13-10. Voltage amplification with a triode.

triode are also shown in Fig. 13-10. We have provided a 400-volt plate supply battery and a 2-volt grid supply battery. The latter is connected to make the grid 2 volts negative with respect to the cathode. The grid resistor, marked R_G in the figure, provides a way for stray space charge electrons, which inadvertently drift to the grid, to get back to the cathode. Such a *grid-leak resistor* is always provided in practical circuits to ensure that the bias is fixed by the supply source and not charge buildup from stray electrons. We have also provided a load resistor, designated R_L, across which the amplified output voltage will be developed. The straight line drawn on the plate characteristics graph, which connects "40 ma" on the current scale with "400 volts" on the voltage scale, represents the effect of this 100 kΩ resistance and is called the *load line.* The intersection of the load line with the -2-volt bias characteristic is a point called the *quiescent operating point,* and shows what's happening in the circuit when there is no input signal to the circuit. In this case, the graph shows that about 14-ma current flows in the plate circuit, and 260 of the 400 volts supplied is dropped across the tube, the remaining 140 volts appearing across R_L. Now let a sinusoid voltage with a $\frac{1}{2}$-volt amplitude (one volt peak to peak) be applied to the grid. Adding this signal to the bias means that the grid voltage "swings" from -1.5 volts, when the signal is positive, to -2.5 volts, when the signal goes negative. Projecting along the segment of the load line between -1.5- and -2.5-volt characteristics in Fig. 13-10, and down to the plate voltage scale, we see that the corresponding variation in voltage drop across the tube (and therefore also across R_L) is about 67 volts. Stop for a moment and appreciate what this point means. Using the triode circuit shown, we have taken a one-volt peak-to-peak sinusoid and produced a 67-volt peak-to-peak sinusoid, a 67-to-1 amplification. What is more, there is no reason why we cannot use the output of this circuit as input to another circuit just like it. The output of that second circuit would then be about $67 \times 67 = 4489$ times amplified over our original one-volt signal.

Thus the triode is basically an amplifier. Note that, in the circuit used, the amplified output voltage is 180° out of phase with the input signal, a characteristic of this circuit. The triode can, of course, also rectify like the diode. In some simple radio circuits, in fact, it is used to do both at the same time, by a suitable choice of operating point. Later we shall see that it can also perform another essential electronic function.

13-9 TUBES OF FOUR AND MORE ELECTRODES

Nothing is so good that it cannot be improved. Even the marvelous triode needs some improvement when we wish to amplify high-frequency signals. The electrodes in a tube are separated by a vacuum, which is a dielectric, and two electrodes separated by a dielectric make a capacitor. The electrodes of a vacuum tube, considered in pairs, therefore act as if they were connected by tiny capacitors. Remember that capacitive reactance decreases with increasing frequency. At low frequencies, this *interelectrode capacitance* in a triode has such a high reactance that it does not noticeably affect operation. At high frequencies, however, the reactance drops low enough to cause problems. Most serious is the grid-to-plate capacitance because this lets some of the input signal "leak" into the output circuit, causing distortion.

The grid-to-plate interelectrode capacitance effect can be minimized by adding a fourth electrode—another grid—between the control grid and the plate. This second grid is called the *screen grid* and acts as an electrostatic shield. It is operated at some positive potential, with respect to the cathode, less then the plate voltage. A tube of this type is called a *tetrode* (from the Greek prefix *tetra-* meaning "four"). A modern, ceramic tube type of tetrode called a nuvistor tetrode is shown in Fig. 13-11 (note size). A cutaway view of this same tube, showing construction, is given in Fig. 13-12.

The schematic symbol and plate characteristics for a typical tetrode are shown in Fig. 13-15*a*. Note the irregular shape of the characteristic curves at low plate voltages. This "hump" is due to secondary emission and, for amplification, is an undesirable characteristic of a tetrode, the penalty paid for a screen grid. Recall that the screen is positive like the plate. In fact, due to the shielding effect, it is the screen voltage rather than the plate voltage whose influence is primary in accelerating electrons through the tube. By the time they reach the plate, they are traveling very fast, fast enough to knock out some secondary electrons. As long as the plate is more positive than the screen, the secondary emission electrons are drawn right back into the plate. At low plate voltages, however, the screen is more positive and it attracts the secondary electrons. The result is a screen current higher than the plate current and the dip in the characteristic curve. This characteristic makes tetrodes useful as oscillators, a function we shall have more to say about presently; but as amplifiers, tetrodes have to be operated at very high voltages, far removed the on characteristic curve from the hump. For this reason, tetrodes are not widely used any longer.

The undesirable secondary emission characteristic of the tetrode can be eliminated by the addition of still another grid between the screen grid and the plate. This grid is maintained at the same potential as the cathode, negative with respect to the plate. It ensures that the direction of the electric field near the plate is always toward the plate; hence any secondary electrons are always recaptured by the plate. Since the purpose of this new grid is

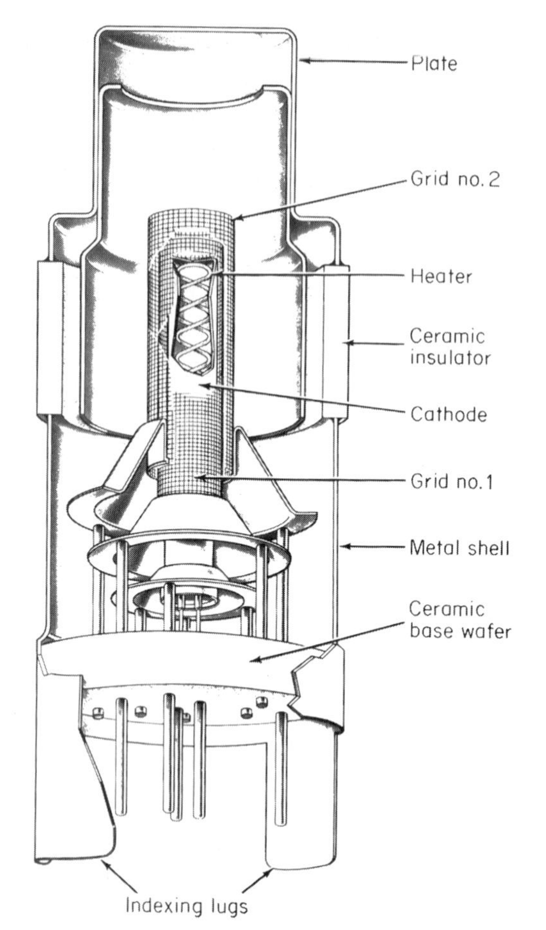

Fig. 13-12. Cutaway drawing of nuvistor tetrode. (Courtesy of Radio Corporation of America.)

Fig. 13-11. A nuvistor tetrode. (Courtesy of Radio Corporation of America.)

Fig. 13-13. A miniature pentode. (Courtesy of Radio Corporation of America.)

to suppress the ejection of secondary electrons, it is called the *suppressor grid*. The five-electrode tube is, of course, called a *pentode* (Greek *penta-* for "five"). Figure 13-13 is a photograph of a modern miniature pentode. Construction details of such a tube are given in Fig. 13-14.

Plate characteristics for a typical pentode, as well as the pentode's circuit symbol, are presented in Fig. 13-15*b*. Comparing with those of the tetrode in Fig. 13-15*a*, we note that the extra grid has changed the characteristic in other ways besides eliminating the secondary emission "hump." For normal operating plate voltages, we see that the plate current is almost a constant, independent of the plate voltage at any given control grid bias.

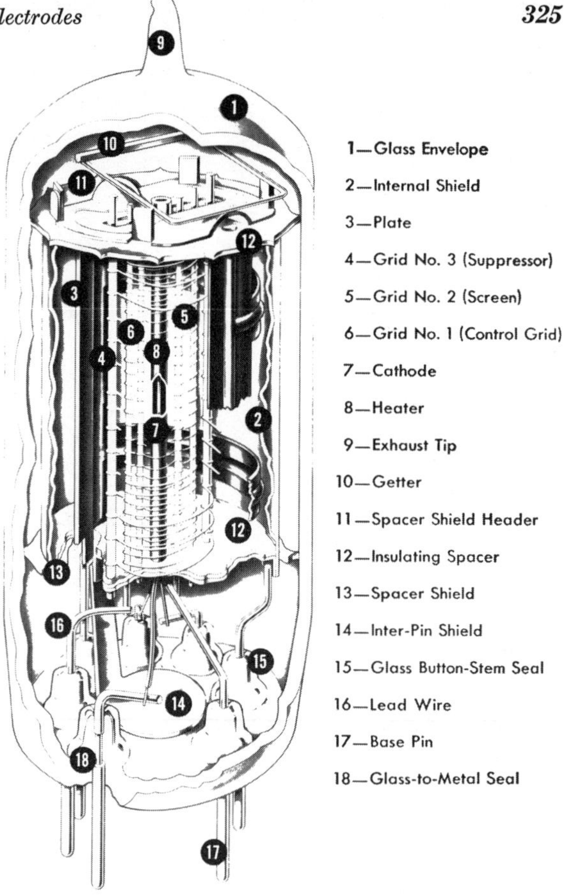

Fig. 13-14. Cutaway drawing of miniature pentode. (Courtesy of Radio Corporation of America.)

Thus the pentode is practically a *constant current* source. We also note that the spacing between characteristic curves is *not* equal for equal steps in grid bias. For this reason, the pentode can not be used to amplify large signals or distortion would result.

Another type of pentode is also very useful. This type is classed as a pentode because it does indeed contain five active electrodes. Their configuration and purpose, however, are quite different from those in the normal pentode, discussed in the previous paragraphs. This special tube is called a *beam power tube,* a typical representative of which is shown in Fig. 13-16. In this type of pentode, the screen grid is wound so that its coils are physically "in line" with those of the control grid, while the suppressor "grid" is not a grid at all but rather two beam-forming plates. The "hidden" screen results in fewer electrons being captured by the screen, that is, less screen current. This effect, in conjunction with the repulsive shaping action of the

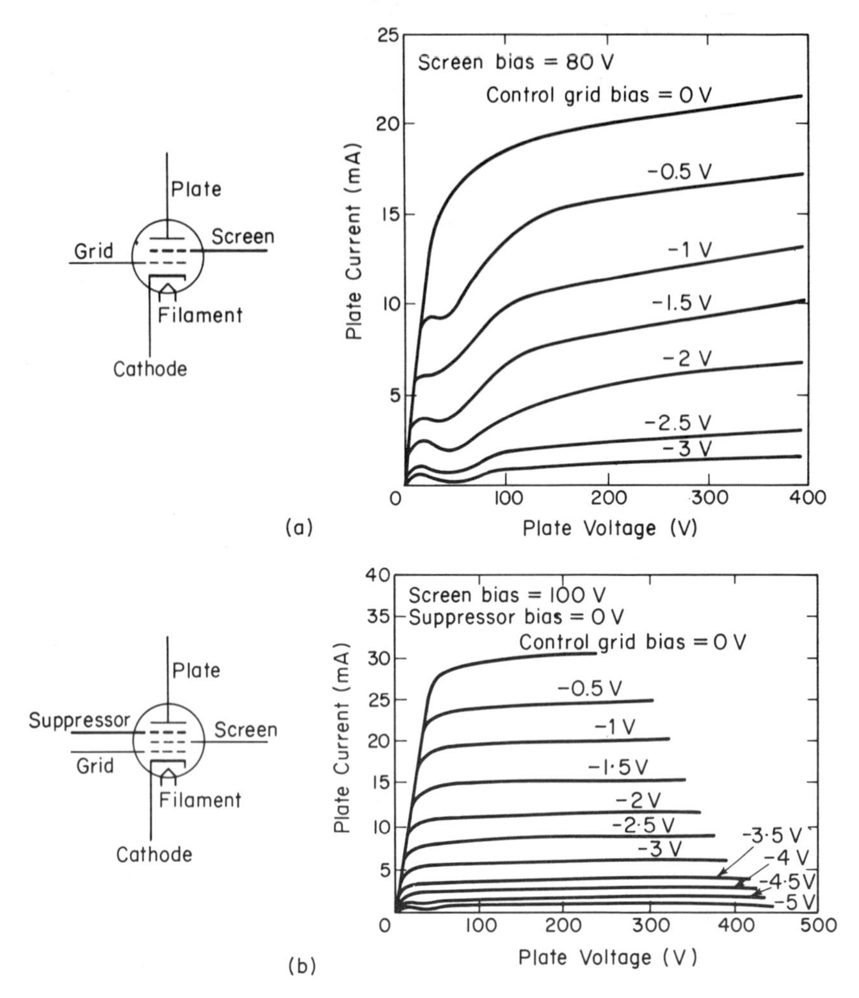

Fig. 13-15. The tetrode and pentode. (a) Tetrode symbol and typical plate characteristics. (b) Pentode symbol and typical plate characteristics.

beam-forming plates, causes the electron space current to "stream" to the plate in "sheets," as indicated in Fig. 13-17*a*. The result is a much higher current than is attainable in an ordinary pentode, tetrode, or even triode (compare the vertical scale of the characteristics in Fig. 13-17*b* with those in Fig. 13-15). The beam power tube is thus extremely useful as a power amplifier.

Certain special-purpose control tubes employ more than three grids, such as the *hexode* (six electrodes, four grids), *heptode* (seven electrodes, five grids), and *octode* (eight electrodes, six grids). Their usage is relatively infrequent.

Fig. 13-16. Beam power tube. (Courtesy of Radio Corporation of America.)

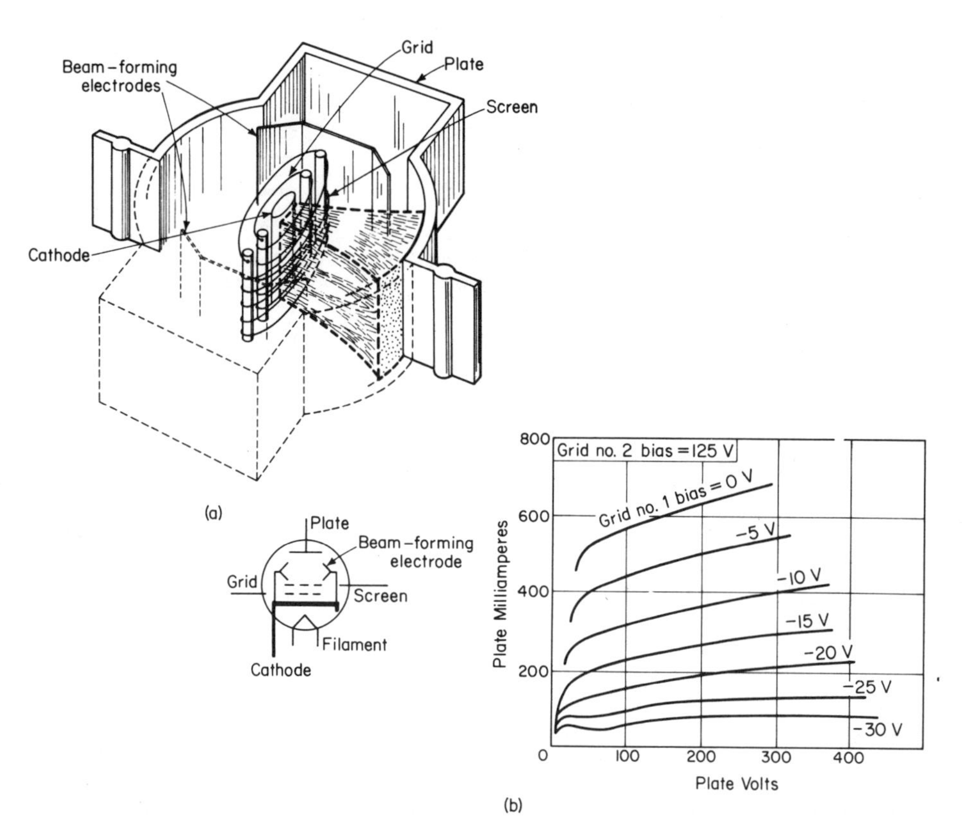

Fig. 13-17. Beam power tube. (a) Structure. (b) Circuit symbol and plate characteristics.

There are also multipurpose tubes that are, in effect, two or even three separate tubes in one envelope. They usually share one common heater and may also have a common cathode. Circuit symbols for several multipurpose tubes are given in Fig. 13-18*a*. Such tubes save space and cost. Note that it is *incorrect* to call a duo-diode a "tetrode" or to call a triode-pentode an "octode."

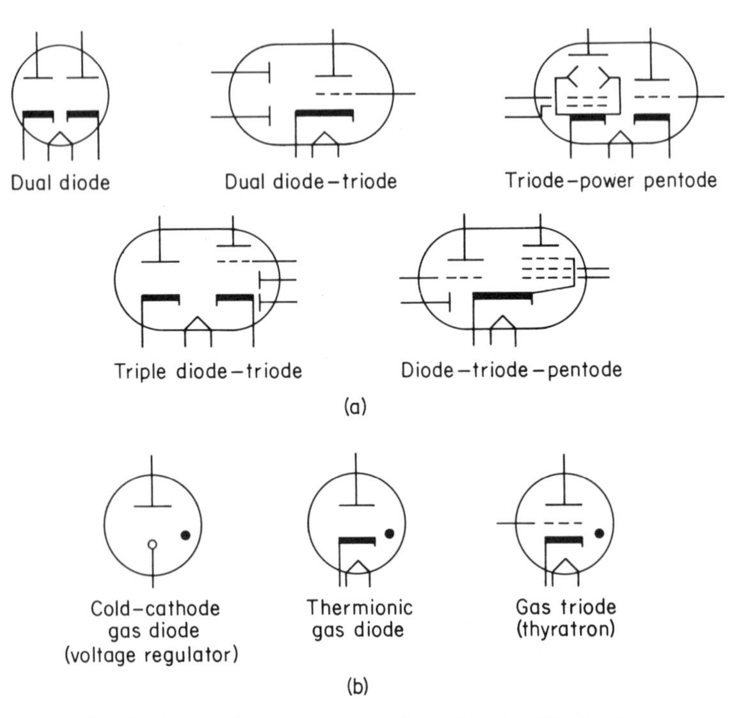

Fig. 13-18. (a) Multipurpose tubes. (b) Gas-filled tubes.

13-10 GAS-FILLED CONTROL TUBES

Gas-containing tubes are more useful than vacuum tubes for certain applications. The significant thing about gas tubes is that when the potential across them is sufficiently high, the gas ionizes and serves as the chief conductor of current through the tube.

Symbols for several gas tubes are given in Fig. 13-18*b*. The cold-cathode diode is, as its name implies, *not* thermionically heated. The operating mechanism is field emission, followed by ionization. When the plate voltage reaches a certain minimum level, a self-sustaining *glow discharge* ion current flows. This current can be increased over a considerable range with little variation in the plate voltage; therefore this type of tube is commonly used as a voltage regulator. The thermionic gas diode serves a different purpose. It is a rectifier, just like its evacuated brother, but can support a much larger current because of the gas ions. Such gas diodes are therefore widely used as high-power rectifiers.

The gas-filled triode has a unique characteristic that puts it into a class by itself. When the plate voltage of this tube, usually called a *thyratron*, is lower than that required to ionize the gas, the tube behaves like a normal triode with the grid controlling current. When the ionization voltage level is reached, however, the ion current literally "swamps out" the space charge and results in a much higher current with lower voltage drop. In this condition, the thyratron is said to have "fired" and the grid has lost control. Grid control can only be regained by shutting off current through the tube and then reapplying a plate voltage lower than the ionization level. The grid bias determines the ionization level. Thyratron circuits are frequently used in controlling machinery.

13–11 CONTROL TUBE PARAMETERS

Control-type vacuum tubes are normally operated under a fairly narrow range of conditions in order to make use of the linear part of their characteristics. For this reason, manufacturers find it unnecessary to give a complete set of characteristic curves for each type of tube produced. Such characteristics are averages for a large number of tubes, anyway, and do not necessarily match exactly those of any particular tube you buy. For the accuracy required in many applications, the plate characteristics of a tube containing one or more grids (i.e., a triode, tetrode, etc.) can be summarized by three numbers that, collectively, are called the *tube parameters*.

One of these tube parameters is called the *amplification factor* and is symbolized by the Greek letter mu (μ). Numerically it is the average change caused in the plate voltage per unit change in (control) grid voltage, keeping the plate current fixed. It is thus a measure of the tube's ability to amplify voltages.

Another tube parameter is the average change in plate current brought about by a unit change in grid voltage, with a fixed plate voltage. Since it is a current-to-voltage ratio, it is a conductance and is called the *transconductance* (or *mutual conductance*) of the tube. The symbol g_m is used for this parameter. Although conductance is normally measured in mhos (see Section 6-6), actual transconductances are quite small and usually measured in micromhos (millionths of one mho).

The third member of the tube parameter team is the *plate resistance*, r_p. Despite its name, it does not represent the resistance of the plate (which is a sheet of metal and hence a good conductor), but rather the change in plate voltage per unit change in plate current when the grid voltage is fixed. Since this is a voltage divided by current, it has the ohms unit and therefore is called a resistance. Comparison of this parameter with the actual load resistance gives an idea of what part of the plate circuit voltage will be dropped across the tube.

The three tube parameters are not independent of each other, but are connected by the simple relationship

Amplification factor = transconductance × plate resistance

Typical values for these parameters in a triode, tetrode, and pentode are given in Table 13-1.

Table 13–1 Typical Tube Parameters

	Triode (Medium-μ)	Tetrode	Pentode
Amplification factor	40	800	4000
Transconductance (μmho)	5000	8000	16,000
Plate resistance (kΩ)	8	100	250

13–12 CRYSTAL DEVICES

It is said that nothing is new under the sun. The development of the transistor is perhaps another justification for this observation. In the infancy of radio, a crystal was used as a detector. The crystal had a wire contact called a *cat whisker*, which had to be continually adjusted to find a suitably sensitive spot on the crystal. Consequently, the old crystal detector was annoyingly unreliable. When Fleming's valve came along, it displaced the crystal detector completely, and in the heyday of vacuum tubes, the poor crystal was fast becoming a mere museum piece.

Then, during World War II, with impetus from the development of radar and other devices, the crystal diode was revived. This time, however, carefully controlled man-made silicon and germanium crystals were used instead of natural galena. Soon the crystal diode was firmly entrenched in electronic technology once again. From there, it wasn't long before Bell Telephone Laboratories scientists had learned how to make a crystal triode, which they called the *transistor*.

Now the cycle is complete and the transistor family seems about to put most of the vacuum tube family in the museum. Perhaps. In any case, let us learn a little about these newcomers and how they are able to usurp the functions of tubes.

13–13 SEMICONDUCTORS

In Chapter 6 we learned that all materials conduct an electric current to some extent. They were grouped loosely as conductors and insulators. We saw that the property of conducting electricity depended on the ability to lose electrons from the valence shell. Remember that the "magic" number of electrons for the valence shell is eight. Atoms having nearly eight electrons will hang on tightly to those they have and will try to borrow from some other atom the one or two they need to make eight, becoming satisfied negative ions in the process. Similarly, atoms having only a few valence electrons will give them up quite readily, being then contented with the next outermost shell, which is complete. This latter type of atom readily becomes a happy positive ion. Caught in the middle, so to speak, are atoms with four valence electrons. They don't know whether to give up the four they have or to acquire four more, so they sit on the fence. Figure 13-19*a*

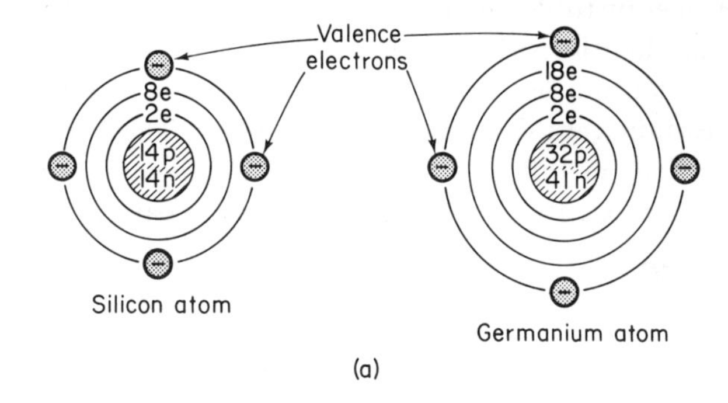

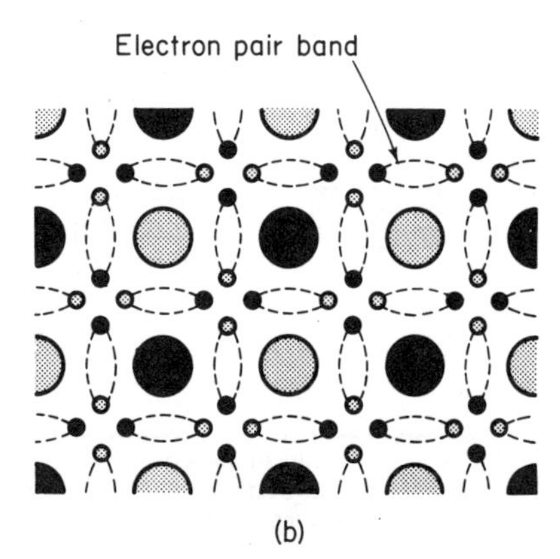

Fig. 13-19. (a) Semiconductor atoms. (b) Semiconductor crystal lattice.

shows two atoms that have four valence electrons: silicon and germanium.

When a large number of atoms form a solid chunk of the material called a crystal, the atoms are arranged in an orderly three-dimensional pattern referred to as the crystal lattice structure. They are held more or less rigidly in this pattern by forces involving their valence electrons. These electron-interaction forces are called *bonds*. Atoms that readily give up or acquire electrons form ionic-type bonds. Those with four electrons, like silicon and germanium, do neither but instead *share* electrons with their neighbor atoms to form *covalent* bonds. Figure 13-19*b* is a simplified conceptual picture of a small piece of a silicon or germanium crystal lattice. We have shaded alternate atoms and their electrons so that it is apparent which electrons belong to which atoms. Each large central circle represents a nucleus and its inner electron shells, while the small dots are the valence electrons. The dashed arcs symbolize covalent electron-pair bonds. Of course, this pattern must be imagined in three dimensions, with some bonds into and out of the page, involving atoms above and below those shown.

Atoms in these covalent-bonded lattices are quite satisfied with the shared-electron situation. Consequently, although they will give up an electron for conduction more easily than an insulator, they require a much larger energy to do so than a conductor. They are *semiconductors*. Understanding their behavior requires a short venture into solid state physics.

Quantum mechanics tells us that the electrons in an atom can only have certain very definite levels of energy. These are shown, conceptually, by the lines at different heights in the left energy diagram of Fig. 13-20. What is more, only *two* electrons can exist in each of these energy conditions.

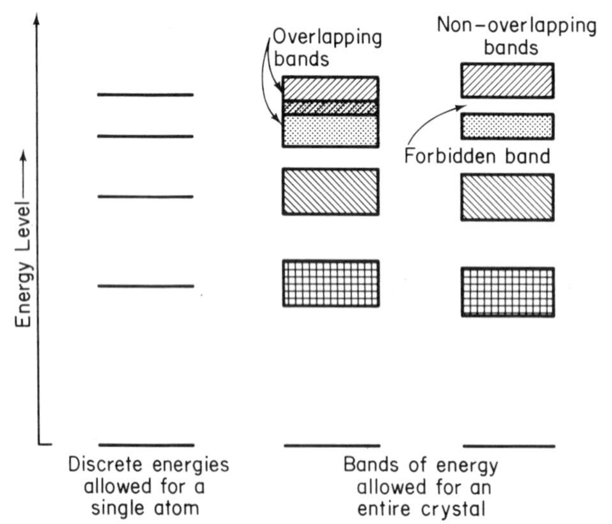

Fig. 13-20. Electron energy levels for atoms and crystals.

Now, when a large number of atoms get together to form a crystal, they are so close together that they, in effect, have a common collection of valence electrons belonging to the crystal as a whole. The principle of discrete energy levels, with only two electrons per level, still holds true, however. The mutual interactions of the atoms adjust these levels slightly so that they are one slightly above the other, rather than all at exactly the same height. The result is an extremely large number of energy levels very close to each other, where we previously had a single level. In other words, each single level for the atom has become "smeared" into an *energy band* for a crystal. Depending on how wide the bands are, those bands corresponding to adjacent levels in the atom may, or may not, overlap.

In a highly conductive metal, the uppermost energy band is (at least partly) unoccupied and overlaps the next lower filled band, as in the central diagram of Fig. 13-20. Consequently, very little energy is needed to "kick an electron up" into the unfilled part of the uppermost band, the *conduction band*. In other atoms, however, there is a gap between the upper filled band (the valence band) and the next energy band above it (the conduction band) that is unoccupied. This gap, illustrated in the right-hand diagram of Fig. 13-20, is a *forbidden region*; electrons cannot have energies in this region. A sizable and definite amount of energy must thus be added to a valence electron to make it "jump" the gap into the conduction band. If this energy requirement is very large, the material is an insulator; if it is comparatively modest, the material is a semiconductor.

13–14 ELECTRONS AND HOLES

Electric charge is conducted through the semiconductor by two effects. One is the *free electron*, which has acquired enough energy by some means to jump into the conduction band. Just the agitation associated with the temperature of the material may be enough to free a few electrons. Thus a piece of silicon, at room temperature, has enough free electrons to conduct a current when a voltage is applied across it.

Now, when the electron leaves the valence band, to enter the conduction band, one covalent bond is broken and there is a "vacancy" in the bonding site from which that free electron has departed, This vacant site is called a *hole* and behaves just like a particle with a positive charge equal to the electron's negative charge. It may be conceived of as a window, through which "one electron's worth" of the nuclear positive charge can be seen. The attraction of this positive charge will do one of two things: (1) It will pull a free electron back from the conduction band, filling the hole and thus "destroying" one free electron and one hole. This process is called *recombination.* (2) It will "steal" an electron from the valence band of a neighboring atom, destroying itself and leaving a new hole at the site from which the electron was "stolen." In this process, the effect is as if the hole had drifted.

We therefore see, as energy is added to the crystal, covalent bonds being broken and free-electron/hole pairs being formed. Some recombine while others drift at random through the crystal lattice. If a voltage is applied across the crystal, the free electrons drift toward the positive side, and the holes drift toward the negative side. Thus both holes and electrons are carriers of charge—that is, the total current through the crystal consists of a free electron current in one direction and a "hole current" in the opposite direction. (Of course, this hole current is also due to electron action, but it is much easier to discuss the "drift of one hole" rather than the multitude of valence electron exchanges among atoms that constitute the "moving" hole. Furthermore, any experimental measurement of hole drift will show that holes "are" positively charged particles. Whether they really are particles is a question for philosophers and beyond the scope of this book.)

To summarize, a crystal of semiconductor contains two types of charge carriers, free electrons and holes. In the *pure* semiconductor, these exist in equal numbers, being continually created in pairs and recombining in pairs. The crystal as a whole is, of course, electrically neutral. Addition of energy to the crystal, as heat for instance, increases the number of electron/hole pairs present. Hence the conductivity of a semiconductor increases as it gets hotter.

13-15 DOPED SEMICONDUCTORS

Figure 13-21*a* shows the phosphorus atom, which has *five* valence electrons. Such atoms are called *pentavalent* and include arsenic and antimony as well as phosphorus. Consider, now, what happens when we add some pentavalent atoms to a melted semiconductor and let the mixture crystallize. The impurity atom fits into the lattice without too much difficulty. However, only four of its valence electrons are tied up in the electron-pair bonds, leaving the fifth one very loosely bound to its parent atom. It takes little energy to move this electron into the conduction band. This situation is shown conceptually in the lattice of Fig. 13-21*a*, where the pentavalent atoms are shaded. Notice that the result of this type of impurity is an abundance of electron carriers; they are in the majority and thus are called the *majority carrier* in this type of material. Since electrons constitute the majority carrier, an impure semiconductor of this nature is clled *N-type*

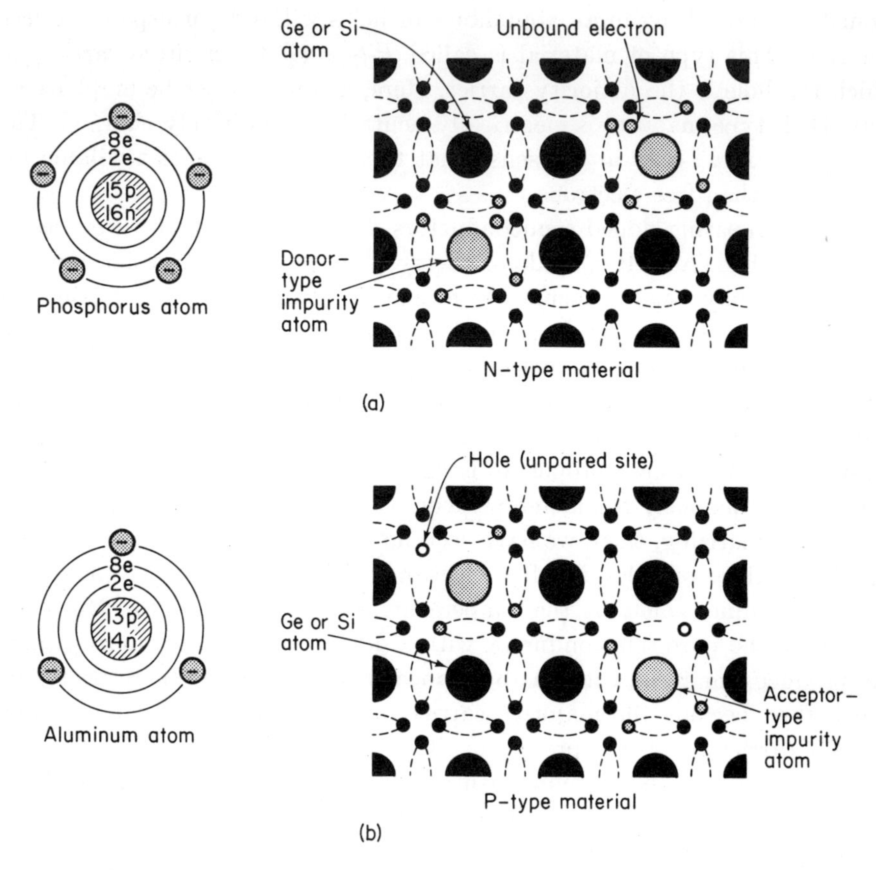

Fig. 13-21. Doping a semiconductor. (a) N-type material. (b) P-type material.

(*N* for *n*egative carrier). The pentavalent atom that provides the "extra" electron is referred to as a *donor* material, and the semiconductor is said to have been *doped* with a donor. Let it be extremely clear that when we say "extra" electron, it is in the sense that we have provided an electron not bound to the valence band without a corresponding hole. This does *not* imply that the N-type material has an electric charge; there is still one electron and no more for each and every proton in each atom of the material.

Consider, alternatively, what happens when a three-valence-electron impurity atom is added to the melt of pure semiconductor. Such *trivalent* atoms include aluminum, shown in Fig. 13-21*b*, as well as boron, gallium, and indium. Again the impurity fits into the lattice, but this time there aren't enough electrons to fill all four covalent bond sites; one is missing. Instant holes! The trivalent atoms are called *acceptor* materials. The lattice shown in Fig. 13-21*b* has been doped with an acceptor material (shaded atoms), thus resulting in an abundance of holes without corresponding free electrons. This type of material is called *P-type* (*P* for *p*ositive carrier) in which the hole is the majority carrier. Here, again, it must be emphasized that the P-type material is electrically neutral, *not* positively charged. The holes are "extra" only in the sense that there are more holes available for conduction than free electrons.

In preparing doped semiconductors, the impurity level is very carefully controlled. A melt is made of a very, very pure silicon or germanium, as the case may be. An accurately measured quantity of the trivalent or pentavalent doping material is added, usually about one impurity atom for every 100 million semiconductor atoms. A piece of previously prepared crystalline material, called the *seed* crystal, is touched to the surface of the melt and then very s-l-o-w-l-y withdrawn. In this manner, a large single crystal with the desired properties is "grown." These operations are carried out in a scrupulously clean, temperature-controlled environment.

Summarizing, then, we have seen how a pure semiconductor contains equal quantities of holes and free electrons, the total number of such pairs being determined solely by the temperature of the semiconductor. However, by doping the pure semiconductor with a donor or acceptor impurity, man can produce a material with an abundance of electrons or holes, respectively, which then become the majority carrier in that material. Of course, both types of carrier are always present in either material, because of the electron/hole pair creation due to temperature, but these thermally produced holes in N material and thermally produced electrons in P material are the *minority* carriers in these materials.

13–16 WHEN P MEETS N

Neither N-type nor P-type material is of any real value by itself in controlling currents. The effect that results in a useful device occurs when P-type and N-type material are brought together, with intimate contact at some surface. This process is illustrated conceptually in Fig. 13-22. In the upper drawing, we have a piece of P material and a piece of N material separate from each other. The little circles in the former and the dots in the latter represent holes and free electrons, respectively. Now let the two pieces be brought into contact, as shown in the center drawing. (In actuality, this would be grown as a single crystal with two different kinds of doping). Some of the free electrons will actually leave the N material, leaving their parent atoms positively ionized. They cross the junction into the P material and there combine with an equal number of holes, making the parent atoms of the holes negatively ionized. Thus we have a bar of combined P- and

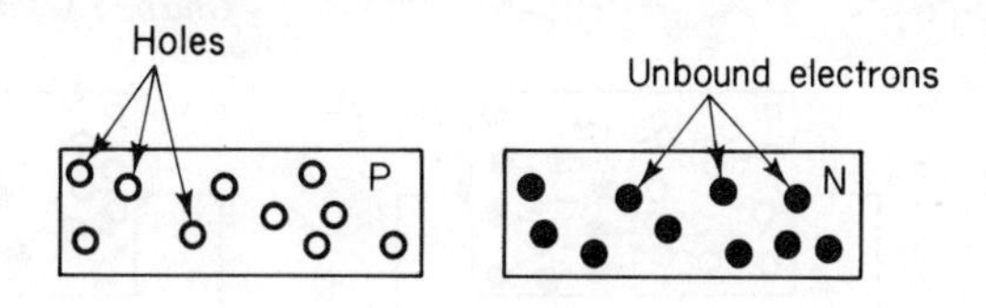

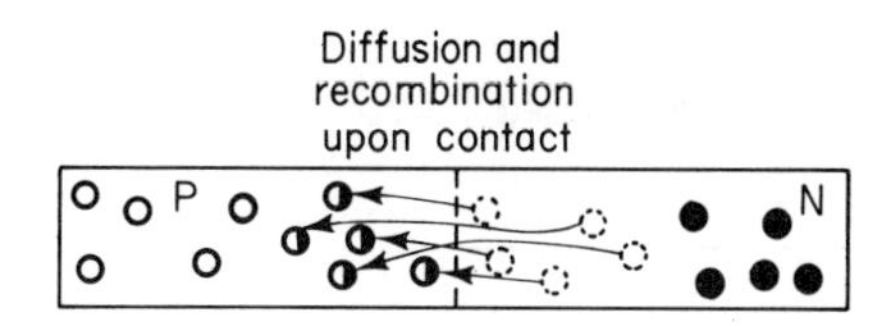

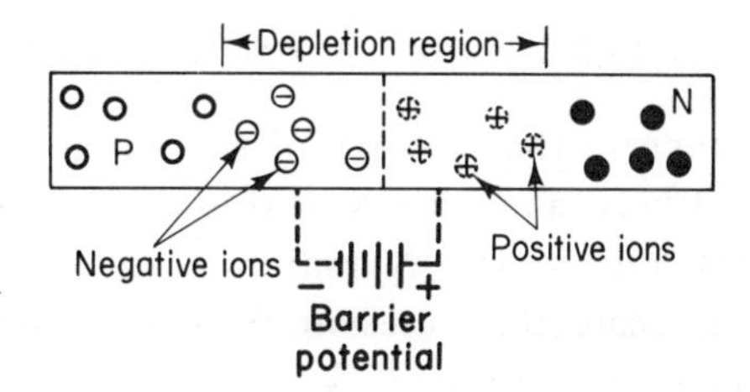

Fig. 13-22. Forming a P–N junction.

N-type material that, as a whole, has no electric charge. Yet, inside the bar, around the area where the P and N materials meet, there is a negative-charge region in the P material and a positive-charge region in the N material, both of which have lost their majority carriers. We have formed a *P-N junction*, and the two ionic areas that have been depleted of their majority carriers are called, appropriately, the *depletion region*.

The significant property of a P-N junction is that it acts like a diode—that is, it conducts a current much better in one direction than in the other. The ionic charges in the depletion region, on opposite sides of the junction, result in a *barrier potential*, shown symbolically as a little battery in the lower drawing of Fig. 13-22. To understand the diode action of the P-N junction better, consider Fig. 13-23. The left half of this figure shows the situation when the P side of the junction is connected to the positive terminal of a battery, and the N side to the negative battery terminal. Free electrons, the majority carriers in the N material, are repelled by the applied negative potential toward and through the junction. Similarly, in the P material, holes are caused to drift toward and through the junction in the opposite direction. In this condition, the junction is said to be *forward-biased* and current flows easily. Thus the forward-biased junction has a low apparent resistance.

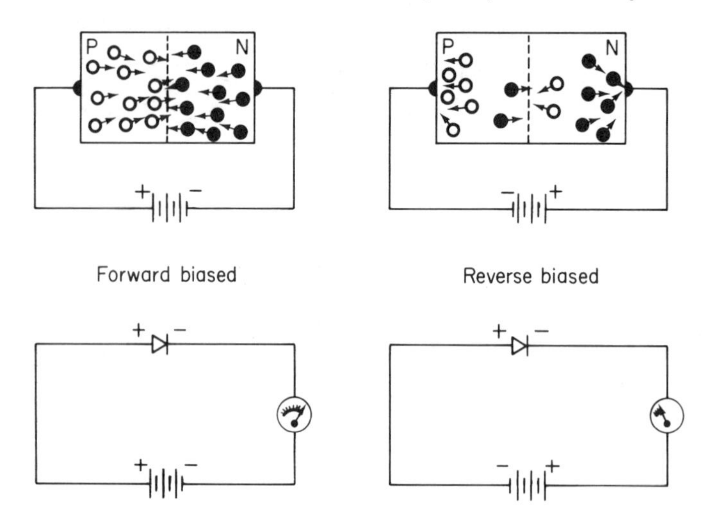

Fig. 13-23. Forward and reverse bias of a P–N junction.

If we reverse the polarity of the battery connections, biasing the P material negatively and the N material positively, the situation is quite different. Electrons in the N material are attracted toward the positive terminal, away from the junction. Similarly, holes in the P material drift toward the negative terminal. As shown in right-hand half of Fig. 13-23, this removes majority carriers from the junction; hence only a tiny current flows due to minority carriers. This situation is referred to as a *reverse-biased* (or inverse-biased) junction, which exhibits a very high resistance compared to the forward-bias condition. Indeed, if there were not the minority carriers present due to temperature, there would be no current at all in the reverse-biased junction and its resistance would appear infinite.

The lower illustrations in Fig. 13-23 show the forward- and reverse-bias conditions, using the proper schematic symbol for a junction diode rather than the conceptual "block of material" shown in the upper drawings. Note that the arrow in the diode symbol points in the direction in which *holes* move easily.

It is easy to remember that a semiconductor diode is forward-biased when the P material is connected to the *p*ositive potential.

13–17 SEMICONDUCTOR DIODES

Many types of semiconductor diodes are in use today. The rectifying effect of a junction between a conductor and the metal selenium (another semiconductor) was discovered about 1880. Yet this property was not exploited to make a commercial selenium rectifier for 50 years. Then, of course, there was the lead sulfide (galena) crystal used in early "crystal set" radios, which was obsoleted by Fleming's vacuum diode.

Until the 1940s, semiconductor diodes were available only as "power" rectifiers. Vacuum diodes, as we know, are limited in current-carrying capacity and have comparatively short lives. To handle large amounts of power, the *area contact rectifier* was developed, also called a "dry disk" or "metallic" rectifier. The semiconductor material in this type of device is either selenium (Se), first used by C. E. Fritts in 1883, or copper oxide (CuO), developed by L. O. Grondahl and P. H. Geiger in the 1920s. Area rectifiers contain a number of "plates" or "cells" of fairly large area (hence the name) in series. Each cell is a diode, consisting of a junction of either Se or CuO with a conductor, which effectively produces a P-N junction. Figure 13-24 shows a selenium stack power rectifier. Such "stacks" can

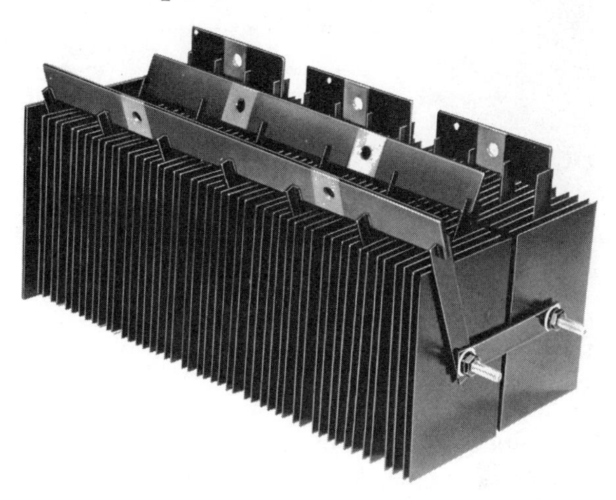

Fig. 13-24. Selenium stack power rectifier. (Courtesy of International Rectifier Corp.)

handle a fairly high current because it is distributed over a large area, which allows good heat dissipation.

The more modern innovation is the silicon diode, which is fast replacing the vacuum diode. A small silicon diode is shown in Fig. 13-25 and a large one in Fig. 13-26. It is the desired current-carrying capacity that primarily determines the size of such a diode. The shape of the characteristic curve for a diode, drawn to a uniform scale, is depicted in Fig. 13-27*a*. The various important regions on the graph are identified. Compare this characteristic with that for a vacuum diode, in Fig. 13-5*a*. To the right of the "avalanche region," the silicon diode's characteristic is similar to that of its vacuum brother, with some small but important differences. Remember that a small reverse bias was required to shut off the vacuum diode; here in the silicon diode, a small forward bias is required to get forward current started. This is due to that "imaginary" battery, the barrier potential (about

Fig. 13-25. Silicon diode, 1.3 amp. (Courtesy of International Rectifier Corp.)

Fig. 13-26. High-power silicon diode, 300 amp. (Courtesy of International Rectifier Corp.)

0.3 volts in germanium and about twice as much in silicon), which must first be overcome. Furthermore, semiconductor diodes do not reach a saturation level. Their forward current does have a limit, however, in that at some point the heat generated will destroy the diode.

When the diode is reverse-biased, it acts like a high resistance and very little reverse current flows. Thus a typical diode might have an effective forward resistance of about 50 ohms, but an effective reverse resistance of 25,000 ohms to several megohms. This small current is due to minority carriers.

If the reverse voltage is made large enough, a point called the *zener potential* is reached, where the minority carriers being hurled into the junction have enough energy to kick valence electrons up into the conduction band. The result is a sort of chain reaction, called *avalanche*, producing an extremely sharp rise in reverse current. If the diode is not designed to operate in this region (and some are, as we shall see), the crystal will be destroyed. Such destruction of a diode is referred to as *avalanche breakdown*. The charac-

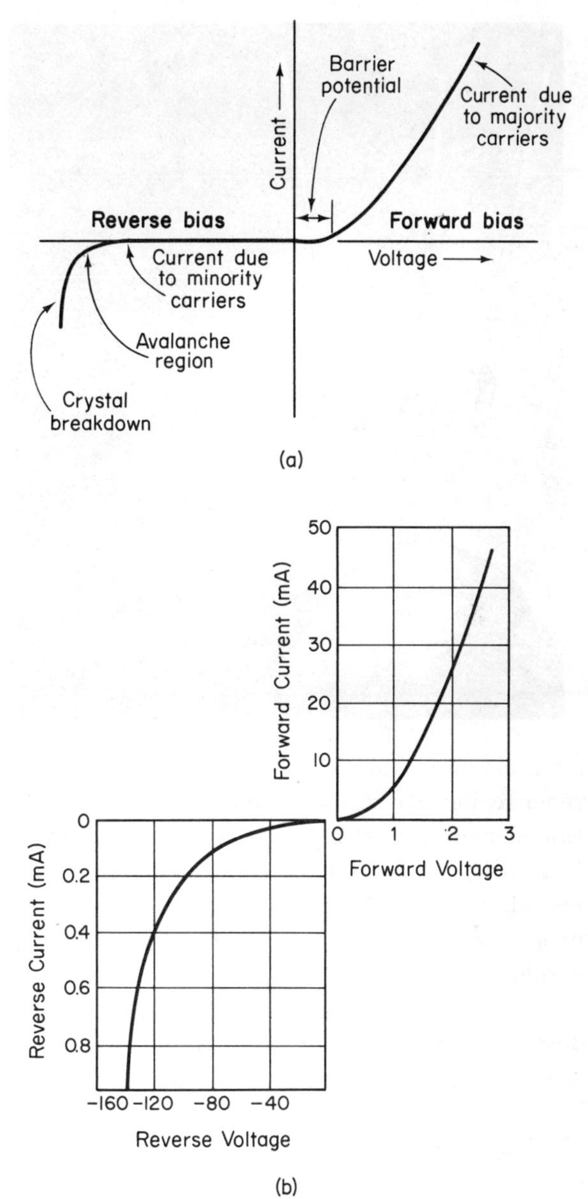

Fig. 13-27. The semiconductor diode characteristic. (a) Shape of the characteristic. (b) Characteristic for a typical diode (note difference in scales for "forward" and "reverse" halves of characteristic).

teristic of a typical diode is shown in Fig. 13-27*b*. Note the two different sets of scales used for forward-biased and reverse-biased conditions; these are necessary to make the reverse current readable.

Two special-purpose diodes are worth our attention for a moment. One of these is the *zener diode*, a one-watt version of which appears in Fig. 13-28. Externally, they look like any other diode of the same power rating.

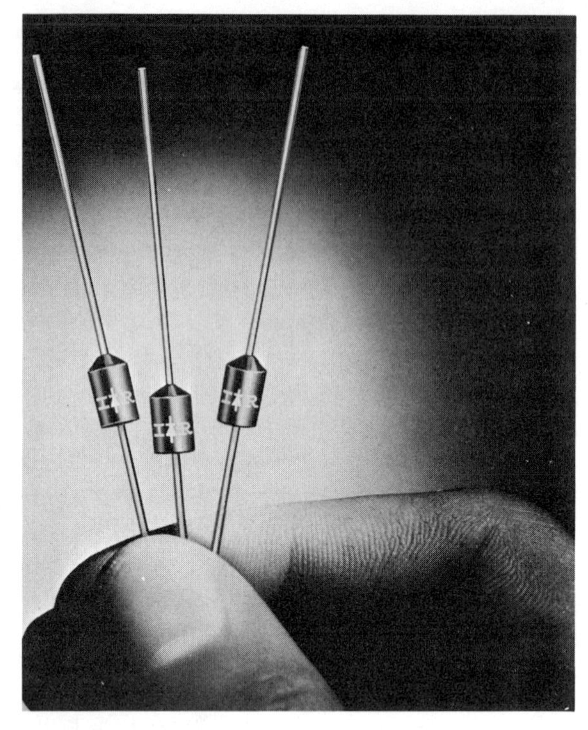

Fig. 13-28. One-watt zener diodes. (Courtesy of International Rectifier Corp.)

As their name implies, however, they are designed to be operated in the zener region. By suitable doping, the avalanche "Knee" of the characteristic can be made very sharp, as in Fig. 13-29*a*, and can be set at a desired voltage. If a suitable zener diode is connected across a dc source, such that is reverse-biased, it will maintain the voltage across its terminals at the zener potential over a wide current range. Therefore it is extremely useful as a voltage regulator, like the cold-cathode, gas-filled tube diode.

A somewhat more unusual characteristic is displayed by the *tunnel diode*. This device, which finds extensive usage in high-frequency circuits, is merely a semiconductor diode in which both the P and N materials are heavily doped and which has an extremely thin depletion region. The tunnel diode symbol and characteristic curve are illustrated in Fig. 13-29*b*. Notice (1) the steep reverse characteristic and (2) the sudden downturn in forward current, followed by a return to a more or less normal diode forward characteristic. This strange characteristic, its "negative resistance" region (the sudden dropoff in forward current with increasing forward bias) somewhat reminiscent of the tetrode, is due to two things going on at once within it. One is the normal diode action of the P-N junction, with which we are already familiar. The other phenomenon is the tunnel effect, which gives the diode its name. This effect is most evident at low forward bias, where we would

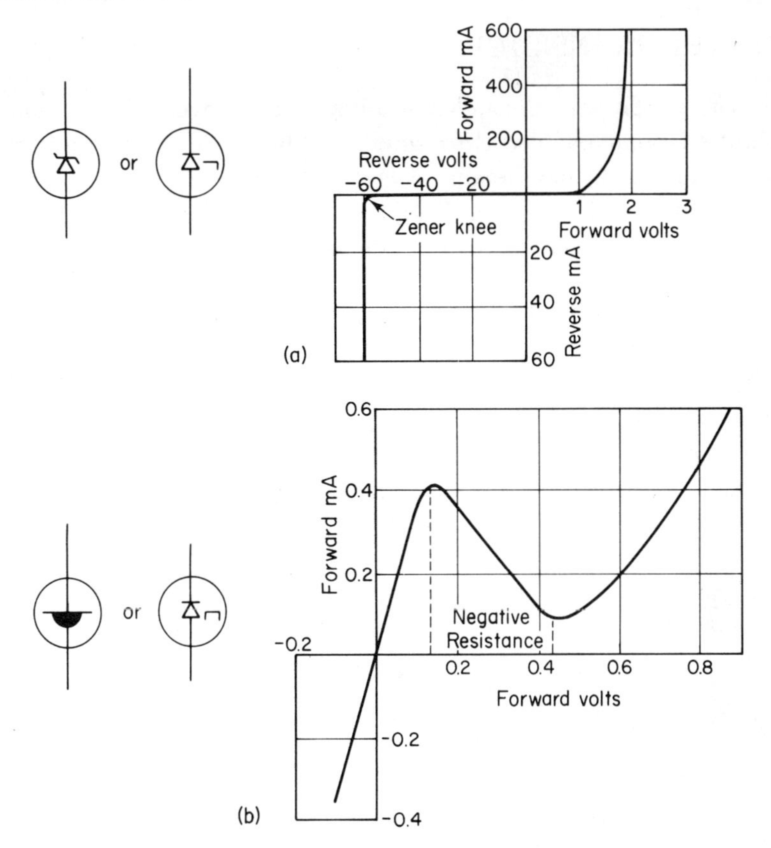

Fig. 13-29. (a) Zener diode schematic symbols and typical characteristic. (b) Tunnel diode schematic symbols and typical characteristic.

expect little current because the potential barrier must be overcome. Instead, amazingly, the carriers are able to cross the very thin junction (less than one-millionth of an inch thick) without "climbing over" the potential barrier. They apparently just disappear from one side of the junction and reappear on the other side, with the speed of light. Leo Esaki, the Japanese physicist who discovered the effect in 1957, said, in a somewhat humorous vein, that it was as if the little carriers had "tunneled under" the potential barrier, and this name stuck to the phenomenon. Its explanation is quantum mechanical and beyond the scope of this book. Suffice it to say that it is the sum of the ordinary diode current and the tunnel current which results in the unusual characteristic. The tunnel current rapidly decreases to zero with increasing forward bias, which explains the "return to normalcy." The negative resistance region makes the tunnel diode very useful for switching circuits and in high-frequency amplifiers and oscillators.

13-18 THE TRANSISTOR

The triode was a wonderful advancement over the vacuum diode; the third electrode enabled it to amplify. When solid state (semiconductor) diodes came into fashion again, electronic experimenters speculated about the possibility of adding a "grid" to a crystal diode. All attempts, however, were unsuccessful until Shockley had mastered the P-N junction in germanium.

William Bradford Shockley, an English-born physicist, joined the staff of the Bell Telephone Laboratories in 1936 upon receiving his Ph.D. from MIT. There, working in conjunction with John Bardeen and Walter Brattain, he discovered the technique of doping germanium and making a P-N junction diode. In 1948 Shockley discovered how to combine two P-N junctions to form a three-element semiconductor analogous to the triode. He called it a *transistor* because it would "*trans*fer" current across a res*istor*. Bardeen and Brattain built the first practical transistor, shown in Fig. 13-30. The active part of this device consisted of a tiny pellet of N-type

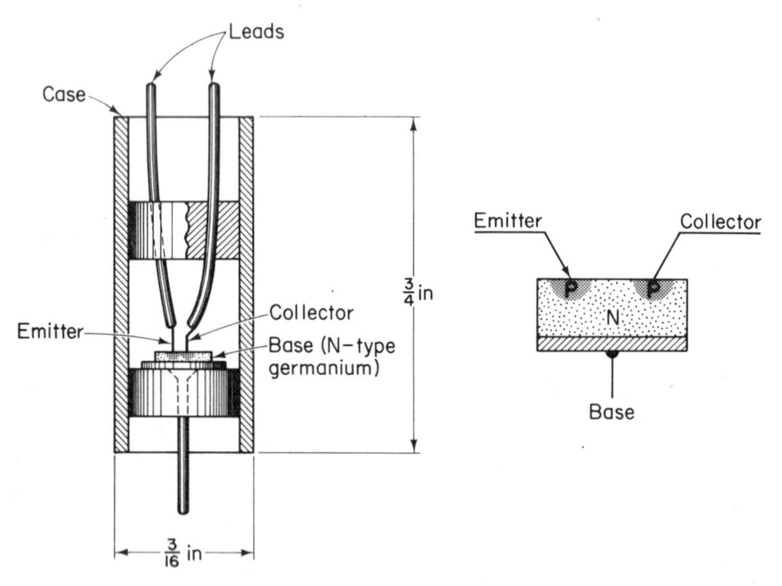

Fig. 13-30. The point-contact transistor.

germanium, no bigger than a pinhead. Two even tinier P-type regions, slightly separated from each other, were produced on one face of the N-type pellet. A very fine wire contacted each of the P regions at a point (whence the name), while a third wire was connected to the N-type substrate material through a contact plate. In use, one of the fine wires injected carriers into the substrate, and so got the name *emitter*. Its function is like that of the cathode in the vacuum tube. Carriers were removed from the substrate through the other fine wire, appropriately called the *collector*, which is analogous to the tube's plate. The current-controlling substrate, which is functionally equivalent to the control grid, is called the *base*. These names have been retained in other transistors that are constructed quite differently.

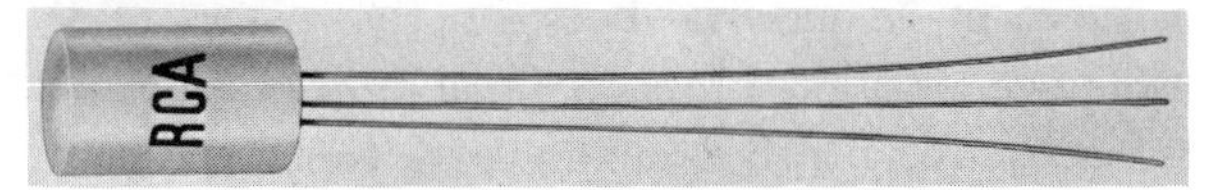

Fig. 13-31. Bipolar junction transistor. (Courtesy of Radio Corp. of America.)

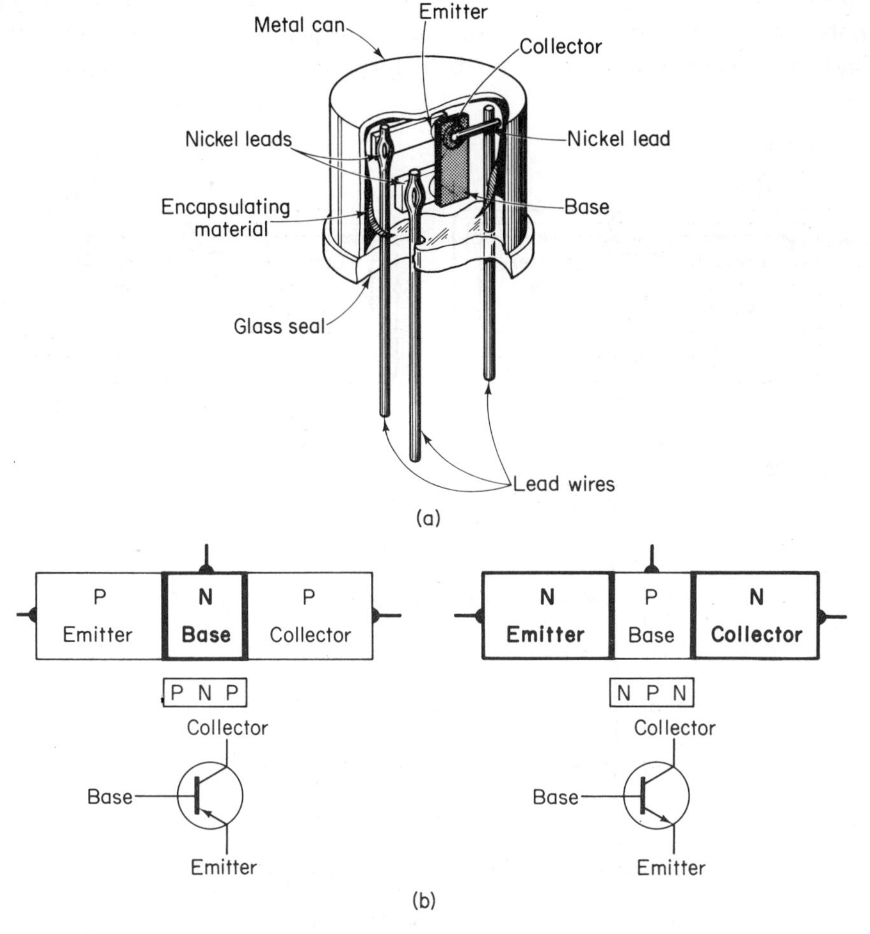

Fig. 13-32. The junction transistor. (a) Typical construction. (b) The alternate types and their schematic symbols.

The point-contact transistor had many drawbacks: it was difficult to make two alike; it was not mechanically rugged; the fine emitter and collector wires could carry only a very limited current. Yet the transistor had proven itself an extremely useful device, and this fact spurred the research that lead to the present types of transistors.

In the transistor most commonly used today, the junctions are formed in sandwich fashion, either a piece of N material between two pieces of P material or vice versa. Such a device is called a *bipolar junction transistor*, or simply a junction transistor. A representative member of the species is shown in Fig. 13-31. Construction of a junction transistor is typified by that shown in Fig. 13-32*a*. Since either type of "sandwich" is possible, we are faced with two alternative constructions: one with a P-type emitter and collector and an N-type base, which is called *PNP*; the other with an N-type emitter and collector and a P-type base, which is, of course, called *NPN*. These two configurations, with their respective schematic symbols, are

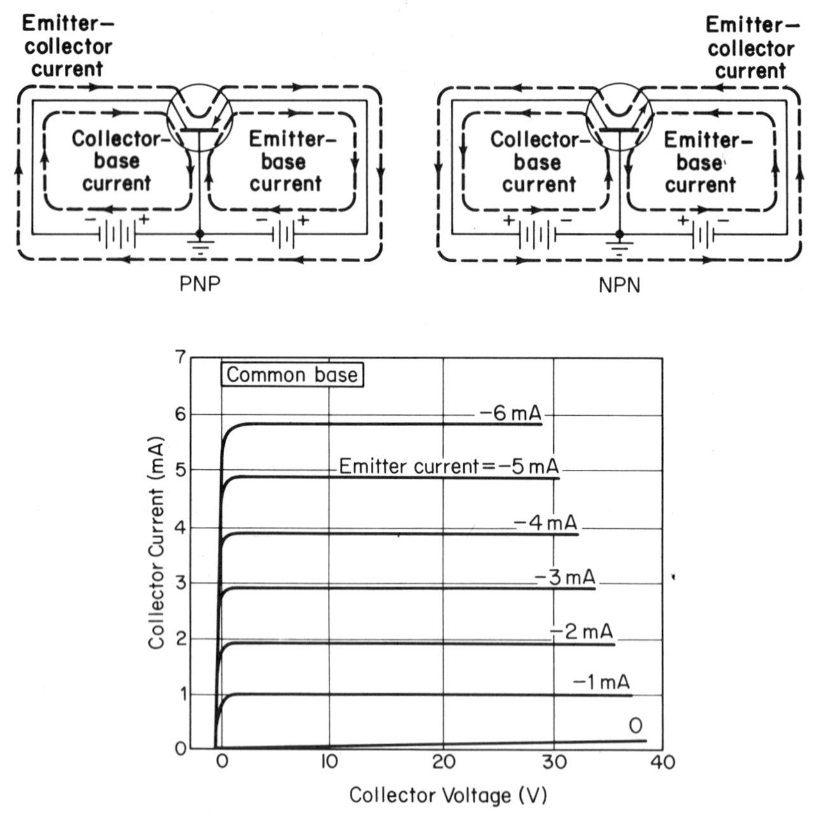

Fig. 13-33. Electron current paths and typical characteristic for common-base circuit.

given in Fig. 13-32*b*. To help recall which way the emitter's arrow goes in each type, remember that the *P*NP's is *p*iercing the base, while the *N*PN's is *n*ot.

The transistor is a low-impedance, current-operated device, while the vacuum triode is a high-impedance, voltage-operated device. Yet both devices can amplify a signal and their operation is analogous. To understand how the transistor operates, we must first clarify a few ground rules of transistor circuits. Remember that a transistor is, in effect, two diodes connected "back to back" (NPN) or "front to front" (PNP). In either case, the *emitter-base junction is always forward-biased and the collector-base junction is always reverse-biased*, when the transistor is to be used as an amplifier. In connecting the bias voltages, any one of the three electrodes can be the common (or "grounded," or "reference") point, but the *common base* and

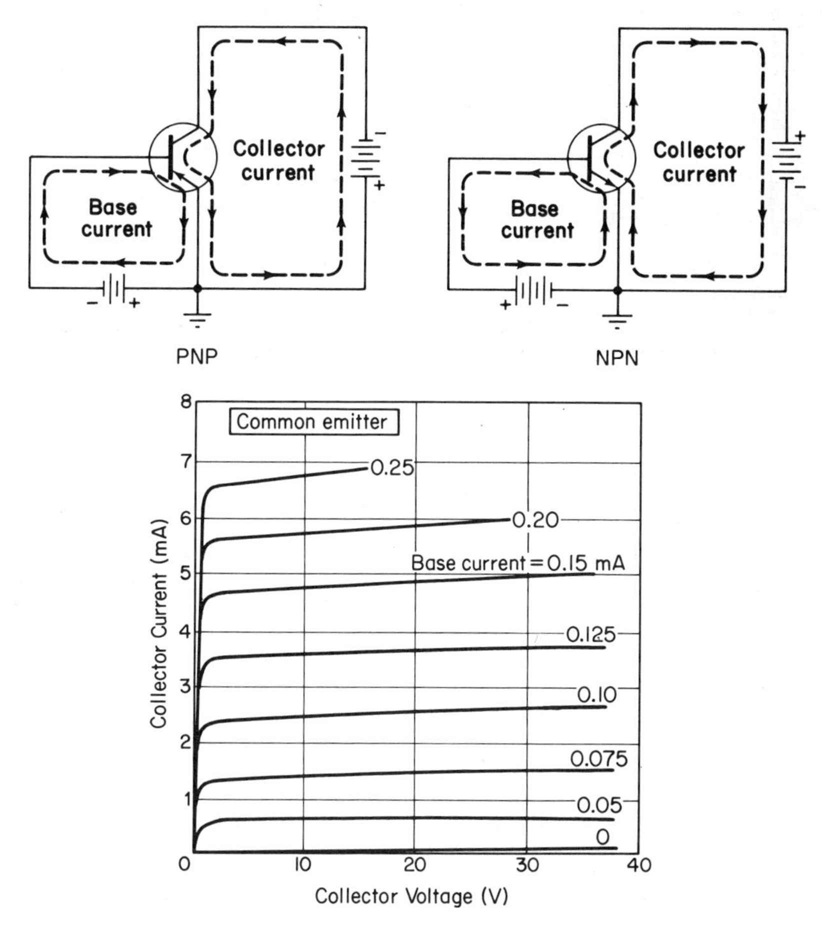

Fig. 13-34. Current paths and typical characteristic for common-emitter circuit.

common emitter configurations are used more often. Figure 13-33 shows the common base configuration, while Fig. 13-34 shows the common emitter setup. Notice how similar the characteristics are to those of a pentode tube.

Let us use the common base circuit and characteristics to explain the transistor action. For sake of definiteness, let us consider an NPN transistor. (The argument is the same for the PNP except that the polarity of the biases and carriers is reversed.) It is the injection of minority carriers into the forward-biased P-N junction that is responsible for transistor action. In our example, the negatively biased emitter injects electrons into the P-type base. This function gives the emitter its name. Inside the base layer, a few electrons recombine with the holes present, but the base is so thin that most of the electrons get through to the collector, attracted by its high

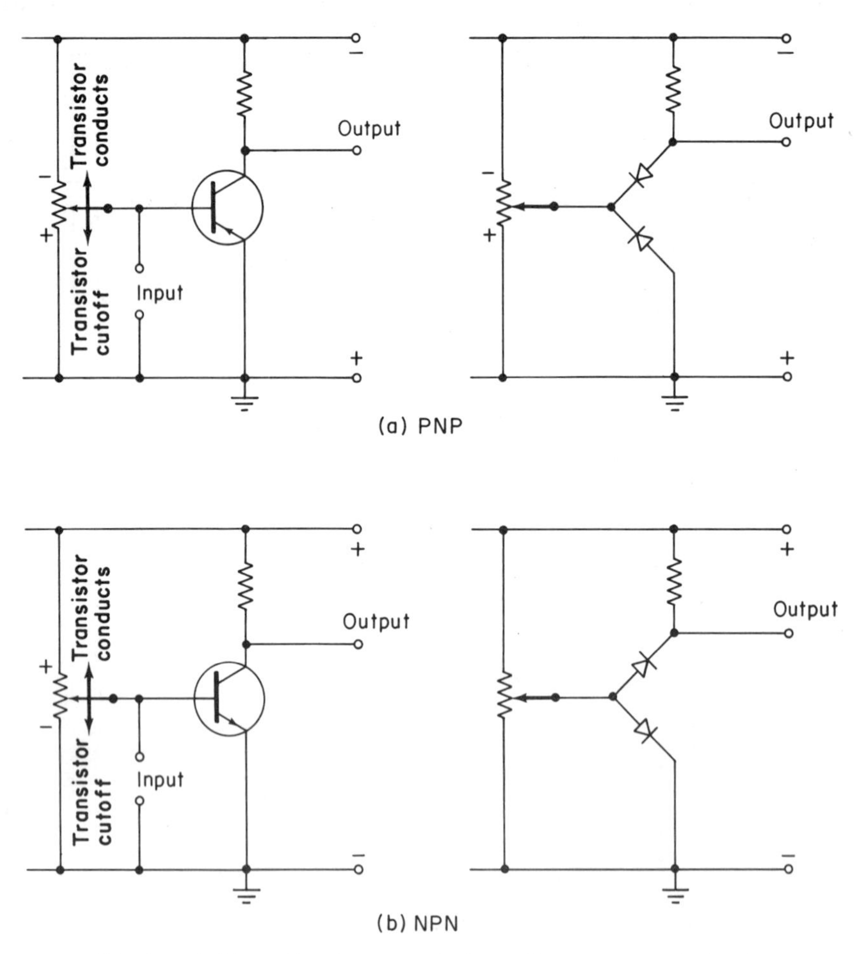

Fig. 13-35. How the base controls collector current in the common-emitter circuit. (a) PNP. (b) NPN.

positive bias voltage. These electrons join with the very small collector-base current. If the emitter voltage were removed, there would be no injected electrons and the collector current would drop off to the low level one expects of a reverse-biased diode. We see, then, a collector current that is almost as large as the emitter current. If the transistor effect were perfect—that is, no electrons recombined with holes in the base—the collector current would exactly equal the emitter current. A useful figure of merit for a transistor is therefore the change produced in the collector current per unit change in emitter current. This parameter is called the emitter-collector current "amplification" factor, the common-base, forward-current transfer ratio, or the *alpha parameter*, since it is usually represented by the Greek letter alpha (α). For an ideal transistor, α is equal to one; for any real transistor, it must be something less than one. Transistors may have α values from as low as 0.85 to as high as 0.999, but the usual value is something between 0.95 and 0.99.

The common emitter circuit of Fig. 13-34 and 13-35 will perhaps seem a bit more familiar, for it is analogous to the most common triode amplifier circuit. As in the common base circuit, minority carriers are injected into

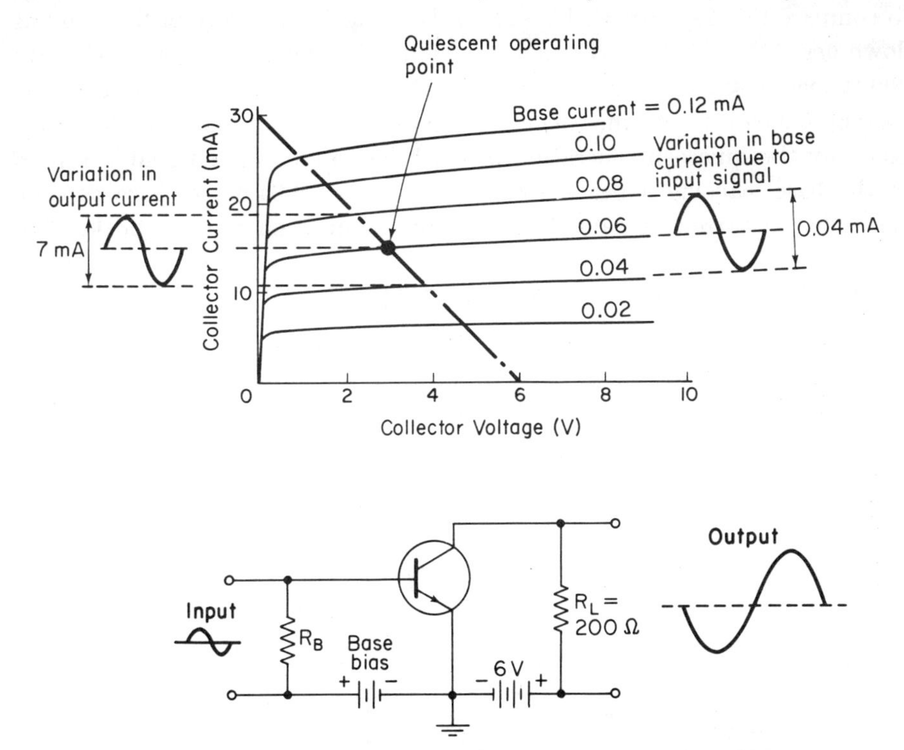

Fig. 13-36. Current amplification with a transistor (common-emitter configuration).

the emitter-base junction, due to the base voltage, and continue on through to produce a large collector current. Here it is the tiny base current that controls the (comparatively) large collector current. For this configuration, then, the appropriate amplification factor is the base-collector current amplification factor, defined as the change in collector current per unit change in base current. This parameter is also called the common-emitter forward current transfer ratio, or simply the *beta parameter*, since it is usually represented by the Greek letter beta (β). Because the collector current is so much larger than the base current, values of β are always greater than one. Actual values range from about 4 to almost a thousand, with typical values of 20 to 100. Since the common emitter circuit is analogous to the common cathode triode amplifiers, such as that shown in Fig. 13-10, the β parameter is analogous to the tube amplification factor, μ. Figure 13-35 is helpful in remembering how each type of transistor tends to behave, in the common emitter configuration, as the base bias is made more positive or more negative.

The amplifying action of an actual transistor circuit is shown in Fig. 13-36. Here we have used a PNP transistor for illustration. It is instructive to compare this figure with Fig. 13-10. Once again the straight line running down across the characteristics is the Ohm's law load line representing the 200 Ω load resistor. With no signal applied to the base circuit, the base current is 0.06 ma, and the collector voltage and current "idle" at the values given by the quiescent operating point. When an ac input signal is applied to the base, causing a peak-to-peak variation of 0.04 ma in base current, the collector current is caused to vary by about 7 ma peak to peak. Thus

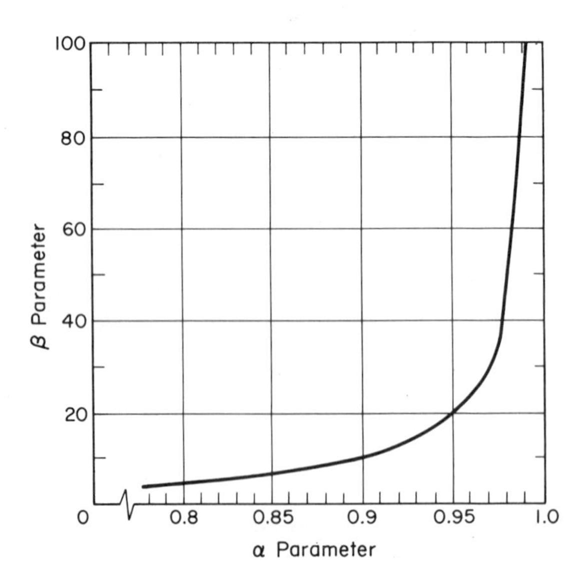

Fig. 13-37. Relationship between α and β parameters.

we have a current amplification for this circuit (which is always somewhat less than β for the transistor alone) of about 175 to 1.

In analyzing transistor circuits, it is heplful to remember that

1. Electron current always flows in the direction oppsite to the emitter's arrow.
2. Emitter current is the sum total of collector plus base current, the latter being small.

The α and β parameters are related to each other, as shown by Fig. 13-37.

13–19 FIELD EFFECT TRANSISTORS

For some applications, the junction transistor is inferior to vacuum tubes. Two of its major disadvantages are (1) low-input impedance, which makes it a poor match for many input devices and (2) low maximum operating frequency, which makes it unusable in high-frequency radio applications. A johnny-come-lately to the semiconductor field, called the *field effect transistor*, goes a long way toward overcoming these difficulties. This device, in its original form, also the brainchild of Shockley, depends on the majority carriers for its operation. It comes in two varieties: the *junction* field effect transistor, called a JFET, and the *insulated-gate* field effect transistor, called an IGFET.

The structure of one type of JFET is represented in Fig. 13-38*a*, together with its schematic symbol. Here a narrow *channel* of P-type material connects two larger P-type regions. On opposite sides of the channel are two N-type regions called *gates*. Through the large P region on one end of the channel, majority carriers (holes in this case) are injected into the channel. This P region is therefore called the *source*. The P region on the opposite and of the channel is negatively biased to withdraw the holes, and is called the *drain*. With no bias on the gates, a majority-carrier current easily flows from source to drain through the channel. When the gates are biased positively with respect to the source, they have reverse-biased junctions with respect to source, drain, and channel. The reverse-biased gates form depletion regions in the channel, cutting down the area through which the holes can drift. This condition increases the effective source-drain resistance and lowers the corresponding current. If the gate bias is made large enough, it will completely deplete the channel and shut off the hole current entirely. Thus the JFET behaves like a triode tube, and can amplify because a small gate bias will control a large source-drain current. In this discussion, the device

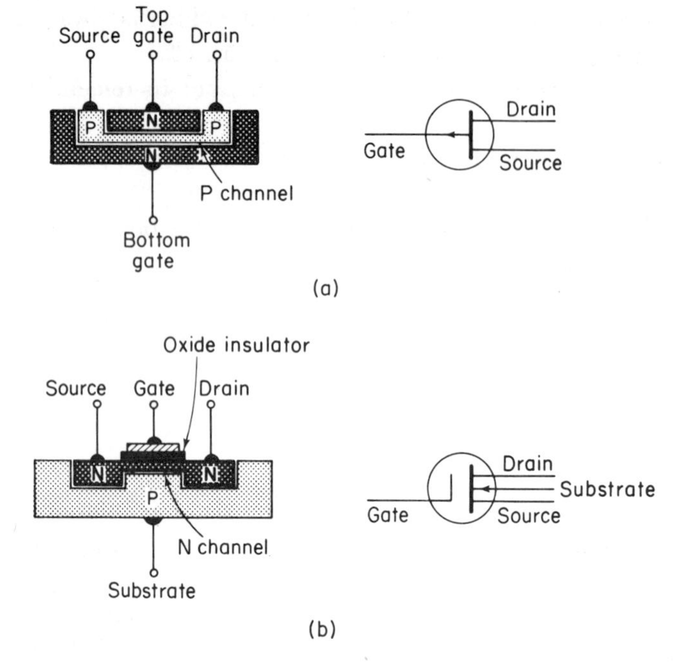

Fig. 13-38. Field effect transistors: (a) Junction type. (b) Insulated-gate type.

was a P-channel JFET. Naturally, the materials can be interchanged to make an N-channel device.

The IGFET, one type of which is illustrated by Fig. 13-38*b*, operates on a similar principle. In this device, however, the lower gate is simply called the *substrate*, while the upper gate (which is now the only gate) is not a semiconductor at all but a small metal plate that is *insulated* from the channel by a thin layer of glass. For the N-channel device illustrated, the drain is biased positively with respect to the source and substrate, the latter being connected together to ground, and electron current flows in the N channel. When a negative voltage is applied to the gate, holes from the P-type substrate are attracted up into the channel, causing a depletion region and reducing the source-drain current. Application of a positive voltage to the gate repels holes into the substrate, thereby widening the channel and increasing the current.

The JFET's gates can only be reverse-biased or it will not function, whereas the IGFET's gate can be biased either way. Moreover, the IGFET's input impedance is higher than that of the JFET because of the insulating glass between the gate and the body of the device. The input impedance of a JFET is typically from one megohm to about 1000 megohms, while that of an IGFET is in the millions of megohms. These considerations make the IGFET adaptable to a wider variety of problems. The glass insulation between the gate and channel in the IGFET is formed simply by oxidizing the silicon itself (silicon dioxide is a glass). For this reason, the popular IGFET is also called the *metal oxide silicon* field effect transistor, or MOSFET, one of which is shown in Fig. 13-39.

Fig. 13-39. Metal oxide silicon field effect transistor, MOSFET. (Courtesy of Radio Corp. of America.)

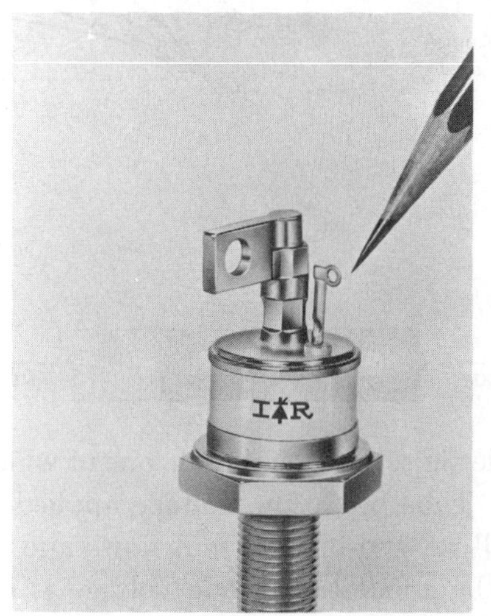

Fig. 13-40. Silicon controlled rectifier. Pencil points to gate terminal. (Courtesy of International Rectifier Corp.)

13-20 THE SILICON-CONTROLLED RECTIFIER

Many other semiconductor devices contain 2, 3, 4, and even more layers of P and N materials. So rapid is the progress in solid state electronics that several new devices will be perfected by the time these words are first printed. It is beyond the intent of this book to attempt to describe or even list all presently used types. There is, however, one more semiconductor control device that deserves a moment of our time. That device is the *silicon-controlled rectifier*, or SCR. Various manufacturers also call it a "thyristor," "thyrode," and "trinistor".

The SCR is a three-terminal device like the transistor, as can be seen from Fig. 13-40. Structurally it contains four layers of alternating P and N material and hence three P-N junctions. Functionally it is the solid state equivalent of the thyratron. The construction and characteristics shown in Fig. 13-41 tell the story. Because of its three junctions, the SCR is like three

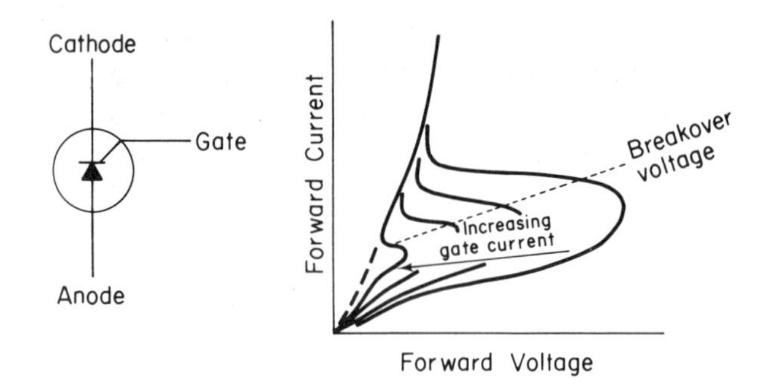

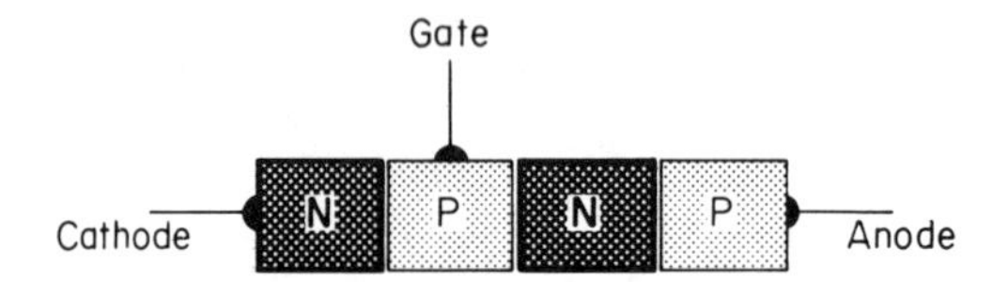

Fig. 13-41. The silicon controlled rectifier. Symbol, characteristic, and structure.

diodes in series, the middle one of which is reversed with respect to the other two. Thus, with low voltage applied and no bias on the gate, only a very small reverse-bias current flows and the SCR is said to be "switched off." As the anode-to-cathode voltage is increased, a critical voltage called the *breakover voltage* is reached where avalanche breakdown occurs at the reversed junction and the SCR conducts. The breakover voltage level is controlled by the gate-to-cathode bias. This fact is understandable in that the SCR can also be conceived of as an NPN and a PNP transistor, mutually interconnected base to collector, with the two emitters forming the SCR's anode and cathode. A positive voltage applied to the gate is thus applied to the collector of the PNP and base of the NPN, which saturates the device in the "switched on" condition.

As in the thyratron, once breakover has occurred and the SCR is switched on, the gate has no control (unless a very large negative potential is applied) and the anode potential must be momentarily removed to restore control.

13-21 TEMPERATURE AND THE THERMISTOR

Heat is the deadly enemy of most semiconductor devices. You will recall that adding heat to the crystal lattice of any semiconductor material, pure or doped, caused the production of pairs of holes and conduction electrons by freeing electrons from the valence band. In transistors and diodes, if enough heat is supplied, the results can be disastrous. These additional carriers add to the small reverse leakage current that always exists in a P-N junction. As the leakage current builds up, the resistance heating effect tends to raise the temperature. If the leakage current, which depends on the temperature of the crystal, reaches the point where it creates heat faster than the physical structure of the transistor can get rid of it, the crystal gets hotter. This process produces more leakage current, which further increases the transistor's temperature and results in still more current, making the temperature still higher, until the transistor literally destroys itself. This condition is called *thermal runaway* and must always be guarded against by proper circuit design and by providing paths for heat to get out of devices handling large amounts of power. Recall the finned plates comprising the selenium stack (Fig. 13-24) and the heavy heat-conducting mounting studs on the power diode and SCR (Figs. 13-26 and 13-40). Since the behavior of semiconductor devices depends on temperature, the characteristic curves must be defined for a *specified* temperature. Normally this temperature is room temperature, 25° C (77° F).

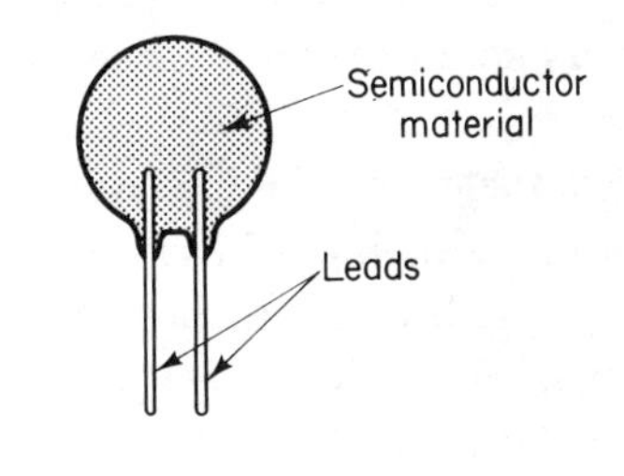

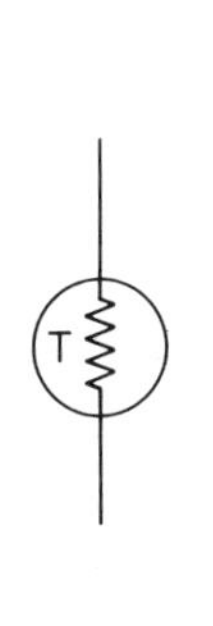

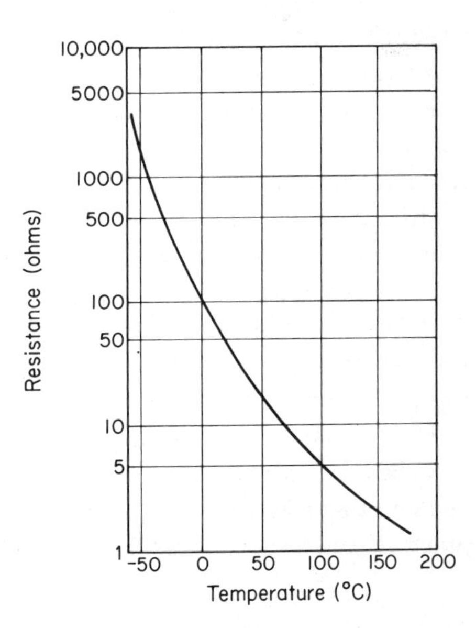

Fig. 13-42. Structure symbol and characteristic of a typical thermistor.

There is one semiconductor device, however, that thrives on heat. It is perhaps the simplest of all semiconductors, containing no P-N junction—the thermistor. A thermistor is basically a resistor whose resistance varies with temperature. Of course, the resistance of all resistors changes with temperature. The thermistor, however, being made from semiconductor material, shows a *larger* variation in resistance with small changes in temperature than conductors, due to the hole-electron pair creation effect we have just discussed. It has no junction to be destroyed and can therefore be used as a temperature transducer. The structure and resistance-versus-temperature characteristic of a typical thermistor are given in Fig. 13-42. Note the logarithmic resistance scale of the characteristic curve, because of the large variation.

13-22 OTHER TUBES AND SEMICONDUCTORS

We will encounter several other important types of tubes and semiconductors as we progress. A few are worthy of brief mention at this point, however.

Photoelectric devices are an important class of control devices. In these units, an incident beam of light, or absence of it, is used to control the current through them, rather than a second bias voltage or current. Earlier we met the photovoltaic cell, a device that *produces* current from the light energy striking it. Now we are interested in photoconductive cells which require a separately supplied bias voltage and pass current (i.e., change resistance) in proportion to the intensity of light impinging on them. Figure 13-43*a* gives the structure, symbol, and a typical characteristic curve for a vacuum tube photocell. Tube types come in both vacuum and gas-filled varieties, each having somewhat different characteristics (the characteristic given in Fig. 13-43 is for a vacuum tube). The analogous semiconductor photocell is called a *photodiode*, its structure, symbol, and typical characteristic being given in Fig. 13-43*b*. Both devices depend on absorption of photon energies by electrons. In the vacuum tube, photoemission takes place. The electrons have absorbed photons of such energy that they can overcome the work function of the cathode material and escape into space, where they are attracted and pulled in by the positive charge on the anode. In the semiconductor photodiode, which is always operated reverse-biased, photons are absorbed by valence band electrons, raising them to the conduction band and creating electron-hole pairs. These minority carriers are thrown into the junction and swept across by the applied bias; hence the current.

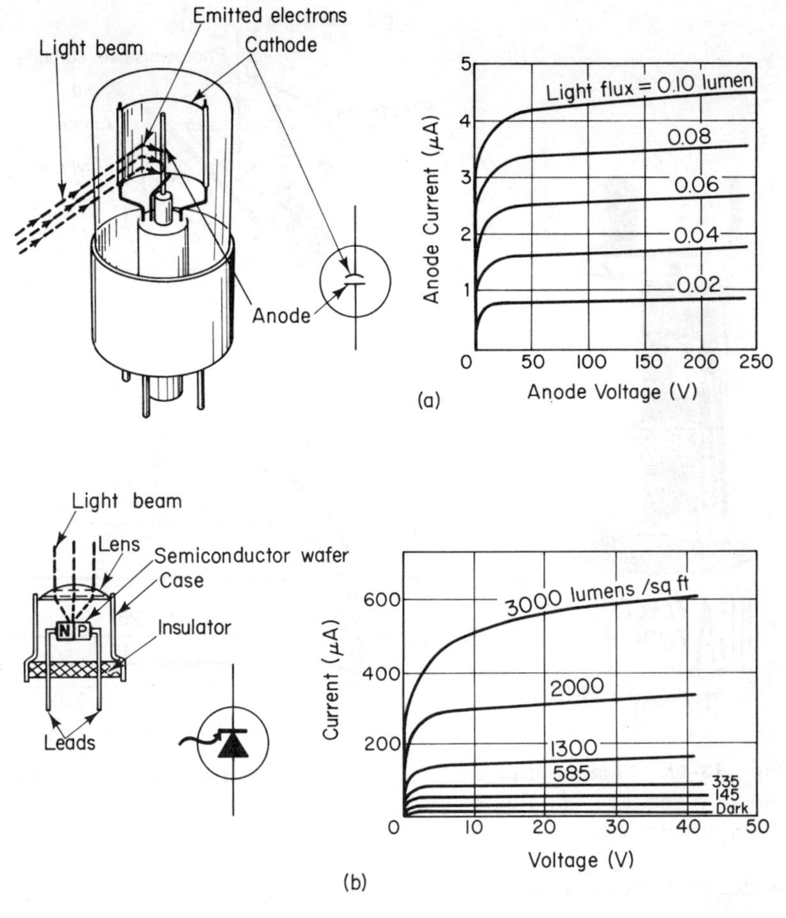

Fig. 13-43. Photoconductive cells. (a) Vacuum tube type. (b) Semiconductor type.

Photocells have many diverse applications. They are commonly used for pickup transducers for the soundtrack in motion picture projectors; tone generating devices in better-quality electronic organs; and in counting moving objects. You may also have seen them in use as an entry-exit notifier in some stores.

An amazing vacuum tube, which is responsive to light, is the photomultiplier tube. As might be implied from its name, this tube is capable of multiplying the effect of the light beam—that is, of getting a relatively strong current from a weak light source. This process is accomplished by employing both photoemission and secondary emission. The tube, a typical specimen of which is shown in Fig. 13-44, contains a photoemissive cathode

Fig. 13-44. Photomultiplier tube. (Courtesy of Radio Corp. of America.)

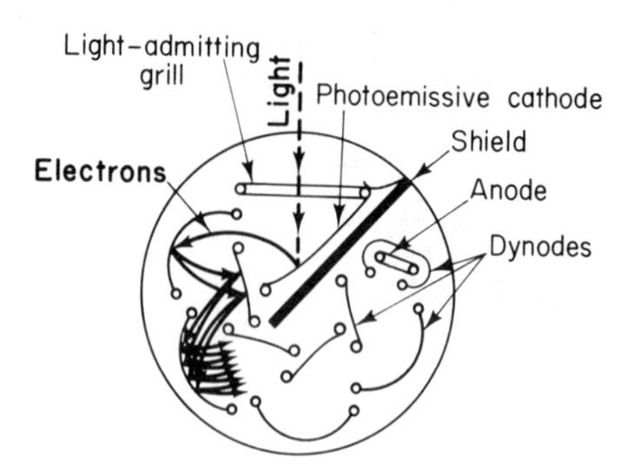

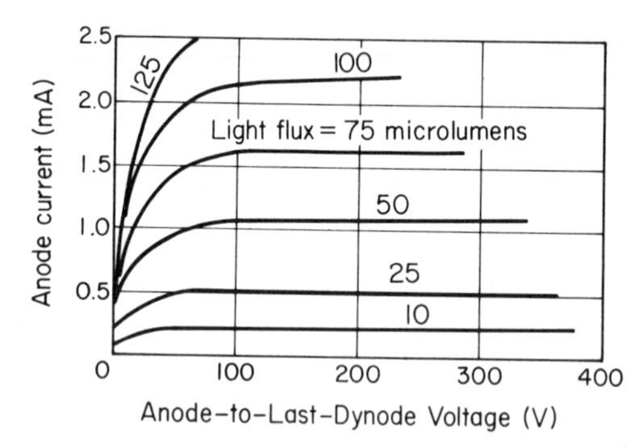

Fig. 13-45. Photomultiplier structure and characteristics.

and anode, just like the photocell. In addition, it contains serveral extra electrodes, called *dynodes*, which are biased to act as "intermediate anodes." In operation, a photon striking the cathode results in the emission of an electron. This electron is attracted by and accelerated to the first dynode which is positively biased relative to the cathode. Upon striking the first dynode, the incoming electron knocks out two (or more) secondary electrons. These secondary electrons are attracted to the second dynode, which has a higher positive bias than the first, where *each* of them now knocks out two (or more) secondaries. These four electrons go on to the third dynode to knock out eight, and so on. The dynodes are placed and shaped so that each will send the secondary emission current toward the next dynode in sequence. The dynodes are biased successively more positive, with respect to the cathode, with maximum positive bias on the anode. A schematic of the structure, plus the first three steps of secondary emission, is shown in Fig. 13-45. Anyone who has ever worked a "double-up" problem knows how fast a number increases if you double it at each step. Thus a photomultiplier that produces just two secondary electrons for each dynode impact, and that has ten dynodes, would deliver 1024 electrons to the anode for each one emitted from the cathode. Notice in Fig. 13-45 that the photomultiplier produces usable currents for light fluxes measured in microlumens (millionths of one lumen) instead of the tenths of lumens required by photocells. Actual tubes can easily multiply the cathode current by factors of ten million and more. High-anode voltages (2000 volts and up) are required, however.

Speaking of high-anode voltages leads us to the final device considered in this chapter, the *X-Ray tube*. At one time the only known sources of X rays were the naturally radioactive substances. In experiments like the Crookes Tube (Section 4-6), it was discovered that X rays could be created artificially. With medical and industrial needs for a reliable and controllable X-ray source in mind, Dr. William Coolidge of General Electric Company developed a practical tube in 1913. A modern version of the Coolidge tube and its schematic symbol are depicted in Fig. 13-46. Electrons are therm-

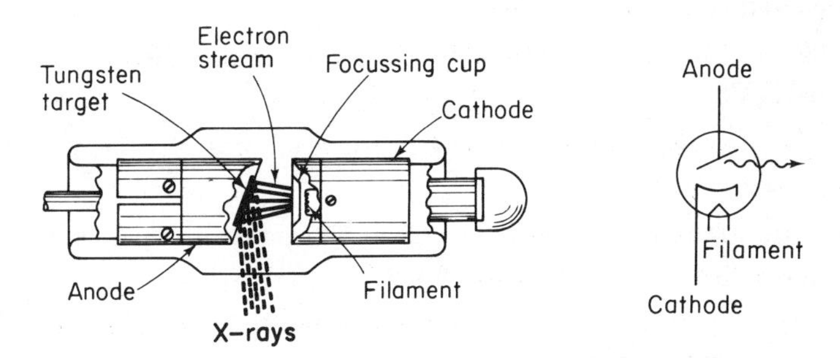

Fig. 13-46. Commercial X-ray tube.

ionically emitted from a heated cathode. In addition, the cathode-emitting surface is cupshaped so that its electric field "shapes" the electron stream into a fairly narrow beam. The tube is highly evacuated to about one billionth of normal atmospheric pressure. A very high anode voltage accelerates these electrons to great energies by the time they strike the *target*, a piece of tungsten set into the anode. These electrons give up energy to the tungsten's valence electrons—not enough to knock them out as in secondary emission but enough to raise their energies to a higher band temporarily. When these electrons "fall back" to their normal state, they emit X-ray photons. The higher the anode voltage, the shorter the wavelength and the higher the energy of the X rays. High energy is required for deep penetration. Typical anode voltages for medical X-ray equipment are in the range of 20 to 100 kilovolts, while some industrial tubes may require as much as 2500 kilovolts.

REVIEW QUESTIONS

1. What obstacle must be overcome to cause an electron to be emitted from the surface of a conductor?
2. Describe the various kinds of emission.
3. If you could travel backward in time, how would you explain the "Edison effect" to Edison himself?
4. Why should ac not be used to heat the filament of a directly heated cathode?
5. What is the most useful characteristic of a vacuum diode? How is it used?
6. Why did DeForest's audion become the heart of the radio industry?
7. Explain how a triode amplifies.
8. Why was the second grid added to the tetrode? The third to the pentode?
9. In what way(s) is a beam power tube different from a garden-variety pentode? When is it useful?
10. Is a diode-triode a pentode? Why?
11. What distinguishes the gas-filled diode from the vacuum diode? The thyratron from the triode?
12. What type of tube would you expect to use if an amplification factor of over 2500 were required?
13. A tube with an amplification factor of 50 will produce what change in plate voltage for a 2-volt swing in grid voltage?
14. What is the significant thing about the atomic structure of semiconductors? About their crystal structure?
15. What is a hole? A free electron?
16. Describe-free electron/hole pair creation and recombination.
17. How does doping affect semiconductor materials?
18. What is a depletion region? Of what use is it?
19. Describe the behavior of a semiconductor diode under (a) forward bias, (b) no bias, (c) small reverse bias, (d) increasing reverse bias. How do these differ from the characteristics of the vacuum diode?
20. Why were area rectifiers developed?
21. What is the connection between avalanche breakdown and zener diodes?
22. What is the peculiarity of Esaki's pet diode?
23. Why doesn't a transistor behave like two diodes in series?
24. There are two varieties of almost every species of transistor. Why?
25. What is the most important rule in connecting a transistor into a circuit?

26. What is the approximate value of β for a transistor with an α of 0.97?
27. Explain the amplifying action in a junction transistor.
28. Draw a parallel between the electrodes in a vacuum tube, junction transistor, and MOSFET.
29. How would you use a thermistor to measure room temperatures?
30. Ten secondary electrons are emitted for each electron striking a dynode in a certain photomultiplier, which has six dynodes. What anode current is produced by a cathode current of 0.002 microampere?
31. Devise a simple "burglar alarm" circuit for your bedroom window, using a photodiode.

14

PLAYING WITH BLOCKS

14-1 PUTTING THE TUBES AND TRANSISTORS TO WORK

Vacuum tubes and transistors are very interesting devices all by themselves and are the subject of a large number of scholarly books. We have said a good deal about them in the last chapter, and we still have several friends to make among their group. As practical people, however, we are even more interested in how these devices are *used* to make the various gadgets in our electronic experience (TV sets, radios, etc,) perform their grand functions. To achieve this kind of understanding, it is necessary to study the behavior of the tube or semiconductor as a member of a small community of components that work together as a unit.

It is not the purpose of this book to get deeply involved in the massive subject of circuit analysis. The plan, in fact, is to use block diagrams to show the major functional units in the circuitry of some rather complex devices we meet a bit later in the text. At this point, however, having gained a passing acquaintance with electronic control devices, we may appropriately take a quick look at a few common circuits that use vacuum tubes and semiconductors. These same circuits, and variants of them, are used over and over as building blocks in more complex circuitry. It is not our intent here to be exhaustive but merely to examine a few typical examples of very common circuits. We will discuss others as we encounter them.

14–2 THE dc POWER SUPPLY

One thing you must have noticed by now about both vacuum tubes and transistors is that they are dc devices. They are used, of course, to produce, control, and amplify ac signals, but they must be supplied with dc at one or more voltage levels. In addition, low-voltage power, either ac or dc, is required by the filaments in vacuum tube circuits. These needs could be, and often are, provided by batteries. More often than not, however, electronic devices are operated from a commercial 120-volt 60-Hz source. Consequently, the first circuit group one encounters in such equipment is designed to take the ac supplied to it and process it to provide the dc voltage level(s) required by other parts of the circuit. Low-voltage ac for tube filaments is also provided by this circuit group, which is appropriately called the *power supply*.

Power supplies come in many varieties, from simple to complex, depending on such factors as how many different dc (and ac) voltages are required; whether any of them must be critically controlled; how much power will be drawn at each voltage; how cheaply must the overall unit be produced, and so on. A typical power supply configuration is shown in Fig. 14-1, in both a vacuum tube and a semiconductor version. In either

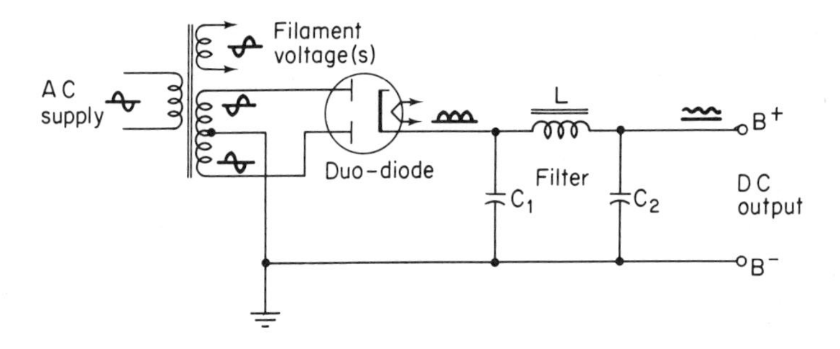

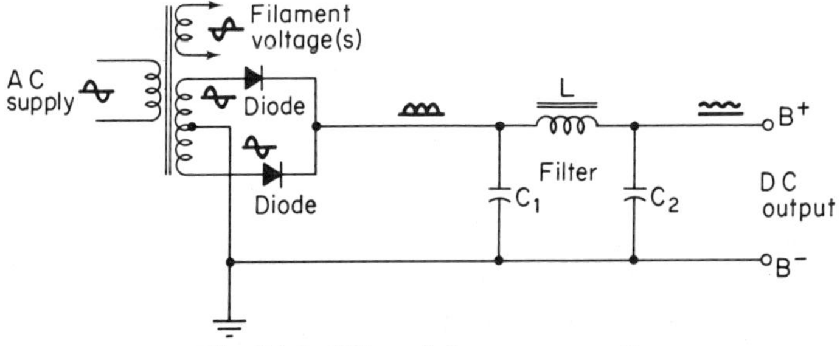

Fig. 14-1. Filtered dc power supplies.

case, the major elements in the power supply are the *power transformer*, two *diodes*, and a *filter* group.

The power transformer has two separate secondary windings. One of these provides the low ac voltage(s) for tube filaments. The other provides a high voltage which will be rectified to dc. The high-voltage secondary winding is usually *center-tapped;* that is, a lead is brought out from the halfway point in this winding. The reason for this center tap will be apparent in a moment. The diodes are used for the actual rectification process. Two separate semiconductor diodes might be used, but in a tube-type circuit, one duo-diode tube would undoubtedly be used and not two individual diode tubes. The output of the diodes is a pulsating dc, as we saw in Chapter 13, and the purpose of the filter group is to smooth out these pulsations into a more or less steady dc voltage level.

To understand the dc supply, pretend for a moment that the filter group is not there; that is, there are no capacitors and the coil L is just a straight-through wire. The active part of the circuit, at any particular instant, includes a path from the center tap of the transformer's secondary winding out through B^- to the load, back through B^+ and one of the diodes, to the center-tap point we started at. The voltages at various points in the circuit are illustrated in Fig. 14-2. The secondary voltage of the transformer is, of course, an ac voltage just like the primary voltage. This means that at any given instant one end of the secondary winding will be positive and the other negative. The significant thing is that the center-tap point is always negative with respect to *whichever* end of the secondary winding is positive. Notice that we have grounded this point, making it the reference level, and that it connects directly to the terminal marked B^-. Now when the upper (in the illustration) end of the secondary winding is positive, and the other end negative, a positive potential appears across the upper diode and a negative potential across the lower diode. Consequently, the upper diode conducts and the lower one is cut off. Current flows from the center tap out the B^- terminal, through the load, into the B^+ terminal, through the upper diode and back through the winding to the center tap. An instant later, the ac voltage across the secondary has reversed its polarity, making the lower end of the winding positive and the upper end negative. This step forward-biases the lower diode and cuts off the upper one. Current still flows out from the center tap, through B^- to the load, and back in through B^+, only this time return path to the center tap is through the lower diode. Thus, as seen in Fig. 14-2*b* and *c*, the actions of the two diodes complement each other, each one being "on" for half a cycle and "off" for the other half-cycle. At the point where the two diodes' outputs are joined, the voltage appears as in Fig. 14-2*d*. In effect, we have "cut off" the negative part of

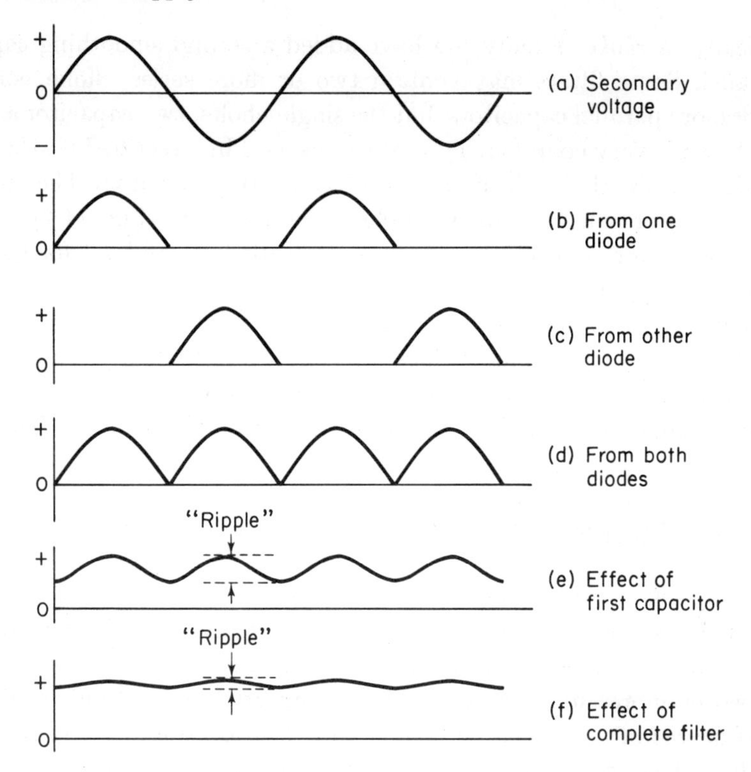

Fig. 14-2. Rectification of an ac voltage.

the sine wave and "flipped it over" onto the positive side of the axis. This is called *full-wave* rectification as opposed to the action of one diode alone, which is called half-wave rectification.

The pulsating dc out of the two diodes is just as good as "pure" constant dc for many applications. In most cases, however, we would like something a little more like a battery voltage, and this is the reason for the filter.

A simple capacitor like C_1, connected across the output terminal (i.e., in parallel with the load), goes a long way toward smoothing the pulsating dc. While the voltages in Fig. 14-2*d* are climbing toward the peak, the capacitor charges, and when the voltages start to decline, the capacitor "dumps" its charge back into the circuit. The result is shown in Fig. 14-2*e*, a "wavy" but more nearly constant dc voltage. This kind of waviness is called *ripple*. Notice that its frequency is twice that of the original ac voltage. (So, for instance, rectified 60-Hz power would contain 120-Hz ripple). The coil, connected in series with the source, also smooths out the ripple by alternately storing energy in and releasing it from its magnetic field. A coil used in this manner—specifically, to oppose the passage of ac—is called, rather graphically, a *choke*. Finally, we have added a second smoothing capacitor in parallel. Some filters may contain two or more series choke coils and three or more parallel capacitors, but the single-choke, two-capacitor arrangement shown is very common. Typical values used in a rectified 60-Hz supply are 10 henries for the choke and 20 to 40 μF for the capacitors. The output of the filter has only a very small ripple, as indicated by Fig. 14-2*f*.

The power supply's dc terminals are called B^+ and B^- for historical reasons. When DeForest invented the audion amplifier, he needed three distinct sources, for which he used batteries. The filament supply he called the *A* battery, the grid bias supply he named the *C* battery, and the plate voltage supply was christened the *B* battery. Since the power suppy's dc voltage is usually used as the plate supply, it is thought of as the *B* supply.

14-3 AMPLIFIERS AND GAIN

At this point, it is no news that transistors and grid-containing vacuum tubes can be used to amplify a weak signal. Circuits for one stage of amplification were shown in Fig. 13-10 and 13-36, and it was mentioned that two or more such stages could be connected end to end to give any desired degree of amplification. When amplification stages are connected so that the output of one stage is the input to the next, the stages are said to be *in cascade*, or *cascaded*. Figure 14-3 presents a tube-type and a transistorized two-stage cascade amplifier. Note that the stages are connected through a capacitor, C_1, which passes the amplified ac signal but blocks the dc supply from stage 1. This is also the reason for the input and output capacitors, C_i and C_2. Connection of amplifier stages in this manner is called *RC coupling* (for *r*esistance-*c*apacitance) and is used at low-signal frequencies. At higher frequencies, *inductive coupling* through air-core transformers is used instead.

In these circuits, we have used the common emitter configuration (common cathode for the tube equivalent). It is most popular because of the large amplification obtainable. Notice, however, the little wave-form drawings, which remind us that the signal, besides being amplified, goes through a 180° phase shift in each stage. Hence, in a many-stage amplifier, the signal out of the first, third, etc., stages is 180° out of phase with the input, while the signal out of the second, fourth, etc., stages is back in phase with input.

We want to point out here that the amplification obtained from a given amplifier stage is *never as great* as the amplification factor (μ) of the tube used or beta (β) factor of the transistor, due to the losses in a real cir-

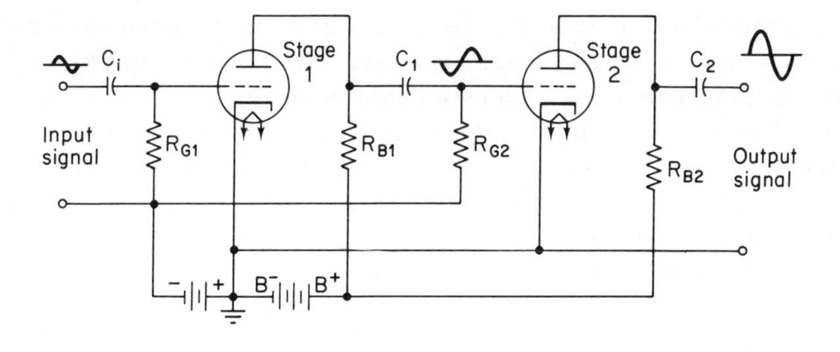

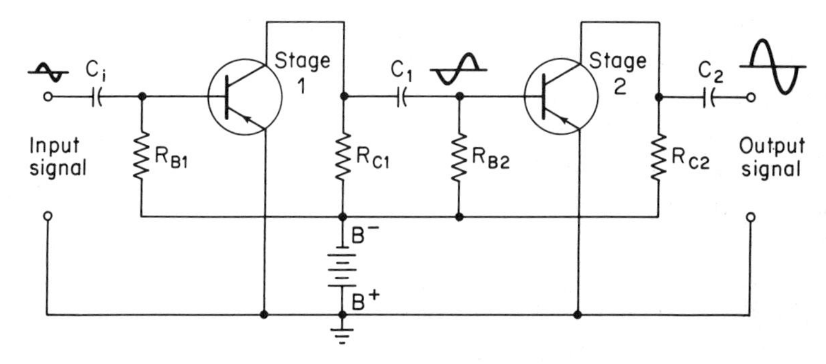

Fig. 14-3. Two-stage cascaded amplifiers.

cuit when a finite current is flowing. We therefore speak of the *gain* of a stage, or of a series of stages, defined as the ratio of output to input. Thus

Voltage gain = output voltage ÷ input voltage

Power gain = output power ÷ input power

For example, an amplifier stage using a tube with an amplification factor of 60 might have an actual voltage gain of about 35.

14-4 MEET THE DECIBEL

Human perception is a strange and wonderful thing. One of its important aspects is that it is not linear. If a certain amount of power pouring into one of the senses causes you to perceive a certain intensity, doubling the power input does *not* result in your feeling that the sensation has become twice as intense. To be specific, this means that although a small glow can be seen in a dark room and a faint sound can be heard when the room is still, it takes fairly intense change in light or sound to catch our attention when the room is already bright or noisy. Nature has evolved this type of response so that we can obtain useful data from our senses over a very large range of input power. This fact was noted in the 1830s by the famous German physiologist Ernst Weber and put into the form of a mathematical "law" by his compatriot physicist, Gustav Fechner, about thirty years later. This "law," which today we know is not as strictly true as Fechner supposed, states that our perception of stimuli is logarithmic. This means that what we perceive as "equal increases" in sensation at two different levels actually corresponds to equal percentage changes in input power. In other words, you perceive the same *change* in loudness between listening to 1 drummer and to 2 drummers, as between listening to 10 drummers and to 20 drummers. Although not exactly accurate over the whole range of any one sense, this law was, nonetheless, a handy approximation.

Now, one of the first uses of electronic amplifiers was to amplify signals that would eventually be converted into sound (the tube, you will recall, was called the *audion*, from the Latin *audire*, to hear). These electric frequencies, corresponding to the sound frequencies that the human ear can hear, are in the range from about 16 Hz to about 20,000 Hz and are called the *audio frequencies*. For the engineer designing an audio amplifier, which is going to power a loudspeaker, the Fechner-Weber law implies that what is important is not so much the power gain itself but rather the percentage gain that must be dealt with. He therefore needed a scale of power gain on which equal intervals represented equal percentage increases. Such a logarithmic unit was defined and named the *bel* in honor of the telephone's inventor, Alexander Graham Bell. This unit turned out to be inconveniently large, so the one-tenth part of a bel, the *decibel* (*deci-* means $\frac{1}{10}$th) became the standard logarithmic unit of gain. The word decibel is usually abbreviated as db. It turns out that the average human ear can just detect a change of 3 db in sound intensity.

Figure 14-4 is a useful chart for converting the power gain ratio into decibels. Notice that if the output power is equal to the input power, so that the power gain is one, the number of decibels gain is zero. This makes good sense. If you get out the same amount that you put in, you have gained nothing. If the output is less than the input, which makes the power gain ratio less than one, the number of decibels is negative, showing a loss (a "negative gain"). If, for example, the power into an amplifying stage is 2 watts and the output power is 40 watts, then the power gain ratio itself is $40 \div 2 = 20$. Looking up this value in Fig. 14-4, we find that it represents a gain of about 13 db.

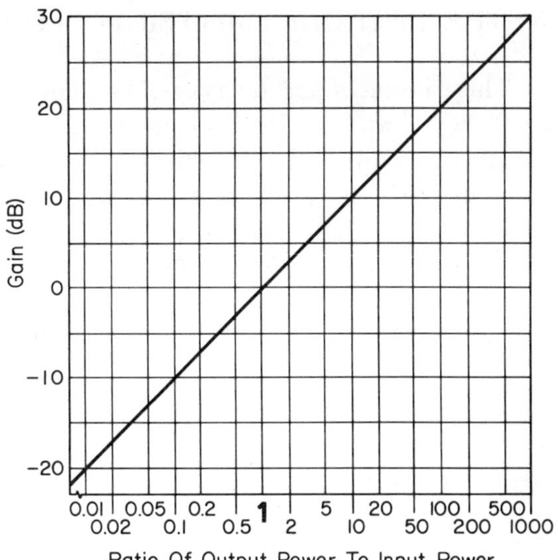

Fig. 14-4. Converting power gain to decibels.

The decibel is a very freely used unit of measure today. As a measure of absolute sound level, the decibel is used in acoustics. What makes the decibel an absolute unit here is that the *zero level* is strictly defined as a certain sound pressure intensity. Without a strict definition of the zero level, however, decibel is a handy *relative* unit that can be used to express the power gain of any device's output relative to its input.

The decibel is also frequently used as a measure of voltage and current gain. In this application, however, the number of decibels is *twice* that for an equal power gain, because of the way voltage and power, or current and power, are related mathematically. We can still use Fig. 14-4 to find decibel gain or loss for voltage or current by interpreting the horizontal axis as the actual voltage gain or current gain ratio and then doubling the number of decibels we read off the vertical scale. Suppose that a certain device has an input voltage of 100 volts and output of 50 volts (i.e., there is a voltage loss). The voltage ratio is $50 \div 100 = 0.5$. From Fig. 14-4 we obtain a value of -3 db, which we must multiply by 2; hence the voltage gain is -6 db, a 6-db loss. When the decibel unit is so used for voltage or current gain, the input and output impedances of the device should be the same.

The characteristic of the decibel unit that makes it so useful outside of acoustics is the way in which successive gains combine to give an overall gain. To find the overall decibel gain of a complicated device, you merely add up the decibel gain figures (subtracting negative ones) for its component stages.

14-5 THE CATHODE/EMITTER FOLLOWER

There is one important type of "amplifier" circuit whose configuration may puzzle you at first. In this configuration, shown in Fig. 14-5, the plate (in the tube version) or the collector (in the transistor version) is the element common to both the input and output circuits. Such an arrangement is therefore properly eligible to be called a common plate, or common collector, amplifier, as the case may be. The popular names for these circuits are, however, the *cathode-follower* and *emitter-follower*, respectively.

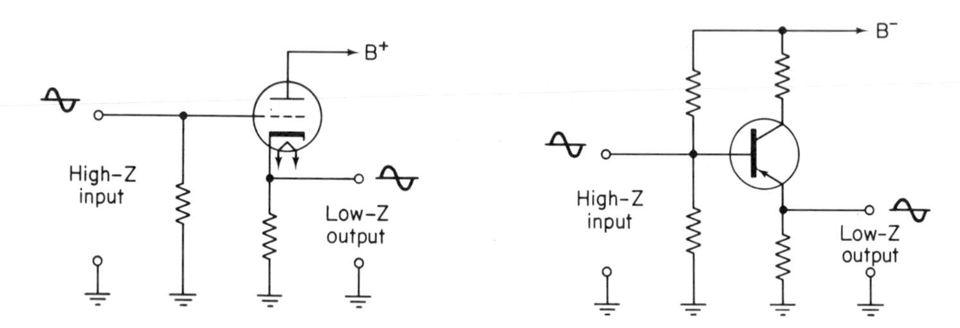

Fig. 14-5. Cathode follower and emitter follower.

The apparently peculiar characteristic of these "amplifiers" is that their voltage gain is less than one. A typical cathode-follower might have a voltage "gain" of −4 db. Why in the world, you may ask, would anyone want to use an amplifier that actually results in a loss? The answer is that voltage and power amplification are not what we are looking for in using a cathode-follower; in fact, we are obviously giving up a little gain for something else. The key characteristics of the cathode-follower and emitter-follower are (1) their input impedance is high and their output impedance is low; (2) their output signal is in phase with their input signal.

To understand the usefulness of these characteristics, we must be aware of the fundamental law of electric power transfer:

> **Maximum power is developed in a load when its impedance is equal to the impedance of the source.**

In high-power devices, like motors and heaters, it is generally not important to "squeeze out every last drop of power efficiency." Electronic circuits, however, are usually low-power sources and this matching of impedances, to obtain maximum transfer of power, becomes more critical. The degree of

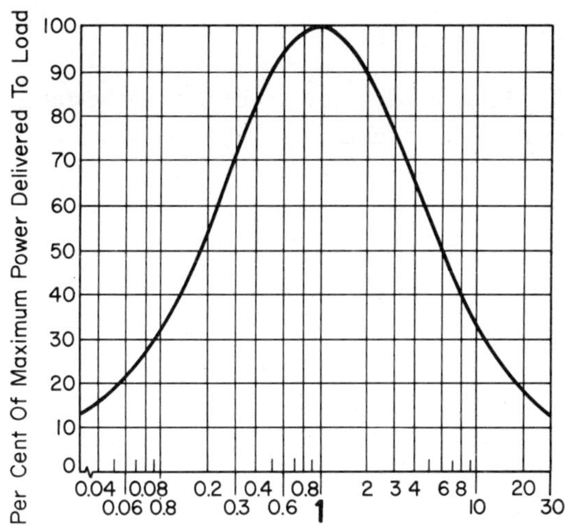

Fig. 14-6. Effect of impedance mismatch on power transfer.

mismatch between load and source is not extremely critical, as demonstrated by Fig. 14-6. Notice that the load impedance can vary from as little as half of the source impedance to as much as twice the source impedance and yet receive 90 percent or more of the maximum power obtainable from that source. Thus impedance-matching is more a question of getting into the right ballpark than of an exact match. This function is accomplished with either a transformer or a cathode- (emitter-) follower.

Returning to the cathode-follower, let us see how it works. Notice that the load resistor, across which the output signal is developed, is in the cathode circuit instead of the plate circuit. The input signal is applied to the grid. If this signal swings more positive, it increases current through the tube and the load resistor. The result is an increased voltage drop across the load resistor, which is our desired output signal, and also a change in the grid-to-cathode bias that reduces the effect of applied signal voltage swing. Similarly, if the applied voltage swings more negative, the tube current decreases and so does the voltage drop across the load resistor. Here, also, the grid-to-cathode voltage change partly compensates for the negative swing of the signal voltage. We therefore see how the change in the cathode resistor's voltage drop "follows" the change in the grid voltage, which is the reason for the name cathode-follower.

This circuit is frequently used to connect an amplifier (a high-impedance source) to a loudspeaker (a low-impedance load). It is also used in test equipment, in cases where the circuits being tested operate at a very low power level, and the power drawn by an ordinary meter (if connected to those circuits) would cause an incorrect reading.

14-6 THE VTVM

In line with the preceding discussion, we now meet a very useful measuring instrument, the *vacuum tube voltmeter*, usually called a *VTVM* for short. In Section 10-8, we saw how a galvanometer connected in series with a high resistance could be used as a voltmeter. Currents of amperes were flowing in the circuits whose voltages we were measuring, and the few milliamperes or microamperes drawn by the meter itself, in making the measurement, were not even "missed." In other words, the energy withdrawn from the circuit under test by the meter did not significantly affect the voltages being measured. In electronic circuitry, however, we are normally dealing with high impedances and low currents—currents measured in milliamperes or even microamperes. Such a circuit often cannot "afford" to give up even the 50 μa required by the best voltmeter, without a drastic change in that circuit's voltages, which makes the measurement useless. For voltage measurements in electronic circuits, we therefore require an instrument with a very high input impedance which itself draws no appreciable current from the circuit under test. The VTVM was developed to meet this need.

The essentials of a typical VTVM voltage-measuring circuit are illustrated in Fig. 14-7*a*. The triodes, Q_1 and Q_2, actually represent the two halves of a dual triode tube. Each half is connected essentially as a cathode-follower (compare with Fig. 14-5) and a microammeter is connected between the two cathode resistors, one of which is variable. With no voltage being measured, the variable cathode resistor is adjusted to compensate for the slight differences in characteristics of the two sets of tube elements, the result being a perfectly balanced condition with zero current through the microammeter. The variable resistor is therefore called the *zero adjust* resistor. When a small dc voltage (of the polarity shown) is impressed on the grid of Q_1, it conducts a greater current, increasing the voltage drop across Q_1's cathode resistor and causing a proportionate current through the meter. The meter scale can thus be calibrated in terms of the voltage across the measuring terminals. The beauty of this circuit is that the circuit being tested only "sees" the high input impedance to the grid of Q_1; the power to drive the meter comes from the B^+ supply, not the tested circuit.

Of course, only small dc voltages can be applied to the grid of the tube; otherwise it would be overdriven into a nonlinear (undesirable!) region of operation, or even burnt out. Use of a *voltage-divider* circuit, such as that shown in Fig. 14-7*b*, permits us to use the VTVM to measure any desired level of voltage. A selector switch enables us to "tap off" a desired fraction

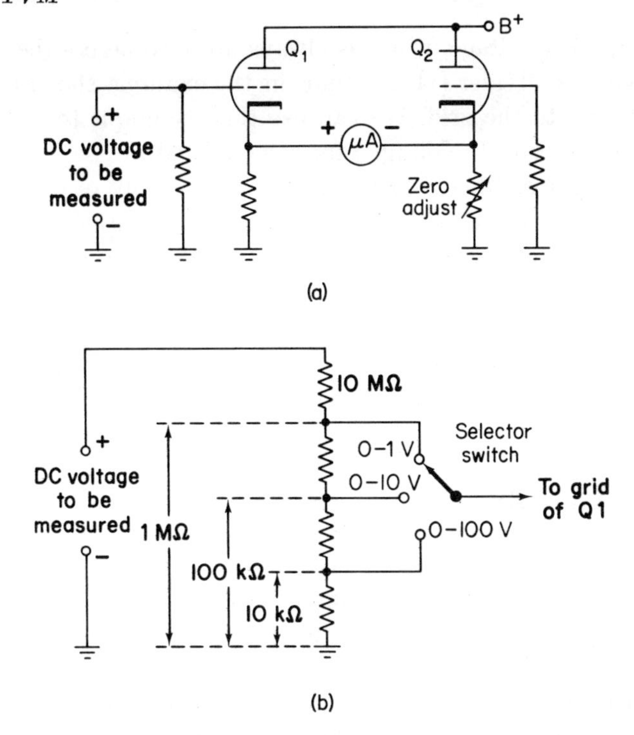

Fig. 14-7. The vacuum tube voltmeter. (a) Simplified circuit. (b) Voltage divider for multiple ranges.

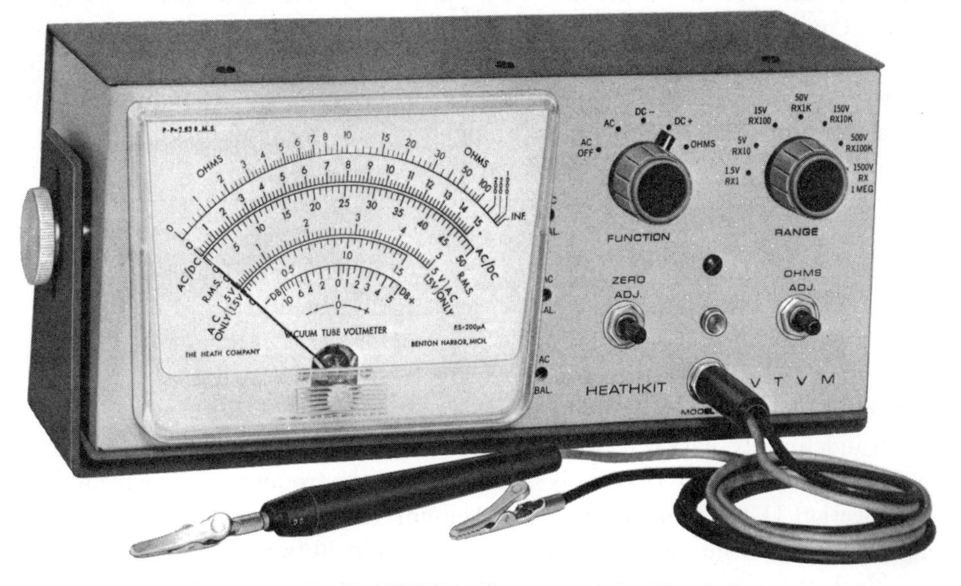

Fig. 14-8. Typical VTVM. (Courtesy of the Heath Co.)

of the total applied voltage, which is always dropped across the total resistance of the voltage divider (11 megohms in the example shown), and apply just that fraction to the grid. We can use this technique to make a VTVM with as many different dc voltage ranges as desired.

By adding a rectifier, which can be switched into or out of the input circuit, we can give the VTVM the capability of measuring ac as well as dc voltages. Provision of a known independent voltage source (usually a battery), which can also be switched into the input circuit as desired, makes our vacuum tube voltmeter also a vacuum tube *ohmmeter*, using the principles discussed in Section 10-8. Figure 14-8 shows a typical VTVM with many ac voltage, dc voltage, and ohms ranges.

14-7 OSCILLATORS

There are many applications where it is necessary to generate ac, of some desired frequency. In these applications, a mechanical alternator is, at the very least, impractical for several reasons. For really high frequencies, it becomes virtually impossible to build a mechanical alternator. Thus we require an electronic circuit to perform this function. A circuit designed to produce a sinusoidal ac signal of a desired frequency is called an *oscillator*.

At the heart of all oscillators is a resonant inductance-capacitance (*LC*) circuit, "tuned" to the desired frequency by suitable choice of components. Consider the simple circuit of Fig. 14-9*a*. Let the switch be in position 1. The capacitor will charge to the voltage of the battery. Now let the switch be thrown to position 2, connecting the capacitor across the inductor. The capacitor immediately discharges through the inductor, building up the magnetic field of the latter. As the capacitor's energy dies away, the inductor tries to maintain the circuit current by releasing the energy stored in its field back into the circuit. This step causes a reverse current and recharges the capacitor, but with the opposite polarity. When the inductor's magnetic field is near total collapse, the capacitor is again nearly charged and it then begins to discharge again through the inductor. This process rebuilds the inductor's field in the direction opposite to its previous direction, as the capacitor discharges once again. In this manner, energy shuttles back and forth indefinitely between the inductor's magnetic field and the capacitor's electrostatic field. The rhythm is sinusoidal and the frequency is the resonant (see Section 11-3) frequency of this *LC* circuit, as indicated by the waveform to the right of the circuit diagram. If we could obtain an ideal inductor, an ideal capacitor, and some zero resistance wire, we could use them to make an oscillator.

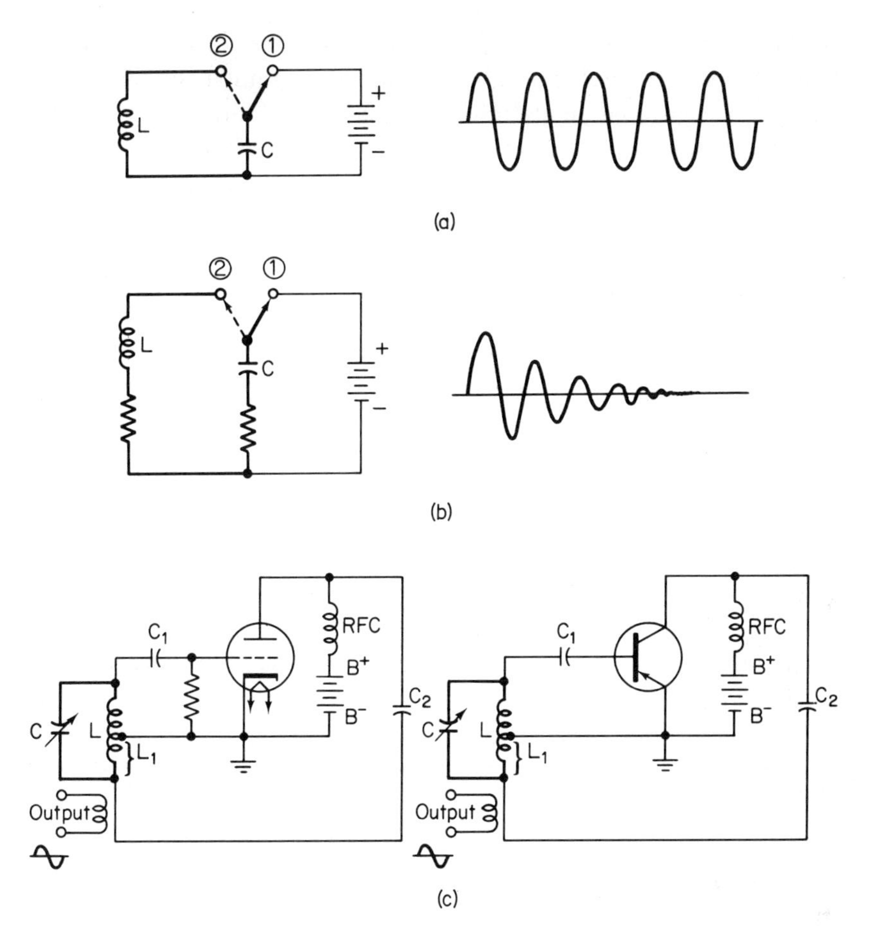

Fig. 14-9. The oscillator. (a) Oscillations in an ideal L–C circuit. (b) Oscillations in a real L–C circuit. (c) Actual oscillator circuits.

Unfortunately, all circuits contain some resistance. If we repeat our previous hypothetical experiment with a real inductor and capacitor, as in Fig. 14-9*b*, we do indeed obtain oscillations at the resonant frequency. Their amplitude, however, quickly dies away to nothing, as shown at the right of the circuit, because energy is continually being lost from the circuit as heat in the resistances.

It would seem, therefore, that we can make an oscillator from a tuned *LC* circuit *plus* some device for adding energy in the right amount, and at the right time, to make up for the resistance losses. This process is accomplished by using an amplifier to amplify the oscillations of an *LC* circuit driving its grid and by *feeding back* a portion of the amplified oscillations from the plate circuit *in phase* with the grid circuit oscillation. The circuits in Fig. 14-9*c* represent tube and transistor versions of one of several popular oscillator configurations. The all-important *LC* circuit is emphasized. The capacitor *C* is variable to permit selection of any desired operating frequency in a band. Note that the inductor *L* has a tap near one end of its winding. The little portion, L_1, is connected into the plate circuit through the capacitor C_2 (whose job is to keep B^+ voltage out of the grid circuit, while passing the ac), and provides the feedback coupling required between the grid and plate circuits. The oscillator is inductively coupled to the circuit requiring its ac signal through the output winding. The inductance marked *RFC* is a *radio frequency choke*, a coil whose function is to prevent the ac from being bypassed to ground through the battery, without interfering with the dc biasing of the plate (collector) circuit.

14-8 FASTEST GUN

Mankind has produced all sorts of guns. Most of them are involved, in some way, in death and destruction. Here, however, is one of the most useful guns ever devised by man—one that does not fire bullets but smaller and more versatile projectiles—the *electron gun.* This gun is the heart of much electronic equipment, notably units having cathode-ray display devices. Without it, the entire television industry would be nonexistent.

An electron gun is basically a device to produce a highly directional beam of electrons, like a flashlight beam, only much narrower. The gun is always enclosed in some sort of vacuum tube or vacuum chamber, depending on what the electron beam is to be used for. The component parts of an electron gun have the same names as those in common vacuum tubes; their shapes and arrangement are quite different, however, because of their purpose. A cross-sectional view of a typical electron gun and the schematic symbols for the elements of the gun are shown in Fig. 14-10. (See how schematic symbols sometimes oversimplify!) It consists of series of elements lined up along a common axis. The filament coil is compact, designed to produce heat in a relatively small area. The cathode is made in the shape of a cylinder closed at one end, forming a small cup, in effect. The filament is inserted into the cup. The forward face of the cathode—the closed end of the cylinder—is the only part of the cathode designed to emit and is oxide-coated. In this case, the name grid is very misleading, for the electron gun's control grid is another cylinder, closed at one end and surrounding the cathode, not a coil of wire. This grid has a hole in the center of its forward face through which the electron beam passes. After the grid comes a group of two or three positively biased anodes. Unlike the plate in a conventional vacuum

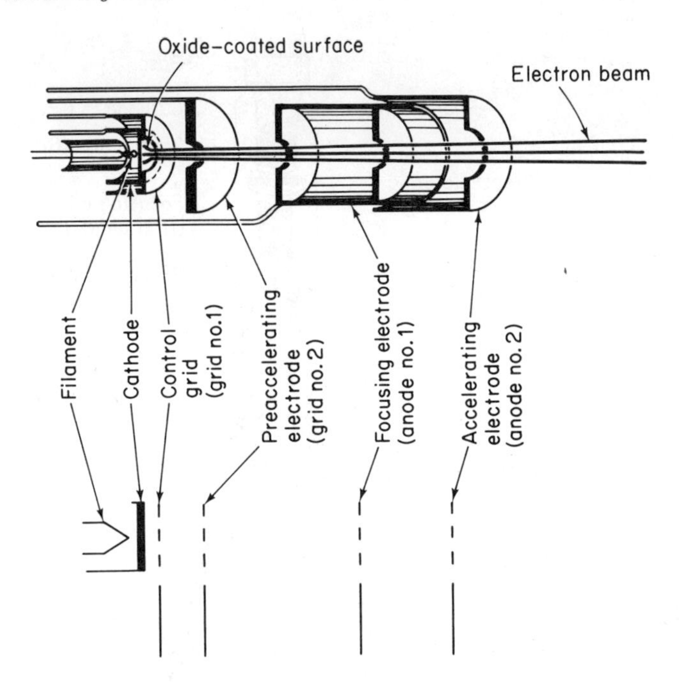

Fig. 14-10. The electron gun.

tube, whose function is to collect electrons emitted by the cathode, these anodes are cylindrical, with central holes to let the beam through. They form a kind of electrostatic "lens," which narrows the beam to a fine "pencil" and accelerates the electrons on their journey to whatever they are intended to strike. Some electron guns, which we encounter a bit later, accelerate electrons to speeds as great as half the speed of light, making them truly the "fastest guns in the West" or anywhere else.

14-9 THE CATHODE RAY TUBE

One of the most prominent uses of the electron gun is in the *cathode ray tube*, a visual display tube. It is so named because an electron beam emitted from the cathode of a tube used to be called a *cathode ray*. The cathode ray tube, or CRT for short, is a funnel-shaped vacuum tube. The narrow end, which is called the neck, is capped by the base, and the electron gun is mounted in this neck, as shown in the cutaway drawing of Fig. 14-11. The flat face of the tube, called the *screen*, is coated on the inside with a *phosphor*, which glows where it is struck by the electron beam. Every CRT is also provided with some means for bending the beam in two directions at right angles to each other, normally vertical (up and down) and horizontal

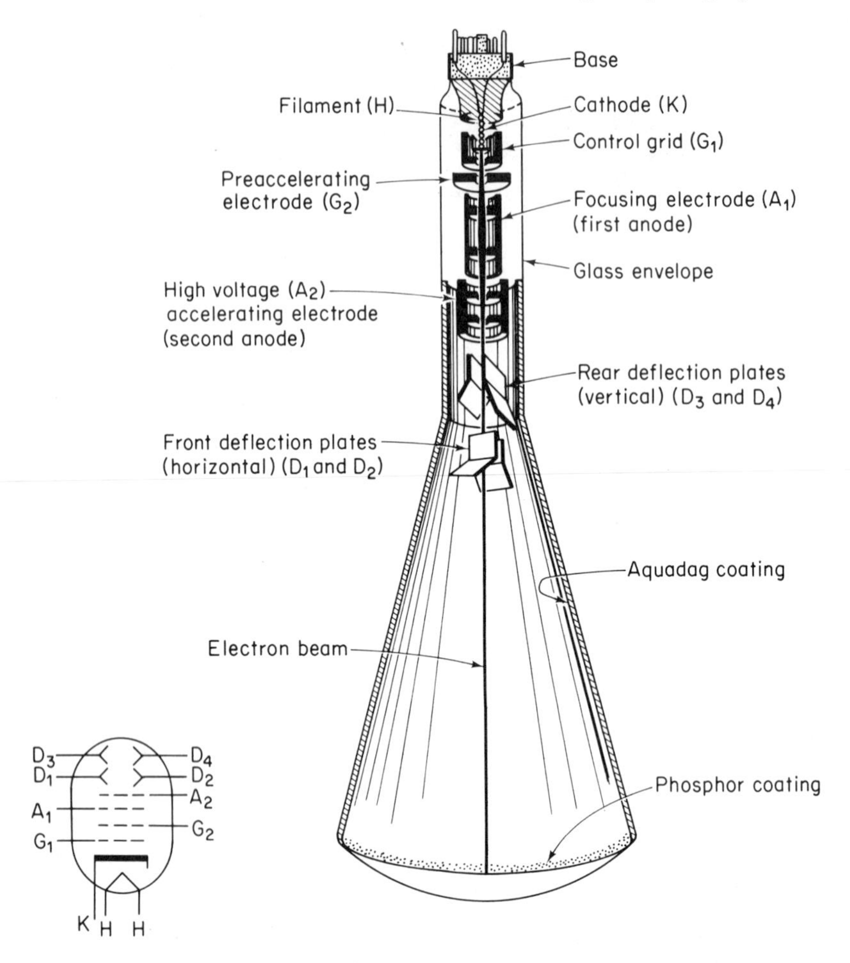

Fig. 14-11. Electrostatic cathode ray tube, with circuit symbol.

(right and left). This step may be accomplished electrostatically, by applying high voltages to two sets of *deflection plates* mounted inside the tube, as in Fig. 14-11. Alternatively, and particularly in the case of large tubes, this process is accomplished magnetically (the motor effect) by two sets of coils contained in a *deflection yoke* which mounts outside the tube, around the neck, as in Fig. 14-12. Tubes using these deflection methods are called electrostatic CRTs and magnetic CRTs, respectively. In either case, the inside of the CRT, between the neck and the face, is coated with a layer of conducting material called *aquadag*. The aquadag is connected electrically to the accelerating (last) anode, acting as a trap for stray electrons and as a shield against external electrostatic fields.

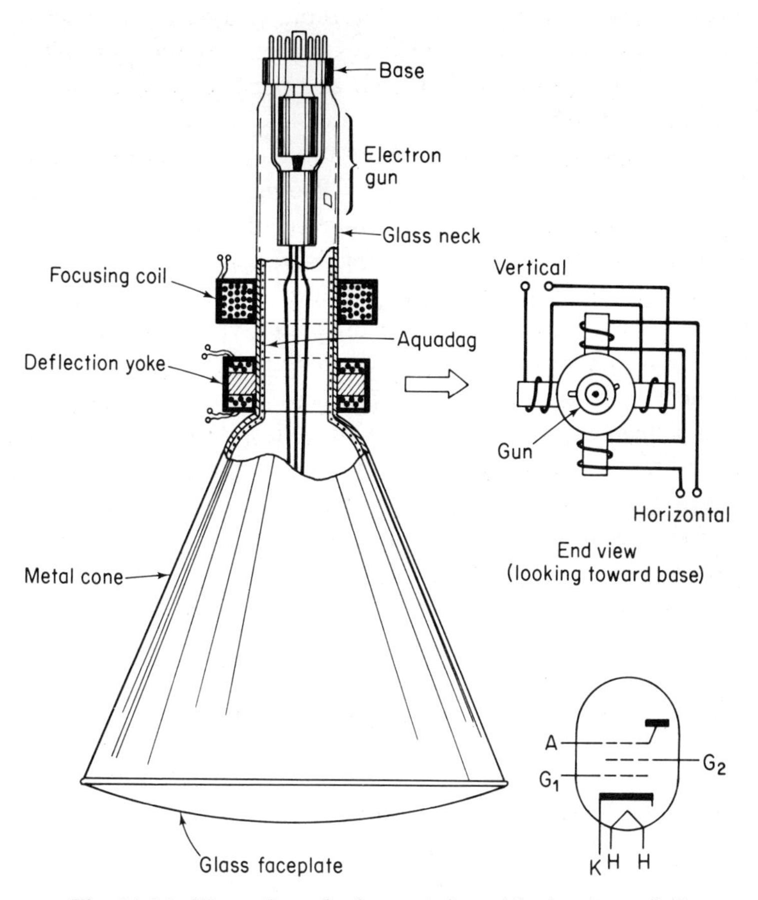

Fig. 14-12. Magnetic cathode ray tube, with circuit symbol.

The phosphor is a chemical compound, or mixture of compounds, which in this instance has the property of absorbing energy from bombarding electrons and reemitting this energy as light. The incoming electrons interact with the valence electrons in the phosphor's crystal structure, "kicking them up" to higher energy levels; upon "falling back," they emit photons of visible light. This light-emission process takes place in two steps: (1) *fluorescence*, light emission while the electron beam is striking the phosphor, and (2) *phosphorescence*, delayed emission after the electron beam excitation is removed. The length of time the phosphorescence continues at a spot after the beam has moved off that spot is called *persistence*, and is a convenient characteristic for describing phosphors. Phosphors whose persistence is between roughly a microsecond and a millisecond are said to have short persistence; between a millisecond and a second or two, the persistence is called medium; and anything greater is long persistence. Phosphors also produce a variety of colors, from red to blue, as well as white. The appropriate choice of phosphor color and persistence depends, of course, on the use of the CRT.

In operation, the CRT's electron gun emits a beam that strikes the screen, producing a small spot of light. The grid controls both the initial "narrowness" of the beam and the number of electrons contained in it (i.e., the beam current). From the viewer's standpoint, varying the negative grid bias changes the intensity of the spot of light on the screen. When a voltage is applied to a pair of deflection plates, the electron beam is bent toward the positive plate and away from the negative plate as it passes between them, and therefore strikes the screen somewhere other than the center. (In a magnetic CRT, the same effect is accomplished by energizing the appropriate coils in the yoke.) Applying a varying voltage to the plates (or yoke) results in a spot of light that moves about the screen. If the persistence of the screen is long enough, the moving spot "draws" a glowing line on the screen. This action is the basis for the CRT's use in television "picture" tubes, radar displays, and in the oscilloscope.

14-10 THE CATHODE RAY OSCILLOSCOPE

After the volt-ohm-milliammeter and the VTVM, the most useful piece of electronic test equipment is the *cathode ray oscilloscope*, or (familiarly) just a *'scope*. It consists of a cathode ray tube of the electrostatic type, a power supply, provisions for applying amplified signals to the CRT's horizontal and vertical deflection plates, and some additional circuitry. A representative laboratory-type oscilloscope is pictured in Fig. 14-13*a*. The block diagram of Fig. 14-14*a* indicates the major functional circuits, other than the power supply. Since the voltages supplied to the vertical and/or horizontal inputs may be quite small, an internal amplifier is provided in each of these circuits. The gain of each amplifier is controlled by an attenuator circuit, something like the volume control on your radio or TV set.

The most common use of the oscilloscope is to study waveforms, as shown is Fig. 14-13*b*. Any voltage that varies with time (or any time-varying physical phenomenon that can be transduced into a voltage; see Section 10-8) can be made to *plot its own graph* on the screen of the 'scope. In order for this to occur, we must move the spot across the screen proportionately to elapsed time, and we must push the spot up and down the screen in proportion to the magnitude of the variable whose waveform we are trying to study. To accomplish the former step, the 'scope has a built-in

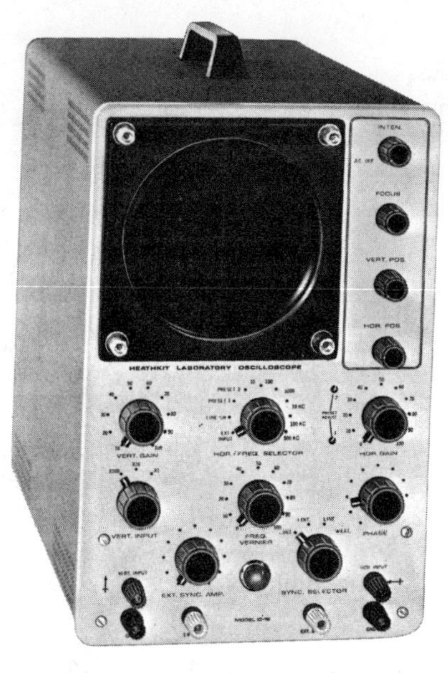

(a)

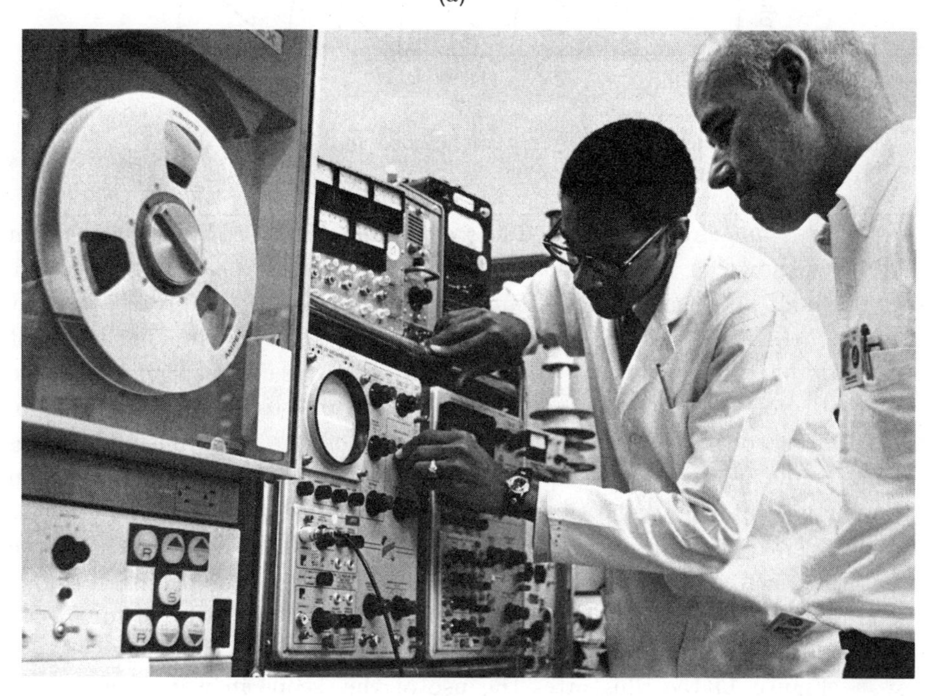

(b)

Fig. 14-13. (a) Cathode ray oscilloscope. (b) Using the oscilloscope as a test instrument. (Courtesy of the Aerospace Corp.)

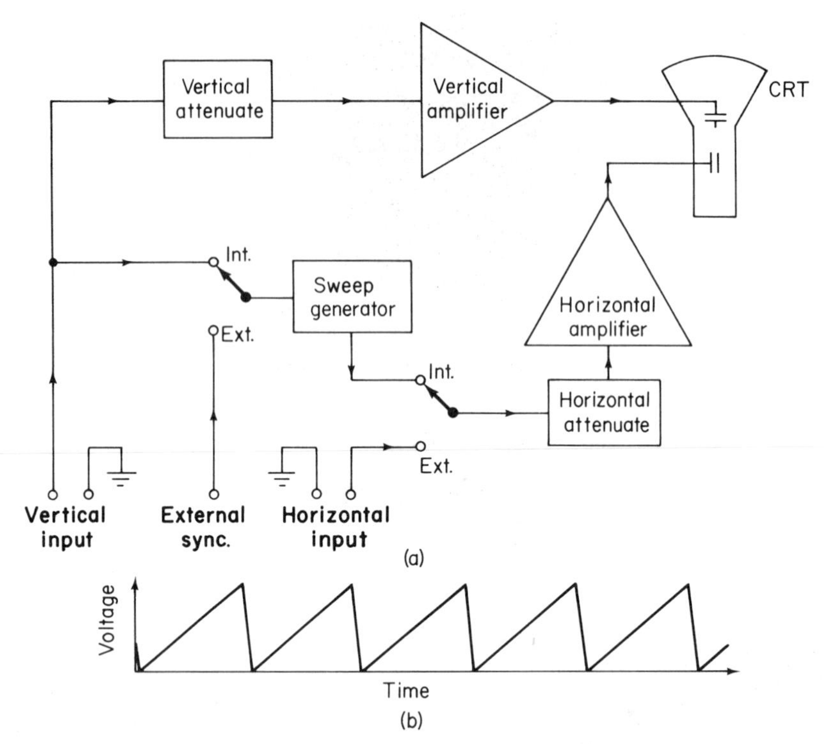

Fig. 14-14. The cathode ray oscilloscope. (a) Functional diagram. (b) Horizontal sweep voltage.

sweep generator circuit that produces a voltage with a sawtooth waveform like that shown in Fig. 14-14*b*. If this waveform were perfect, the voltage would build up linearly to a maximum (moving the spot from left to right across the screen) and then drop instantly back to zero (moving the spot instantly back to the left side of the screen). In any real sweep generator, the sawtooth voltage takes some small but measurable time to return to zero. For this reason, a *blanking* circuit that shuts off the beam entirely during (what would be) its return trip is provided. A coupling is provided between the vertical input and the sweep generator to *synchronize* their frequencies exactly, once they are set closely enough by hand. This step "locks" the pattern in a fixed position on the screen; otherwise it would tend to move around. Provision is made for switching to an external source for sweep or synchronization.

Figure 14-15*a* illustrates the use of the 'scope in waveform study. The figure on the left is the graph of the voltage to be studied and, plotted on the same axes, the graph of the appropriate sweep voltage. When these are applied to the vertical and horizontal plates, respectively, the pattern

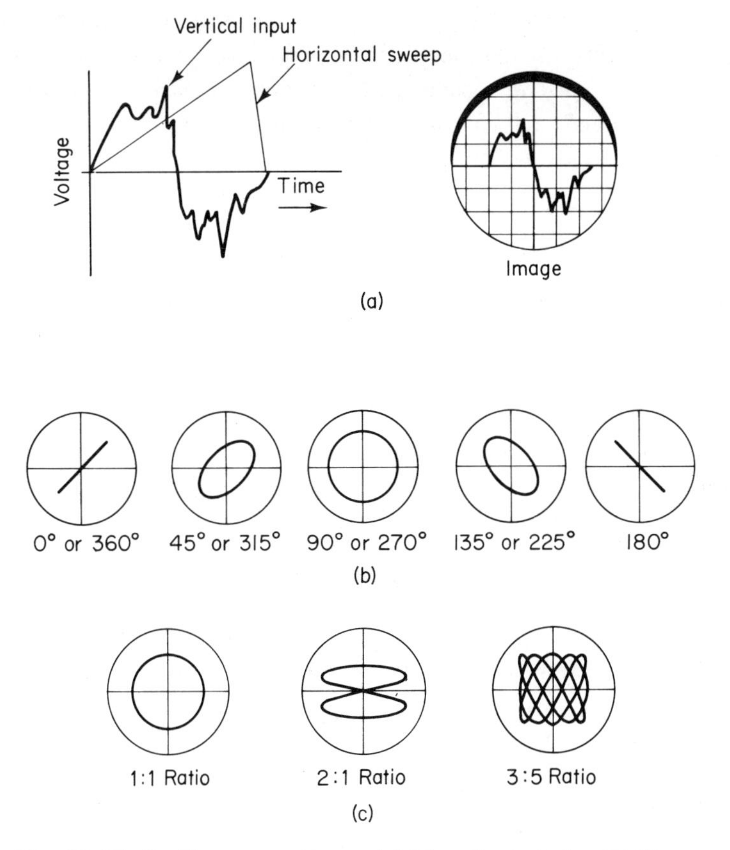

Fig. 14-15. Oscilloscope patterns. (a) Waveform study. (b) Phase difference Lissajous figure. (c) Frequency ratio Lissajous figures.

on the screen faithfully reproduces the voltage waveform, as shown by the left figure. This usage of the oscilloscope is called time-base display.

Besides studying waveforms, the oscilloscope is also useful for determining the phase shift of an ac voltage in passing through a device and for determining an unknown frequency. Both measurements depend on forming patterns of a type called *Lissajous figure*, after the French scientist who discovered them. When sinusoidal signals are applied to both the vertical and horizontal inputs (internal sweep not used), a Lissajous figure pattern results. If both are of the same amplitude and frequency, a single-loop pattern results, which varies from a circle to an ellipse to a very "skinny" ellipse (a diagonal line), depending on the phase angle between the two signals. This effect, illustrated in Fig. 14-15*b*, is used for the phase difference measurement. If the two frequencies are not the same, a many-loop Lissajous pattern results, the number and position of loops relating directly to the frequency ratio. This effect can be used to measure an unknown frequency applied to one axis, by connecting the other axis to an oscillator whose output frequency can be varied and is accurately known at any setting. When the Lissajous pattern shows a frequency ratio of 1:1, the unknown frequency corresponds to the oscillator's frequency.

14-11 ELECTRON-BEAM WELDING

An interesting direct use of the electron gun principle commercially is in the *electron-beam* (E-B) *welder*. Here the energy of the beam is used to weld two pieces of metal together, that is, to melt and fuse together the contacting surfaces of the metal pieces. If the CRT contains an electron "gun," then the E-B welder uses an electron "cannon" by comparison, consuming many kilowatts of power. The setup is shown schematically in Fig. 14-16. The vacuum chamber, in this case, is large enough to contain

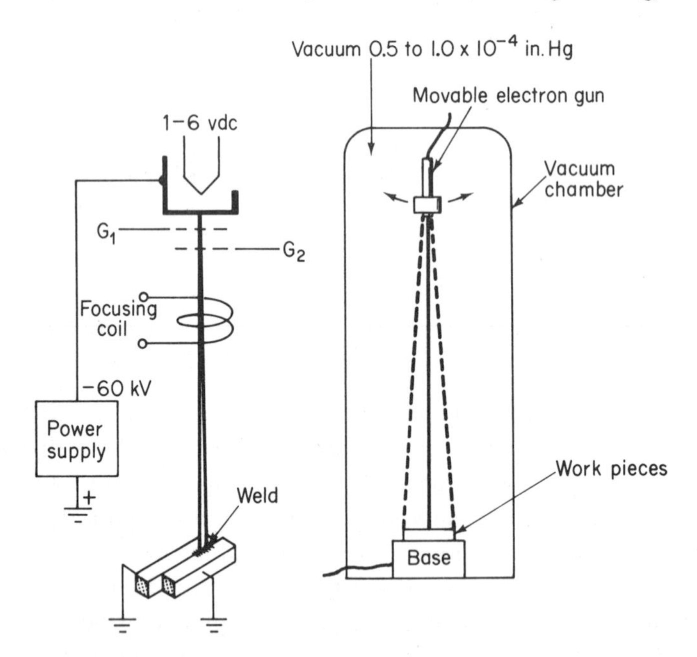

Fig. 14-16. Principle of the electron-beam welder.

the gun and the items to be welded. A vacuum pump keeps the pressure inside below 10^{-4} in. of mercury. The welding gun is mounted on a movable joint to "track" the weld, instead of using deflection plates or yoke. The work pieces are grounded, as is the positive side of the power supply, while the cathode operates at high negative potential, typically −60 kilovolts.

E-B welding is becoming increasingly popular. It is especially suited to light work, where a clean weld is required with minimum heating of the parts.

14-12 THE ELECTRON MICROSCOPE

Another interesting and important application of the electron gun is the electron microscope. To understand this application, it is well to spend a moment first on the ordinary optical microscope.

The key to the microscope, and to many other optical devices as well, is the lens. A lens, loosely speaking, is a piece of transparent material, with suitably shaped curved surfaces, which refracts (i.e., bends) light rays to form an image. The refraction of light rays by a simple lens is illustrated in Fig. 14-17*a*. A microscope, in its simplest form, is a two-lens device. An

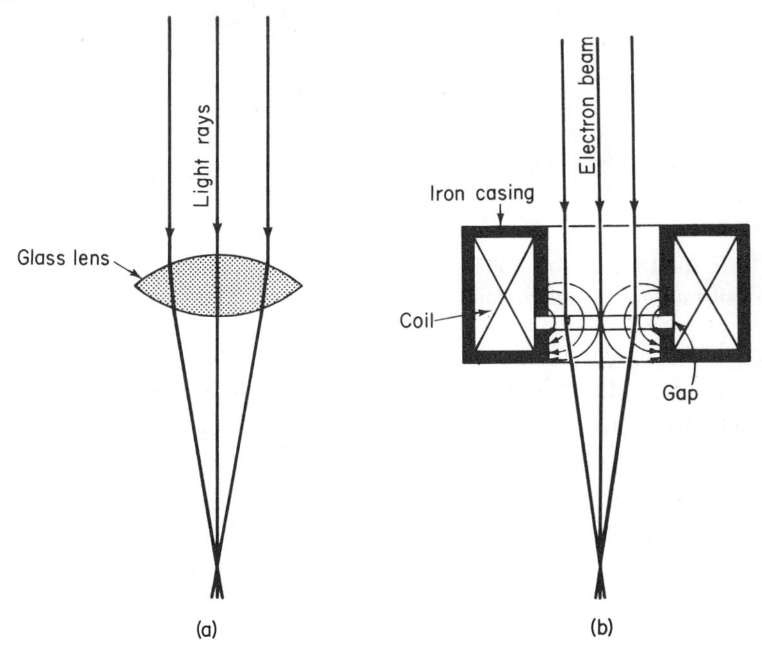

Fig. 14-17. Comparison of (a) optical, and (b) magnetic lenses.

image is formed of the illuminated specimen by one lens; a second lens is then used to magnify the image formed by the first. In high-powered microscopes, an additional lens, called a condenser, is used between the light source and the object studied to provide a concentrated uniform illumination. This configuration is shown in Fig. 14-18*a*.

You might think that, with suitably designed lenses, an optical micro-

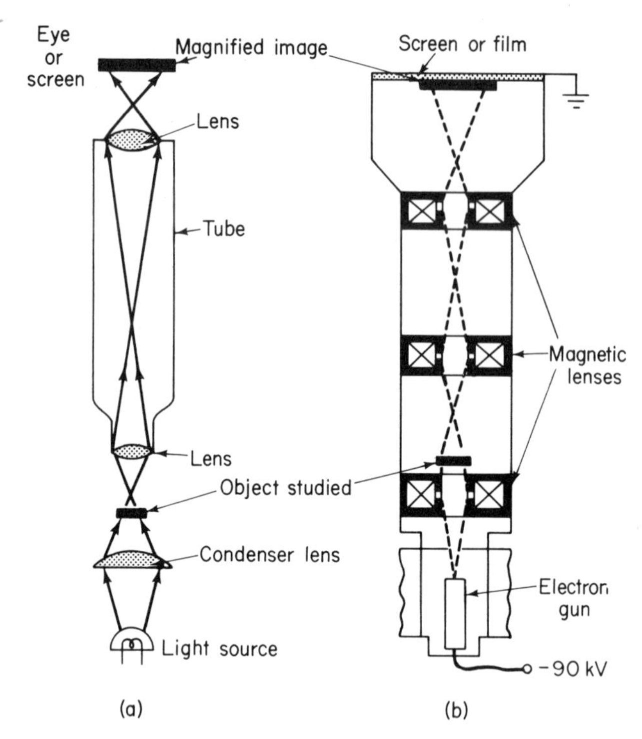

Fig. 14-18. One form of electron microscope (b), and comparison with an optical microscope (a).

scope can be designed to give any desired amount of magnification. This is true to a point. Magnification alone is not all-important, however. Much more important in an optical instrument like the microscope (or telescope, or binoculars) is *resolution*, the distance between points that are so close together that the eye just barely sees them as two points (rather than as one big blurry point). The reason that a limit of resolution exists is that light consists of waves, as explained in Section 8-6, and waves passing through lenses do *not* form perfectly sharp images. The blurriness of the image is due to a phenomenon called *diffraction*, which pertains to how waves "spread" around corners. The unaided human eye can just resolve two points separated by about 0.0003 in. at normal reading distance. It can be demonstrated that the resolving power of the best microscope is about 400 to 500 times as good as the eye. Therefore it is not practical to magnify much more than 500 times with the microscope, for there is no more detail to see beyond this limit and we lose image brightness to boot. Since, however, larger images are easier to study, even if they don't show any more detail, practical microscopes are constructed with magnifications up to two thousand diameters (2000x). What if greater resolution is desired—for instance, to make it possible to see viruses?

The most important factor in determining the limit of resolution is the wavelength of the illumination on the object being studied. In practice, the smallest length that can be resolved is taken to be one-third of the wavelength of the illumination. If we take 6×10^{-7} meter as the average wavelength of light, the smallest dimension visible with a microscope is therefore about 2×10^{-7} meter (about eight millionths of an inch), which leads us to the $500 \times$ maximum magnification we discussed in the previous paragraph. To get greater resolution, and higher magnification, the answer is simple: use shorter wavelength illumination! For a long time, however, this was easier said than done.

The beginning of the breakthrough came in 1924 when Louis De, Broglie announced his theory that elementary particles also had a wave nature. He gave a formula for the wavelength associated with a particle of given mass and velocity. In particular, if an electron is accelerated through a potential of 60 kilovolts, it has an associated wavelength of about 5×10^{-12} meter, only 1/100,000th that of light. In theory, therefore, it would be possible to use electrons as illumination in some sort of microscope that could magnify details 100,000 times. The trouble was that no one knew how to make an "electron lens" to refract an electron beam, like the ordinary lenses do to the light beam in the optical microscope. This problem was first solved by H. Busch, who in 1926 published a paper giving the theory of both electrostatic and electromagnetic electron lenses. By the time a year had passed, he had successfully tested a magnetic lens.

The concept of a magnetic lens is illustrated in Fig. 14-17*b*. In essence, it is a doughnut-shaped (toroidal) coil, completely encased in ferromagnetic material, except for a slit running around the inside of the "hole in the doughnut." This slit is the gap, on opposite sides of which the magnetic poles form. An intense magnetic field exists across this gap, which "refracts" the electron beam like a lens.

Electrostatic lenses were also developed slightly later, and the science of electron optics was established. Many different forms of electron microscope were actually developed through the splendid efforts of equally many workers, both too numerous to discuss in detail here. Perhaps worthy of special historical note was the rapid development in the early 1940s of practical high-voltage electron microscopes by E. Ruska and co-workers in Germany, and by Vladimir Zworykin and his associates at the Radio Corporation of America laboratories in the United States.

The electron microscope represented schematically in Fig. 14-18*b* is a magnetic-type *transmission* microscope, which is most directly analogous to the optical microscope. Its use is limited to very thin specimens, for electrons are easily absorbed. Electrons from a gun are focused by one magnetic lens on the specimen. The beam from the specimen is processed by two more magnetic lenses to form a final image, either on a phosphor-coated screen for direct viewing or on a photographic plate.

Zworykin and others went on to develop the relatively compact electron microscopes we know today. Assembly of one such unit is shown in Fig. 14-19. Using present electron micrograph techniques, followed by photographic enlargement, usable magnifications of 100,000× to 500,000× are attainable today.

Fig. 14-19. Assembly of an electron microscope. (Courtesy of Radio Corporation of America.)

REVIEW QUESTIONS

1. Why does a "power supply" circuit group appear in most electronic equipment?
2. How does the filter section work?
3. Explain the difference between half-wave rectification and full-wave rectification.
4. An amplifier circuit uses a triode whose amplification factor is given as 80 in the specifications. The circuit only gives a voltage gain of 40. Is something wrong? Why?
5. Express the following gains in decibel units: voltage gain of 5; power gain of 0.1; power gain of 200; voltage gain of 0.5.
6. Where would you use a cathode follower?
7. An amplifier with an 8-ohm output impedance is used to operate a 16-ohm loudspeaker. Is this practical? Why?
8. Explain how a VTVM measures a dc voltage. In what way is it superior to the VOM?
9. What is a voltage divider? Why is it used in the VTVM?
10. What determines the frequency of the signal generated by a sine-wave oscillator? How?
11. Name and state the function(s) of the component parts of an electron gun.
12. How do electrostatic and magnetic CRTs differ?
13. What characteristics are desirable for a phosphor to be used in a lab oscilloscope CRT? A TV picture tube (black and white)? A display CRT whose images will be photographed?
14. State three uses for the cathode ray oscilloscope. How is it set up for each of these jobs?
15. What major theoretical and practical developments were required to produce the electron microscope?

15

A PANTRY FULL OF SOUNDS

15-1 THE MIRACULOUS VOICE FROM THE WIRE

Today we take communication technology so much for granted that few of us ever stop to realize how truly marvelous it would have seemed to our forefathers to have a human voice come from a wire. Nevertheless, each time we turn on a radio, TV set, phonograph, or tape recorder, or use the telephone, a human voice is, in effect, traveling through a wire. Perhaps even more mysteriously, in the case of radio and TV, the voice and picture spend most of their journey traveling through space. This wireless transmission process we consider in the next chapter. A second wonder is also associated with the phonograph and tape recorder, however. We refer to the ability to *store* the sound of the human voice, and any other sound as well, in a thin plastic disk or a little roll of plastic tape. These "canned" sounds can be kept indefinitely, just like canned preserves, and can be taken out whenever we want to hear them. Moreover, we can "play back" these preserved sounds as often as we like.

In this chapter, then, we will see how sounds are recorded and played back, the inventions that made the process possible, and the degree of realism (*fidelity*) that is achieved.

15-2 SOUND AND HOW WE HEAR IT

Like the fish in the sea, we live in an ocean of air—our atmosphere. On the average, this large mass of air presses against every square inch of our body surface with a force of about $14\frac{1}{2}$ lb (give or take a little, depending on how close you live to the bottom of Death Valley or a Himalayan mountain top). Now, whenever an object moves through or vibrates in the air, small local fluctuations occur in the air pressure. These fluctuations spread out from the point at which they are generated, like the ripples on a pond spread out from the point where a stone hits. Nature took advantage of these small pressure variations as a possible warning device by giving most animals, humans included, a sense for detecting them. This sense we call *hearing* and our perception of the pressure variations is called *sound*.

The sense organ for sound is, of course, the ear. Inside the ear is a flexible membrane, like a drum head, that is free to vibrate. This membrane, the eardrum, is set vibrating by the sound waves (the pressure variations) reaching our ear. The vibrations of the eardrum are then transmitted, by three small bones, to an organ in the inner ear that contains the auditory

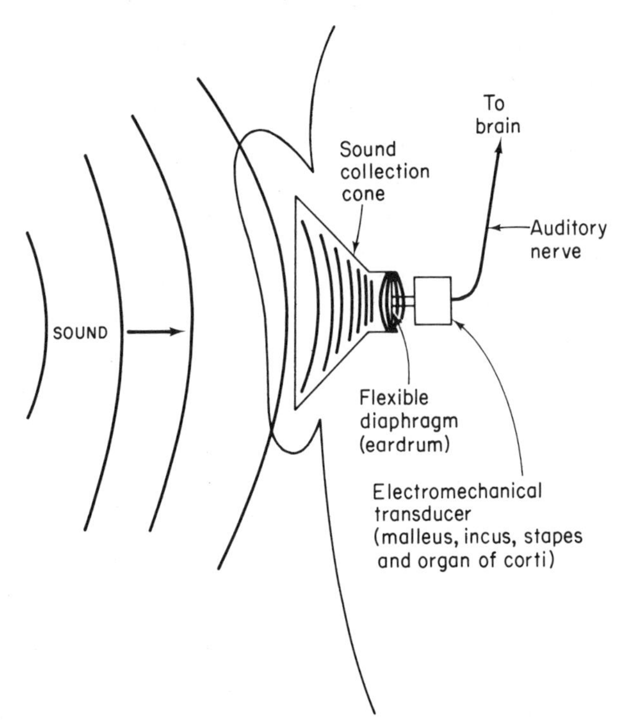

Fig. 15-1. The hearing apparatus.

nerve endings. There the vibration is transduced into electrical impulses, which travel up the nerve to the brain. These electrical impulses are what the brain interprets as sound. The process is represented in Fig. 15-1, in highly schematic form.

It should be noted that hearing is neither the only function of the ear nor its most important function. The ear evolved as the organ for determining orientation (up and down) and motions. It "picked up" the additional job of hearing along the evolutionary road. Some animals, like the tortoise, have well-developed balance organs but no hearing.

15-3 IT BEGAN WITH BELL

It is quite a step from the telegraph, which we have discussed earlier, to the *telephone* The word telephone comes from two Greek roots meaning "sound at a distance." Any device that could produce a controllable signal at a distance could be called a telegraph. The telephone, on the other hand, is intended to reproduce more or less faithfully at the "receiving" end the *actual sound* put into the "sending" end. The problem, stated in modern terms, was to develop two electromechanical transducers: one that would use sound waves to produce a varying electric current; the other, to use the electric current to recreate the sound waves. Although Alexander G. Bell is usually credited with the invention of the telephone, it is fair to say that other inventors, Bell's contemporaries, were also working on the problem of electric speech transmission and had made considerable progress. Professor Elisha Gray filed notice of telephonic work at the U.S. Patent Office only a few hours after Bell's patent application was submitted.

The problems of speech and hearing were very much a part of the life of the Bell family of Edinburgh, even before that day in 1847 when Alexander was born. Both his father and his grandfather were students of sound. Bell's father was among the poineers in teaching speech to the deaf, the vocation that Alexander himself took up. A siege of tuberculosis drove the Bells from their native Scotland to Canada in 1870, taking the lives of Alexander's two brothers in the process. After a year spent regaining his own health, Alexander came to the United States. He became interested in the electromechanical reproduction of sound, while holding a professorship in vocal physiology at Boston University.

In 1874 he announced the principle of telephony "If I could make a current of electricity vary in intensity as the air varies in density during the production of sound, I should be able to transmit speech telegrapically." In his quest, Bell had three invaluable aids: the developed technology of the telegraph, the theoretical work of Hermann Helmholtz, and the personal acquaintance of Joseph Henry. Vacationing in Canada in 1874, he described an electromagnetic device to his father, which, conceptually at least, would produce the desired "undulating current" from sound waves. A year later he directed his assistant, Thomas Watson, in building the first practical telephone transmitter (microphone). On March 10, 1876, while working on his "acid cup transmitter," Bell accidently spilled battery acid on his trousers He cried out "Watson, please come here. I want you." His assistant, working near the other end of the circuit on another floor, heard the words from the instrument beside him. Thus the telephone was born.

It is ironic that Bell's first successful transmitter, the "acid cup" device, was comparable to Elisha Gray's design and did not resemble the devices sketched in Bell's patent application.

The telephone was exhibited at the 1876 Centennial Exposition in Philadelphia. "It talks!" exclaimed a distinguished visitor, the Emperor Pedro II of Brazil, and the newsmen present immortalized that remark, along with Bell's invention. His patent claims did not go uncontested, however. Some 600-odd lawsuits, on behalf of Gray and others, were fought out in the courts until Bell was finally established as the inventor. He was rich and famous at thirty, but continued his inventive career.

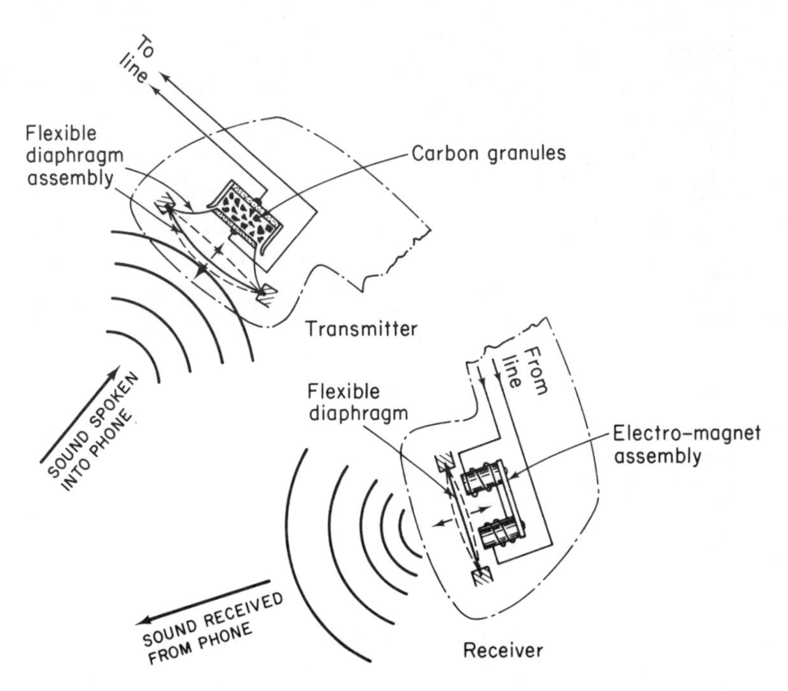

Fig. 15-2. Telephone transmitter and receiver units.

The telephone transmitter and receiver, in essentially their modern form, are shown in Fig. 15-2. The transmitter unit is a carbon button microphone, which was invented by Edison, not Bell. The heart of this device is a little box full of carbon granules fitted with a piston at one end. When the piston moves in and out, it compresses the column of granules to a varying degree. The greater the compression, the lower is the resistance of this column. The piston is attached to a flexible diaphragm, which vibrates like the eardrum when sound waves impinge upon it. Thus the resistance of the column of carbon granules is made to vary with the sound pressure. A potential impressed across the column of granules, through a connection to the piston and a contact plate at the opposite end of the column, produces a varying current as its resistance changes.

Fig. 15-3. Crossbar switches, part of Central Telephone Co. System, Las Vegas, Nevada. (Courtesy of International Telephone and Telegraph Co.)

The receiving unit is, in its simplest form, nothing more than an electromagnet placed near a flexible ferromagnetic diaphragm. When current from the transmitter flows through the receiver's coil the, varying strength of the resulting magnetic field causes the diaphragm to flex in and out. In this manner, the vibrating diaphragm "disturbs" the air in rhythm with the current in the coil, reproducing the original sound wave. Thus we have the early *earphone* and ancestor of the loudspeaker.

In addition to the millions of transmit-receive units, with the associated dialing and alerting units, our modern American telephone system comprises a vast network of miles of wire bundles, amplifying equipment, and complex switching equipment, like that shown in Fig. 15-3. Such equipment automatically shuttles your call to the party you are calling. Today telephone conversations are multiplexed, like telegraph messages, to permit sending a large number of conversations over the same wire.

15-4 THE VOICE ON FOIL

The first successful attempt to solve the problem of sound recording and reproduction was made by Thomas Edison in 1876. The Edison phonograph was a simple, strictly mechanical device that acted as both the recording and playback unit. It consisted of a drum covered with tinfoil which could be rotated by a crank, as shown in Fig. 15-4*a*. As the drum turned, a needle connected to a flexible diaphragm traced out spiral groves in the tinfoil. The diaphragm was mounted at the end of a horn whose function was to direct sound waves.

To record, you merely spoke or sang into the horn while turning the crank. Sound waves, concentrated by the horn, made the diaphragm vibrate and the attached needle *cut* the groove in the tinfoil. To play back the sound, you placed the drum with the needle at the beginning of the groove and turned the crank. This time the needle *followed* the existing groove pattern, the little "hills" and "valleys" at the bottom of the groove causing the needle to move up and down, as in Fig. 15-4*b*, which, in turn, caused the diaphragm to vibrate. Vibration of the diaphragm produced a sound wave in the column of air in the horn, which was recognizable as the original sound. The reproduced sound was, however, a very poor copy of the original. Furthermore, it didn't take too many "replayings" to make the recording unintelligible, for the coarse steel needle soon flattened out the hills and dales at the bottom of the groove. Nonetheless, Edison's phonograph, his own favorite invention, was at the beginning of the rapid evolution of "talking machines."

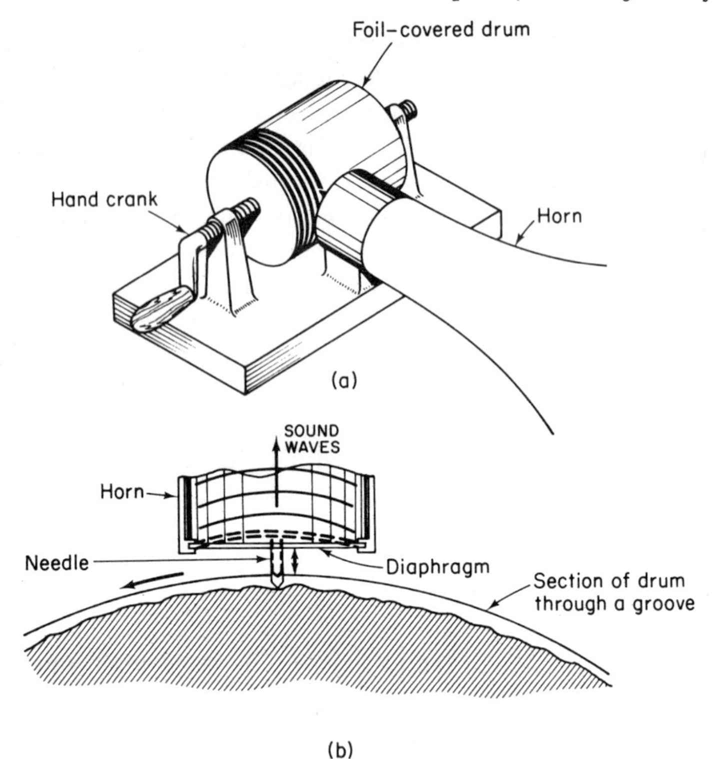

Fig. 15-4. The Edison phonograph. (a) Configuration. (b) Section showing hill-and-dale recording.

15-5 TOWARD THE MODERN PHONOGRAPH

Shortly after Edison's phonograph was announced, others followed. One great improvement was realized in giving up the hill-and dale sound track on the tinfoil drum in favor of "side-to-side wiggles" on a flat disk of resin or shellac. This *lateral* type of disk recording became the modern phonograph record.

The major problem, however, with all the early recording techniques was the feeble air pressure forces available in sound waves. Even with a horn to "concentrate" these forces, they make a dim impression at best, with most of their energy spent in flexing the stiff metallic diaphragms connected to the cutting needles. The playback devices were somewhat improved, too, but it took a rather massive weight (by modern standards) on a sharp needle to get a respectable sound out of the shellac disks. Some of us still remember the old wind-up Victrolas,[1] whose sharp steel needles

[1] *Victrola* was the trade name of early machines of the Victor Talking Machine Co. (now RCA), which became a household synomym for "disk record player."

had to be changed frequently. Of course, these sharp needles, tracking the relatively soft grooves at pressures of more than 25 *tons* per square inch (would you believe?), rapidly wore out the recordings. Their sound quality degraded badly in a few playings. Nevertheless, these battered old disks, spinning at 78 to 80 rpm, brought us pleasure for many years. Like Aladdin commanding a genie, we could summon Nelson Eddy, Al Jolson, or Enrico Caruso to sing at our whim, and we didn't mind if their voices suffered a little in the acoustic-mechanical journey between their throats and our ears.

15-6 WITH ELECTRON VOICES

Mechanical recording got a new lease on life after 1926, when electronics took a hand. Through the invention of several important types of devices, it became possible to use the feeble air forces to produce powerful electric currents that *electromechanically* recorded and reproduced the sound.

One of these was the microphone, early forms of which were invented by Bell and Edison, and which we have already met as the telephone receiver unit. There are many types of microphones, but they all have a common job—to produce an electric voltage or current that fluctuates in the same pattern as the sound waves striking them. Depending on the type of microphone, they either generate a varying voltage themselves or they cause a steady current supplied them by another source to pulsate by changes in their internal impedance. Several types are illustrated conceptually in Fig. 15-5.

The Edison-type carbon button microphone is shown in Fig. 15-5*a*. As explained earlier, it acts like a variable resistance when its granules are alternately compacted and loosened by the impinging sound waves. This type of microphone is relatively inexpensive and finds widespread use where faithfulness (fidelity) of reproduction is not critical.

Another variable-impedance microphone is the *capacitance* microphone, first perfected by E. C. Wente of Western Electric and represented in Fig. 15-5*b*. Once again the key element is a vibrating diaphragm, which performs a function analogous to the human eardrum. This metallic diaphragm serves as one plate of a capacitor, held away from a rigid plate by an insulating ring. A voltage is applied between the two plates. As the one plate vibrates, in rhythm with the incoming sound waves, it alternately widens and narrows the air gap between itself and the rigid plate, thus varying the dielectric thickness. The changing capacitance results in a changing current "molded" to the pattern of the sound waves.

A third type of microphone is the so-called crystal type. This micro-

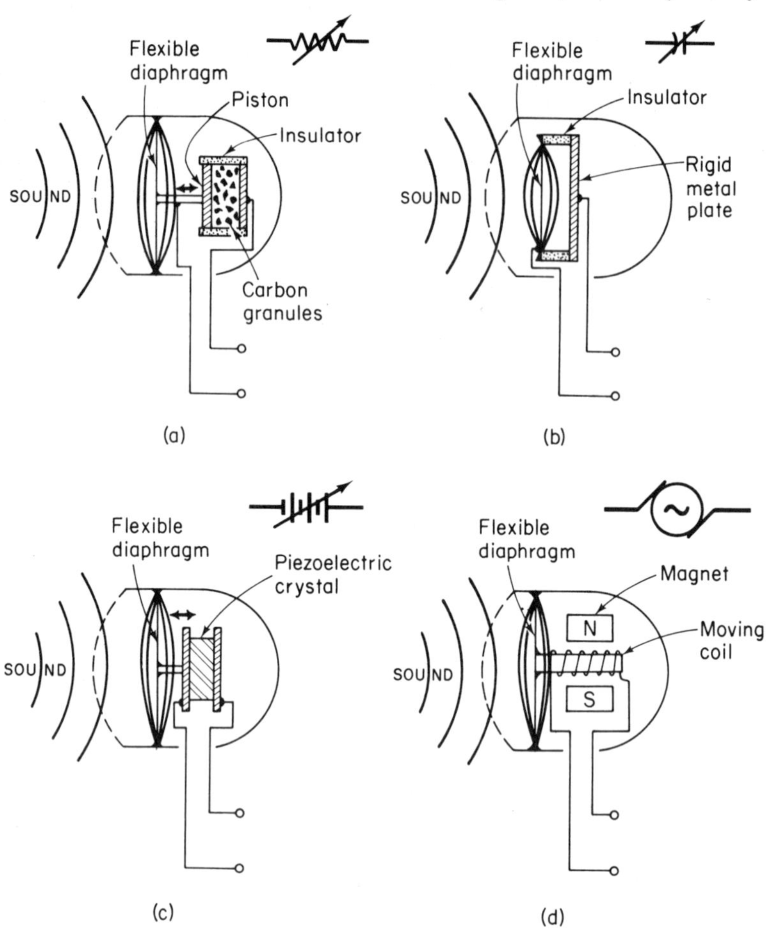

Fig. 15-5. Microphones. (a) Carbon button. (b) Capacitance. (c) Crystal. (d) Dynamic.

phone, shown diagrammatically in Fig. 15-5*c*, exploits the principle of *piezoelectricity*. Crystals of certain compounds when subjected to mechanical stresses will produce a feeble dc voltage between the stressed faces of the crystal, the emf depending on the amount of stress. In the crystal microphone, the crystalline material (once generally rochelle salt or ammonium dihydrogen phosphate, now commonly barium titanate) is sandwiched between two contact plates from which the varying voltage is picked off. One plate is rigidly mounted, while the other one is actuated by a flexible diaphragm, alternately "squeezing" and releasing the crystal. This is also a fairly inexpensive type of microphone, with reasonable fidelity, and is commonly supplied with low-cost recording devices.

A fourth class of microphones is the *dynamic* microphone, which uses electromagnetic induction to generate the varying voltage. The design of one type of dynamic mike is represented schematically by Fig. 15-5*d*. The key mechanical element is, as usual, a flexible diaphragm, to which a small coil is attached in this case. The coil is moved through the field of a permanent magnet by vibration of the diaphragm, inducing a voltage in the coil. Dynamic microphones completely dominate commercial broadcasting and recording applications. A typical commercial unit is illustrated in Fig. 15-6. Such microphones translate the programmed sound into electrical impulses with maximum accuracy.

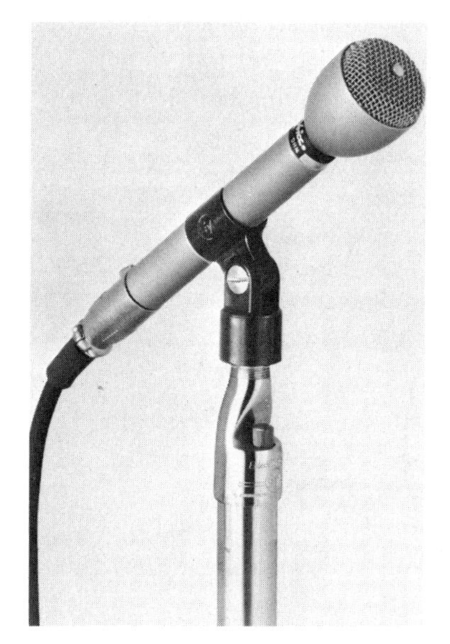

Fig. 15-6. Omnidirectional dynamic microphone. (Courtesy of Electro-Voice, Inc.)

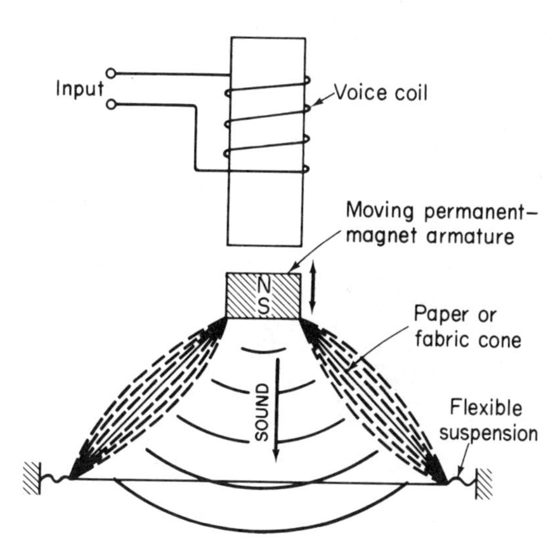

Fig. 15-7. Principle of inexpensive magnetic loudspeaker

At the "other end of the line" from the microphone is a device that must perform the opposite function, the translation of electrical impulses back into sound waves. Such a device is called a *loudspeaker*, or simply a *speaker*. The heart of the simplest (and cheapest) loudspeakers is an arrangement much like the telephone receiver—an electromagnet that is mounted near a flexible iron diaphragm. Variation in the intensity of the electromagnet's field alternately attracts and releases the diaphragm. To get more air moving, a suspended cone of paper or fabric is usually flexed by the diaphragm itself. This cone acts as a coupling device with the air in the room.

Since all loudspeakers are essentially microphones operated "in reverse," most microphone types (except carbon button) have their speaker equivalents. There are crystal, capacitance (more usually called electrostatic), and electromagnetic types of loudspeakers. The high-power requirements and crystal-size limitations of crystal "loudspeakers" limit their design to small units used in earphones. Electrostatic types are occasionally used in sound systems but are not popular because of their power requirements and special voltage conditioning requirements (usually met by circuits incorporated in the "package"). By far the most common loudspeakers are electromagnetic types. The low-cost version, shown in Fig. 15-7, has a moving permanent magnet armature that flexes the cone. The electrical signal is supplied to the rigid voice coil electromagnet. In contrast, most high-quality loudspeakers are of the dynamic type, wherein the voice coil moves within a rigid, massive magnet structure, as in Fig. 15-8. The moving coil, in this case

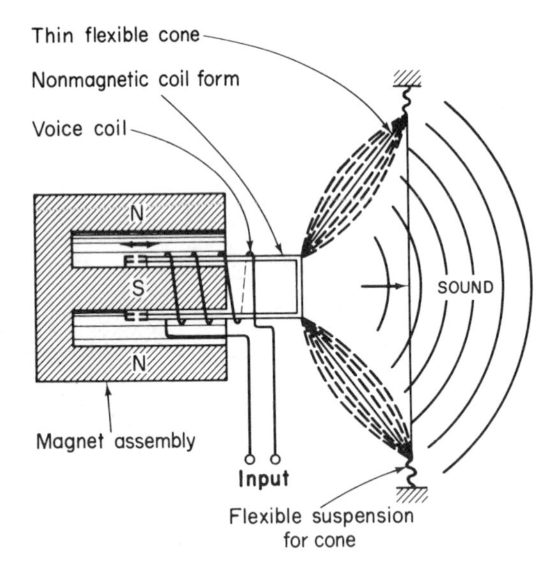

Fig. 15-8. The moving-coil dynamic loudspeaker.

wound on a form of nonmagnetic material, causes the cone to flex and produce the sound. A typical quality dynamic speaker is illustrated in Fig. 15-9. To get more faithful reproduction of the higher audible frequencies, this relatively large speaker has a separate smaller cone (often called a *whizzer*) mounted coaxially with the large cone. Mechanical design is such as to transfer most of the flexure energy to the small cone at high frequencies.

Between the microphone and the loudspeaker are the devices that put the impulses on the record and those that pick it up again. The former evolved from Edison's horn-loaded diaphragm with attached needle into a

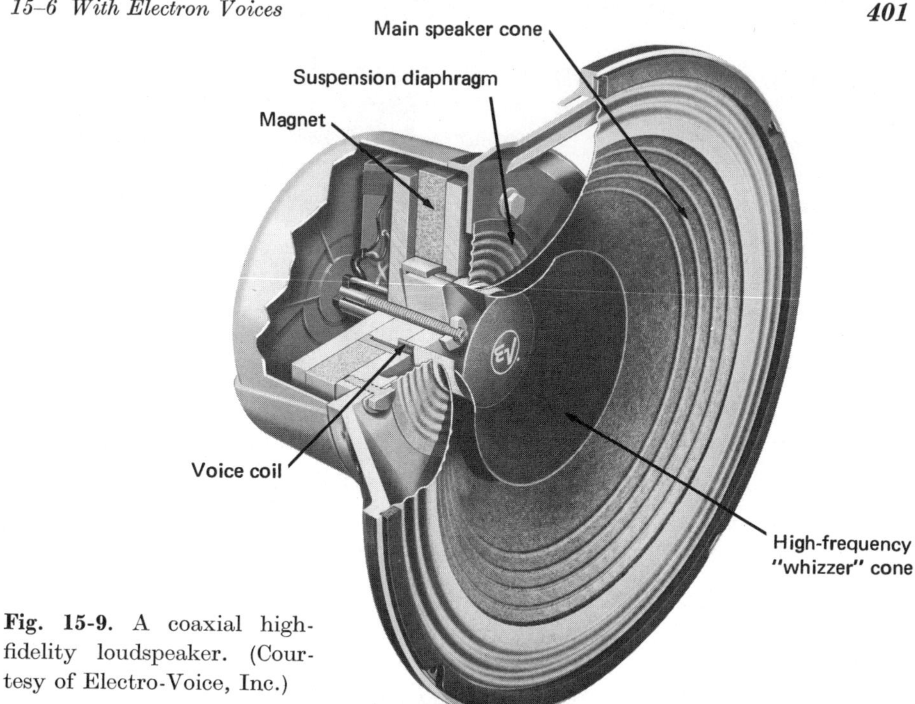

Fig. 15-9. A coaxial high-fidelity loudspeaker. (Courtesy of Electro-Voice, Inc.)

sophisticated electromagnetic cutting head, in principle, a dynamic speaker element with a cutting stylus. The pickups, too, went through a lengthy evolution, paralleling that of the microphone. As a result, there are crystal, capacitance, and electromagnetic phonograph pickup cartridges. Once the principles of such cartridges were physically realized, the major improvements were in reduction of dynamic mass and required tracking force. The former is important because dynamic mass—the total mass of those parts of the cartridge that must vibrate as dictated by the wiggles in the record—is the key to reproducing faithfully the most difficult sounds; the smaller the dynamic mass, the better. Tracking force, on the other hand, is the major issue in record wear. Remember that the 100 to 200 grams required in pre-1935 pickups produced pressures of tons per square inch at the sharp needle point. After 1935, pickup and record design improvements reduced the required force to one-tenth that value. Modern high-fidelity cartridges track at forces of one gram and less.

Of course, none of these devices would be of any value without the electronic amplifier. Microphones and pickup cartridges generate insignificant amounts of power, while cutting heads and loudspeakers require large amounts of power. DeForest's triode, and its successors, therefore played a central role in the development of sound recording and reproduction.

15-7 TODAY'S PHONOGRAPH

The phonograph (or gramophone, as the British call it) of today is a conceptually simple but sophisticated piece of equipment. The essential components are given in the block diagram of Fig. 15-10. The record is

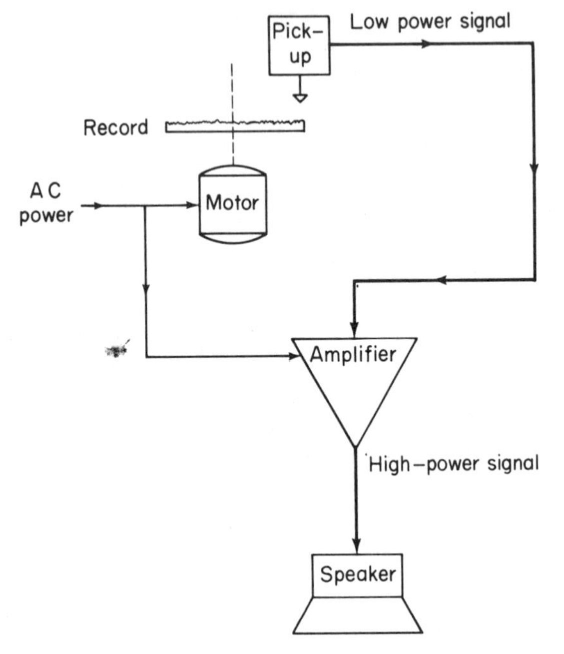

Fig. 15-10. Components of a modern phonograph.

rotated by a motor-driven *turntable* at a specific design speed. A stylus tracks the record's grooves, generating a voltage in the pickup cartridge. The signal from the cartridge is amplified by the amplifier section, which usually has provisions to adjust its power level (volume control) and quality (tone control) and is then fed to one or more loudspeakers. Power to drive the motor and supply the amplifier is taken from either the ac line or from batteries in the case of a portable unit.

The record-playing mechanism, a typical example of which is shown in Fig. 15-11, is worth a little study. Generally three or four playing speeds are provided by the motor drive assembly. A rotational speed of 78 rpm was used on early popular "singles" and is still provided for the convenience of persons who have collections of these records. Scientific study of the reproduction process showed, however, that there was an optimum speed for maximum playing time with a given record size, range of groove diameters, and stylus configuration. H. C. Harrison and J. P. Maxfield published the results of such a study in 1926, in which $33\frac{1}{3}$ rpm was selected for the 16-in. record used for early sound movies. The first commercial 12-in.

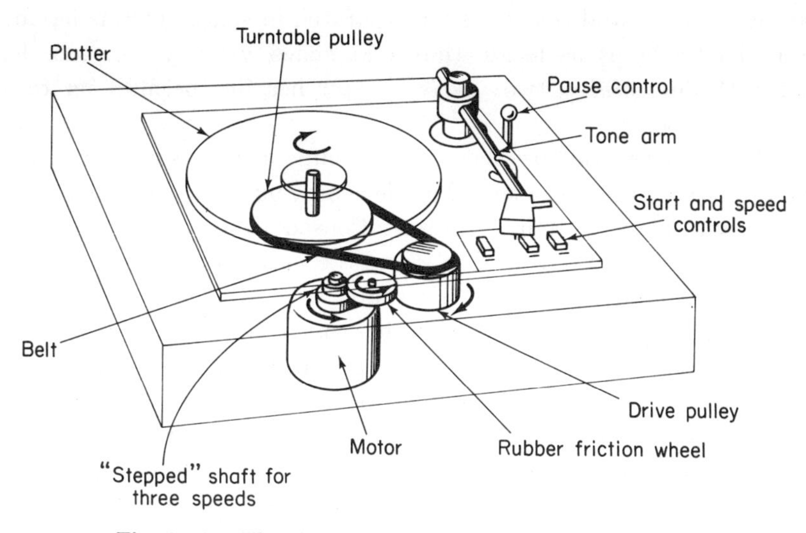

Fig. 15-11. The three-speed record playing assembly.

$33\frac{1}{3}$ rpm record was offered by the RCA-Victor Company circa 1931. Columbia followed suit with its own 15-minute per side long playing record, or *LP*. To provide better fidelity to the "singles" market, RCA introduced the 7-in. 45 rpm record in 1949. Hence all our modern drive mechanisms provide the 78, 45, and $33\frac{1}{3}$ rpm speeds. Where a wide range of tones was not required, such as in readings of Biblical and other literary material, it was realized that recording space could be economized by using a still slower speed, $16\frac{2}{3}$ rpm, and this speed is also provided on most phonographs. The multitude of speeds is normally achieved by a stepped shaft arrangement in conjunction with a constant speed motor of either the four-pole or synchronous variety. The torque is transmitted, through an intermediate idler, either to the rim of the turntable directly or to a main drive pulley and then to the turntable through a belt, as in Fig. 15-11.

The cartridge is mounted in a *tone arm*, which is designed to hold the cartridge for optimum tracking across the entire record. Tracking force is adjustable in better quality units. An automatic record-changing mechanism may be provided, in which case the playing assembly is called a *changer* rather than a turntable. Other frequently provided controls are a pause control and speed adujstment.

15-8 TWO EARS ARE BETTER THAN ONE

Each of us has *two* ears, one on either side of our heads. "That's obvious," you say, "so what?" What we don't often realize is how our comprehension of what we hear is greatly influenced by the fact that we do have more than one ear and that they are separated in space. Just as having two separated eyes helps us locate things in space visually, so, too, having *binaural* (Latin: *bini* = two, *auris* = ear) hearing permits us to place sounds in space.

At sea level and ordinary room temperatures, sound waves move along at a brisk 1130 fps, a considerable speed, but slow in comparison with light and with many of today's aircraft. Consequently, if we are not looking directly toward a nearby source of sound (i.e., our nose is not pointed at it), then a given sound wave from that source arrives at one of our ears later than the other by a small but detectable time difference. In evaluating the impulses coming over the auditory nerve, our brains have learned to interpret this time difference as an "off-center" location in space. We may turn our heads to look at the source or may simply note its eccentric placement. Thus one major factor in the sense of "realism" or "presence" in a hearing experience is that the sounds from sources placed differently in space reach our ears with different relative time lags.

In the ordinary *monaural* recording process discussed so far, all the recorded sound enters a single microphone and appears as a single set of lateral wiggles in the record grooves. The pickup converts this pattern back to electrical impulses, which, after amplification, are fed to the phonograph's speaker or speakers. Thus we may sit in front of a musical "combo" and hear the piano on our left, trumpet in the center, and drums on our right; but when we hear the recording, all of their sounds seem to come from the same place, the speaker. Even if there are two or more speakers, separated by some distance, they still do not recreate the placement of the original instruments in space, for they all receive the identical musical information from the amplifier.

The key to recording sound that recreates the binaural sensation is to have *two* (at least) separate and distinct *channels* of music. Each channel is recorded from a different vantage point in space, and they are played back separately. In terms of our previous example, we might have a mike at the left that records on one channel a loud sound from the piano, a moderately loud and slightly delayed sound from the more distant trumpet in the center, and a softer and more delayed sound from the drums at the right. At the same time, another mike at the right records a second channel on which the loudness and delays in the sounds of the instruments are just the opposite, the nearby drums being loudest with no delay and the piano being faintest with maximum delay. These two recorded channels are played back separately through two amplifiers and two sets of speakers that are physically separated by 8 to 10 ft in the listening area. The listener then hears properly delayed sounds at each ear and gets the spatial sensation of the instruments.

For this reason, a two-or-more-channel system is called *stereophonic*, meaning "spatial sound" and popularly shortened to just *stereo*.

In principle, stereophonic music could be recorded on two separate records having some sort of a synchronized playback mechanism. In practice, however, it is much simpler(and cheaper) to record both channels on the same disk. To accomplish this, our old friend, the hill-and-dale recording of Edison, was revived. On a modern vinyl stereo record, one channel of music is cut in a lateral pattern, and the second channel in a hill-and-dale (i.e., up and down) pattern. The pickup stylus therefore also has both up-and-down and side-to-side motions, as shown in Fig. 15-12*a*. These two motions are communicated by the stylus to crystal or magnetic elements in the stereo cartridge. The cartridge elements are designed so that each produces a signal from the motion in one plane only, as indicated in Fig. 15-12*b*.

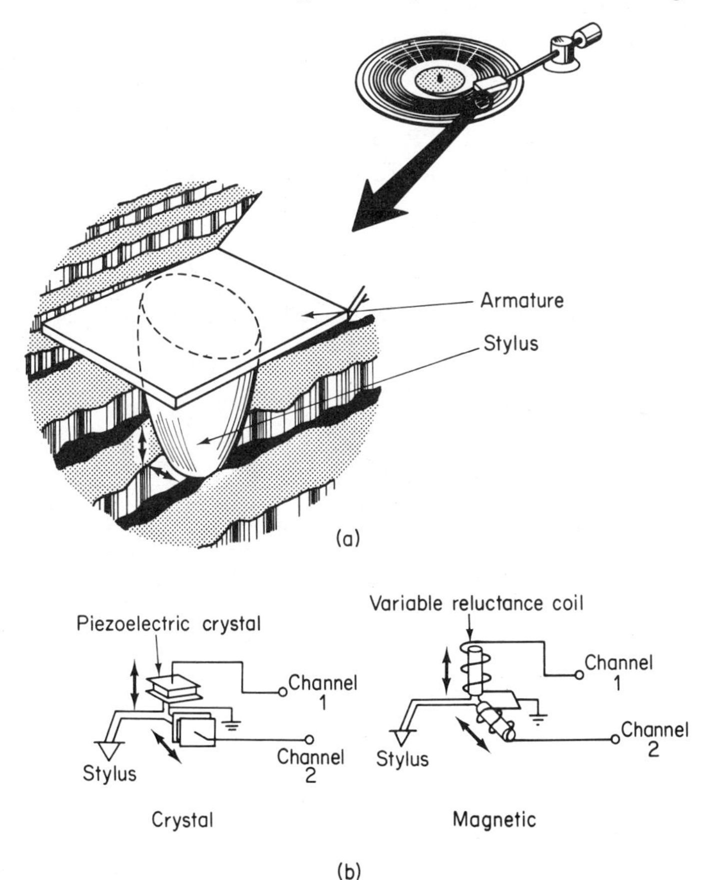

Fig. 15-12. Stereo record playback. (a) Motions of the stylus in a groove. (Adapted from a drawing, courtesy of Shure Bros. Inc.) (b) Principle of crystal and magnetic pickups.

Notice that we haven't yet said anything about fidelity; sterophonic sound is *not* necessarily high-fidelity sound. A true two-channel stereo system can be made (many are, in fact, sold today) with very low quality components, giving poor fidelity. The reverse is, of course, also true: a single high-fidelity channel of music *cannot* give a stereo effect. We shall have more to say about fidelity presently.

15-9 MAGNETIC RECORDING

We have seen the difficulties involved in early mechanical recording of sound, prior to the electronic amplifier, due to the weak forces available for cutting the impression in the drum or disk. An alternative to mechanical recording is magnetic recording.

A Danish telephone engineer, Valdemar Poulsen, is credited with the first invention in this field. In 1898 he patented the first *wire recorder*, after five years of experimentation. Poulsen, often called the "Danish Edison," took his *telegraphon*, as he called it, to the 1900 Paris Exposition. It won the Grand Prix. This first practical magnetic recorder is illustrated in Fig. 15-13.

Fig. 15-13. Poulsen's wire recorder. (Courtesy of Magnetic Products Div., Minnesota Mining and Manufacturing.)

Its operating principle is simple: A steel wire, 10 mils[2] in diameter and wound on a drum, is drawn past an electromagnet, the current in which has been caused to vary with the desired sound waves. The result is an induced pattern

[2] A mil is one thousandth of an inch.

of residual magnetism on the wire, a magnetic memory of the sound. Running this wire by a pickup coil (in Poulsen's device, the same electromagnet that recorded the pattern) produces small fluctuating currents in that coil by electromagnetic induction which are used to "drive" a headphone, reproducing the recorded sound. Although crude by modern standards, Poulsen's wire recorder worked and offered the major advantage of easy erase for reuse.

The coming of DeForest's amplifier tube in 1912 provided one element necessary to the evolution of magnetic recorders. Another, five years earlier, was the discovery that the quality of recordings could be improved greatly by using a fixed bias current to the recording head. This bias is required to force the recording process into a suitable operating region of the magnetization curve for that recording medium. Prior to World War II, most magnetic recording was on wire and the Germans developed the bulk of the technology. Steel tape, as well as wire, was used successfully as a recording medium.

The first nonmetallic tapes to be used as a magnetic recording medium had powdered iron coatings and were made by a German named Pfleumer around 1927. These tapes were very rough, often producing a fine spray of powder as the tape moved past the recording head. By 1939, however, the Germans had developed a good plastic tape. In the United States, meanwhile magnetic recording technology was developing at a more leisurely pace. The war gave a considerable "shot in the arm" to the industry, primarily because of Navy contracts. By 1946 the wire recorder had reached the peak of its development in the United States, when its death knell was sounded by the development of a high-quality magnetic oxide tape that could be produced at a reasonable cost. This breakthrough caused the meteoric rise of the tape recorder as we know it.

15-10 THE TAPE RECORDER

The basic processes involved in recording on tape and playing back the result are illustrated by Fig. 15-14. The heart of the tape recorder is the *head*, a small horseshoe electromagnet with a very small and precise gap between the poles. This gap is typically $\frac{1}{4}$ mil ($2\frac{1}{2}$ ten-thousandths of an inch) wide. The core of the head is a "magnetically soft" material; it has little residual magnetism when current to the coil is shut off. By contrast, the tape coating must be a "magnetically hard" material that retains its magnetization indefinitely.

The tape-transport mechanism does for the tape what the turntable does for records. It is provided with two spools, or reels, upon one of which the tape is stored, and to the other of which it is transferred, in the course of

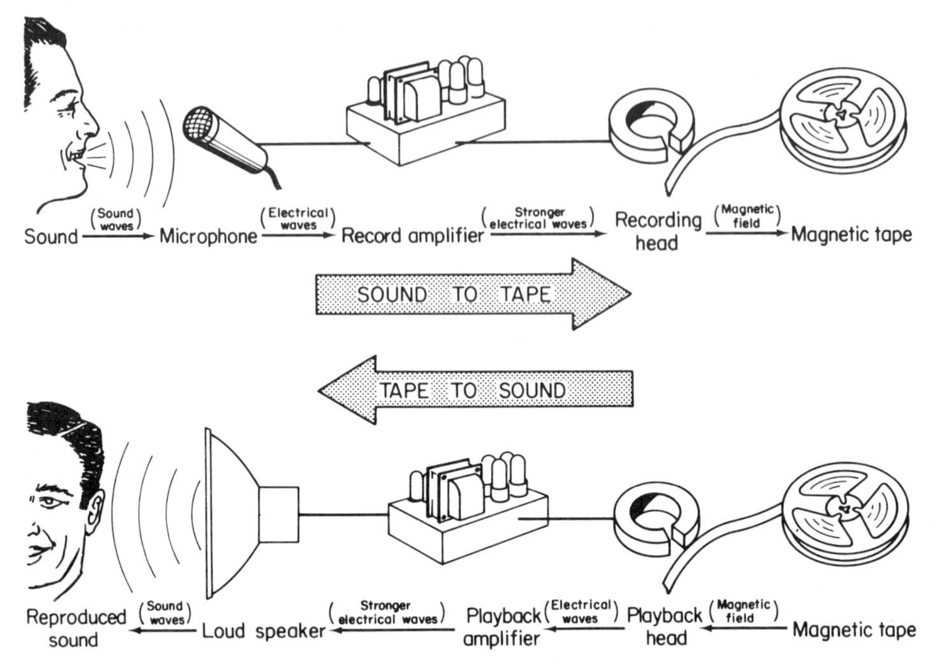

Fig. 15-14. Recording on magnetic tape. (Courtesy of Magnetic Products Div., Minnesota Mining and Manufacturing.)

recording or playback. One or more motors drive the reels and a *capstan*, a pulley that draws the tape by the heads at a precise speed. In a high-quality tape machince, such as the one shown in Fig. 15-15, individual heads are used for the record, playback, and erase functions; In more inexpensive units, two of these functions may be combined in a single head of compromise design. These components, together with an oscillator to produce an ac bias, recording amplifier (s), and the first stage of amplification for playback (playback preamplifier), make up what is called a *tape deck*. If the unit also includes the power amplifier(s) and speaker(s) necessary to make it a self-contained recording and playback unit, then it is called a *tape recorder*.

The magnetic tape itself is a thin plastic foil, either $\frac{1}{2}$, 1, or $1\frac{1}{2}$ mils thick, and is commonly made from cellulose acetate or Mylar.[3] The coating is red iron oxide (γ-$Fe_2 0_3$) dispersed in a *binder*, a complex mixture of resins and plasticizers. The pulverized oxide, shown in Fig. 15-16, is mixed with the binder and then spread evenly on rolls of backing 24 in. wide. The tape then moves at precise speed through a scrupulously clean drying oven, as in Fig. 15-17. During this journey, under exactly controlled temperature

[3] E. I. DuPont Co. trademark for polyester film.

Fig. 15-15. Studio tape recorder. (Courtesy of the Ampex Corp.)

Fig. 15-16. Red oxide of iron used for magnetic tape. (Courtesy of Magnetic Products Div., Minnesota Mining and Manufacturing.)

and humidity conditions, solvents evaporate from the binder, leaving a uniform oxide coating only about one-sixth the thickness of a human hair. Finally, the tape is cut into strips of the desired width—$\frac{1}{4}$ in. for amateur sound recording, $\frac{1}{2}$ to 2 in, for commercial and videotape use—and packaged.

To get a better "feel" for the tape recording process; suppose that we wanted to record a 1-kHz signal. This means that the current in the recording head reverses itself 2000 times each second. For each maximum current condition, the magnetic field of the head produces a tiny permanent bar magnet in the tape coating, its north-south polarity depending on the direction of the current in the recording head and its width determined by the head's gap. In this example, we would therefore be creating 2000 little bar magnets on the tape each second. These 2000 little magnets are laid out end to end with nonmagnetized spaces between them corresponding to the points where our 1-kHz sinusoidal signal goes through zero, as depicted in

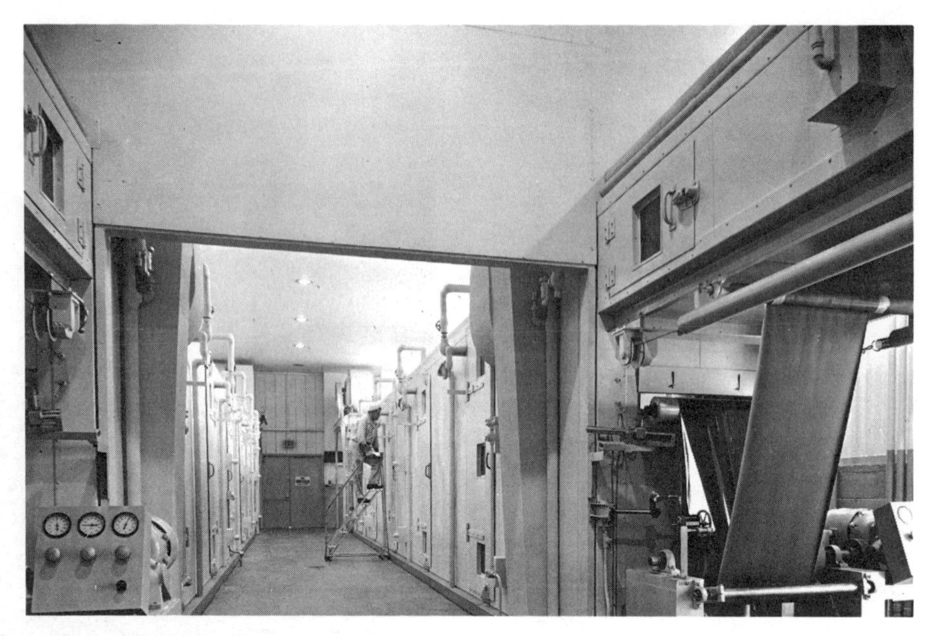

Fig. 15-17. Manufacturing magnetic recording tape. (Courtesy of Magnetic Products Div., Minnesota Mining and Manufacturing.)

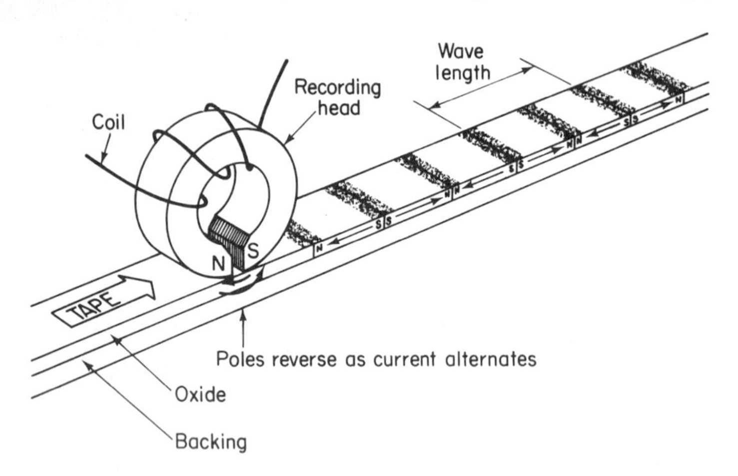

Fig. 15-18. The recording process. (Courtesy of Magnetic Products Div., Minnesota Mining and Manufacturing.)

Fig. 15-18. A magnified section of an actual tape, with the recorded pattern made visible, is presented in Fig. 15-19. You can see that, for a given gap size, to "lay down" the required number of little magnets in a second, there is some minimum speed at which the tape must move past the head. The higher the tape speed, the higher the frequency that can be recorded and

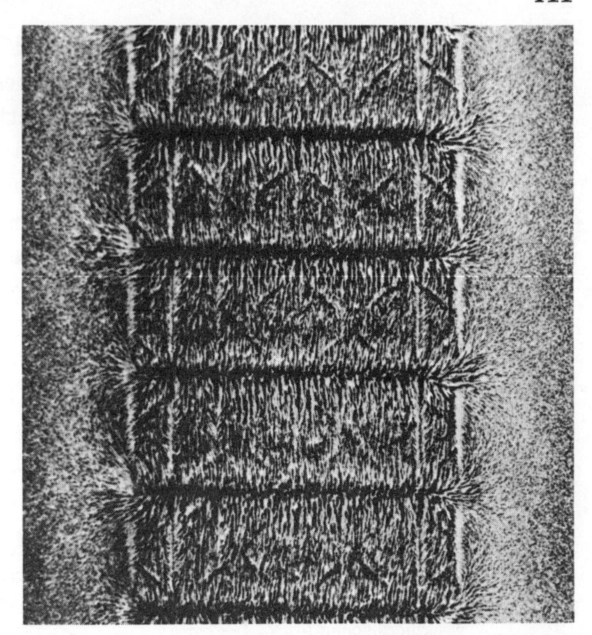

Fig. 15-19. Recorded pattern on magnetic tape. (Courtesy of Magnetic Products Div., Minnesota Mining and Manufacturing.)

played back. Studio recording equipment operates at a tape speed of 15 inches per second (ips). Amateur equipment usually provides a speed of $7\frac{1}{2}$ ips for high-fidelity music recording, with $3\frac{3}{4}$ ips and occasionally $1\frac{7}{8}$ ips for less-critical applications, such as readings and dictation.

Most present tape equipment is capable of recording and playing back two or four channels simultaneously, one above the other, on a $\frac{1}{4}$-in wide tape. This provides an excellent medium for stereophonic recording; in fact, widespread exploitation of stereophonic techniques began with tape, disk recordings following. Bell Telephone Laboratories had demonstrated stereo recordings (on steel tape) at the 1939 World's Fair, but the first commercial stero machine was a tape recorder produced in 1949.

A magnetic coating, called *striping*, is also put on photographic film for synchronized sound movies. This technique is particularly popular for amateur movies.

15-11 OPTICAL SOUND RECORDING

In sound movies, the cart actually came before the horse. Although he cannot be credited as sole inventor, Thomas Edison was responsible for the crucial breakthrough (using perforated film with drive sprockets) that made motion pictures practical. He was driven to accomplish this feat by a strong desire to have recorded pictures to go with the sound recorded on his phonograph. The first *talkies* used synchronized sound from a disk-type record in a process called *Vitaphone*, introduced by Warner Brothers in 1926. The synchronization problem was tricky, at best, and a way was sought to put the sound right on the film with the picture.

The problem was solved with the optical sound track, a narrow strip of light-and-dark pattern on the film that represents the sound photographically. The process is as simple in principle as making a phonograph record, but in this case the idea is to make a light beam vary in intensity with the sound pattern, rather than moving a cutting stylus. In one method of optical recording, the amplified signal is applied to a glow discharge lamp, causing its brightness to vary with the sound pattern. (An ordinary incandescent bulb can't be used because its filament takes too long to heat up and cool down to follow the high-frequency sounds.) Light from this lamp is focused by a lens system on a narrow strip of the film frame about one-tenth of an inch wide, the rest being devoted to the picture. This step is done on the

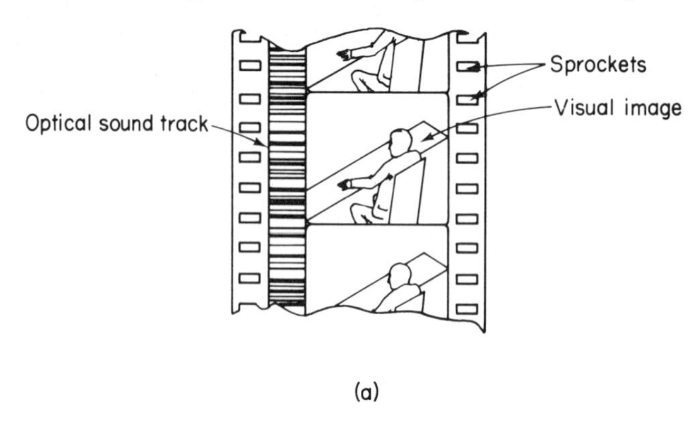

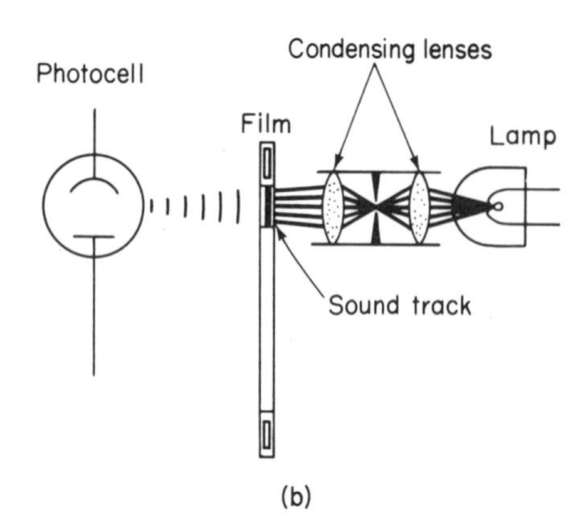

Fig. 15-20. Optical sound recording. (a) Film layout. (b) Scanning an optical sound track.

master negative film, from which the prints to be screened in theaters are made. The result, on the final print, appears as the series of dark and light bands illustrated in Fig. 15-20 *a*. To play back this sound track, an "optical pickup" system is built into the projector, shown schemaically in Fig. 15-20*b* looking along the strip of film. Light of fixed intensity from an incandescent source is projected by a lens system through the sound track to a photocell. As the film moves by, the band pattern lets varying amounts of light through to the photocell, thereby resulting in a current in the photocell that fluctuates like the original sound pattern. An amplified version of this current is fed to the theater's loudspeaker system.

An alternative optical recording technique uses a constant light source and instead drives a rotating mirror galvanometer with the amplified signal. The result is a film pattern consisting of bright stripes of variable width instead of fixed-width stripes of different brightnesses. The general principle is the same, however.

In today's motion pictures, both optical sound tracks and magnetic striping similar to that used in ordinary recording tapes are in widespread use. Magnetic striping seems to be winning out, because it is absolutely essential in certain stereophonic processes that require more than one channel of sound, notably Todd-AO and Cinerama. Magnetic striping also lends itself better to the low-cost amateur sound movie field.

15-12 HOW HIGH THE "FI"?

If anyone can be accused of prostituting the use of scientific terminology, it is the advertising profession. One good example of this fact is the term *high fidelity*, or *hi-fi* for short. There is scarcely a music-producing box on the market—phonograph, tape player, TV, or radio—that does not claim to have high fidelity. We would be indeed remiss, within the scope of this book, if we did not provide our reader with at least some background on the subject, to protect him from the morass of suspect or fraudulent equipment claims.

To begin with, the name high fidelity is a little redundant. Fidelity means *faithfulness* and, strictly speaking, the reproduced sound is either a faithful copy of the original or it is not. However, accepting the idea that any reproduction is to some degree faithful to the original, we go along with the idea of hi-fi as a *nearly perfect reproduction* of the original sound. The general idea is that, in listening to the reproduction, you can close your eyes and believe that the performers are there in the room before you. Now a good part of this illusion is, as we have already seen, the stereophonic effect, which gives the placement of the performers in space. While essential to the total sense of realism, the stereo effect is separate from fidelity; we can have sterophonic sound of very low fidelity. The other part to the aural sensation of realism is the characteristic of the reproduction that makes the violin music sound like it is coming from a real violin and the footsteps sound like they are coming from the impact of a Flamenco dancer's boot on the floor rather than from a loudspeaker. To understand it, we must first study briefly how musical instruments produce their characteristic sounds, and here we include the human voice as an instrument.

The sound of an instrument is caused by the vibration of some type of body within the instrument: a membrane in the drum, a column of air in the pipe organ, or a cord in the case of the human voice, violin, and harp. The discussion we are about to begin concerns a vibrating string. In principle, with slight modifications, it applies just as well to membranes, air columns, and other vibrating bodies.

Consider a string tightly stretched between two rigid attachment points. If this string is "plucked"—that is, grasped near its midpoint, stretched away from its undisturbed position, and released—it will vibrate back and forth about that undisturbed position, the amplitude of the vibration gradually dying away to nothing. A string constrained in this manner can only vibrate in certain specific ways, or *modes*. These modes represent *stand-*

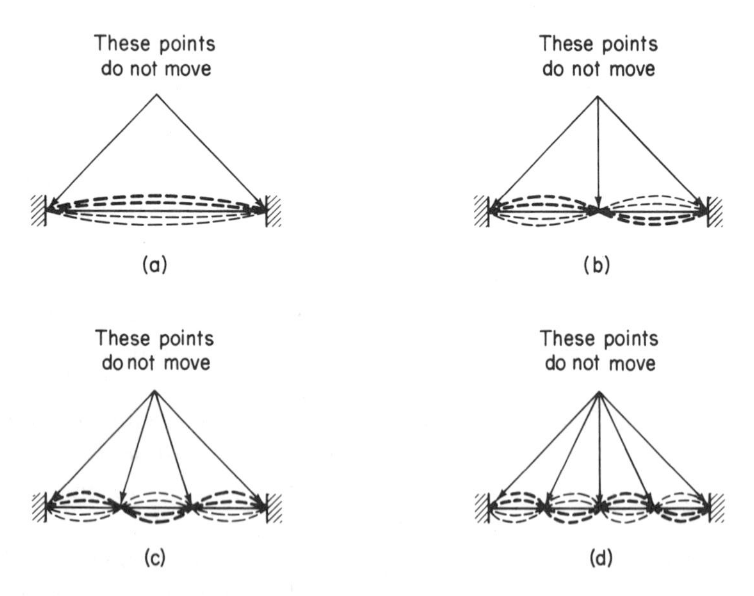

Fig. 15-21. Some natural modes of vibration in a string. (a) Fundamental. (b) Second harmonic. (c) Third harmonic. (d) Fourth harmonic.

ing waves in the string at natural resonant frequencies. The first four modes of vibration for the string are depicted in Fig. 15-21*a* through *d*. In any mode, there are certain points on the string called *nodal points*, or just *nodes*, which do not move from their undisturbed position. The fixed endpoints are, of course, always nodal points. Between any two nodes, every point on the string moves up and down sinusoidally with time, but the pattern remains fixed in space, hence the name *standing wave*. In the simplest mode of vibration, only the endpoints are nodes, as in Fig. 15-21*a*. This mode is called the *fundamental* mode. The next simplest mode of vibration, called the *second harmonic* mode, has a node midway between the endpoints, per Fig. 15-21*b*. Figure 15-21 *c* and *d* illustrates the *third harmonic* and *fourth harmonic* modes, which have two and three nodes, respectively, equally spaced between endpoints. The distance between nodes corresponds to half a wavelength for the standing wave produced. From these first four modes, we see that the permissible modes of vibration for the string are only those for which the half-wavelength is divisible into the length of the string a whole number of times. In other words, the string's length is some exact multiple of (i.e., equal to, two times, three times, four times, etc.) the half-wavelength of all its possible standing waves.

Each of these modes also has a characteristic frequency, which, you will recall, is inversely related to wavelength; the smaller the wavelength, the greater the frequency and vice versa. Since the fundamental mode has the greatest wavelength (i.e., the length of the string itself), it also has the lowest frequency. The second harmonic is twice the frequency of the fundamental, the third harmonic is three times the fundamental, and so on. The sound we attribute to each of these pure sinusoidal modes of vibration (providing the frequency is within the range of human hearing) is a *pure musical tone*, or *pitch*. For instance a pure sinusoidal vibration whose frequency is 440 Hz is heard as *A* above middle *C* on the piano keyboard, while middle *C* itself is a frequency of about 256 Hz. When an audible frequency is doubled, the new frequency is heard as the same note, one octave higher. Thus 880 Hz is also an *A*, because it is twice 440 Hz.

When we actually pluck the string, we do not get a simple, pure sinusoidal vibration but a complex periodic vibration that is a mixture of the fundamental and some of the harmonics (which musicians call *overtones*). A great French mathematician, Jean Batiste Joseph, the Baron de Fourier, showed in 1807 that any periodic motion, no matter how irregular the waveshape, can be considered the sum of the fundamental of the same frequency plus all the harmonic modes (of appropriate amplitude and phase). The plucked harp string or bowed violin string, which has been adjusted to produce a fundamental corresponding to middle *C*, will vibrate in a complex way, giving the desired middle C tone along *with some additional harmonics.* It is these additional harmonics that give the instrument its characteristic *tone quality.* Figure 15-22*a* shows the middle C waveforms from several instruments, as they would appear on an oscilloscope, compared with a pure 256-Hz sinusoid, such as an oscillator might produce. This is, in effect, a "picture" of why a French horn sounds different from an oboe playing the "same" note. All three patterns have a fundamental vibration frequency of 256 vibrations per second (remember, we can only use the term hertz for sinusoids), but they have different harmonic contents. Figure 15-22*b* shows how the same fundamental mode can be combined with just one or two harmonics of different amplitudes and phases, to produce dramatically different waveshapes.

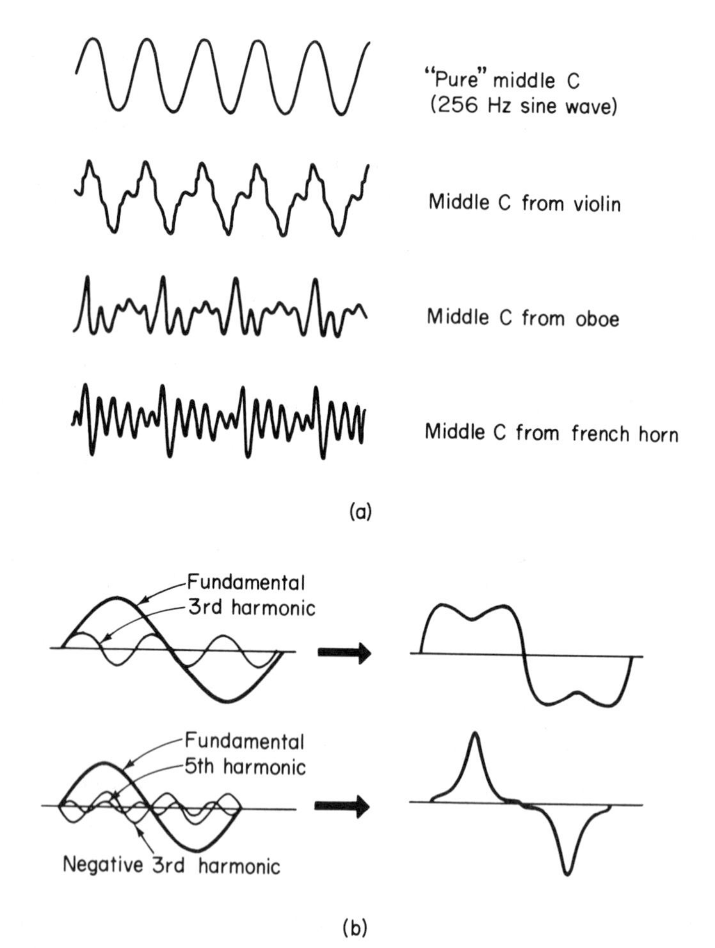

Fig. 15-22. (a) Middle C tone from various instruments. (b) Addition of harmonics to the fundamental.

Here, then, is key to high fidelity. To reproduce the sound of any instrument faithfully, the record/playback devices must be able to reproduce all the audible harmonics contained in the sound without changing the amplitude or phase relationships. This, of course, is just a longwinded way of saying that the hi-fi system must reproduce the sound's waveform without distortion.

The question arises as to what range of frequencies we need to consider for hi-fi reproduction of sound. We already know that the average human ear can hear sound frequencies in the range from about 20 to 20,000 Hz. Is this full range really required to cover the sounds of musical instruments and other sounds to be recorded? Figure 15-23, a bar chart showing frequency range of various sources of sound, tells us emphatically Yes!. A hi-fi system really worthy of the name should be capable of reproducing all frequencies between 20 and 20,000 Hz without distortion of relative amplitude or phase.

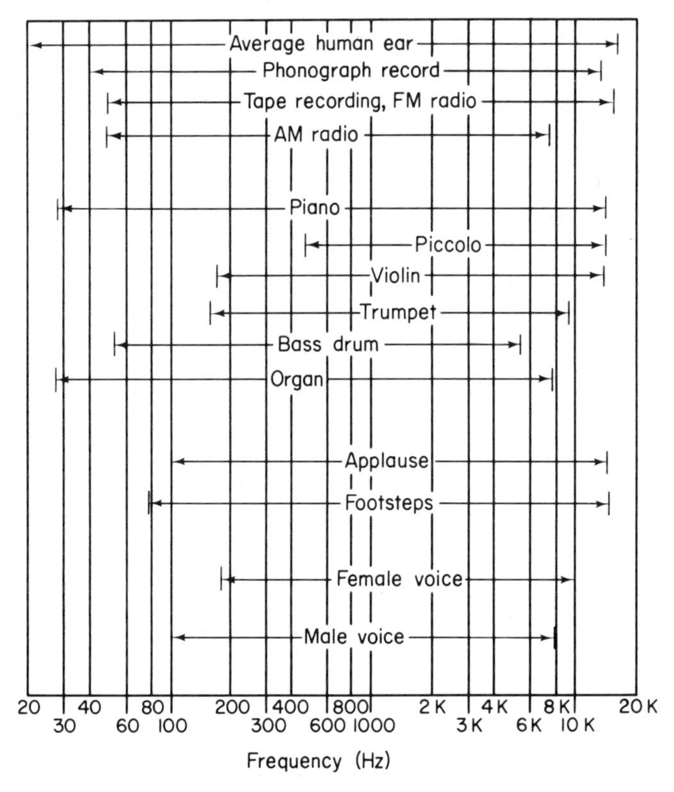

Fig. 15-23. Range of fundamentals and audible harmonics for various sound sources.

15-13 WHERE HARDWARE FAILS

From the previous discussion, it is evident that all of the components in a hi-fi system should treat all of the audio frequencies (those in the audible range, from 20 Hz to 20 kHz) in the same way. Therefore the phonograph pickup cartridge should put out the same voltage for the same amount of travel across the groove, regardless of how fast the wiggle; the amplifier should produce the same output voltage for a given input voltage, irrespective of its frequency; and the speaker should produce the same air pressure variation for a given current, whatever its frequency. Unfortunately, things don't quite work out ideally in real life.

We introduce the idea of a *frequency response curve* for a component. This is a graph that, as the name implies, shows how a given component responds to different frequencies. The audio engineer has adopted the frequency 1000 Hz as a standard of reference, because 1000 is a nice round number which (on a logarithmic scale) is right "in the middle" of the band of audio frequencies. The frequency response curve is a graph of how the output of a component *differs* at any given frequency from its output at 1000 Hz, its settings or adjustments (if any) being kept fixed. The output variation is customarily measured in decibels, so a typical frequency response curve has a vertical scale of "response" (difference in output from that at 1000 Hz) marked off in even increments of decibels, while the horizontal axis is the "squeezed up" (logarithmic) frequency scale. Because of the way we have defined response, the response will always be zero decibels at 1000 Hz. In terms of the frequency response curve, then, an ideal component would have a graph that was a horizontal straight line at zero decibels, meaning that its output is the same at all audio frequencies as at 1000 Hz.

Frequency response curves for some typical *low-fi* components are given in Fig. 15-24. Since the human ear, under normal circumstances, can just detect a change of 3 db, the response curve may be considered *flat*, from the practical standpoint, so long as it stays between +3 db and −3 db. For the components represented in Fig. 15-24, we see that the RC coupled amplifier's response is flat from about 60 Hz to about 13kHz; the inductively coupled amplifier's is flat from about 140 Hz to 3 kHz; while the poor little speaker's response is only flat in the range from about 900 Hz to 6500 Hz.

The reasons for these dropoffs at low and high frequencies lie in the nonideality of the parts used to make these components. In the RC coupled amplifier, for instance, the coupling capacitor begins to look like a high impedance at low frequencies, while the stray capacitances to ground (as between tube elements) start to look like a short circuit path to ground at high frequencies. In the inductively coupled amplifier, the low impedance of

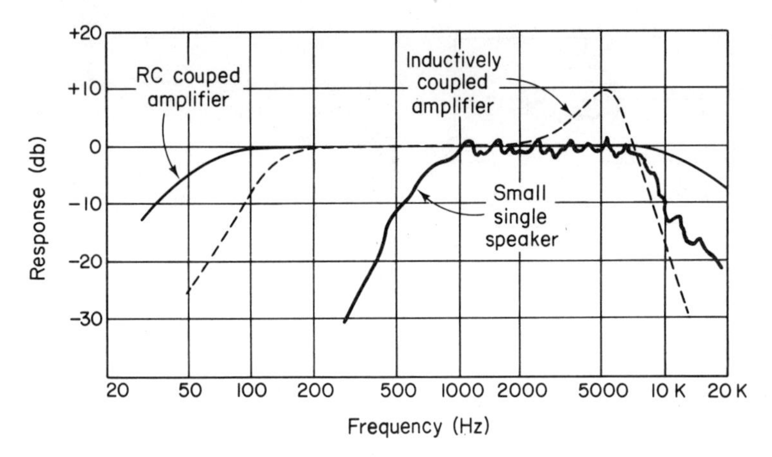

Fig. 15-24. Frequency response curves for some typical components.

the transformer at low frequencies does not permit high amplification, while at high frequencies the stray capacitance to ground is the culprit once again. In addition, the inductively coupled amplifier has a response "peak" (around 5000 Hz in the example shown) due to resonance of the stray capacitance with leakage inductance. The speaker's problems are mechanical rather than electrical, relating to its ability to move large enough quantities of air, and the inertia of the moving cone assembly. Although not illustrated here, record cutting heads, phono cartridges, and tape heads have their frequency limitations, too, of course.

15–14 THE HI-FI SYSTEM

Most adults cannot actual hear frequencies as low as 20 Hz or as high as 20 kHz. In particular, as we grow older, our *high*- frequency response falls off badly. A sound system having a flat (within ± 3 db) response from, say, 50 Hz to 15 kHz would therefore be a high-fidelity system to all but the most sensitive and discriminating ears. It is indeed possible to achieve this goal at a reasonable cost and possible to surpass it if the audiophile (i.e., the hi-fi buff) has money to burn.

The easiest thing to "fix" is the amplifier. By careful choice of components, and by using feedback and some other little tricks, the audio design engineer gets extremely flat response across a frequency range broader than the audio range, but at some sacrifice in power gain. Since considerable power is needed—to drive the multiple speakers required—a final stage of undistorted power amplification is required. Thus we arrive at a clever circuit called a *push-pull* amplification stage, shown in its tube and transistor

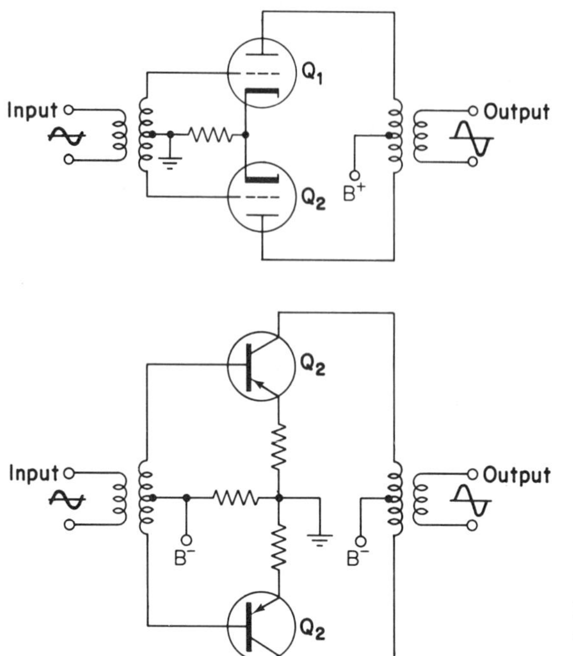

Fig. 15-25. Tube and transistor versions of a push-pull power amplification stage.

versions in Fig. 15-25. This stage is transformer-coupled to the previous voltage-amplification stage, the secondary of the transformer being center-tapped. The connections and biasing are such that the tube or transistor, Q_1, conducts only through half a cycle[4] and is cut off during the other half, and Q_2 operates during the other half-cycle. The outputs of Q_1 and Q_2 are combined in the output transformer, whose primary is also center-tapped. The result, per Fig. 15-26, is that the circuit can supply twice the current, and four times the power, provided by a conventional single-tube (or single-transistor) stage, with each tube doing only half the work. This yields maximum power with minimum distortion of the signal.

The response of the phonograph pickup is improved by careful design, using very light and delicate moving parts. Most high-fidelity cartridges are the magnetic type, usually called "variable reluctance" cartridges. Of course, the turntable drive and the tone arm must also be correctly designed to minimize noise and other distortion.

A lot can be done, by careful design and a few tricks, to broaden the response of a speaker. No amount of design effort to date, however, has produced a single-element speaker that can cover the whole audio frequency

[4] This is called "Class B Operation" as opposed to Class A, where conduction is over a whole cycle.

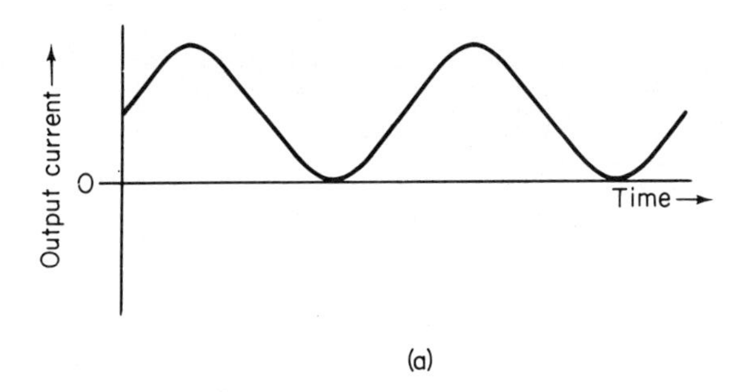

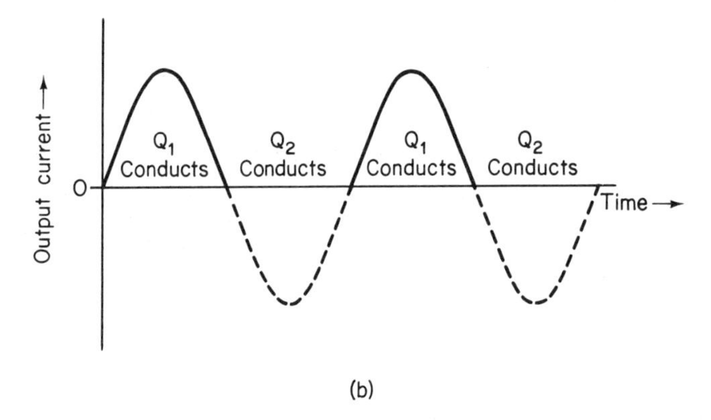

Fig. 15-26. Comparison of the output of a push-pull stage (b), with that of a single-tube stage (a).

range. As a result, we have the hi-fi speaker *system*, containing two or more speakers of different sizes (sometimes mounted on a common frame and called a *coaxial* or *diffaxial* speaker). One is always a fairly large speaker, called a *woofer*, whose function is to reproduce the low frequencies. Another is always a small speaker, often with a horn instead of a simple cone, which generates the high-frequency sounds and is called a *tweeter*. More elaborate systems contain a third *midrange* speaker for the middle frequencies. The various frequencies contained in the signal out of the amplifier are sorted out and directed to the proper speaker by an electrical network called a *crossover network*. In hi-fi reproduction we therefore deal with the frequency response of the speaker system as a whole.

Figure 15-27 gives response curves for some typical hi-fi components. Although no tape-recording components are illustrated here, it should be pointed out that tapes are, potentially, the highest fidelity program source. Tape response is determined primarily by the quality and gap size of the head, together with the tape speed, the latter being under the control of the

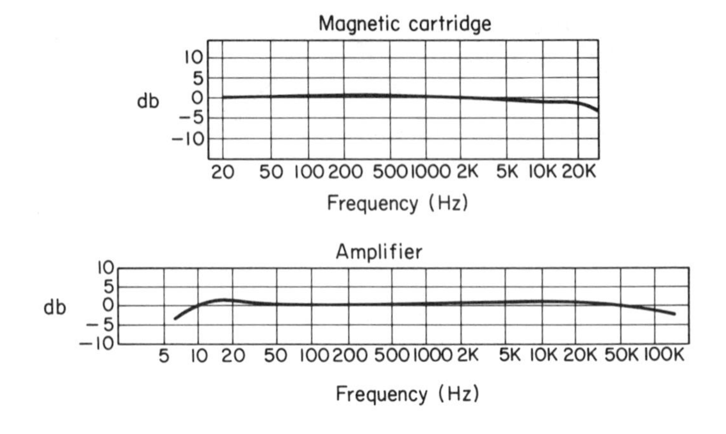

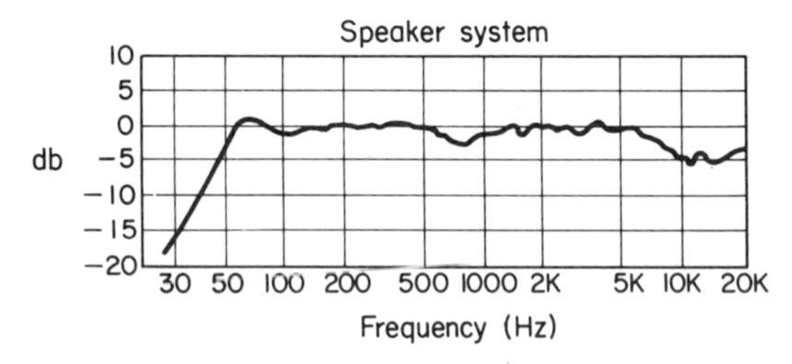

Fig. 15-27. Response curves for typical high-fidelity components.

operator. Our 50-to-15,000-Hz goal is easily met with a good tape machine at the $7\frac{1}{2}$ inches per second (ips) tape speed, and can be bettered at 15 ips speed.

The message to be taken from the foregoing discussion, if there is one at all, is that on a dollar-for-dollar basis the weakest link in the hi-fi system is the speaker system, for it is there that the flatness of frequency response is hardest to achieve.

The true present-day hi-fi system is both stereophonic and of high reproductive fidelity. A representative system would have three sources of program material, a phonograph, a tape deck and a *tuner* (an FM or AM/FM radio without the amplifier and speaker). The layout of such a system is shown in block diagram form in Fig. 15-28 and pictorially in Fig. 15-29. (See if you can identify the major components in this picture.) Note the two independent signal channels all the way through Fig. 15-28, as required for true stereophonic effect.

The heart of the system is the *integrated amplifier*, which consists of two sets of preamplifiers and actual power amplifiers built on a single chassis.

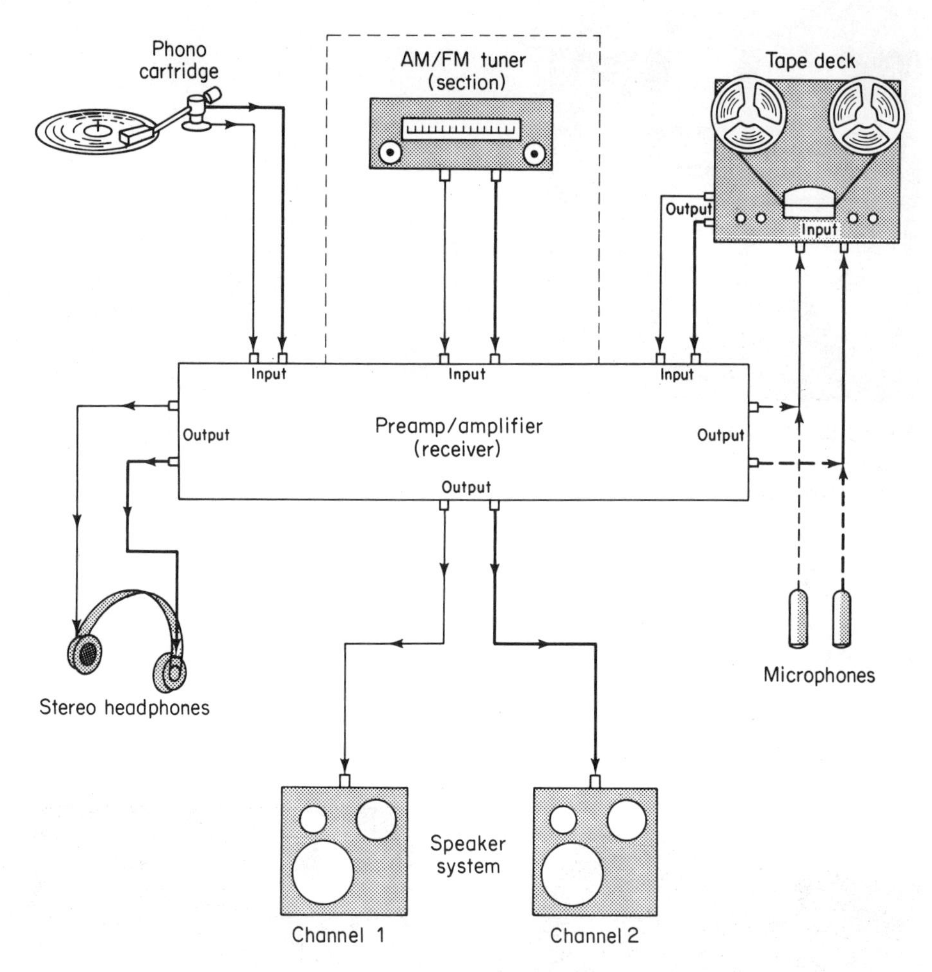

Fig. 15-28. A complete stereophonic hi-fi system.

The preamplifier circuits contain adjustments (which appear as control knobs and switches on the front panel of the unit) permitting the listener to control volume (or loudness) of both channels, bass and treble emphasis (tone control), and to compensate for certain other vagaries of the recording/reproduction process. In addition to the outputs to the speakers, most amplifiers provide separate outputs for tape recording and for private listening with headphones. The popular tendency now is also to mount the tuner on the same chassis as the preamplifiers and power amplifiers, indicated by the dashed line in Fig. 15-28, in which case the combined unit is called a *stereo receiver*. A representative receiver unit is illustrated in Fig. 15-30, this particular one being an AM/FM unit. Note the array of controls. Taking this

Fig. 15-29. Home high-fidelity music system. (Courtesy of Allied Radio Corp.)

Fig. 15-30. Typical stereo receiver. (Courtesy of the Heath Corp.)

Fig. 15-31. Compact stereo music system. (Courtesy of the Heath Co.)

Fig. 15-32. Typical high-fidelity tape deck. (Courtesy of the Ampex Corp.)

"combination unit" business even one step farther, a recent entry into the market is the *stereo compact,* in which the stereo receiver and a phonograph changer mechanism are mounted together in one compact unit. Figure 15-31 shows a stereo compact.

A breathtaking variety of tape recorders and tape decks is available in all degrees of quality and price. A quality unit, such as the one in Fig.15-32 offers three speeds and the capability of playing both monaural and two-and four-track stereo tapes.

The components of a typical high-quality speaker system are shown in Fig. 15-33. These "active" components alone, however, do not make a

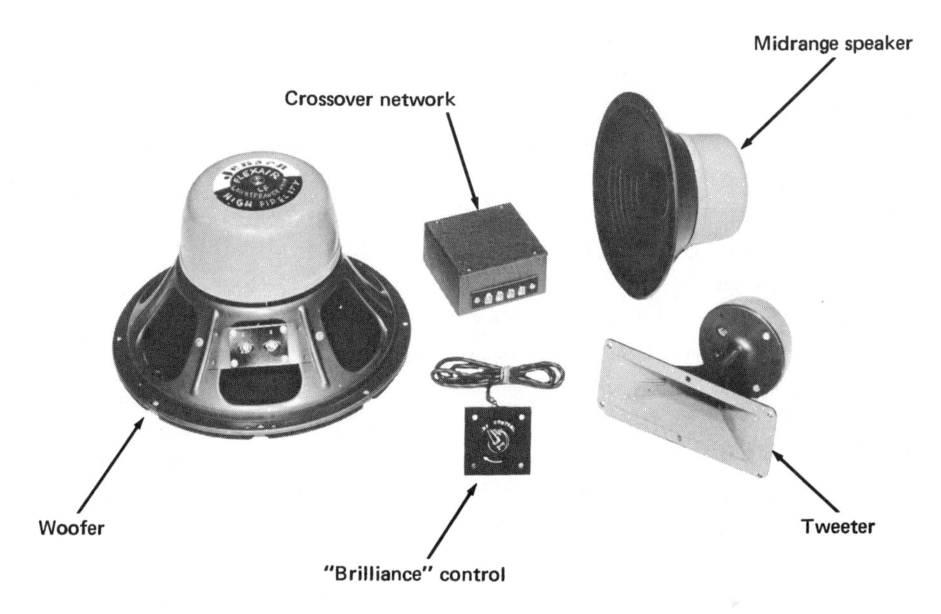

Fig. 15-33. Components of a quality high-fidelity speaker system. (Courtesy of Jensen Manufacturing Div., The Muter Co., Chicago, Ill.)

speaker system. The cabinet into which they are mounted, usually called the *enclosure*, has a very strong influence on the sound quality and is, in fact, an integral part of the speaker system. The enclosure is really a musical instrument that must be adequately matched to the speakers which excite its contained air volume.

The author sincerely hopes that this section will be of assistance to the reader who is considering the purchase of a hi-fi system. The reader is cautioned, however, that after all the preliminary screening has been done on the basis of manufacturers' specifications and frequency response curves, the ultimate decision has to be made by the ears (in conjunction with the pocketbook). Always have the salesman demonstrate the exact combination of components you intend to purchase. They still may sound slightly different in your home because of the differences in room acoustics.

15–15 SOUND IN THE SEA

One final application of electronics to sound, which is worth a little of our time, is *sonar*. The word *sonar* is an acronym that is derived from the words *so*und *n*avigation *a*nd *r*anging." Basically it is a technique for locating objects in the sea by means of sound echoes. The term was originally applied to echo-ranging equipment for submarine detection, but is now used for underwater sound devices in general.

Sonar systems are of two general classes: *active* and *passive*. The passive system is nothing more than a listening device to detect sounds generated by the located object. An active system, on the other hand, generates its own sound, alternately sending out a pulse of sound and listening for the return of echoes.

The idea behind sonar is not new. It has long been recognized that sound waves offer the only means for "seeing" any significant distance underwater, since light is badly absorbed and dispersed over short distances. Even with sonar, underwater "seeing" is limited to a few miles range. Since the velocity of sound in seawater averages about one mile per second (compared with the 186, 200 mile-per-second velocity of light in air), sonar search is rather slow. The principle of sonar is simplicity itself. A simplified form known as a *hydrophone* (Greek: "sound in water") was used in World War I to warn ships of the presence of submarines. The hydrophone is a passive system equivalent, in simplest terms, to a submerged microphone whose signal is amplified and monitored with earphones.

The strategic pressures of two world wars, and collaboration between the governments of Great Britain and the United States, brought about the evolution of sonar to its several present forms: *Searchlight* sonar produces a narrow beam of sound and swings this beam by moving the transducer like an optical searchlight. Many small ships carry a simple straight-down-only searchlight-type sonar, called a *fathometer*, for depth measurement. *Scanning* sonar produces a beam that spreads out in all directions, but the echo-detecting elements are highly directional and these are scanned for bearing information. Helicopters use an adaptation called *dipping* sonar, which employs a spherical transducer on a long cable that is dipped into the water and drawn along as in trolling. Similar to dipping sonar is *variable depth sonar* or *VDS*. Here, too, a cable-suspended transducer is used, this time hung from a ship, the purpose being to probe deeply by lowering the transducer below troublesome thermal layers in the sea. The *underwater telephone* uses a beam of sound waves as a carrier of ship-to-ship messages underwater, like radio waves are used in the air. Sonar is the basis of a highly sophisticated navigation system for our nuclear submarines, and a more mundane system for locating schools of fish for our fishing fleets.

We illustrate a typical active sonar system, the Azimuth-Scanning System. The principle, as in all active systems, is to emit a pulse of sound into the water and listen for returning echoes, as in Fig. 15-34*a*. The sound pulse is referred to as a *ping*. A block diagram of the major components of such a system is given in Fig. 15-34*b*. Under control of the operator, who is

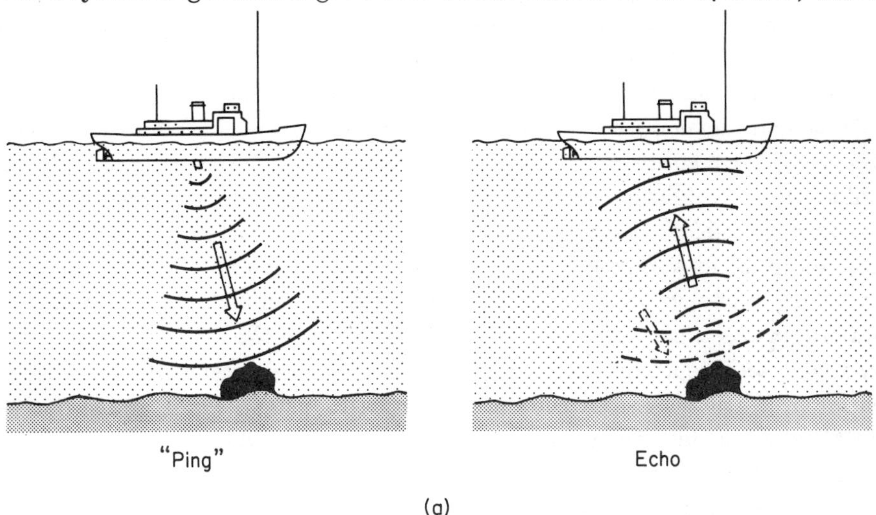

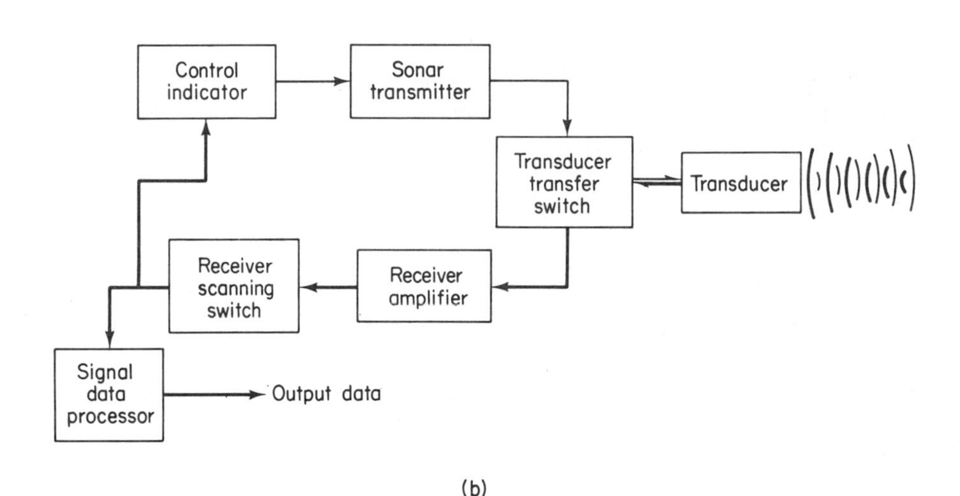

Fig. 15-34. Sonar. (a) Principle. (b) Block diagram of typical azimuth-scanning sonar system.

both visually and audibly monitoring the processed echo data, the transmitter generates an audio pulse of desired pulse width and repetition rate. Typical transmitter frequencies are near the upper audio limit of 20 kHz. The pulse width is typically adjustable from 5 to 100 msec and the repetition rate can be anything from one lone pulse to something like 60 pulses per minute. The repetition rate selected depends on the range desired; enough off time must be allowed for all echoes of interest to get back before the next pulse. A representative average power level for the transmitter is 8 kw with instantaneous peak pulse power levels of about 160 kw.

The transducer, you will note, is functionally related to the loudspeaker (as a sound generator) and to the microphone (as a sound detector). Both functions are combined in one transducer. A piezoelectric crystal device could be used, but far more common is a transducer that operates on the principle of *magnetostriction* (see Section 9-10), the change in volume of a ferromagnetic material in a magnetic field. The essential element of the transducer is a tube of nickel wound with a coil, the tube being connected to an immersed diaphragm. As a sound generator, the coil receives the transmitter's pulse, producing an intense field in the nickel tube, which shortens and deflects the diaphragm. The release of the diaphragm disturbs the water and produces the ping. As a detector of sound, the diaphragm is deflected by the echo wave, compressing the nickel tube and thus varying the flux through the coil in the manner of a variable-reluctance phonograph pickup. The transfer switch alternately connects the transducer to either the transmitter or receiver, as appropriate.

The signal from the received echo is amplified by the receiver amplifier and passed on to the receiver *scanning switch assembly*. Here the audio and visual display signals are separated by ingenious motor-driven switches and other circuitry. The visual signal, suitably conditioned, is fed to a special-purpose, spirally scanned cathode ray tube display device called a *plan-position indicator*, or *PPI*. This device gives a continuous sonar map, much like a radar map. The audio signal is monitored through phones and speakers.

Data from the scanning switch assembly are fed to the signal *data processor*, together with data on the search ship's position and course. From these data, the data processor computes range, speed, bearing, and course of the detected object.

REVIEW QUESTIONS

1. Explain the hearing process.
2. What are the functions of the principal transducers required to make a telephone? Where are they used in the sound recording/reproducing process?

3. Given enough time, could Edison have gotten high-fidelity sound from a purely mechanical phonograph? Why?
4. What electromechanical transducer principles are common to microphones, phonograph pickups, and loudspeakers?
5. Why do we have so many different record-playing speeds?
6. What is the primary requirement in producing stereophonic music? How is this reflected in the record? In the phono pickup cartridge?
7. What is drastically different about tape recording, relative to disk recording? Which has the potential for highest fidelity? Why?
8. What is the nature of the recorded pattern on tape? How is it produced?
9. With regard to tape recording, explain the terms capstan and bias.
10. What is the difference between a tape recorder and a tape deck?
11. Why is a gaseous glow discharge lamp used in optical sound recording?
12. What would you say about a friend's hi-fi which has two 6-inch speakers?
13. What is the major requirement of a true high-fidelity sound system?
14. What function do harmonics perform in the hearing process?
15. Why is a push-pull amplifier stage so named?
16. Explain what is meant by a frequency response curve.
17. How does an active sonar system work? Describe its major components.

16

FAR OVER THE WAVES

16-1 RADIATED ENERGY

Back in Chapter 6 we learned that energy could be lost from a circuit as heat, because of interaction of the electric current with the "resistance" elements of the circuit. By now, we are all at home with resistive heating losses. It happens that there is a second way a circuit may lose electric energy to its surroundings, by emitting it as *waves of electromagnetic radiation*. The discovery of this phenomenon, and of the means to control it, has revolutionized the world, perhaps more so than any other discovery. It was the breakthrough that led to radio, television, and all other known forms of wireless transmission of information.

The way was paved by the work of that experimenter *par excellence*, Michael Faraday, who "invented" electric and magnetic lines of force to explain the things he discovered about electric charges and magnets. To Faraday, these lines of force were very real. How else could one charged body or magnet affect another some distance away, with no visible interaction between them? Faraday wondered about them a good deal. Were lines of force, in some way, like light? Thinking that they might be, Faraday performed a series of experiments and was finally able to change the polarization of a ray of light slightly with a very powerful electromagnet. He thus showed that light had some electromagnetic character. At this stage, however, Faraday had to drop the matter. An important theoretical step had to be taken to point the way to the next group of experiments, and mathematical physics was not the forté of Michael Faraday. Onto the stage stepped Maxwell.

16–2 THE GREAT THEORETICIAN

James Clerk Maxwell was as great a theoretical scientist as Faraday was an empiricist. Descended from a family of Scottish peers, the Clerks of Penicuik, James made his earthly debut in Edinburgh in 1831. His genius was recognized early, in a misguided sort of way, by his classmates, who hung the nickname "Daffy" on him. More appropriate recognition came to young Maxwell when, at 15 years of age, he wrote a technical paper on the construction of Cartesian ovals. His professor submitted this original paper to the Royal Society of Edinburgh, which could scarcely believe that the boy was its author. The next year, "Daffy" entered the University of Edinburgh. Through acquaintance with William Nicol, one of the pioneers in proving the wave nature of light, Maxwell became interested in the nature of light and color and distinguished himself in that area while still at Edinburgh.

Maxwell's father sent him to Cambridge in 1850. His accomplishments there almost stagger the mind. He laid the foundation of the kinetic theory of gases and statistical thermodynamics, contributed to astronomy a theory (held to this day) of the nature of the rings around the planet Saturn, gave a new outlook to the study of heat, and did basic work on color perception. The work that made him world famous, however, was on electromagnetic theory.

Maxwell had long puzzled over lines of force and the question of "action at a distance." Around 1855, he published a monograph entitled *On Faraday's Lines of Force*, giving his initial thoughts. He recognized the fact that electric and magnetic fields are closely related for a moving charge, whether it be a static charge on some object that is being moved mechanically or the moving charges that make up an electric current. In his finest work, carried out between 1864 and 1873, Maxwell developed a simple and elegant electromagnetic theory. All the phenomena of electricity and magnetism he wrapped up in a few simple equations, which are called *Maxwell's equations* in his honor. Using his theory, Maxwell showed that electricity and magnetism cannot exist independently, but are two different aspects of a single phenomenon—electromagnetism. He further showed that an oscillating electric charge produced an oscillating electromagnetic field, that is, electromagnetic waves, which radiated away from the source at a constant speed. This speed could be calculated from measurable physical constants. Maxwell made some unusually precise measurements and calculated the speed to be about 300,000 kilometers per second, a value equal to the then known speed of light. Maxwell seized on this fact as certainly no coincidence and concluded that light was just another form of electromagnetic radiation. Furthermore, reasoned Maxwell, since charges can oscillate at any desired rate, there should be a whole spectrum of electromagnetic radiation, of which light is just a small part.

Maxwell himself never devised a circuit that could produce the waves, and the theory that is his greatest triumph remained unproven until nine years after his untimely death from cancer in 1879.

Maxwell's equations are of a mathematical type called *differential* equations, which treat the *rate at which quantities are changing* rather than the values of such quantities themselves. By applying certain mathematical techniques, such general equations are "integrated" (solved) for the actual values of the changing quantities in a given set of circumstances. It is therefore difficult, in a nonmathematical book like this one, to give an appreciation of Maxwell's equations. Basically they are relationships between the electrostatic and magnetic fluxes through a circuit and the moving charges producing them. Roughly translated into English, the equations in one form state that (1) the rate of change of magnetic field passing through a circuit is reflected in the work done in moving an electric charge around that circuit; and (2) the rate of change of the electrostatic field passing through a circuit is reflected in the work done in moving a unit magnetic pole around that circuit. There are also certain conditions imposed on radiation in empty space.

16-3 HERTZ MAKES WAVES

In 1857, while Maxwell was still speculating about the properties of lines of force, Heinrich Rudolf Hertz (for whom the frequency unit is named) was born in Hamburg. He began an engineering career, but was captivated by the appeal of pure science and went on to become a physicist. In Berlin he studied under, and became the lifelong friend of, the great Helmholtz. At the University of Kiel, in 1883, Hertz became interested in Maxwell's electromagnetic field theory. The Berlin Academy of Science was offering a prize for confirmation of the theory, and his old friend, Helmholtz, suggested that the bright young Hertz try his hand at it.

In the period between 1885 and 1889, by then a professor of physics at Carlsruhe Polytechnic, Hertz studied and experimented. Finally, using

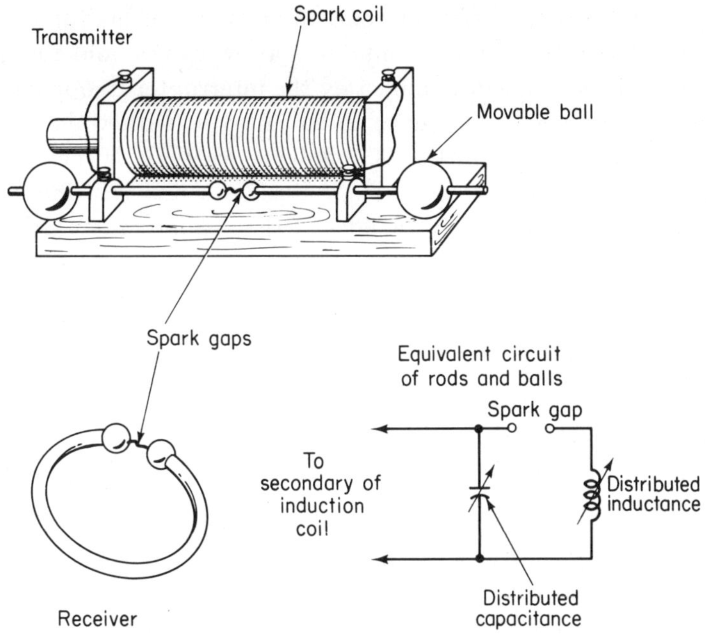

Fig. 16-1. Hertz' transmitter and receiver.

apparatus like that in Fig. 16-1, he was able to demonstrate the correctness of Maxwell's theory.

You must remember that Hertz did not have a triode to produce oscillations, so he used what was available in his day. The induction coil in his transmitter is already familiar to us (see Section 10-13). The secondary winding of this induction coil is connected across an adjustable spark gap formed by two rods, upon which two movable metal balls are mounted. The inset to Fig. 16-1 shows the equivalent circuit for the rods-and-balls assembly. At high frequencies, the distributed capacitance and inductance of this assembly are significant, and it becomes a resonant LC circuit.

In operation, the induction coil provides surges of voltage of comparatively low frequency (the frequency of the interrupter). These voltage excursions build up a charge on the distributed "capacitor," up to the voltage level where a spark will jump the gap. At that point, the "capacitor" discharges back and forth across the gap at high frequency, the resonant frequency of the distributed "inductor" and "capacitor" of the rods-and-balls assembly. It is the rippling electromagnetic field produced by this oscillatory discharge that spreads out into space in waves like the ripples on a pond. Of course, these high-frequency waves die out very quickly, damped by the high resistance of the spark gap, but the next surge of voltage in the ignition coil's secondary starts another train of such waves. So we see that the output of Hertz' transmitter consists of bursts of high-frequency electromagnetic waves, each burst containing a few cycles, and the number of such bursts per second is determined by the interrupter frequency.

Hertz' receiver, or *resonator* as he called it, with which he detected the waves from his transmitter, is simplicity itself. It is merely a loop of heavy wire with a spark gap. The loop itself is a one-turn inductance and the conductor's ends, on opposite sides of the insulating gap, make a small capacitor. Thus it is also an *LC* circuit. Hertz would bend the loop, adjusting its gap distance precisely with a micrometer, and move it around near the transmitter in the dark, watching for a small spark across this receiver gap.

Hertz could adjust the high frequency produced by his spark coil transmitter, within reasonable limits, by moving the balls on the rods and somewhat by adjusting the gap. The size of the balls and the gap determined the capacitance, while the position of the balls determined inductance.

16-4 WAVES IN THE ETHER

We know a good deal today about how electromagnetic waves (hereafter we'll say *radio* waves, for short) behave, but we do not really know what they *are*. Waves of sound energy, for instance, travel through air or some other material. The "swells" in the ocean travel through water. And when we shake one end of a rope that is rigidly attached at the other end, as in Fig. 16-2*a*, the waves of energy imparted by our hand travel through the rope. Radio waves, on the other hand, have no known medium and, for that matter, don't need one. They travel quite happily through empty space. At one time scientists believed that there "must" be a medium for radio waves. They named this medium the *ether* (also spelled "aether") and it had to have amazing properties: it was supposed to be mechanically like a very stiff elastic solid that permeated all space and matter; yet it was invisible, intangible, and ordinary matter moved through it without friction.

The ether all but died as a result of a famous experiment to detect it and whose results were totally negative. That was the famous Michelson-Morley experiment of 1887, which set Einstein on the road to the theory of relativity. More recently, some nuclear physicists have advocated reviving the ether for other reasons. It has also been said that the medium for radio waves is "the electromagnetic field itself," and also "the very fabric of space," whatever those two statements may mean.

For our purposes, it is completely unimportant whether there is or is not an ether or other medium for radio waves. Whenever we come across the expression "going out over the ether," we may treat it as a literary relic in the same sense as "the sun rises."

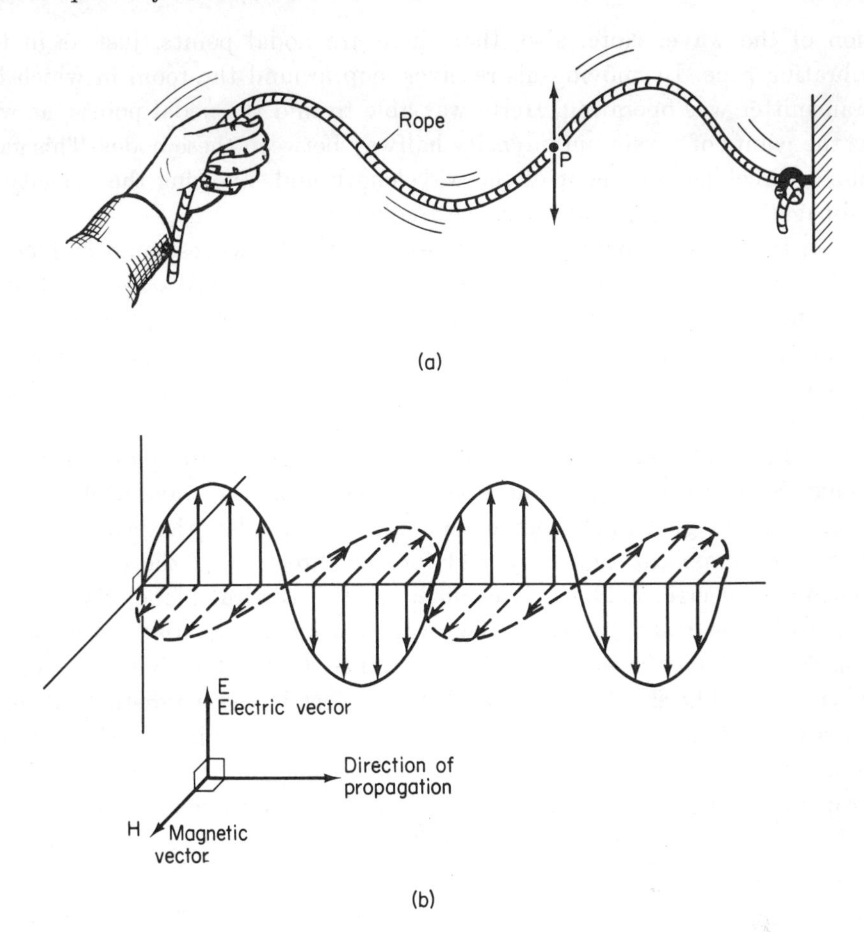

Fig. 16-2. (a) Energy wave in a rope. (b) An electromagnetic wave (vertically polarized).

16-5 PROPERTIES OF RADIO WAVES

In a somewhat oversimplified way, we may think of the radio wave as like the wave in the rope of Fig. 16-2*a*. The oscillations are at right angles to the direction in which the energy is propagated. Thus, in the rope, the individual little segments of rope like P move only up and down, while the wave of energy passes from the hand on the left to the wall on the right. In the radio wave, it is the intensity (amplitude) and direction of the electric and magnetic fields that fluctuate at right angles to the direction of propagation. This condition is shown, somewhat schematically, in Fig. 16-2*b*, which is a perspective view, at some instant in time. Note that the electric and magnetic fields are at right angles to each other, as well as to the direction of the wave. Note, also, that there are nodal points, just as in the vibrating rope. By moving his receiver loop around the room in which his transmitter was operating, Hertz was able to find the nodal points, as well as the points of maximum intensity halfway between these nodes. This gave him a direct measurement of the wavelength and, knowing the velocity of propagation, he could calculate the frequency.

In his experiments, Hertz showed that radio waves could be focused and reflected like light, completely bearing out Maxwell's theory. Unlike light, however, the reflecting surface used for focusing or simple reflection of radio waves has to be *metallic*; otherwise the waves pass right through as though there were no obstruction. This is why we can use indoor radio and TV antennas.

Like light, radio waves can be *polarized*. The word polarized, in this usage, refers to a special condition imposed upon the wave such that its electric and magnetic field components have a particular orientation in space. The type of polarization is named for the direction of the electric field. When the electric field, E, is confined to the vertical plane, the wave is said to be vertically polarized. The wave shown in Fig. 16-2*b* is vertically polarized. When the electric field is confined to the horizontal plane, the wave is called horizontally polarized. If the electric and magnetic fields were interchanged in Fig. 16-2*b*, the wave would be horizontally polarized. Sometimes the electric field's direction is made to spiral around the direction of propagation, like a corkscrew, as in Fig. 16-3. Such a wave is called *circularly polarized* because, looked at along the direction of propagation, the tip of the electric field vector traces out a circle. In any case, the magnetic field,

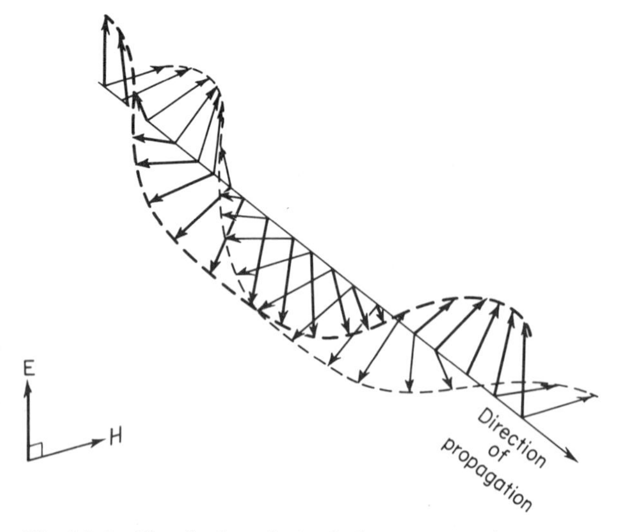

Fig. 16-3. Circularly polarized electromagnetic wave.

H, always remains perpendicular to the electric field, whatever its local orientation. Common practice in the United States is to produce vertically polarized AM radio waves and horizontally polarized FM and TV waves. Circular polarization is limited to space satellites and other specialized applications.

16–6 THE ELECTROMAGNETIC SPECTRUM

Hertz succeeded in demonstrating Maxwell's contention that light was an electromagnetic wave. As Maxwell supposed, charges can be made to oscillate at any frequency, from zero to infinitely[1] fast. Therefore electromagnetic waves can (at least in principle) have any frequency in this range.

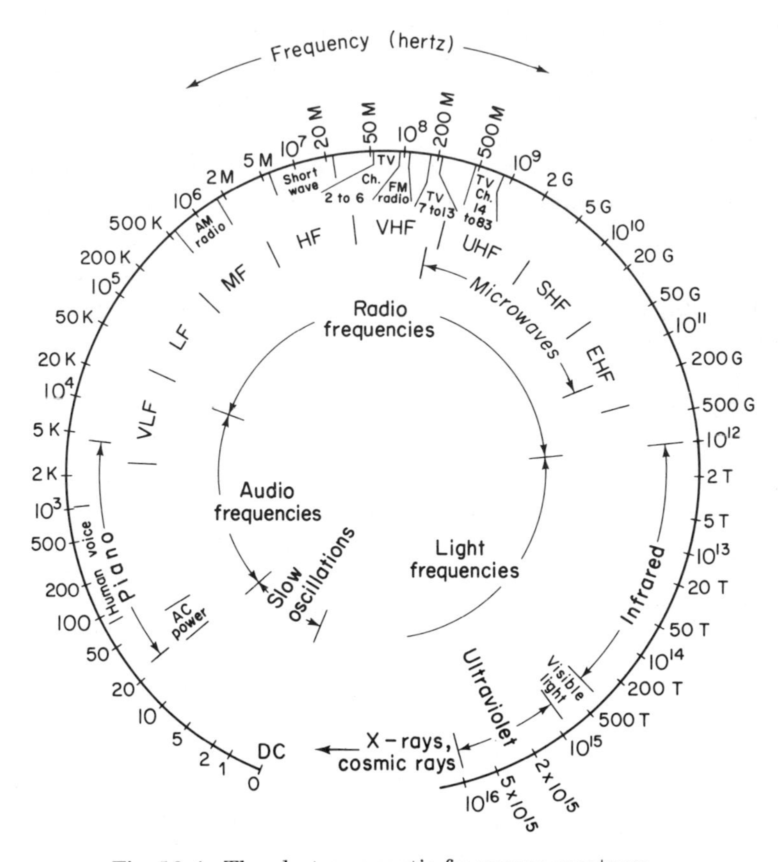

Fig. 16-4. The electromagnetic frequency spectrum.

[1] Certain abstruse relativistic theory indicates that there may be an upper limit of about 2×10^{43} Hz, corresponding to the "frequency of fluctuations in the space-time metric. "For practical purposes, we may assume it is infinite.

Figure 16-4 depicts what we may call the "familiar" part of the electromagnetic spectrum, that part whose frequencies cover the range from 0 to 10^{16} Hz. Although this figure is marked off only in frequency units, the spectrum is also often discussed in terms of wavelength (λ). These two characteristics of a wave are related to each other, as you may recall, in a very simple way:

Wavelength (meters) = 300,000,000 ÷ frequency (hertz)

So a radio wave whose frequency is one megahertz (= 1 million hertz) has a wavelength of

$$\lambda = 300{,}000{,}000 \div 1{,}000{,}000 = 300 \text{ meters}$$

Common practice is to refer to the frequency of a wave of one megahertz or less, to refer to the wavelength of a wave whose frequency is 10^{11} Hz or more ($\lambda = 3$ mm or less), and to talk about either the frequency or wavelength of a wave in between.

Notice that this segment of the spectrum is broadly divided into: *slow oscillations*, from no frequency at all (dc) to 20 Hz; the *audio frequencies* (af), from 20 Hz to 20 kHz, corresponding to the frequencies of sound waves audible by human beings; *radio frequencies* (rf), from 20 kHz ($\lambda = 15$ km) to 1 terahertz, or 10^{12} Hz ($\lambda = 0.3$ mm); and *light frequencies*, from 10^{12} to 10^{16} Hz ($\lambda = 0.00003$ mm). Within these categories, the spectrum can be further broken down in a number of ways. The radio frequencies of most interest to us here are subdivided into the following bands:

Very low frequency, vlf	3 kHz–30 kHz
Low frequency, lf	30 kHz–300 kHz
Medium frequency, mf	300 kHz–3 MHz
High frequency, hf	3 MHz–30 MHz
Very high frequency, vhf	30 MHz–300 MHz
Ultrahigh frequency, uhf	300 MHz–3 GHz
Superhigh frequency, shf	3 GHz–30 GHz
Extremely high frequency, ehf	30 GHz–300 GHz

Note that the the vlf band spills over into the audio frequencies on the low end. Figure 16-4 indicates where AM radio, short wave, TV, and FM radio broadcast frequencies fall within this portion of the spectrum. Most of the uhf, together with the shf and ehf bands, make up what is often called the *microwave* region, because the wavelengths are smaller than 1 meter. We have more to say about microwaves later. For the present, however, we'll limit ourselves to talking about the rf region below the microwave frequencies.

16–7 WIZARD OF WIRELESS

Hertz' transmitter and receiver were nothing more than laboratory experiments. This is not to belittle the greatness of Hertz' achievement. However, he was a scientist trying to substantiate a scientific theory about electromagnetic fields, not an engineer trying to develop a communication system. It therefore remained for Hertz' successors to devise a way to use his waves to transmit intelligence over a distance without wires, that is, to invent *wireless communication.*

The word radio, as used in the United States today, is short for *radiotelephony*, which must be distinguished from radiotelegraphy or wireless. In radiotelegraphy, the earliest wireless communication, the electromagnetic waves were used only to carry a coded signal—the familiar Morse-type "dit-dah-dit"—which was received as a series of beeps or clicks. In radiotelephony, on the other hand, it is a human voice, music, or other sound that is transmitted, and a reasonably faithful replica of that sound is supposed to be produced by the receiving device. Many men contributed to the development of these two fields. The most significant contributions, however, were made by DeForest, Fessenden, Lodge, and, perhaps the best known of all, Marconi.

If anyone ever "had it made" right from the beginning, it was young Guglielmo Marconi. He was born into money, in 1874, his father being a member of the landed gentry of Italy, and his mother's family being distillers of some of the finest Irish whiskey produced in County Wexford. Consequently, the young Bolognese never lacked for anything. He was privately tutored, studying physics under the finest Italian professors without ever actually enrolling at a university. He also had the advantage of being bilingual, speaking English with his Scottish-Irish mother.

In 1894 Marconi came across an article by Crookes suggesting the potential of Hertz' waves for communication, and this stirred his imagination. With financial aid from his father, he equipped a laboratory on the family estate at Pontecchio and proceeded to experiment. His transmitter was a Hertzian induction-coil type. His receiver, however, was an antenna connected to a *coherer*, a device invented by the French physicist Edouard Branly. The coherer was something like a carbon button microphone in reverse. It comprised a glass tube containing metal filings, whose resistance was normally high until they were compacted by the received waves. This change of resistance could be used to operate an earphone. Unfortunately, the filings liked to stick together, once compacted, so Marconi had to invent another device to keep them "unstuck." He called it a *decoherer*, and it was essentially a little buzzer that tapped the coherer tube.

Hertz had used an antenna that today we call a horizontal *dipole*, both segments of which are suspended in space insulated from the ground. Such a dipole must be some exact multiple of one *half-wavelength* in length. One of Marconi's great contributions was turning the antenna to the vertical position and connecting the lower end to ground. Such an antenna is still called a *Marconi antenna* and is usually one *quarter-wavelength* long but may be some multiple. If both types of antennas were broadcasting the same frequency, the Marconi antenna only needs to be half as long as the Hertz antenna. In actuality, Marconi found that his antennas had to be quite long, because the increased capacitance in the antenna circuit due to grounding resulted in much lower broadcast frequencies and therefore larger wavelengths. It was this longer antenna length that both literally and figuratively got Marconi off the ground. It soon gave him a broadcast range of almost 2 miles. At this point, Marconi was ready to show it to the Italian government, but they weren't ready for Marconi or wireless. Undaunted, Marconi went to England in 1896, at his mother's suggestion. There the way had already been prepared somewhat by the efforts of Crookes, Rutherford, and Sir Oliver Lodge. Through a series of lectures and demonstrations, Marconi gradually gained support in England and elsewhere. He formed the Wireless Telegraph and Signal Co., Ltd., which eventually became the Marconi Wireless Telegraph Co., Ltd.

The first honest-to-goodness, paid-for *Marconigram* was sent in 1898 from Lord Kelvin to Stokes. The resourceful Italian was now riding high on a wave of support from the British post office, War, and Navy departments, as well as from such people as Lodge and Fleming. Lodge himself had done considerable work in wireless telegraphy and held several patents on receiver and transmitter tuning. Marconi's real moment of triumph came on December 12, 1901, when he was able to receive the signal for the letter S in Newfoundland from a transmitter in Cornwall, England.

Marconi continued to develop improved wireless apparatus. Large ships carried Marconi wireless transmitters. The last message from the ill-fated *Titanic* came over a Marconi transmitter. The security requirements of strategic communications during World War I led him to experiment with and recognize the value of using waves of much smaller wavelength (i.e., *short wave*). In 1919 Marconi invested some of his considerable wealth in a yacht, which he had fitted out as a floating communications laboratory and which he appropriately named the *Elettra*. In the eighteen years preceding his death, he spent much of his time at sea experimenting, particularly with short wave communication.

Marconi did not suffer from lack of recognition for his work. Among other honors, he was made a Knight of Italy, shared a Nobel Prize, was named to the Italian senate, received the title of marchese, and was elected to the presidency of the Royal Italian Academy.

16-8 FIRST DISC JOCKEY

In direct contrast to the famous Marconi, we come now to a man who is relatively obscure and yet who was both the real father of radio as we know it and second only to Edison in the number and kind of inventions patented. That man was Reginald Aubrey Fessenden, a Canadian-born immigrant physicist who had worked for both Edison and Westinghouse.

Fessenden was not satisfied with broadcasting the monotonous "dah-dit-dah-dit-dit" of telegraphy. He wanted to transmit actual sounds. His greatest discovery was how to *modulate* and *demodulate* an rf wave—that is, how to impress an audio signal on top of an rf wave, broadcast the combined wave, and then get back the original audio signal (or at least a reasonable facsimile) at the receiver.

Thus it was, on Christmas Eve in 1906, that Fessenden became the world's first disc jockey by sending Christmas music and a Scriptural reading over the wireless from his experimental transmitter on the Massachusetts Coast. Shipboard radio operators, from Cape Cod to Cape Hatteras, receiving Fessenden's Christmas Special instead of the customary dit-dah-dit, wondered if they'd been at sea too long.

Fessenden's radiotelephone did not become practical until a reliable source of *continuous* rf waves (not bursts, like Herts' transmitter put out) was available. The first successful source was an ingenious 100-kHz alternator by Dr. Ernest Alexanderson of General Electric. The coming of DeForest's audion, however, saw the Alexanderson alternator supplanted by the simpler, cheaper, and much more flexible vacuum-tube oscillator.

16-9 MODULATION AND RADIO COMMUNICATION

The sequence of events in radio broadcasting, then, is: (1) generating an appropriate radio frequency called the carrier; (2) modulating it with the desired audio intelligence; (3) radiating this combined signal as an electromagnetic wave; (4) reproducing the combined signal in the receiver when the wave reaches it; and (5) demodulating it to get back the audio intelligence, discarding the carrier in the process. There are several ways of doing all these steps.

You may wonder, why bother with the radio frequency carrier at all? After all, Maxwell's theory tells us that oscillations of any frequency will radiate electromagnetic waves, doesn't it? Why not just transmit and receive audio frequencies? In principle, this could be done, but it is not practical for a number of reasons. For one, all transmitters would operate in the same frequency range and would mutually interfere with each other. So, unless each geographic area were limited to one radio station, the receiver would be flooded with a meaningless jumble of noises. Secondly, at the low frequencies and long wavelengths involved, some of the actual hardware required becomes impractically large. This factor is particularly true of the inductors and capacitors that would be used in the tuned (i.e., resonant) circuits, and even more so of the antenna. If, for instance, we take 1000 Hz as the middle of the band of frequencies to be broadcast, a quarter-wavelength Marconi antenna would have to be 300 km (about 186 miles) high! It would also be out of tune at the lowest and highest frequency in the band, since the latter would be about a thousand times as great as the former. Finally, the poor efficiency of an audio-frequency antenna would severely limit the range of the transmitter.

All of these transmission problems are solved by using an rf wave to carry the message, hence the name *carrier*. The information to be transmitted can be combined with this carrier in several ways. The two techniques of major interest to us are *amplitude modulation* (AM) and *frequency modulation* (FM), which are illustrated conceptually by Fig. 16-5. The rf carrier

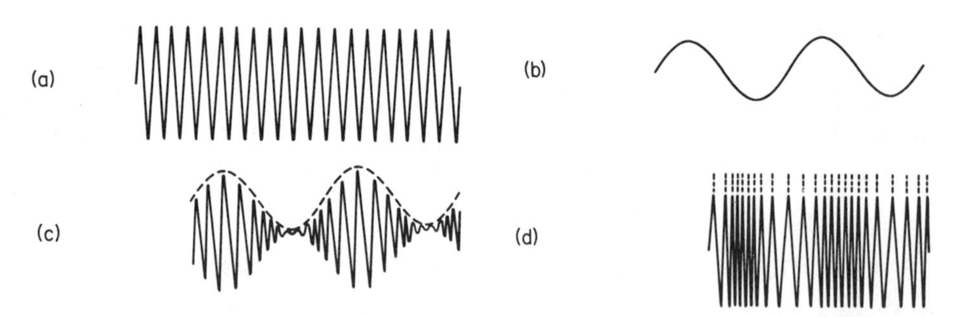

Fig. 16-5. Modulation. (a) rf carrier. (b) af signal to be transmitted. (c) Amplitude-modulated carrier. (d) Frequency-modulated carrier.

is shown in Fig. 16-5*a*. It is a pure sine wave. The actual audio signal to be transmitted, the modulating signal, is represented by the sinusoid in Fig. 16-5*b*. This sinusoid is used for simplicity in understanding the other two parts of this figure. An actual bit of speech or music would have a more complex waveform. No attempt is made here to show the audio and carrier frequencies to scale; in practice, the carrier is at least 100 times the modulating frequency.

In amplitude modulation, the modulating signal is used to produce changes in the amplitude of the carrier while its frequency stays fixed. The result, shown in Fig. 16-5*c* is that the *envelope* of the carrier (an imaginary line connecting all the peaks of successive cycles of the carrier) looks like the modulating waveform. Thus the AM modulating device produces a "mold," so to speak, which shapes the rf carrier.

In frequency modulation, on the other hand, the modulating signal is used to produce changes in the frequency of the rf carrier without affecting its amplitude. The effect here, as indicated by Fig. 16-5*d*, is to "bunch up" and "stretch out" the carrier, in proportion to the amplitude of the modulating signal, the carrier's frequency being increased by positive peaks and decreased by negative peaks. The FM modulating device is therefore figuratively like an "accordion" in which the carrier's frequency is compressed and expanded.

We discuss AM and FM in more detail in the sections to come.

16-10 HOW THE WAVE GETS THERE

Upon being radiated by the transmitter, the modulated wave may travel to the receiver by a number of paths. Part of the radiated energy travels straight out from the transmitter and makes up what is called the *ground wave*. The remainder travels upward and becomes the *sky wave*.

If you could see the electromagnetic field around a vertical transmitting antenna, it would look something like a giant doughnut lying on the ground, with the antenna at the center of the hole in the doughnut. The energy propagated through the lower half of the doughnut is the ground wave, while that sent through the upper half is the sky wave.

The ground wave itself must be considered in two parts: the surface wave and the space wave. The surface wave "hugs" the curved surface of the earth, while the space wave travels the straight line path called the line-of-sight. The surface wave soon loses strength, however, giving up energy by inducing eddy currents in the earth (which is a fairly good conductor). A range on the order of 100 miles for the ground wave, and on the order of 40 miles for the space wave, is representative for the average transmitter.

Yet we know that radio signals are sent over very long distances, even around the world, by transmitters of modest power. This fact was first noticed when Marconi had the letter S sent from Wales to Newfoundland. The answer lies in the upper region of the earth's atmosphere and how it affects the sky wave. One of the major functions of our atmosphere, besides providing the oxygen we need to breathe, is to act as a protective blanket in shielding us from harmful radiations coming from the sun and elsewhere in space. In absorbing the energies of these radiations, the atoms of oxygen and nitrogen in the upper atmosphere become ionized, and this region has therefore been named the *ionosphere*. The ionosphere acts like a partial mirror to rf, reflecting much of the sky wave energy back to earth. In 1902 two brilliant theoreticians, the Anglo-American electrical engineer, Arthur E. Kennelly, and the self-taught British genius, Oliver Heaviside, quite independently predicted the existence of an electrically charged layer in the upper atmosphere which could reflect radio waves. Consequently, in their honor, it has been called the *Kennelly-Heaviside layer* although it really involves several ionospheric layers. The degree to which a radio wave penetrates the Kennelly-Heaviside layer before it is turned back depends on three factors: the frequency of the signal, the angle at which the beam enters, and the immediate condition of the ionosphere (which, in turn, depends on the time of the day and year, sunspots, etc). The frequency is important in that very high frequency sky waves, of 30 MHz and above, go right through the ionosphere and are not turned back, while very low frequencies, which have most of their energy in the ground wave anyway, are severely attenuated (absorbed). Between the extremes of frequency, most signals are reflected back, with some absorption, providing the angle and ion density are right. For each frequency, there is some angle called the *critical angle*, illustrated in Fig. 16-6. Any portion of the sky wave that strikes the Kennelly-Heaviside layer at an angle greater than the critical angle will pass on through.

The ionosphere's properties vary drastically in the course of a day and to a lesser extent with the season, sunspot cycles, and other phenomena. In the daytime, sunshine promotes a high degree of ionization, resulting in a lower altitude limit for the Kennelly-Heaviside layer of perhaps 20 miles. At night, the lower portions disappear while the upper portions coalesce, in effect raising the lower boundary of our ionic umbrella to about 200 miles.

A receiver located between the range limit of the ground wave and the point where the sky wave returns to earth cannot receive the transmitter's signal at all. Thus a certain part of the intervening distance is skipped over, as shown in Fig. 16-6, and is called the *skip distance*. A wave can make two or more trips from ground to ionosphere and back again before it is received by a particular receiver. The signal received by receivers A, B, or C in Fig. 16-6 is termed a *one-hop* transmission, while that arriving at receiver D is a *two-hop* transmission.

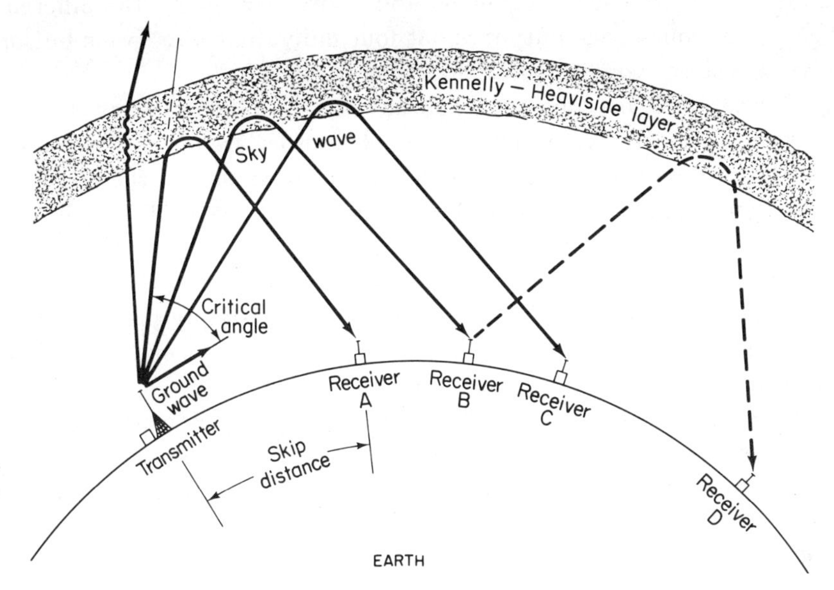

Fig. 16-6. Propagation of radio waves.

Since the "ceiling" moves up at night, the nocturnal skip distance is greater, giving a larger nighttime transmission range for the standard broadcast and short-wave frequency bands. Commercial television, FM, and other high-frequency broadcasts must rely on the line-of-sight ground wave, so their range is fairly constant day or night.

16–11 HETERODYNING

A very useful principle, applied in several ways in electronic communication, is the *heterodyne* principle. The word heterodyne comes from two Greek roots meaning "other power," a not too informative name. It means to make a wave of a given frequency interact with another wave of a different frequency (you guessed it! that's the "other power"). The wave mixture that emerges from this interaction contains not only the two original frequencies but also two new frequencies,[2] which are called the *beat frequencies* when talking about sound waves and *sidebands* when we speak of electrical frequencies. These beat frequencies, or sidebands, are the *sum* and *difference* of the two original frequencies. So, for example, if we heterodyne a 500-kHz signal with a 2-kHz signal, the resulting wave also contains

[2] Depending on how the heterodyning is accomplished, several other unwanted frequencies are also produced, which must be suppressed.

a 502-kHz signal (the sum frequency) and a 498-kHz signal (the difference frequency). Of course, the output is not four individual waveforms but one complex waveform containing all four components.

Furthermore, if a constant amplitude signal of one frequency is heterodyned with a varying amplitude signal of another frequency, the amplitude of the sum and difference sidebands is in proportion to the amplitude of the varying amplitude signal. This principle, as we shall see, is what makes communication by radio waves work.

Electrically, heterodyning is accomplished by feeding the two signals into a *nonlinear device*, a device whose output is not simply proportional to its input. This device is usually a vacuum tube or semiconductor circuit designed to operate in a nonlinear manner.

In communications work, the word *heterodyning* is usually reserved for the interaction of two frequencies close together, while the control of a high (rf) frequency by a low (af) frequency is actually the "modulation" process.

16-12 AM COMMUNICATION

We first met amplitude modulation, or AM, in Section 16-9, where we learned that it involves changing the amplitude of a constant frequency rf carrier in proportion to the amplitude of the desired audio signal. Now, let's take a look at the actual AM transmitter and receiver.

The circuit shown in Fig. 16-7*a* is a greatly oversimplified AM transmitter. You recognize the part of the circuit between the antenna on the left and the transformer, T, on the right as an oscillator circuit, similar to that discussed in Chapter 14. This oscillator provides the pure sinusoidal rf carrier, its frequency being determined by the tuned circuit consisting of the inductance L_1 and the capacitance C_1. This oscillator-amplifier circuit is a nonlinear device. Sound striking the microphone produces a pulsating dc current in the primary of transformer T. This action induces an ac audio frequency in the secondary, which heterodynes with (i.e., modulates) the oscillator output to produce an amplitude-modulated signal. The capacitors C_2 and C_3 bypass the radio frequencies around the power supply and transformer. Through inductance L_3, the antenna is coupled to the circuit, causing electrons to rush up and down the antenna as the carrier swings from positive to negative and back, the intensity of the surge being governed by the modulating signal. These oscillating charges in the antenna radiate their energy as radio waves.

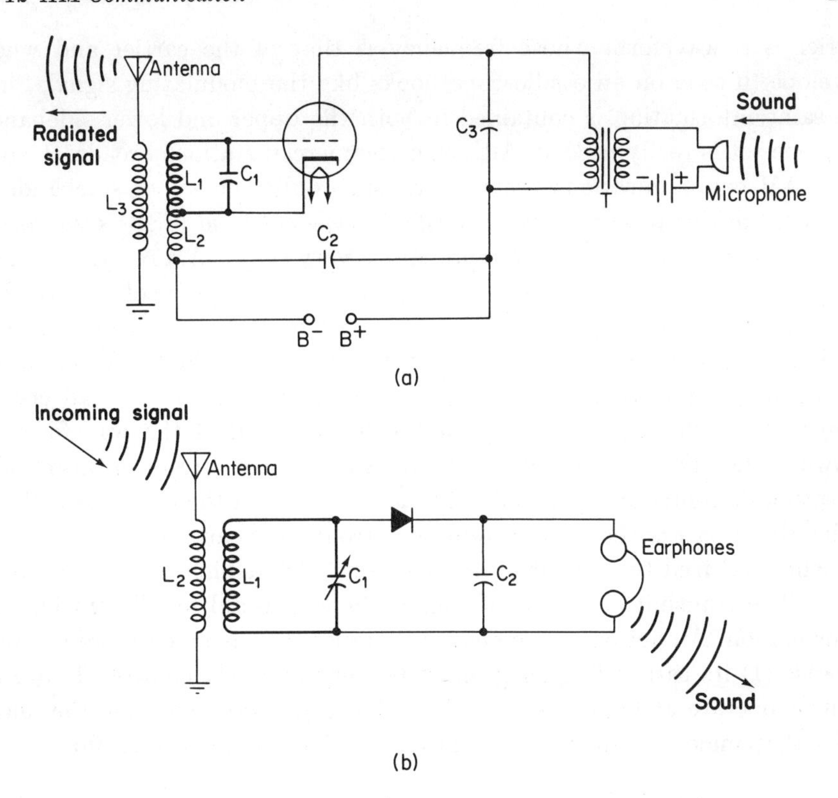

Fig. 16-7. (a) Simple AM transmitter. (b) Receiver.

If the sound striking the microphone is a pure sinusoidal "note," which has only one frequency, the result of the modulation is the carrier plus two sidebands. These are, of course, the sum and difference frequencies. So if a pure 1-kHz note is played into a 700-kHz oscillator circuit, the output signal contains the frequencies 699 kHz, 700 kHz, and 701 kHz. The tuned RF circuit comprising L_1 and C_1 offers almost no impedance to the 1-kHz audio signal, so it develops no voltage in the output and is lost.

Actual broadcast material is, of course, no simple sine wave but a complicated audio waveform. Recall (Chapter 15), however, that Fourier's theorem tells us that any complicated waveform is merely the sum of a large number of sine waves of different amplitudes and frequencies. Each of these component audio sine waves *independently* heterodynes with the rf carrier; each one produces its own pair of sideband frequencies whose amplitudes are proportional to its own amplitude. It is important to note that all the transmitted intelligence is contained in the sidebands; there is no information in the carrier. The combination of all these sidebands with the carrier is a waveform whose frequency is that of the carrier and whose envelope (if seen on an oscilloscope) looks like the modulating signal. Since the same information is contained in both the upper and lower sidebands, only one set is really needed. Although commercial stations broadcast both, some AM communications systems transmit only one set of sidebands to save transmitter power. These are called *single sideband* (SSB) systems.

The circuit in Fig. 16-7*b* represents a very simple AM receiver, similar to the old crystal sets and even some novelty radios being sold today. The tuned circuit containing L_1 and C_1 rejects all frequencies except the relatively narrow band containing the desired carrier and its sidebands. The diode (which was a galena crystal with a cat-whisker contact in the old crystal sets) rectifies the signal, cutting off the bottom half of the waveform, as shown in the first two parts of Fig. 16-12. The average current in this rapidly pulsating dc reproduces the audio frequency sound in the earphones. Meanwhile, the rf part of the signal is bypassed around the earphones by capacitor C_2, which is sized to offer little impedance to rf but high impedance to af.

The transmitter circuit of Fig. 16-7*a* was useful for illustrating AM transmission. An actual commercial AM transmitter is much more complex because (1) its carrier frequency must be very precisely controlled, and (2) it must operate at high power levels. A block diagram, showing the major units that make up an AM transmitter, is given in Fig. 16-8. Most trans-

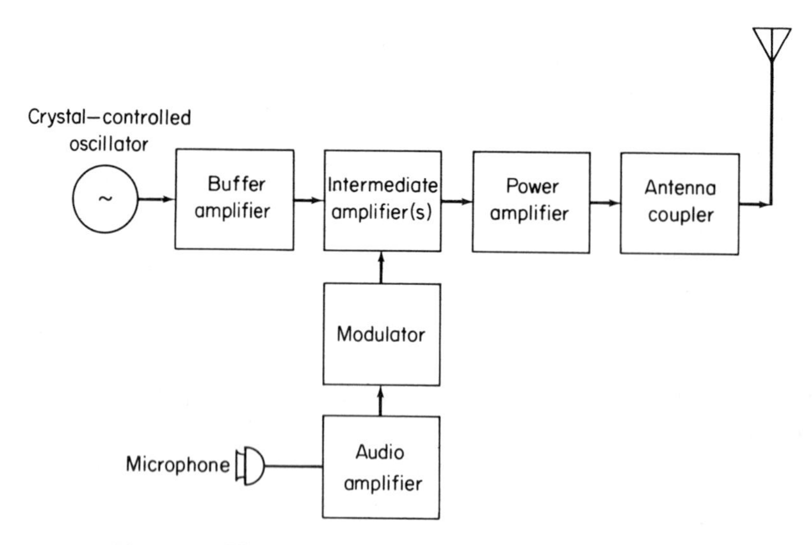

Fig. 16-8. Block diagram of an actual AM transmitter.

mitters operate on a single frequency. This frequency is assigned by the *Federal Communications Commission* (FCC). In most cases, a single-frequency transmitter will use a crystal-controlled oscillator to produce the carrier. A quartz crystal, or other piezoelectric crystal, which has been cut to such a thickness that its natural frequency is the desired frequency, is placed in the grid circuit of such an oscillator. A crystal-controlled oscillator can hold the developed frequency fixed to one part in a million, or better. The output of the oscillator feeds an amplifier called the *buffer* amplifier, whose function is to isolate the oscillator from loading by the later amplifier stages, as well as to amplify. The oscillator frequency is then fed into intermediate amplifiers, which may also be used to multiply the frequency if the required rf carrier is of higher frequency than can practically be obtained from a crystal (about 15 MHz) directly. Meanwhile, the intelligence that is to be transmitted is fed from the microphone (or tape head, or phono pickup) through an audio-frequency amplifier to the modulator, which prepares the modulating signal. The latter is then combined with the carrier in one of the intermediate amplifiers, called the *modulated amplifier*. Its output, in turn, goes to the final power amplifier, or *driver*, as it is often called. Between the driver and the antenna, an impedance-matching circuit, called the *antenna coupler*, must be inserted to ensure maximum power transfer to the antenna. A typical radio broadcasting station, in a remote area, is shown in Fig. 16-9.

Fig. 16-9. Typical remote broacasting station. (Courtesy of International Telephone and Telegraph Co.)

The allowed power output of radio transmitters in the United States is also determined by the FCC. Commercial stations vary from about 250 watts to over 50 kilowatts, depending on the area they serve. Amateur stations are severely restricted in power level, depending on their assigned frequency and time of day, and generally are held to under one kilowatt.

16-13 BETTER AM RECEIVERS: TRF AND SUPERHET

AM receivers evolved rapidly after DeForest introduced his audion tube. For some time, the most popular AM receiver was the *tuned radio frequency* (TRF) receiver, shown in block diagram form in Fig. 16-10*a*. It

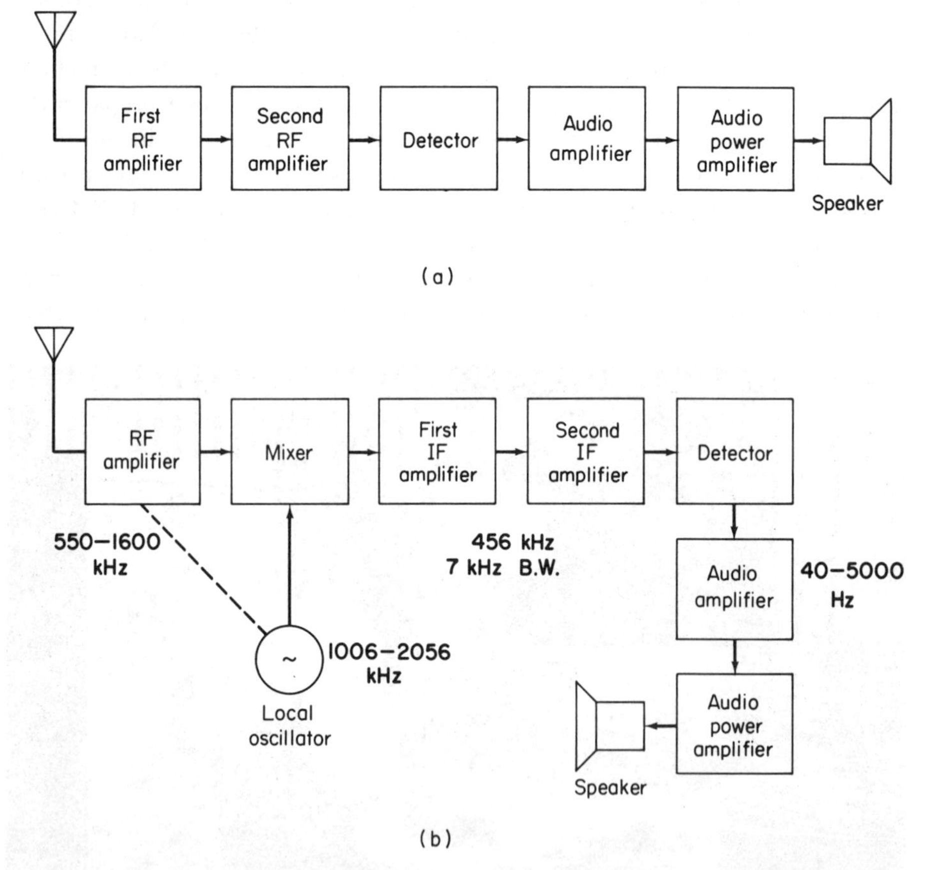

Fig. 16-10. AM receiver. (a) Tuned radio frequency (TRF). (b) Superheterodyne.

consists of an antenna, two or more tuned radio-frequency amplifier stages (whence the name), a detector, two stages of audio amplification, and a loudspeaker. The key to the operation of the TRF receiver is *ganged tuning* of the resonant circuits in the rf stages. The TRF stages are shown, in a somewhat simplified fashion, in Fig. 16-11*a*. Note that the stages are induc-

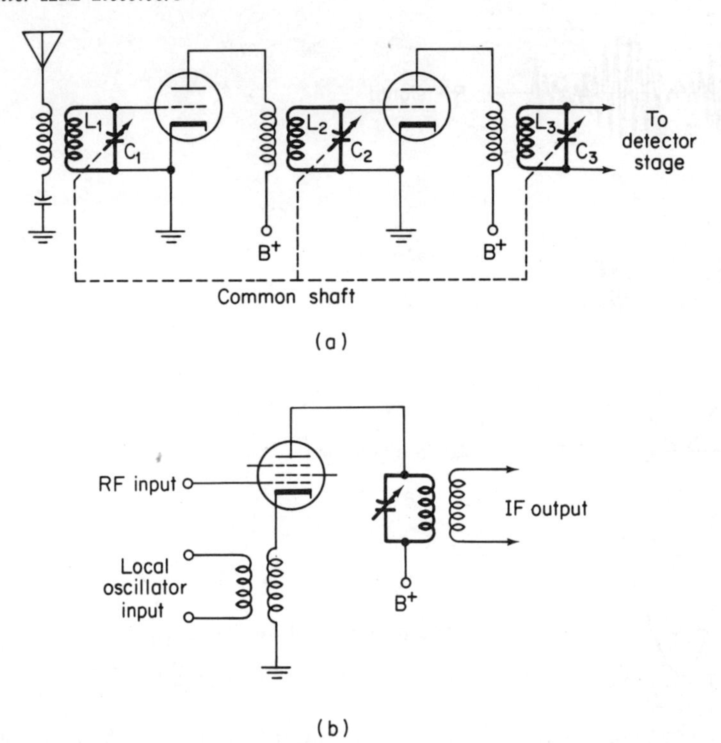

Fig. 16-11. (a) Tuned rf amplifier stages. (b) Typical mixer circuit.

tively coupled and that the input to the rf stages and the following detector stage is a tuned circuit. The variable capacitors, C_1, C_2, and C_3, are all physically mounted in a single unit, with their movable plates on a common shaft. Rotation of this shaft, via the tuning knob, adjusts all three capacitors simultaneously, making all three *LC* circuits resonant to the same frequency. The TRF arrangement provides high sensitivity, for the incoming rf signal is amplified two or more times before detection. It also provides good selectivity, for there are three tuned circuits "in a row" to reject unwanted frequencies. The nature of the signal waveform, as it moves through the TRF receiver, is illustrated in Fig. 16-12. The symmetrical output waveform from the rf amplifiers has its bottom shaved off in the detector, just as in the simple receiver discussed previously. At the detector output, the rf component is bypassed to ground, while the af signal is passed on to the audio amplifier. Its output, the variable dc voltage shown, is then fed to a final audio-power amplifier, a push-pull stage in better quality receivers, which provides the power needed to drive the speaker.

Despite the once widespread popularity of the TRF receiver, today it has been all but supplanted by the *superheterodyne* receiver. Perhaps you've seen this impressive-sounding word in advertisments for radios, or in the

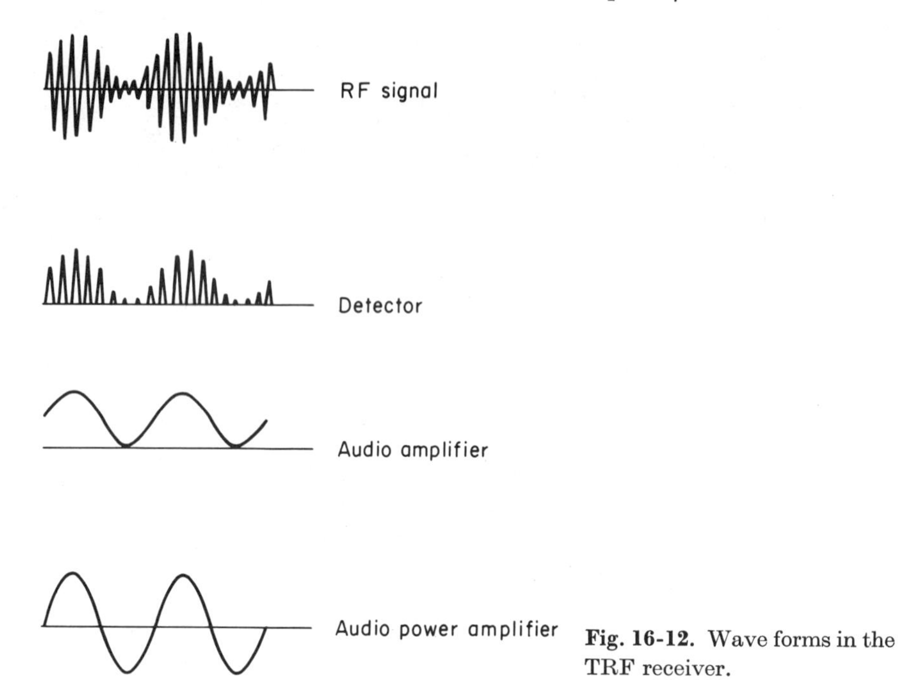

Fig. 16-12. Wave forms in the TRF receiver.

literature that came with a radio you've bought. This invention, which made radios "easy to use" and resulted in there being at least one radio in almost every household, was the brainchild of an American electrical engineer, Edwin H. Armstrong.

Armstrong, a graduate of Columbia University, had been a radio "ham" since his teens. World War I found him, now an Army major, attached to the Signal Corps in France. He became interested in the problem of locating enemy bombers for anti-aircraft batteries. Locating the bombers was then being accomplished by listening for the sounds of the aircraft with a highly directional listening device, but Armstrong believed better accuracy could be achieved by detecting, instead, the electromagnetic waves produced by the planes' ignition systems. Since these waves are in the rf range, they have to be lowered to be audible, and Armstrong devised a receiver that used the heterodyne principle to reduce the rf to af. His device came too late to be of any strategic value in World War I, but it was the prototype for the final form of the AM receiver.

A superheterodyne receiver is outlined, in block diagram form, in Fig. 16-10*b*. The fundamental characteristic of the superhet receiver is that it transfers the intelligence from the incoming rf signal, whatever its carrier frequency, to a new and fixed-frequency carrier generated in the receiver. This fixed-frequency carrier, usually 456 kHz, is called the *intermediate fre-*

quency, or *i-f*. The modulated i-f is obtained by heterodyning the modulated rf signal coming out of the rf amplifier with a fixed amplitude signal produced by a separate oscillator circuit in the receiver. The frequency of this *local oscillator* is set exactly 456 kHz higher (or lower) than the rf carrier, so that the difference frequency is the i-f. The heterodyning is done in a circuit called a *mixer*,[3] one form of which is shown in Fig. 16-11*b*. The sidebands are effectively moved from around the carrier frequency to around the i-f, resulting in a modulation envelope on the i-f that looks just like the one on the rf signal received from the transmitter. The i-f signal then goes through several stages (depending on the quality of the radio) of amplification, followed by detection and audio processing just like that in the TRF receiver.

The waveforms at various stations in the superhet receiver are shown in Fig. 16-13. The local oscillator is automatically set to the right frequency for heterodyning with the input rf signal, because its tuning capacitor is ganged with that of the rf amplifier so both are adjusted simultaneously.

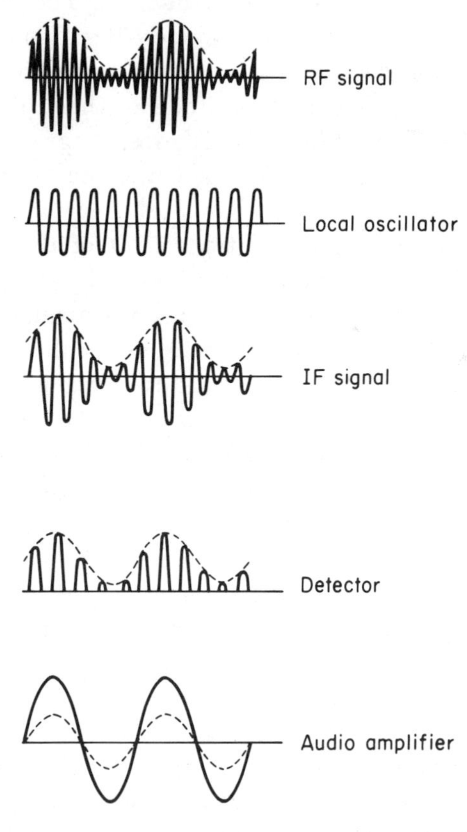

Fig. 16-13. Wave forms in the superheterodyne receiver.

[3] Sometimes the local oscillator and mixer functions are combined in a single tube called a *converter*.

All frequencies produced by the heterodyning, other than the difference frequency (and its sidebands), are rejected by the mixer's tuned output circuit.

The superhet receiver is simpler and therefore more economical than a TRF receiver, for only the rf input and local oscillator circuits must be continuously tunable. Throughout the i-f circuits, only one predetermined frequency is used; thus, once adjusted, no further tuning of these circuits is required. Furthermore, the i-f stages can be designed for optimum performance at that frequency. All in all, the superhet gives better selectivity, more uniform gain, and more performance per dollar than the TRF receiver.

16-14 STATIC AND LO-FI IN AM

AM communication suffers from two major drawbacks: susceptibility to interference (*static*) and poor audio fidelity. The sensitivity to static is a consequence of amplitude modulation. The amplitude of a signal is the easiest to change of all its characteristics. Things like lightning and automobile ignition systems radiate component frequencies that may happen to match certain sidebands and add to them, if already present, or appear in place of missing sidebands. The resulting sound from the speaker is distorted or noisy. Today this is not a serious problem except in some rural areas.

The low fidelity is not the fault of the receiver (although most of them do not use high-fidelity audio hardware) but is due to bandwidth. The FCC has squeezed as many stations as possible into the allowed broadcast band, from 550 to 1600 kHz, each being allowed a bandwidth of only 10 kHz, or 5 kHz on either side of the carrier. For instance, a station assigned a broadcast frequency of 1000 kHz is allowed to use all the frequencies between 995 and 1005 kHz. Since these two frequencies are the most extreme sidebands allowed, this means the highest audio frequency that can be transmitted on AM is 5 kHz. Since we have already seen (Chapter 15) that true high-fidelity reproduction requires frequencies up to at least 15 kHz, even the finest AM radio cannot deliver high-fidelity sound.

Despite the problems of static and fidelity, AM is the most popular modulation technique. Besides the commercial broadcasts—whose "pop" disc jockeys are the reason that most teen-agers seem to have a transistorized superhet growing from one side of their heads—AM is also the darling of the amateurs. In the last decade, the citizen's band transceiver (a *trans*mitter and re*ceiver* mounted in one unit) has been appearing in autos, trailers, pleasure boats, etc. in growing numbers. A typical mobile unit is shown in Fig. 16-14.

Fig. 16-14. Mobile 50 MHz (6-meter) transceiver. (Courtesy of Allied Radio Corp.)

16–15 FM BROADCASTING

The problem of static in radio transmission was attacked, and finally licked in 1939, by Armstrong, the same Edwin Armstrong who invented the superheterodyne receiver. After six years of effort, he succeeded in devising a way to transmit an audio signal by modulating the *frequency* rather than the amplitude of the carrier. This technique is frequency modulation, or FM. The receiver is, of course, detecting the variation, in the frequency of the carrier and doesn't give a hoot (or any other sound) for amplitude changes in the carrier. Since static is essentially a random amplitude disturbance, FM neatly eliminates static.

Figure 16-15*a* illustrates conceptually the FM transmitter. In it, you will recognize once again a basic oscillator circuit which is producing the carrier frequency. This time, however, the tuned-circuit capacitance which, together with the inductance L_1, determines the output frequency, actually consists of two capacitances connected in parallel. One is the large capacitor,

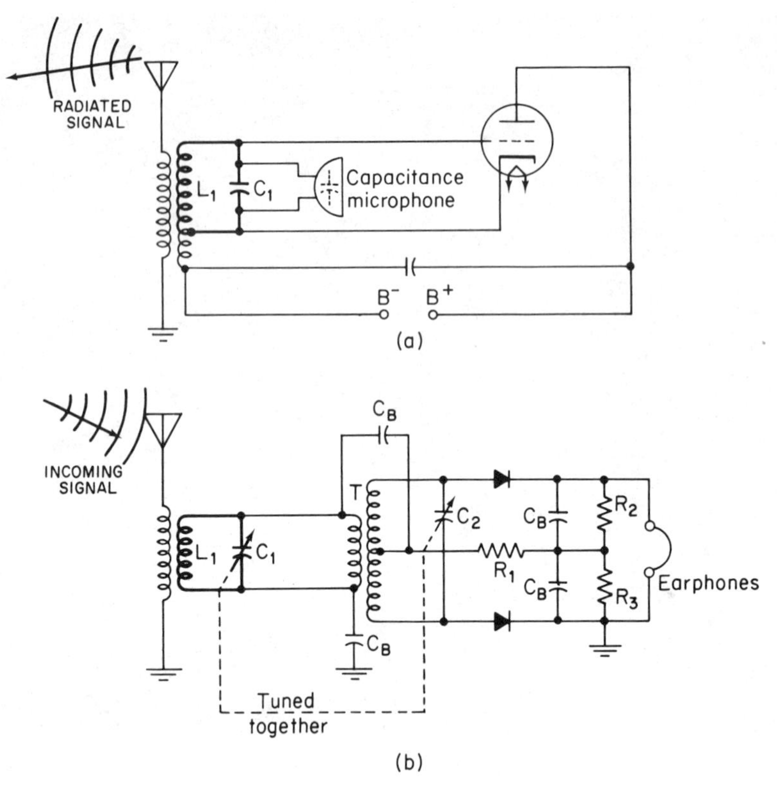

Fig. 16-15. Simplified FM equipment. (a) Transmitter. (b) Receiver.

C_1; the other is the small varying capacitance of the microphone. When no sound is reaching the microphone, it has a fixed capacitance and the circuit puts out an unchanging carrier like that pictured in Fig.16-5*a*, called the *resting frequency*, or *center frequency*. Sound waves striking the microphone element produce small changes in its capactiance proportional to the amplitudes of those sound waves. These small changes in the tuned circuit's total capacitance result in proportionate small changes in the frequency of the carrier. Thus a pure sinusoidal note, like that in Fig. 16-5*b*, would result in an output that looks something like Fig. 16-5*d*. In FM transmission, the difference between the resting frequency and the maximum (or minimum) frequency produced by the modulation is called the *frequency deviation* of the signal.

In actuality, a transmitter like that in Fig. 16-15*a* is not practical for a number of reasons, the major one being that it cannot produce sufficiently large frequency deviations. A typical commercial FM transmitter uses a circuit called a *reactance-tube modulator* to vary the frequency of the oscillator circuit. This is simply a vacuum-tube (or semiconductor) circuit whose effective capacitive reactance varies with the amplitude of the applied audio signal, and which is connected across the tuned circuit of the oscillator just like the simple capacitance microphone.

The principle of the FM receiver is illustrated by the circuit in Fig. 16-15*b*. The part of the circuit between the tuning capacitor, C_1, and the earphones is called a *discriminator* circuit. This is an FM "detector," equivalent in function to the single diode in the simple AM receiver. Its job is to take in a signal of varying frequency and put out a voltage whose amplitude is proportional to the deviation from the resting frequency. The key to this circuit is the transformer, T. The capacitor, C_2, across the secondary of the transformer, is ganged with the tuning capacitor, C_1, so that the secondary of the transformer is always resonant at the center frequency of the band tuned in by the $L_1 C_1$ circuit. The capacitors marked C_B are all high-frequency bypass capacitors; that is, they act like simple conductors at the FM signal frequencies. The primary voltage of the transformer T is therefore also fed into its secondary circuit at the center tap. Meanwhile, the current in the primary has induced a voltage in the secondary. The secondary voltage, in turn, causes a current to flow in the secondary circuit. This secondary current either leads, is in phase with, or lags the secondary voltage, depending on whether the instantaneous signal frequency is lower than, the same as, or higher than the center frequency, respectively. The reactive voltage drop across the capacitor, C_2, which is produced by this secondary current, lags this current by 90°. Half of the drop appears on each side of the center tap. The voltage across either diode is the combination of the primary voltage, introduced at the center tap, and half the reactive drop across C_2. The rectified diode voltages appear across the load resistors, R_2 and R_3. When the signal frequency is the resting frequency, the transformer's secondary current is in phase with its induced voltage, causing the voltages out of the diodes to be equal, but of opposite polarity. Consequently, they cancel each other, and the circuit's net output voltage is zero. When the instantaneous signal frequency is higher or lower than the transformer secondary's resonant frequency, more voltage is applied first to one diode and then to the other. This step causes the output voltage to increase first in one direction, then in the other, depending on the amount and direction of deviation from the resting frequency. In this way, our discriminator circuit, called a Foster-Seely discriminator after its inventors, produces a varying amplitude audio voltage from a constant amplitude, varying frequency rf voltage.

In actuality, the circuit of Fig. 16-15*b* could not be used. The discriminator requires a truly constant amplitude input or its output will be distorted. Furthermore, it is not intended to be operated at the high FM carrier frequencies, 88 to 108 MHz. A block diagram of a real FM receiver is given in Fig. 16-16. Comparison with Fig. 16-10*b* shows that it is very similar to the superhet AM receiver except that it has a *limiter* and a discriminator stage in place of the AM detection stage. The limiter's function, as its name implies, is to limit the amplitude of its output voltage. It literally *clips* the tops off signals whose amplitudes are too great, thereby giving the discriminator the constant amplitude signal it requires. Note, also, that the FM signal has first been heterodyned down to a fixed intermediate frequency (10.7 MHz in the receiver shown), so that the discriminator is operating on this fixed lower frequency and does not have to be tuned to the carrier frequency of the desired station.

There are other types of FM detectors. One that is becoming increasingly popular is the *ratio detector*. This circuit, unlike the discriminator, is not sensitive to the amplitude of the incoming FM signal and does not require a limiter.

Before leaving FM transmission, a few words on bandwidth are in order. We have seen that, in AM, each pure audio frequency in the modulating signal produces a pair of sidebands, one on either side of the carrier frequency. In FM transmission, it can be shown mathematically, each pure audio frequency creates an *infinite* number of sidebands, at least in theory. These sidebands represent the sum and difference frequencies of the carrier with the modulating frequency and *all of its harmonics*. FM broadcasting would not be practical if it were required to reproduce an infinity of sidebands for each tone. Fortunately, all but an insignificant fraction of the energy is contained in a finite and relatively small number of sidebands. Their number

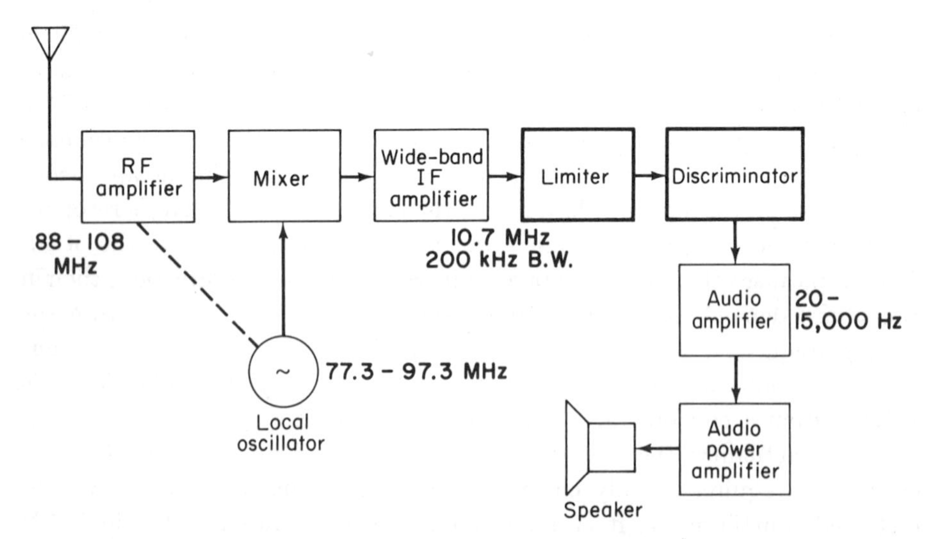

Fig. 16-16. Block diagram of an actual FM receiver.

is determined by the maximum frequency deviation and the highest audio frequency to be broadcast. For commercial FM broadcasting, there are eight important sidebands. The FCC limits the maximum deviation to 75 kHz on either side of the carrier, while the decision to transmit high-fidelity audio dictates a maximum audio frequency of 15 kHz.

Note that although FM is static-free by its very nature, it is hi-fi only because of this agreement to provide response to 15 kHz. There is plenty of room in the FM broadcast band, 88 to 108 million Hz, to provide the mere 120 thousand Hz bandwidth required for each station. Such a bandwidth requirement would not permit even nine stations to operate at once in the 540 to 1600 thousand Hz AM band.

REVIEW QUESTIONS

1. How would you describe an electromagnetic wave? How are its components related?
2. What properties of "light" did Hertz find in radio waves?
3. Can a radio wave ever be both horizontally and vertically polarized? Why?
4. What is the wavelength of a radio wave whose frequency is 600 kilohertz? 20 megahertz?
5. What were Marconi's major contributions to the field of radio?
6. Why is a high-frequency carrier wave used in broadcasting?
7. Describe the difference between amplitude modulation and frequency modulation.
8. Of two parties listening to a specific radio stations, why might the one living nearer the transmitter get poorer reception? Give several explanations.
9. A "ham" radio enthusiast tells you he can hear farther at night than during the day. Is he pulling your leg? Why?
10. What is meant by a sideband?
11. If an 800-kilohertz carrier is mixed with a 3-kilohertz signal in a nonlinear device, what frequencies appear in the output?
12. What functional units do you need to make the simplest possible AM transmitter? AM receiver?
13. What advantages are gained by going from the crystal set to the TRF receiver? To the superhet?
14. Why is even the best AM radio a "lo-fi" device?
15. What makes FM static-free? Is this what makes it a high-fidelity program source?

17

MORE PATTERNS IN THE ETHER

17-1 BEYOND THE RADIO

In the last chapter we saw the historical development of radio communication out of electromagnetic wave theory. By now we should be somewhat comfortable with oscillators, modulators, detectors, and mixers. It is our intention to continue with the more recently developed areas of radio wave applications. The trip will take us from the "boob tube" in the living room to laser beams and radio telescopes.

17-2 PICTURES THROUGH SPACE

One of the most outstanding (and perhaps also one of the most diabolical) inventions of our age is television. It is the electronic realization of that magic box, given to the princess in the fairy tale, in which little people acted out plays for the princess' enjoyment. The princess was fortunate in that her little folk apparently had no sponsors, so she was not belabored with commercials.

From what we already know about radio broadcasting, it should not be difficult to imagine how the electronic information representing a picture can be sent surging through space by a transmitter and picked up by a receiver. The sixty-four-dollar questions, at this point, revolve around a couple of transducers and their associated circuitry: (1) How is an image converted into electrical signals at the transmitter? (2) How are these electrical signals converted back into a visible image at the receiver? We deal first with monochrome (black and white) TV and later with color TV.

Perhaps the first workable television scheme was proposed by G. R. Carey of Boston in 1875. His transmitter was a photosensitive surface in which a gridwork of intersecting wire was embedded, equivalent to an array of tiny photocells. The receiver consisted of a bank of small lamps, each of which was operated by a relay from the photocell element in the corresponding position on the transmitter plate. A light pattern projected on the photosensitive transmitter plate thus caused a similar pattern to be formed by the lamp bank in the receiver. This scheme, with its hundreds of lamps and maze of wires, could never be a practical television system. It did, however, serve as the basis from which certain animated advertising signs were developed, some of which you may see in operation today.

Modern television is based on a *scanning* system and, like the motion picture, works only because of two imperfections of the human eye:

1. There is a limit to how fine a detail the eye can resolve.
2. When the eye is presented with two discrete images in a period of time shorter than about $\frac{1}{25}$th of a second, they are "blended" together into one continuous image due to *persistence of vision.*

The limit of resolution permits using a reasonable number of discrete *elements* (i.e., bits of blackness and whiteness) to form a picture. The persistence of vision allows the eye to see a complete picture when the individual elements are really being presented in sequence, but very rapidly. A TV system of the scanning type thus consists of a device that looks at the scene to be transmitted, one element at a time, and generates an appropriate electrical pulse in proportion to the brightness of the element; the radio transmitter circuits that broadcast the pulses; the radio receiver circuits that receive the pulses; and a device that traces out the image, one element at a time, so rapidly that the eye sees a picture.

Mechanical scanning was tried first, using motor-driven disks with a spiral hole pattern, invented by Paul Nipkow in 1884. Used in conjunction with a phototube on the transmitting end and a neon lamp on the receiving end, the whirling Nipkow disks could indeed transmit a visual image, if properly synchronized. This unwieldy and difficult-to-synchronize mechanical approach was soon supplanted by an all-electronic scanning system. A camera tube is used at the transmitter and a *kinescope* tube (a close relative of our old friend, the cathode ray tube) is used to display the image in the receiver.

17-3 CAMERA TUBES

Camera tubes are of several types, but they have certain common characteristics:

1. A photosensitive surface, upon which the desired scene is focused by a lens, and which converts the light pattern into a pattern of electric charge.
2. A means of storing the charge pattern between scans.
3. An electron beam executing the scan pattern.
4. A means of producing an output signal proportional to the charge encountered at the element scanned.

Historically the oldest, and now all but obsolete, is a pot-shaped camera tube called an *iconoscope*. It was invented around 1933 by Zworykin, the same V. K. Zworykin of electron microscope fame. A sectioned view of an

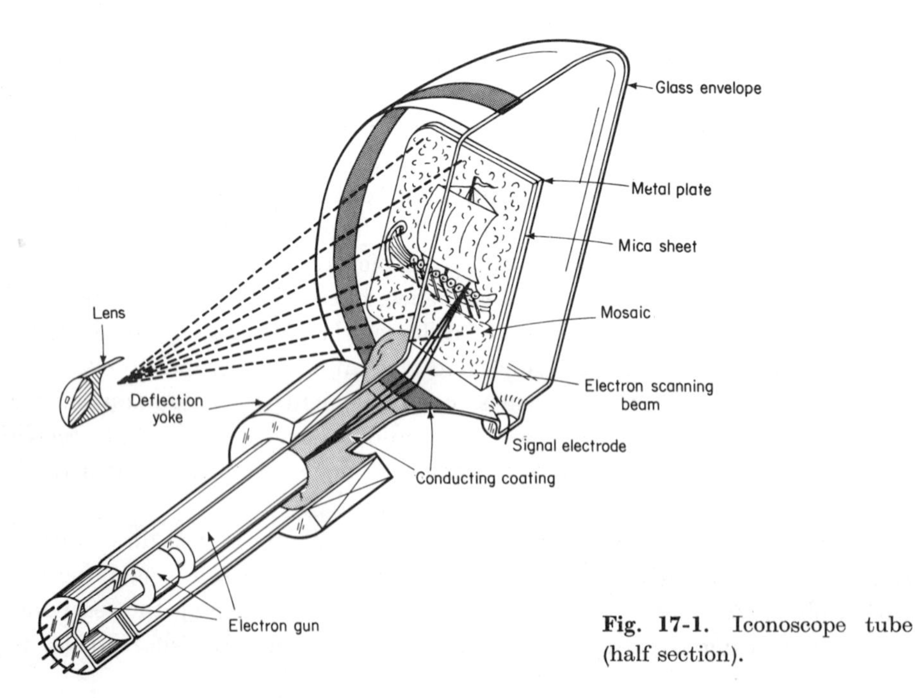

Fig. 17-1. Iconoscope tube (half section).

iconoscope is shown in Fig. 17-1. Its major components are an electron gun with a magnetic beam-deflection yoke (forming the "handle" of the "pot") and a photosensitive "sandwich" called the *mosaic* (contained in the "pot"). The mosaic consists of silver globules, sensitized with cesium, deposited on one side of a mica sheet, the other side of which has a metallic backing. Each little silver globule forms one plate of a tiny capacitor, the mica forming the dielectric and the backing—or *signal plate* as it is called—serving as the second plate common to all the globules. When a light pattern is focused on the mosaic by a lens, as in Fig. 17-1, photoemission takes place at each globule, the charge depending on the intensity of the illumination. Each little capacitor retains its charge through the scan until the electron beam reaches it. The impinging electrons then suddenly replace those lost by photoemission, causing a proportionate current pulse to flow in the signal plate. The amplitudes of the pulses from the signal plate circuit thus represent the relative brightnesses of the mosaic elements as the beam scans across them.

The iconoscope requires very strong illumination and is subject to erroneous "shading" effects because of redistribution of the emitted electrons. These weaknesses are overcome in the *image orthicon*, the standard commercial camera tube of today. Figure 17-2*a* shows the structure of the image orthicon. The image to be televised is focused on a thin, transparent photocathode at the front of the tube. Photoemission takes place at each point in proportion to the amount of light reaching that point. An accelerator grid increases the velocities of these photoelectrons, hurrying them on to strike the target, a very thin piece of glass of the same potential as the cathode in the electron gun. The incoming high-speed electrons knock secondary electrons out of the target. These secondaries are immediately collected by a screen placed near the target. The result is a pattern of charge on the target that looks just like the light pattern on the photocathode. As the beam from the electron gun sweeps over the target, it is turned back in the direction from which it is coming, since the target is of essentially the same potential as the cathode in the gun. At the uncharged places on the target (corresponding to dark areas in the image), the whole beam is turned back, while at charged points some electrons are absorbed from the beam to neutralize the charge. The returning beam strikes a disk-shaped electrode, knocking out secondary electrons. These secondaries are directed to a set of dynodes, similar to those in a photomultiplier tube, which surround the gun and provide amplification of the varying return-beam current. The amplified signal is then drawn off through the signal electrode.

The image orthicon is an excellent camera tube. It has a resolution of 670,000 elements per square inch and can produce a usable signal from a

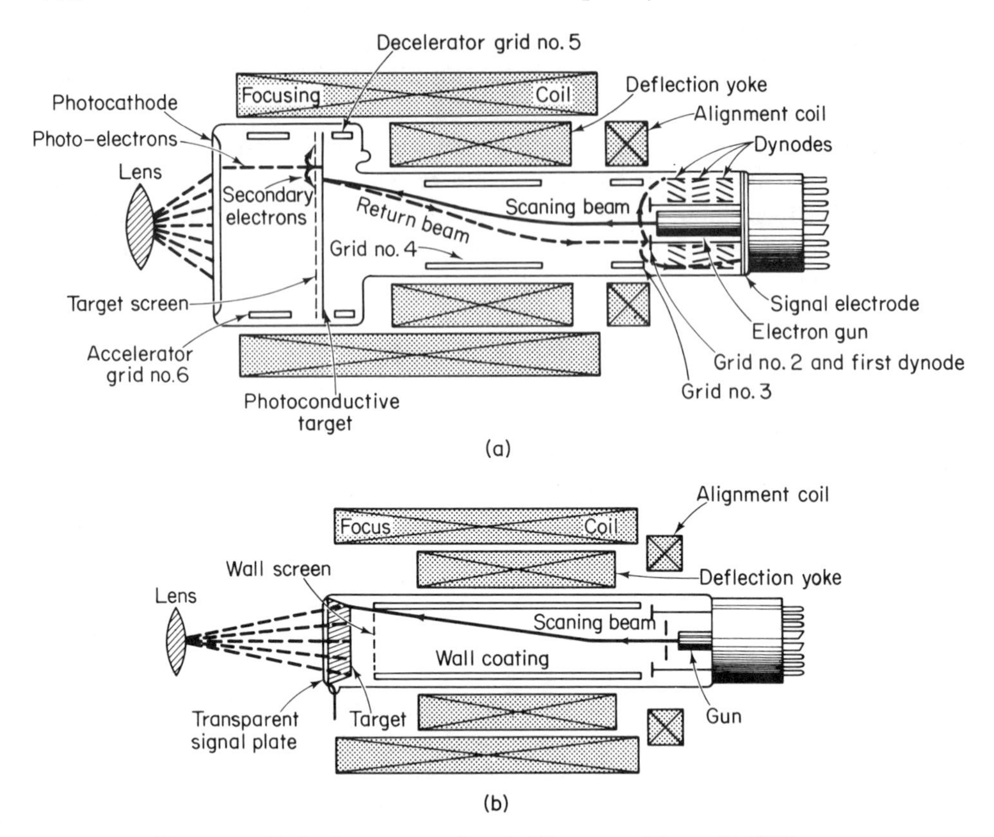

Fig. 17-2. Today's camera tubes. (a) Image orthicon. (b) Vidicon.

scene illuminated by bright moonlight. Unfortunately, it is also a temperamental tube, which requires careful temperature control and many adjustments. An image orthicon camera is bulky and very expensive.

A much smaller, simpler, and more rugged camera tube, which is becoming increasingly popular, is the *vidicon*. It differs from the image orthicon in several ways, perhaps the most significant of which is that its sensitive target depends on photoconductivity rather than photoemission. Figure 17-2*b* depicts the structure of a vidicon. The image is focused, through a transparent conductive film that acts as the signal electrode, onto the photoconductive target, which is biased slightly positive. In darkness, this layer acts like an insulator. An electron gun provides a beam that is slowed down by the wall coating and screen to moderate velocity and deflected magnetically. Wherever the scanning beam strikes the back of the target, it neutralizes the charge there, giving up just enough electrons to make up for those that have leaked through the partially conductive coating since the last scan. The brighter the illumination, the greater the conductivity of the target, the greater the leakage current, and the greater the number of electrons robbed from the beam. Once again, therefore, we get "bursts" of current at the signal electrode in proportion to the brightness of the spot scanned.

The vidicon's resolution is extremely high, about a million elements per square inch, limited only by the electron beam width. Consequently, a very small vidicon can deliver quite a respectable picture. The standard vidicon is one inch in diameter and half-inch vidicons are used in many applications. A portable TV camera system, using a vidicon tube, is shown in Fig. 17-3.

Fig. 17-3. Portable vidicon television camera system. (Courtesy of General Electric Co.)

17-4 MONOCHROME TV BROADCASTING

To understand the TV broadcasting process, let us stop for a moment and consider what goes into one single picture. Remember that the scene televised is "taken apart," one element at a time, by the camera tube's scanning beam and is "redrawn," one element at a time, on the screen of the receiver's picture tube. These two scanning operations must, of course, be perfectly synchronized.

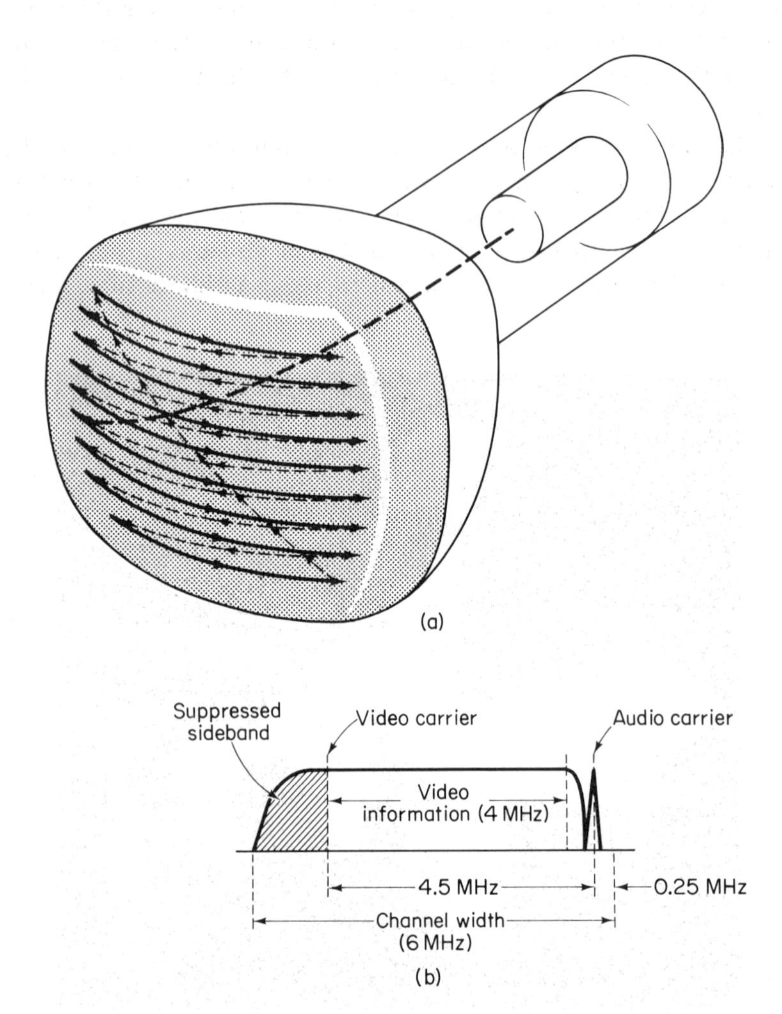

Fig. 17-4. (a) Field scanning pattern. (b) Distribution of information in the standard 6-MHz television channel.

The scanning action is illustrated in Fig. 17-4*a* for the receiver tube (it is the same in the camera tube). The beam traces a *line* of elements from left to right. When the beam reaches the right-hand edge, the magnetic-deflecting field for the horizontal direction reverses sharply. This step would jerk the beam back across the face of the tube, drawing a right-to-left line, except that the beam is momentarily turned off during the retrace by a *blanking* signal. At the same time, the magnetic-deflecting field in the vertical direction has increased to move the starting point on the left down one "notch." The blanking signal is turned off and the next left-to-right line is drawn. This pattern repeats until the bottom right corner of the image is reached; then the beam is deflected sharply back to the top edge, as well as to the left side, during blanking. The picture is scanned along all the odd-numbered lines first, providing one *field*, then along all the even numbered lines, producing a second field, the two being meshed together to make one complete picture or *frame*. This standard scanning pattern is called two-to-one interlaced scanning. The total number of lines allocated per frame is 525. However, only about 485 to 490 actually appear in the image, the rest being lost during the vertical retrace. Thirty total frames are scanned each second so that the eye cannot see flicker. Because each frame consists of two fields, the field rate is therefore 60 per second, which permits use of the

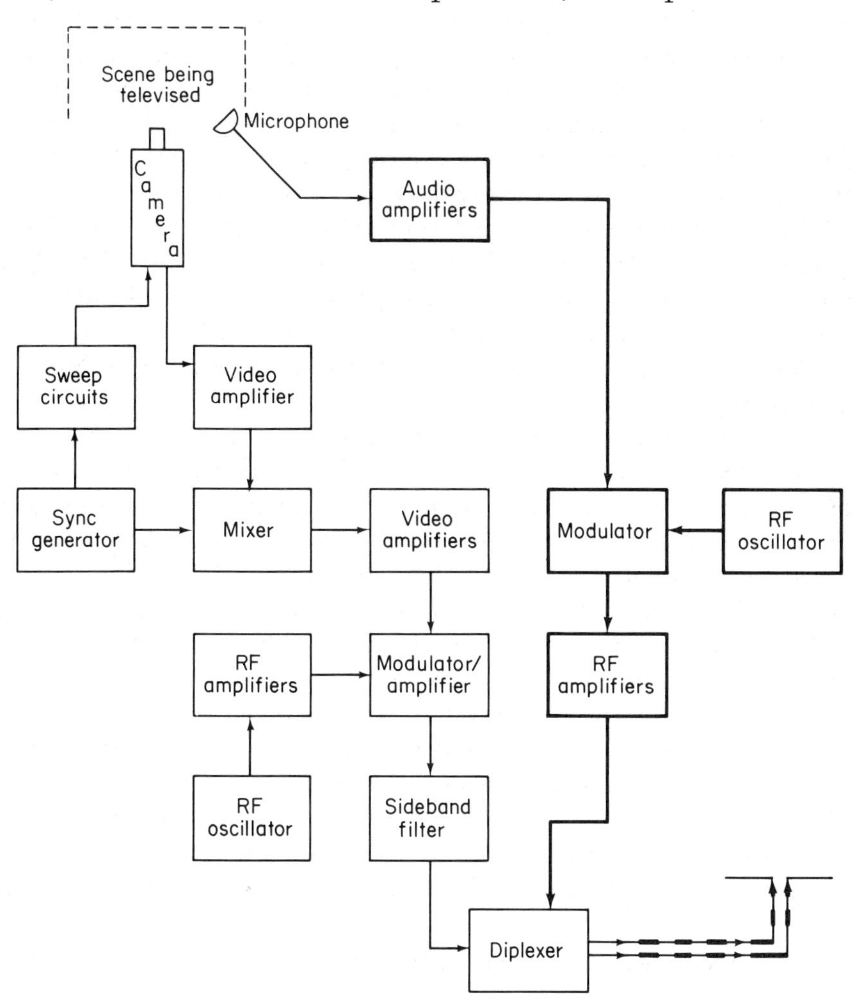

Fig. 17-5. Diagram of typical monochrome television transmitter.

carefully controlled 60-Hz power line frequency for synchronization.[1] Since thirty 525-line frames are scanned per second, the horizontal scanning frequency must be 15,750 lines per second. The shape of the picture is roughly rectangular, $1\frac{1}{3}$ times as wide as it is high. These proportions correspond to the frame size in commercial 35-millimeter movie film. (See, they knew right from the beginning that they were going to show us a lot of old movies!)

The TV broadcast channel carries a lot of information. In addition to the video signal itself (the brightness-variation signal generated by the camera tube during scanning), synchronizing pulses, equalizing pulses, and blanking pulses are also transmitted. The synchronizing pulses, as their name implies, are used to "lock" the movements of the receiver picture tube's beam with that of the camera tube. The equalizing pulses help the receiver discriminate the vertical sync pulses. The combination of video signal proper with sync, equalizing, and blanking pulses is the *composite video* signal, which is impressed on the video carrier frequency by amplitude modulation. The same channel must also carry the sound that goes with the picture, and this sound is frequency-modulated onto the audio carrier. All this information requires a channel bandwidth of 6 MHz, compared with only 120 kHz (0.12 MHz) required by FM radio and a mere 10 kHz (0.01 MHz) for AM radio. The utilization of the 6-MHz bandwidth is illustrated by Fig. 17-4*b*. The sidebands of the video carrier are 4 MHz wide, so only a single sideband can be used effectively and the lower one is almost completely suppressed (i.e., cut off).

The TV transmitter is given, in block diagram form, in Fig. 17-5. It is really two separate transmitters—an AM video transmitter and an FM audio transmitter—side by side. Both feed a common antenna through a *diplexer*, a special circuit that prevents the audio signal from getting into the video circuits and vice versa. The audio transmitter needs no further comment. The heart of the video transmitter is the sync generator, which provides all the critical timing functions to both the local camera's sweep circuits and to the mixer for transmission to the receiver's sync circuits. The output of the camera tube is mixed, after amplification, with the sync signals to form the composite video signal. The rest of the video transmitter is like any other AM transmitter, except that it operates in the TV frequency band and has a sideband filter to suppress the lower video sidebands. Figure 17-6 shows the placement of camera and microphone in the studio.

The television receiver is diagrammed in Fig. 17-7. Up through the

[1] In Europe, where the power line frequency is 50 Hz, the frame rate is 50 per second.

Fig. 17-6. Television sequence being shot. (Courtesy of Minnesota Mining and Manufacturing.)

video detector stage, it is simply a wideband superhet receiver. The rf amplifier, local oscillator, and mixer together form a unit called the *tuner*. In the video detector, the video signal is separated from the audio. The audio is passed on to what amounts to a standard FM receiver. The composite video signal is fed to both the control grid of the picture tube to control the brightness of the scanned elements and to a sync separator stage. The horizontal and vertical sync pulses are isolated in the sync separator and used to drive the receiver's sweep circuits, which, in turn, drive the beam-deflection coils of the picture tube. The flyback transformer circuitry uses the sharp transient produced by horizontal retrace to produce high voltage for the picture tube.

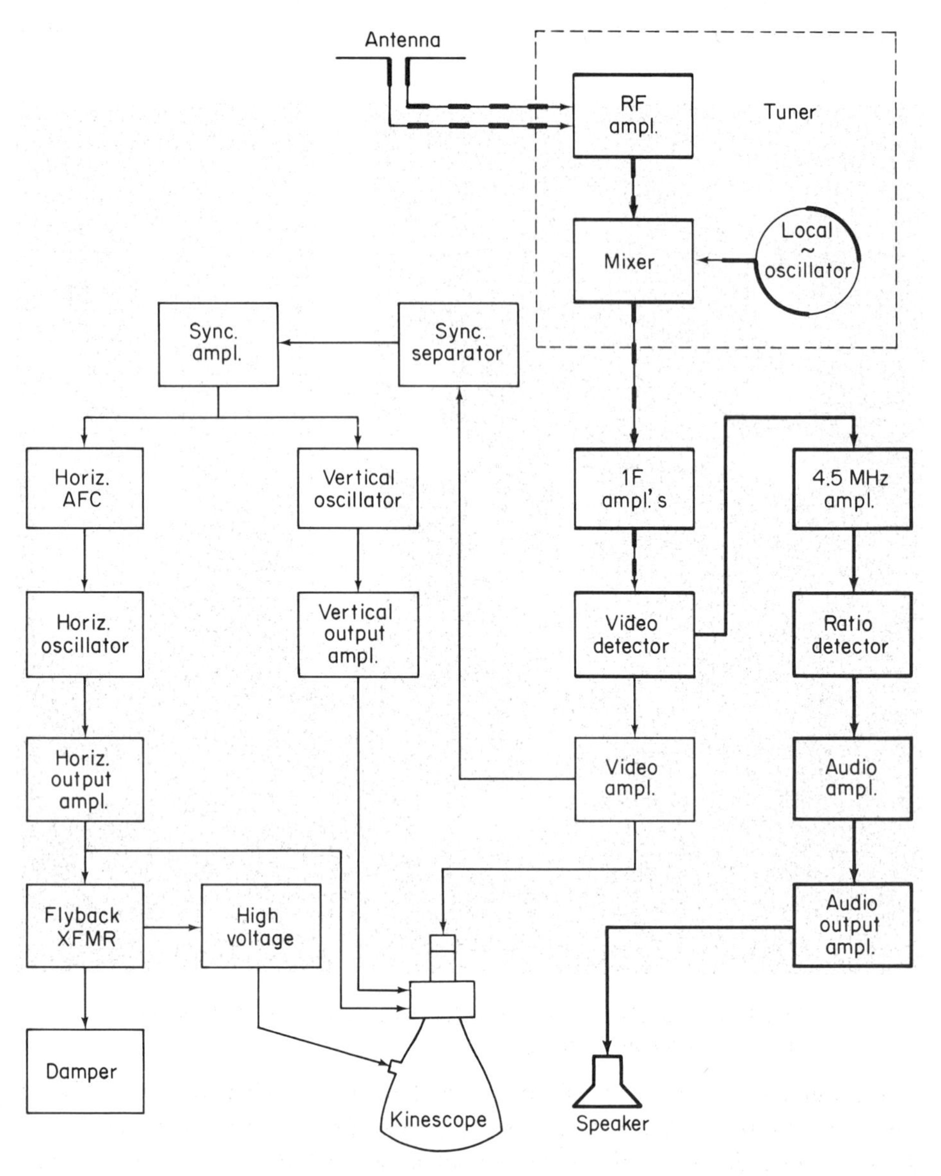

Fig. 17-7. Standard monochrome television receiver circuits.

17-5 COLORS THROUGH THE ETHER

Given an understanding of black and white TV, it is not too big a jump to color. In many ways, there is a close similarity to the transition from black-and-white photographs to color photographs. To begin with, we must consider for a bit the property of "color" and how it is seen.

If you haven't thought about it before, it may surprise you to realize that color is a three-dimensional quality. By that we mean that three separate, independent characteristics of a colored image must be given in order to specify its color accurately. These characteristics are *hue* (or chroma), *saturation* (or purity), and *brightness*. The hue is the basic characteristic that gives the color its name, such as yellow and purple. Saturation is a measure of how much the color is diluted with white. For instance, catsup, blood, and the alternate stripes in the American flag are fairly saturated reds, while strawberry ice cream, ladies' pink lingerie, and flamingo feathers are examples of reds of low saturation. Brightness refers to the brilliance of the color, compared to a scale of neutral shades running from white through progressively darker grays to black. Any system of color television (or photography) must deal with these three dimensions.

White is not a color but a mixture of all the rainbow's colors. Almost everyone has seen the classic experiment in school, where a narrow beam of white light is made to fall upon one face of a glass prism, and the beam leaves from another face of the prism, broken up into all the component hues. Traditionally, we have grouped these hues into six bands with the names: *red, orange, yellow, green, blue,* and *violet*. These are the *spectral* colors. By mixing the end colors, the red and violet, in various proportion, we obtain purples and magentas that are *nonspectral* colors. There are, of course, a large number of separate hues in each of these loose color names. Black is not a color, but the absence of all color. Grays are simply weak whites. Browns are reds, oranges, and yellows of low brightness.

It is believed that the eye sees color through the combined action of three different types of receptors (little sensors) in the eye, each sensitive to a different band of colors, which, for simplicity, we may call red, green, and blue receptors. It can be shown that if we select any three colors that are sufficiently far from each other in the spectrum, we can reproduce almost any color by a suitable mixture of these three. Any three colors forming such a set are called *primaries*. Although the choice is somewhat arbitrary, it has become standard practice to select a certain red, a green, and a blue-violet (which we will hereafter call simply blue) as primaries.[2]

Figure 17-8 shows three projectors whose beams of white light pass through filters, each of which absorbs most of the colors in the beam and passes only one primary color. The beams intersect as they reach a screen. If their relative intensities are adjusted properly, the area illuminated by a portion of all three beams will appear white. Under these circumstances,

[2] Specifically, we are referring to *additive* primaries here.

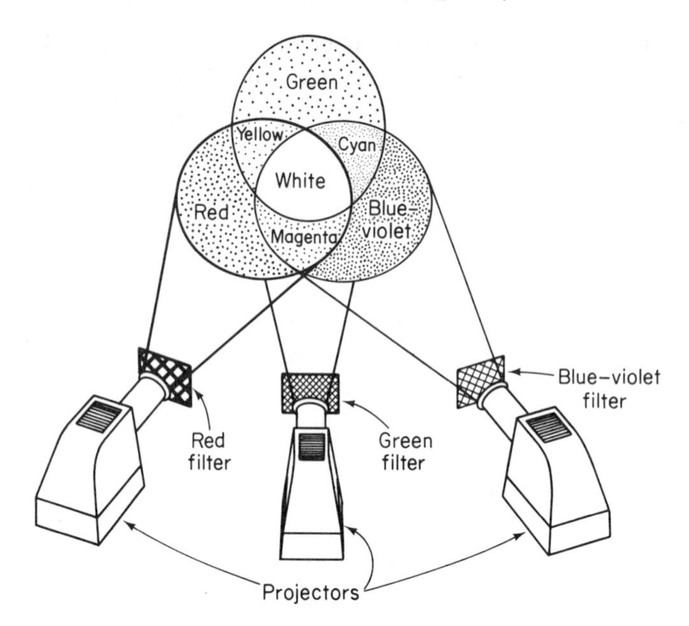

Fig. 17-8. Additively mixing light beams of the primary colors.

the overlap of the red and green beams produces a yellow area; red and blue-violet mix to form a magenta; and the blue-violet with green combination produces an aqua color, which is called *cyan* in the jargon of the business. By suitable adjustment of the relative brightness of the three beams, any color can be adequately reproduced in one of the areas of beam overlap. This additive process is the fundamental basis for color television.

The color transmission technique involves two opposite processes: *color separation* at the transmitter and *color synthesis* at the receiver. It would be perfectly feasible to transmit a color picture in the following way: Place three cameras side by side, connected to three video transmitters on different carrier frequencies. Let each camera photograph the scene through a color separation filter, one red, one green, and one blue. Then one transmitter will transmit only the red portion of the scene; another, only the green portion; and the third, only the blue. The receiving unit consists of three standard television receivers mounted in one cabinet. Each is tuned to receive one of the three color signals, and the phosphor in its picture tube is selected to fluoresce in that color. A system of lenses and mirrors combines the three separate single-color images into one full-color picture. Isn't that simple enough? Although this technique is not used for economic reasons, it does show, in principle, how color TV works. Let us now consider the actual process.

17-6 COLOR TV BROADCASTING

Historically monochrome TV came before color. Millions of black and white sets were already in use when color broadcasting techniques were finally perfected. As a result, the FCC decreed that commercial color broadcasts had to be "compatible"; in other words, color broadcasts had to be so structured that existing monochrome receivers could receive the color broadcast, in black and white, without modification. This ruling meant that established practices like 525 lines, 30 frames per second, and 6-MHz bandwidth had to be kept "sacred." The squeezing of three-dimensional color information into the standard black-and-white broadcast channel is truly an example of human ingenuity.

The specialized circuits of a color TV transmitter are represented in the block diagram of Fig. 17-9. The color camera does indeed use three camera tubes, either image orthicons or vidicons. Instead of three separate systems of lenses and filters, however, one lens system and *dichroic mirrors* are used. A dichroic mirror is a beam-splitting mirror that reflects the color for which it is named and passes all others. The light coming through the taking lens encounters a blue dichroic mirror, so the blue component of the scene is split off and sent to one camera tube. The remaining red and green components go through the blue dichroic mirror and encounter next a red dichroic mirror. This process splits off the red component, which is reflected to a second camera tube. What remains, after passing through both dichroic mirrors, is merely the green component, which goes directly to the third camera tube. The signals from the camera tubes are then processed for transmission. First, 59 percent of the green signal, 30 percent of the red signal, and 11 percent of the blue signal are added together to produce a *luminance* signal, which corresponds roughly to brightness. It is this luminance signal that is used for the black-and-white picture in monochrome receivers and that makes the broadcast technique "compatible." The camera output signals are also processed in the matrix unit to produce two signals, each of which contains some hue information and some saturation information. These two signals are called *chrominance* signals, and we have called them the "orange-cyan" and "bluegreen-magenta" signals.

Now we come to the question of how to fit this information into the 6-MHz channel. Careful study showed that the monochrome video signal consisted of narrow groups of sidebands, clustered in steps of 15,750 Hz (the line scanning frequency, remember?) around the video carrier, and with unoccupied gaps between the clusters. It was found that if the chrominance signals were used to modulate another frequency, called the *subcarrier*,

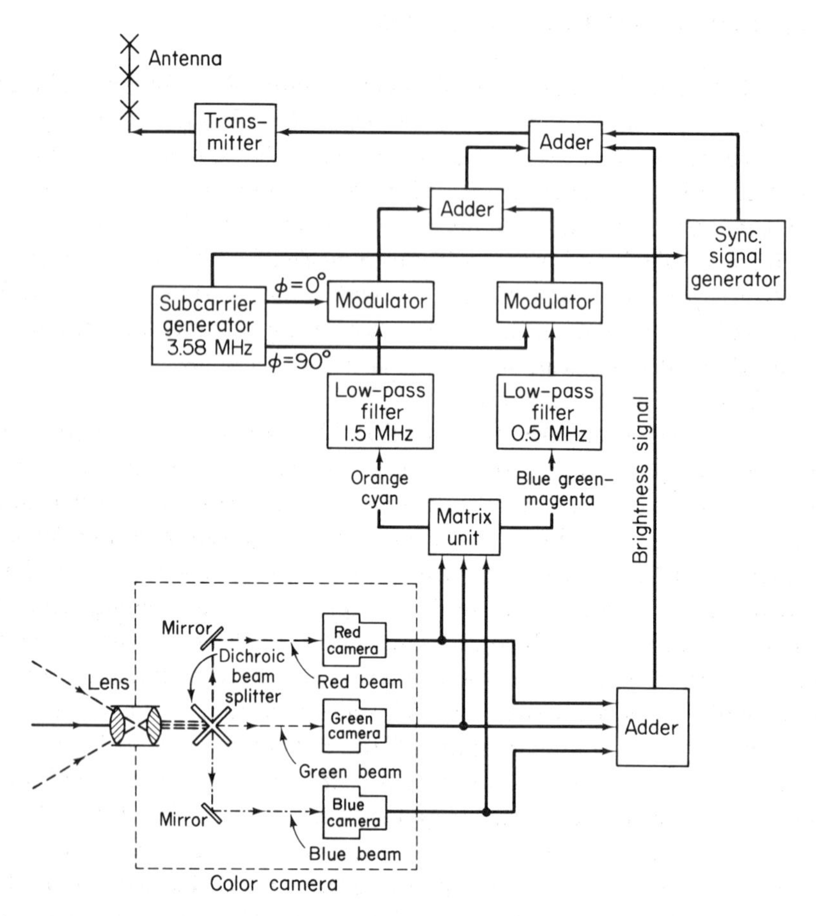

Fig. 17-9. Simplified block diagram of a color television transmitter.

just about 3.58 MHz above the video carrier frequency, the resulting sidebands would "sandwich in" nicely between the monochrome sidebands. This is exactly what is done. The subcarrier itself is filtered out and not transmitted. It must therefore be generated again by an oscillator in the receiver. The two chrominance signals are blended with the subcarrier, 90° out of phase with each other, and the subcarrier is removed, leaving only the sidebands. The sidebands are added and act as a *phase modulation* for the carrier. The amplitude of this combined chrominance signal carries the saturation information, and its phase angle conveys the hue. The combined chrominance signal, together with the luminance signal and some additional sync pulses (needed to keep the subcarrier oscillator in the receiver "locked" to the one in the transmitter, with respect to phase angle), then modulates the video carrier. The rest of the transmitter is like a monochrome transmitter.

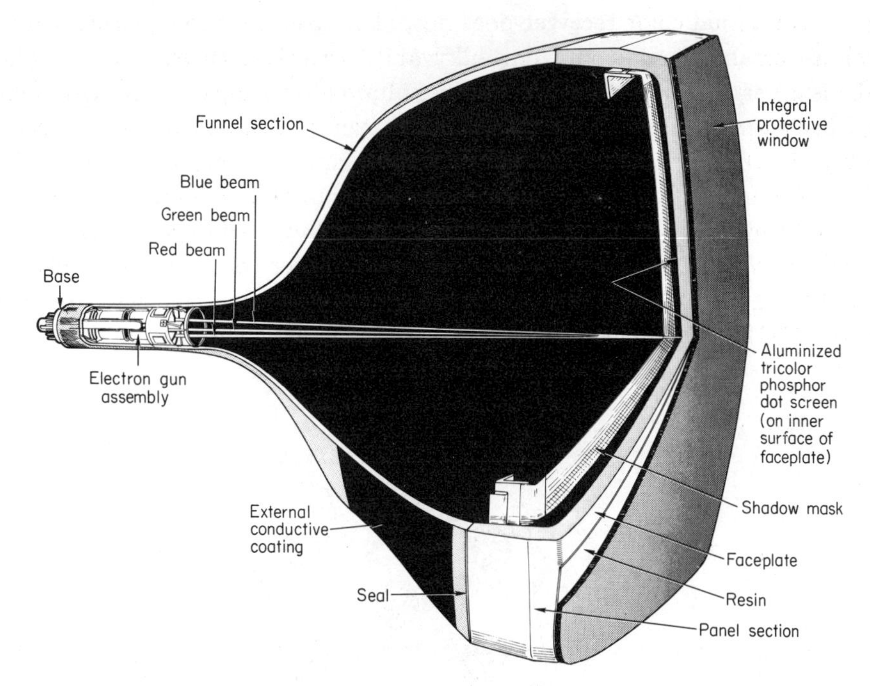

Fig. 17-10. Cutaway drawing of color picture tube. (Courtesy of Radio Corporation of America.)

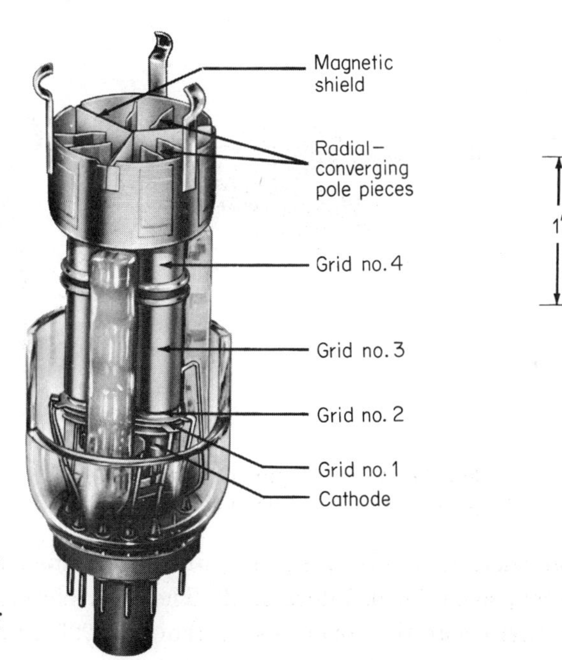

Fig. 17-11. Tri-gun assembly for color picture tube. (Courtesy of Radio Corporation of America.)

An actual color receiver does not, of course, use three picture tubes; such an arrangement would be bulky and expensive. Instead, the display tube is a very complex kinescope that contains three separate electron guns and has a special face construction. The structure of this tube is illustrated in Fig. 17-10, and a detail view of the three-gun assembly is shown in Fig. 17-11. The inside face of the screen is coated with three different phosphors, not uniformly, but in a three-dot pattern. A thin, perforated metal sheet, called the *shadow mask*, is placed near the phosphor-coated screen, between it and the guns. A portion of the mask and the phosphor coating are shown in Fig. 17-12*a*. The shadow mask has about 250,000 holes in it, and for each

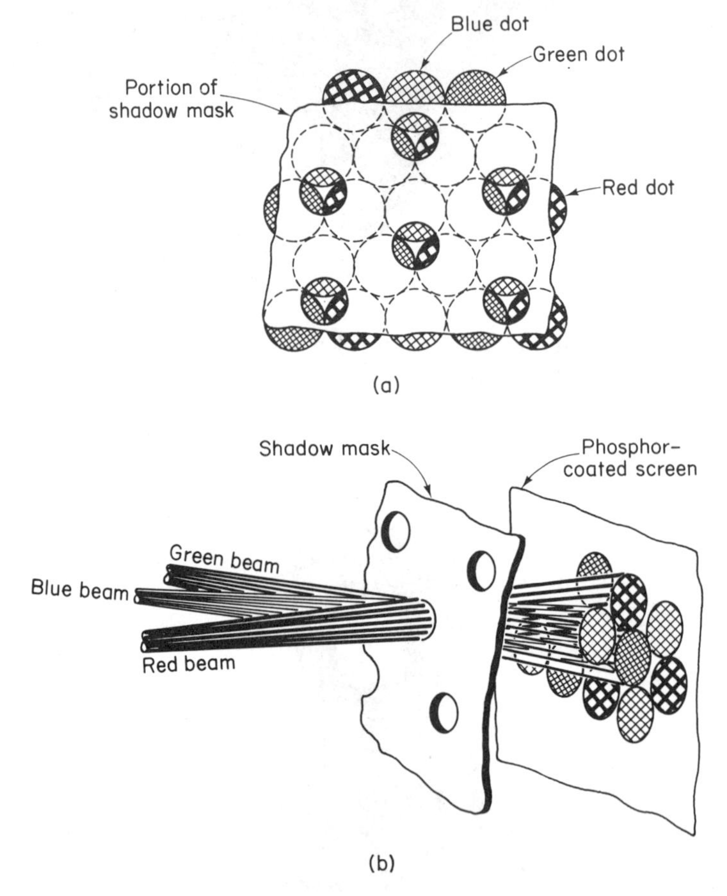

Fig. 17-12. (a) Alignment of shadow mask with phosphor pattern. (b) Beam paths through shadow mask.

hole there is a red- , a green- , and a blue-fluorescing phosphor dot on the screen, some 750,000 dots in all. The color picture tube has an extra *convergence electrode* that operates at about 11,000 volts and that forces the red, green, and blue signal beams to go through the same shadow mask hole. After passing through, the blue signal beam strikes the phosphor dot that fluoresces blue, and the other two beams also strike their matching dots, as shown in Fig. 17-12*b*. These tiny glowing dots of color are too small for the eye to resolve under normal viewing conditions. Thus, if all three dots are illuminated with equal intensity, the eye sees the three dots as a white element; if only the red and green dots are activated, the eye sees a yellow

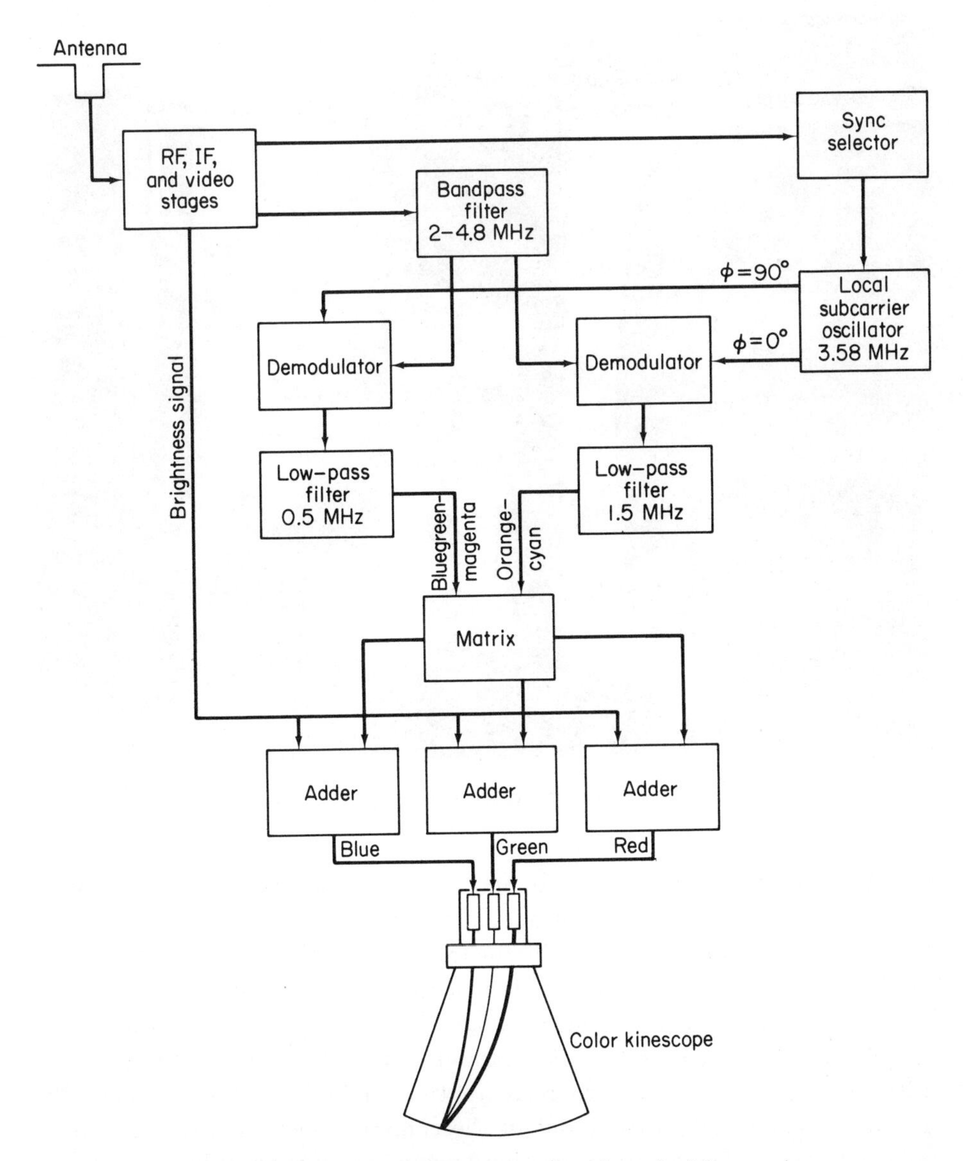

Fig. 17-13. Simplified block diagram of a color television receiver.

element; and so on. The effect is somewhat similar to the way color comics are created in the Sunday newspaper, out of innumerable tiny dots of colored printer's ink.

The circuits of a color receiver that differ from its black-and-white counterpart are given in simplified block diagram form by Fig. 17-13. An actual receiver is shown in Fig. 17-14. In the rf and i-f stages, the video and

Fig. 17-14. Solid-state home color television receiver. (Courtesy of Motorola Inc.)

the audio signals are separated. In the video detector, the luminance signal is split off from the combined chrominance signal. The latter is demodulated by heterodyning with a local oscillator frequency 3.58 MHz above the video carrier and synchronized with that produced in the receiver. This process recreates the individual orange-cyan and bluegreen-magenta chrominance signals, which are fed into the matrix and adders. There they are combined with the luminance signal to recreate the original red, green, and blue signals, proportional to those out of the camera tubes. Finally, these pure color signals are fed to the respective guns of the picture tube.

17–7 MICROWAVES

As we go up the frequency ladder (or down the wavelength ladder) of electromagnetic radiation, we eventually come to the *microwave* region. This terminology covers those radio waves whose wavelength lies below one meter and is greater than one-tenth of a millimeter, which is the same as saying waves whose frequency is between 300 MHz and 3 terahertz (3×10^{12} Hz). In this frequency region, the radio waves behave somewhat like light. They can be focused and reflected (if the surfaces used are conductors), and they travel in straight lines. Unlike light, however, they can penetrate clouds and fog. These properties led to the development of radar, which we discuss presently.

One of the most popular reflectors for microwaves (and light) is the *parabolic* reflector shown in Fig. 17-15. The paraboloid is a "dish" shape

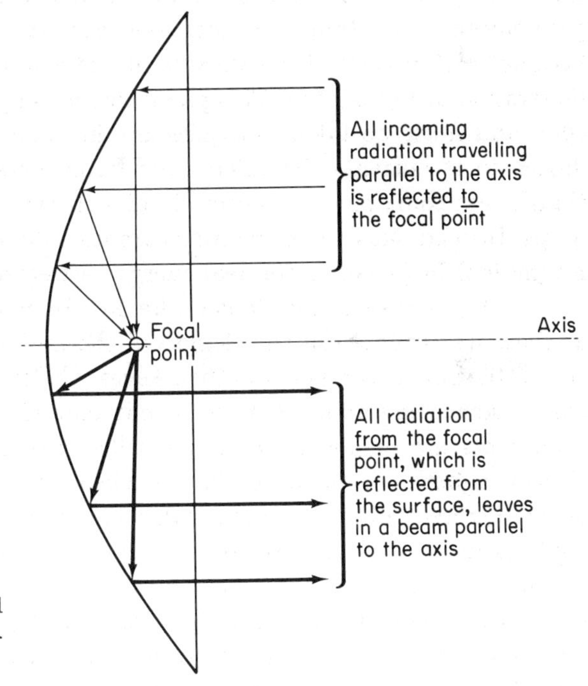

Fig. 17-15. Microwave and optical properties of a parabolic surface.

obtained by rotating a certain geometric curve about its own axis. Associated with every paraboloid is a point in space, on the axis of the paraboloid, which is called the focal point. The significant property of the paraboloid is that any radiation reaching it which is traveling parallel to its axis, is made to converge at the focal point. Conversely, if a small source is placed at the focal point, all radiation striking the dish is reflected out in a beam parallel to the axis.

These properties make parabolic reflectors very useful in microwave antennas.

When we begin to talk about microwave circuitry, we must be prepared to abandon some of our old, cherished circuit concepts. Consider, for instance, the distribution of current in a wire. In our usual Ohm's-law way of thinking, we more or less tacitly assume that the electron flow is distributed essentially uniformly through the cross section of the wire, as in the left illustration of Fig. 17-16*a*. This is true for dc, and for ac until we reach the radio frequencies. For rf, the picture begins to change. The magnetic field associated with the electrons moving near the center of the wire is almost all contained within the wire, the lines of force being "bunched up." The field associated with electrons moving near the surface, however, is largely in the space (or insulator) around the wire, and these lines are spread out. The result is that there is greater inductance, hence greater inductive reactance and less current, in the center of the wire, as shown in the center illustration of Fig. 17-16*a*. This effect becomes more and more pronounced as the frequency increases, and at high frequency, the current becomes a thin shell at the *surface* of the conductor only, as in the right illustration of Fig. 17-16*a*. This phenomenon is appropriately called the *skin effect* and must be taken into account in determining the resistance of a device to rf currents. At microwave frequencies, it is impractical to use Ohm's law thinking to analyze circuits in terms of currents and voltage drops. Instead, Maxwell's equations are used directly to analyze the electromagnetic field (which is the real energy carrier) associated with the circuit.

The hardware, too, looks different for microwaves. Remember that at these frequencies the wavelength of the radiation is of the same general size as the circuit components themselves. Under these conditions, we might expect something unusual. It turns out that some of the most satisfactory conductors are hollow metal tubes called waveguides, through which the electromagnetic waves are conducted. These tubes may be circular or rectangular in cross section, the rectangular ones being more common. The waveguide is not considered to carry a "current," like the wire carrying low-frequency ac, but merely to act as a *boundary* to confine the electromagnetic wave in an enclosed space. The signal is fed in at one end by means of a tiny radiating antenna or a single-turn inductive coupling loop. The waveguide then directs the signal to the receiving end by reflections against its inner wall. No energy leaks out through the walls because of skin effect.

For any waveguide, there is a lowest frequency (longest wavelength) that it will propagate. The major dimension of the waveguide, at right angles to the direction of propagation, must be slightly longer than half of this longest desired wavelength. The frequency to which it corresponds is called the *cutoff frequency* of the waveguide. Going in the other direction, there is

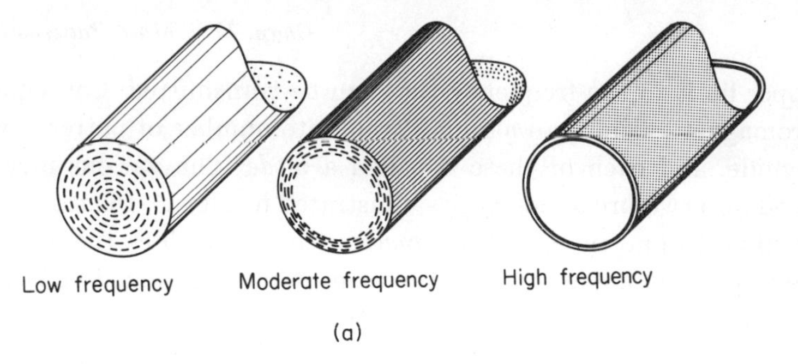

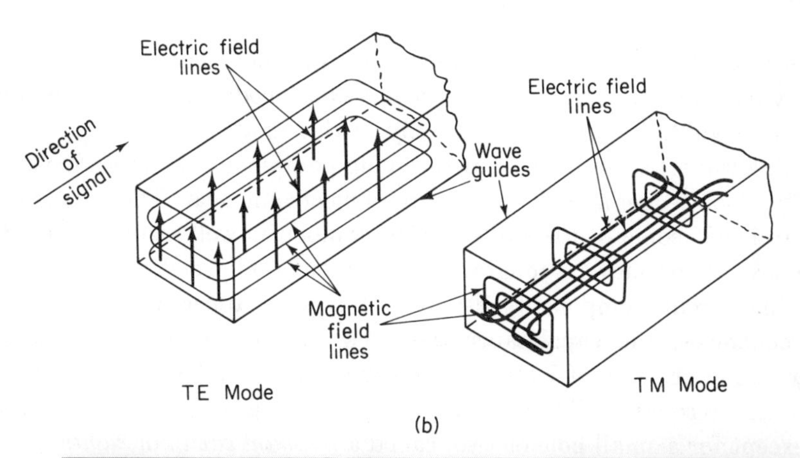

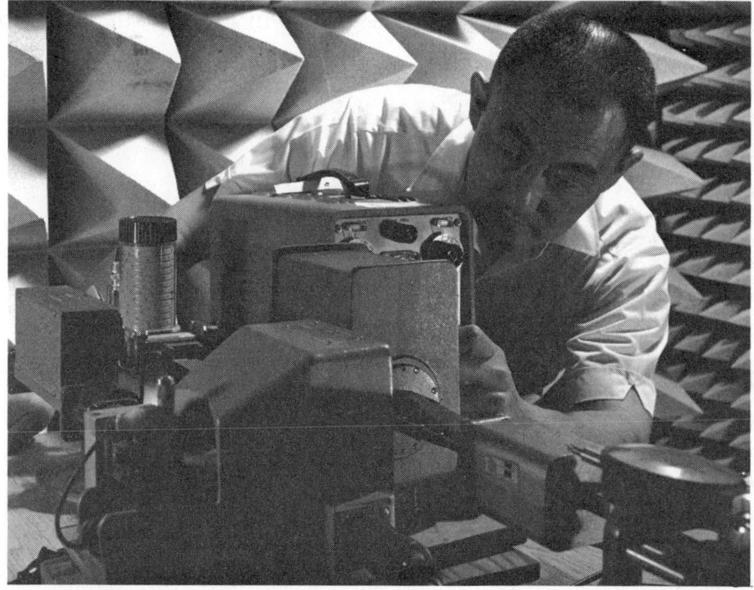

Fig. 17-16. (a) Skin effect in a conductor as frequency is increased. (b) Wave propagation modes in a rectangular waveguide. (c) Microwave apparatus being tested in an anechoic chamber. (Courtesy of the Aerospace Corp.)

no upper limit to the frequency that can be transmitted. Consequently, the electromagnetic fields can assume an infinite number of patterns within the waveguide, and each of these is called a *mode*. The various modes can be grouped into two broad categories, illustrated in Fig. 17-16*b* for a rectangular waveguide. In one type, called *transverse electric*, or TE, the electric field is entirely at right angles to the direction of propagation, but the magnetic field is partly in the propagation direction. In the other type of mode, called *transverse magnetic*, or TM, the magnetic field is completely at right angles to the direction of transmission, but the electric field has a component in the direction. The number of possible modes for a given frequency increases, the higher that frequency is above the cutoff frequency. The lowest possible frequency has only one mode, called the *dominant mode* for that waveguide, and the waveguide is generally used only in this mode.

Microwave transmitting and receiving equipment therefore looks as much like a bootlegger's "still" as a piece of electronic gear. The microwave technician needs some of the skills of the plumber as well as of the electrician. Figure 17-16*c* shows a microwave apparatus under test.

Other circuit components also take on a novel appearance. At microwave frequencies, the capacitance (between the successive turns) of an ordinary inductor, and of even a wirewound resistor, becomes a serious problem. The resonant *LC* circuit for microwaves is a little box, completely closed except for a small hole or two, called a *resonant cavity* or *cavity resonator*. Consider the behavior of the device shown in Fig. 17-17*a*. At microwave

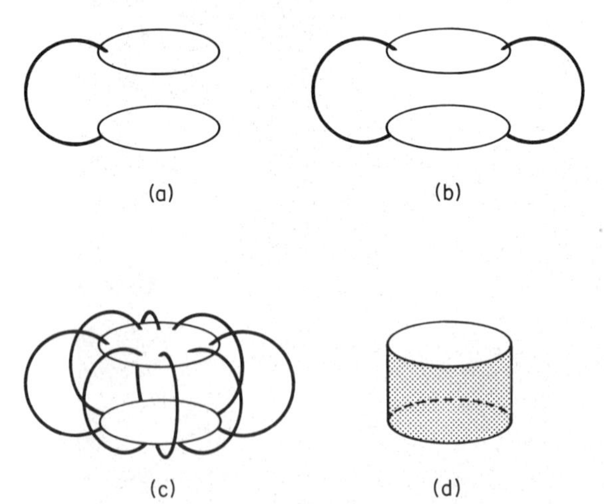

Fig. 17-17. Development of the cavity resonator.

frequencies, any little old loop of metal is a significant inductance, and any two bits of metal close, but not touching, form a capacitor. This simple little device is therefore a parallel *LC* circuit, which is resonant at some frequency. If we raise the frequency at which we wish to resonate, the inductance of this single turn is too great and we lower it by connecting a second inductor (loop) in parallel with it, as in Fig. 17-17*b*. To raise the resonant frequency still higher, we must add more inductive loops in parallel, and our *LC* circuit soon begins to look physically like Fig. 17-17*c*. In the limit, we end up with an infinite number of loops in parallel, that is, a solid wall. Our resonant *LC* circuit has become the metal "pill box" of Fig. 17-17*d*, which is the cavity resonator. It may be a cylinder, as shown, or a thin square box, or a sphere. Effectively, it is a piece of a waveguide sized so that a wave of a particular wavelength can be maintained inside. Resonant cavities can easily provide Q values of 1000 and greater, compared to values of 100 and less, which are typical of low-frequency *LC* circuits.

The tubes and semiconductors used at microwave frequencies must also be of radical design. Two characteristics of ordinary vacuum tubes make them unusable with microwaves: interelectrode capacitance and transit time. As we are well aware by now, every little capacitance is significant at these high frequencies. In an ordinary tube, the interelectrode capacitances would act as short circuits, bypassing the signal around the tube. To appreciate the significance of transit time, consider a 600-GHz microwave signal. Such a signal goes through 600 billion cycles each second. In the time it takes to complete one cycle, light (which is traveling 186,000 *miles* per second) can only cover a distance of about 20 thousandths of an inch. If the spacing of electrodes in the tube is this great, the electrons from the cathode don't get a chance to reach the plate before the grid field reverses again, and the higher the input frequency, the worse this problem becomes. The interelectrode capacitance problem and the transit time problem work against each other. The easiest way to decrease the capacitance is to move the electrodes farther apart, but this step increases the transit time from cathode to plate.

The basic triode concept can be used up to about 10 GHz, in what is called a *lighthouse* tube. This is a triode in which the cathode, "grid," and plate are closely spaced disks, the external envelope resembling a lighthouse. Above 10 GHz, the transit time of the triode becomes intolerable and new types of tubes are required. Some of these tubes are shown in Fig. 17-18.

The magnetron, Fig. 17-18*a*, is a distant relative of the cyclotron-type nuclear particle accelerator. It consists of a ring-shaped anode surrounding an indirectly heated cylindrical cathode. Several resonant cavities, for the desired wavelength, are cut into the anode. Sizing of the device and the intensity of the magnetic field are such that electrons emitted by the cathode are forced to travel in completely circular paths, just grazing the anode and returning to the cathode. The time required to complete the circular path

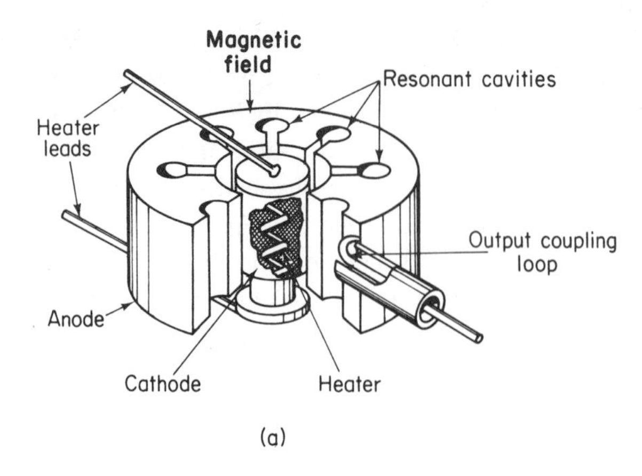

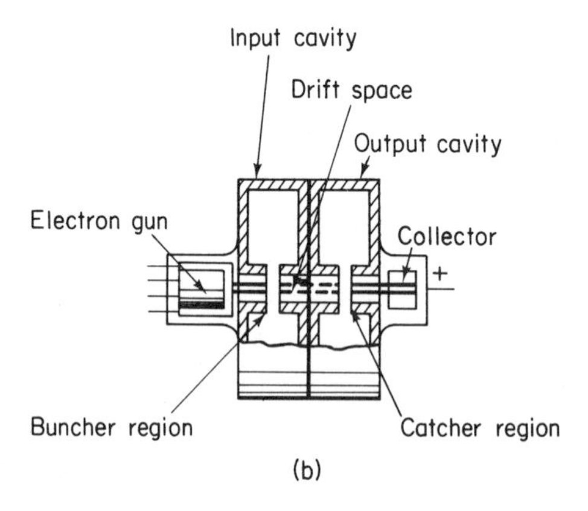

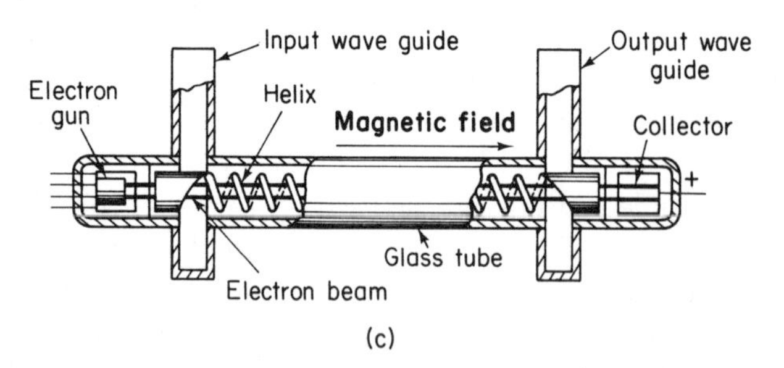

Fig. 17-18. Microwave tubes. (a) Magnetron. (b) Klystron. (c) Travelling wave tube.

corresponds to the period of the desired frequency, setting up strong oscillating fields in the resonant cavities. The energy of these fields is tapped off through a coupling loop.

Another microwave tube making use of resonant cavities is the *klystron*, illustrated in Fig. 17-18*b*. An electron gun fires a beam past two cavity resonators to a collector electrode. The first resonator (nearest the gun) acts like a grid to which a high-frequency oscillator is connected. Its field alternately tends to accelerate or decelerate the electrons in the beam. In the free space between cavities, the electrons drift at whatever speed they left the first resonator. There, the "speeded-up" electrons catch up with the "slowed-down" ones, which passed through an instant earlier. The overall effect is that the electrons in the stream are "bunched up" into little clusters by the first cavity resonator, which is therefore called the *buncher*. The second cavity is tuned to the same frequency as the first, so that an intense field is set up in it. The amplified energy is drawn off through a waveguide from the second, or *catcher*, resonator. By interconnecting the two resonators, the klystron can be made to work as a high-frequency oscillator. Another version of the klystron, called a reflex klystron, uses only one cavity by sending the beam back through again with a reflector electrode.

An interesting microwave amplifier tube that does not use a cavity (and can therefore handle a wider band of frequencies) is the *traveling wave tube* (TWT), depicted in Fig. 17-18*c*. In this tube, the signal is fed into a metallic helix (spiral) by a waveguide. The microwave field prefers the metal of the helix to the vacuum in the tube and travels the longer helical path. Meanwhile, an electron beam from a gun is shot down through the center of the helix to a collector at the other end. The pitch of the helix and the voltage accelerating the beam are such that the signal and the beam electrons travel along "side by side" at the same speed. The cumulative effect is a transfer of energy from the beam to the field. The latter is withdrawn through an output waveguide at the other end of the tube. Power amplification a million times, or more, may be obtained from a TWT.

Fig. 17-19. X-band microwave mixer. (Courtesy of Sylvania Electric Products Inc.)

Semiconductors used in microwave work are characteristically very tiny, having very thin depletion regions. The problem is transit time, just as in vacuum tubes. By way of example, Fig. 17-19 shows a semiconductor mixer unit for X-band (5.2 to 10.9 GHz) microwaves. The major semiconductor elements are tiny Schottky barrier diodes, thinner than a human hair, which appear as "dark spots" or "gaps" in the printed circuit board in the figure. A greatly magnified view of one of these little diodes is given in Fig. 17-20.

Fig. 17-20. Schottky barrier diode, used in X-band mixer. (Courtesy of Sylvania Electric Products Inc.)

17-8 USES OF MICROWAVES

Microwaves were actually discovered by Hertz, since very short wavelengths are characteristic of his spark coil transmitter. The first significant use of microwaves, however, and the one that really gave the proverbial "shot in the arm" to the development of microwave technology, was the use of *radar* in World War II.

The word radar is an acronym—that is, it is made up from the first letters of the words "*ra*dio *d*etection *a*nd *r*anging." The principle is simplicity itself and is much like sonar except that radar uses electromagnetic microwaves instead of sound waves. Figure 17-21*a* shows conceptually how the radar transceiver sends out a pulse of microwave energy and receives back a portion of that pulse reflected by a conductive target. The direction of the beam is the direction of the target, and the distance of the target is auto-

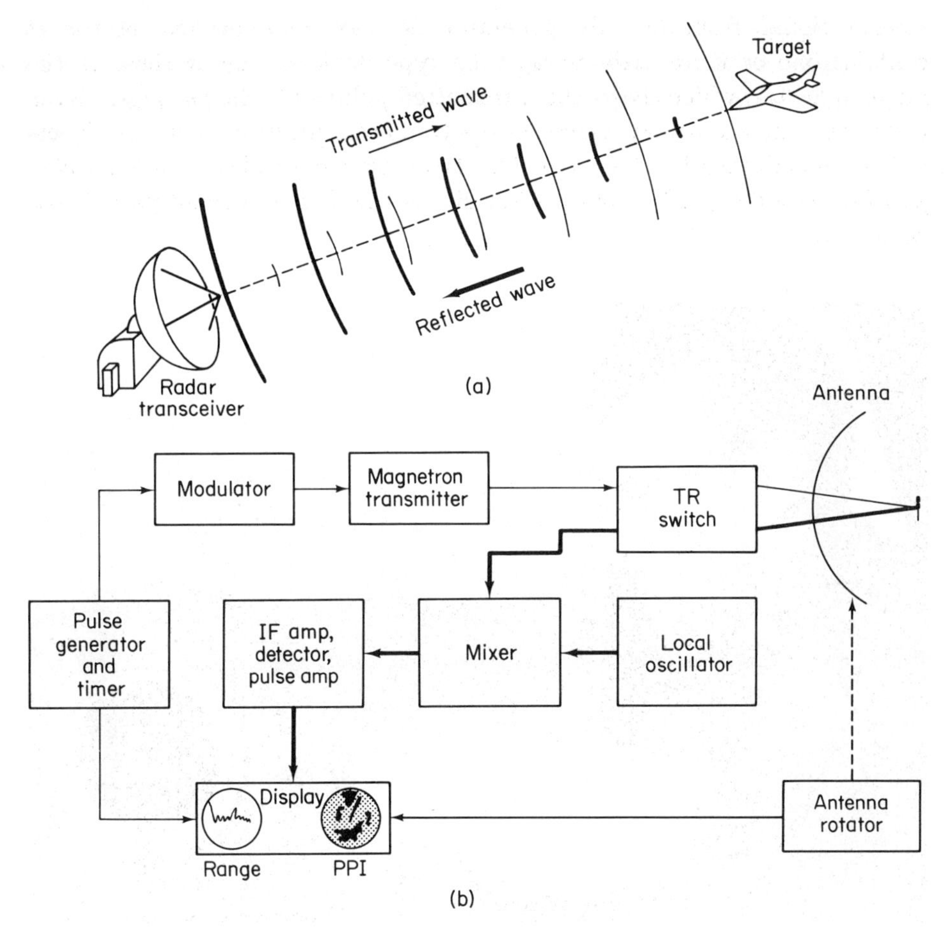

Fig. 17-21. Radar. (a) Basic principle. (b) Block diagram of pulse radar unit.

matically computed from the time between the pulse and its echo and from the known speed of the waves.

A block diagram of the typical pulsed radar transceiver is given in Fig. 17-21*b*. The heart of the unit is the pulse generator and timer which produces the pulse to be transmitted and then shuts down the transmitter to "listen." The pulses produced are impressed by the modulator on the microwave carrier which is generated by a magnetron. This signal is fed to the antenna through a *transmit-receive*, or TR, switch assembly that connects the antenna alternately to the transmitter or receiver circuitry. The returning "echo" signal is directed by the TR switch, in its turn, to the mixer stage of what amounts to a superhet receiver. A motor drive unit rotates the antenna, for scanning action. The output section is a display unit that receives signals from the pulse generator, detector, and antenna rotator. It contains one or more cathode-ray-tube type devices. One of these is the range indicator, which shows the transmitted pulse and echo as "pips." Even more common is a display where the electron beam draws a "map" of objects in the field scanned by the beam. This latter type of display is called a *plan position indicator* (PPI). Figure 17-22 shows the PPI of a small patrol boat radar unit.

Fig. 17-22. A marine radar system at work. (Courtesy of Raytheon Co.)

With the coming of the space age, microwave frequencies were adopted for communication between ground stations and spacecraft like those shown in Fig. 17-23. One type of ground-tracking antenna is illustrated in Fig. 17-24.

Another application of microwaves to space exploration is the *radio telescope*. In 1931, while working on the problem of static in ship-to-shore radio, Karl Jansky of the Bell Telephone Laboratories discovered radio emissions from outer space. Thus was born the science of radio astronomy. There was not much professional interest at first. In fact, apparently the only man in the world whose imagination was stirred was a student at Illinois Institute of Technology named Grote Reber. In 1937 Reber built

Fig. 17-23. Experimental communications satellites. (Courtesy of Raytheon Co.)

the world's first dedicated radio telescope, a 31-ft dish,[3] in his backyard in Wheaton, Illinois, and for several years he alone "listened" to the sky. Radio astronomy really came of age when Sir Alfred Charles Bernard Lovell, of the University of Manchester, directed the construction of the 250-ft "steerable" dish at Jodrell Bank Experimental Station in England. Today radio astronomers are making some of the most significant contributions to their science, working at the very frontiers of the visible universe. In addition to the pioneering Jodrell Bank facility, there are now large radio telescopes all over in the world, such as at National Radio Astronomy Observatory in Greenbank, W. Va.; at Parkes, New South Wales, Australia; and the massive nonsteerable 2000-ft dish constructed in a bowl-shaped valley in Puerto Rico.

[3] Now belonging to the National Bureau of Standards.

Fig. 17-24. Microwave tracking antenna. (Courtesy of General Electric Co.)

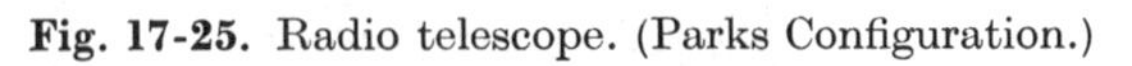

Fig. 17-25. Radio telescope. (Parks Configuration.)

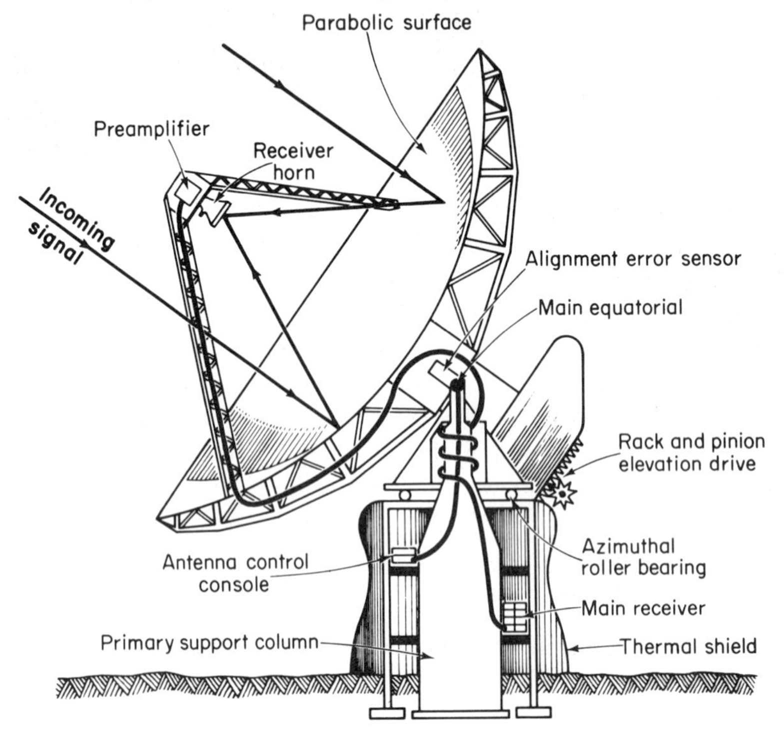

In principle, the radio telescope is nothing more than a sensitive superhet microwave receiver with a very large parabolic antenna, as shown in Fig. 17-25. Just as in case of an opitcal telescope, the bigger the reflector, the better the resolution of the telescope, hence the effort to build bigger and better dishes. Radio astronomical observations are made at frequencies throughout the microwave spectrum, but one of the most frequently studied is 1.42 GHz, which corresponds to a characteristic emission frequency of hydrogen gas, the most abundant element in the known universe.

A more homely application of microwaves arises from the fact that they will produce *dielectric* heating in vegetable and animal tissue, since it is a relatively poor conductor. This property leads easily to the development of a microwave oven, such as the one shown in Fig. 17-26, which will heat

Fig. 17-26. Radarange ® microwave oven. (Courtesy of Raytheon Co.)

a meal in seconds. The same principle, involving milder "doses" of microwaves, forms the basis for medical *diathermy* treatments, where warmth is produced within muscles to relieve spasms and pain.

17-9 MASERS AND LASERS

One of the most glamorous recent scientific developments is the *laser*, which has been touted as the science-fiction ray gun come true. While it is a fact that this device has wide applicability, supposedly including some potential as a death ray, it was originally developed out of the continuing search for amplifiers of ever-higher frequencies. Its invention increased the upper limit of amplifiable frequencies from about 120 gigahertz to over 1200 *tera*hertz, an increase of ten thousand times.

To start at the beginning, we should first understand the difference between coherent and incoherent radiation. Naturally occurring radiation is normally incoherent, which means that it consists of energy distributed over a wide band of frequencies, and even in that portion delivered at a single frequency, the waves are not in phase with each other. Coherent radiation, on the other hand, is energy radiated at a single frequency, and all the wave trains are in phase and polarized in the same plane. This difference is illustrated diagrammatically by Fig. 17-27.

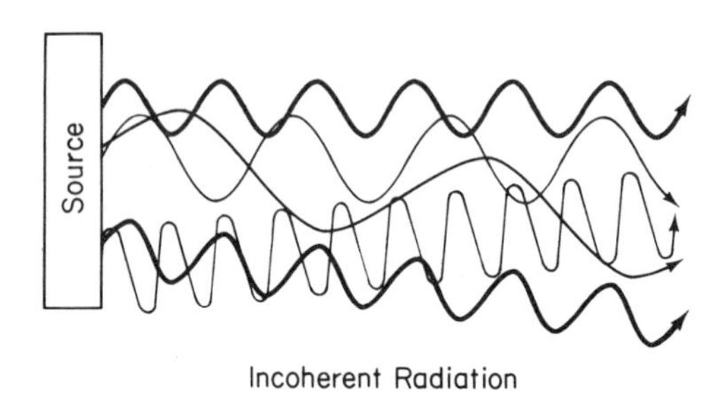

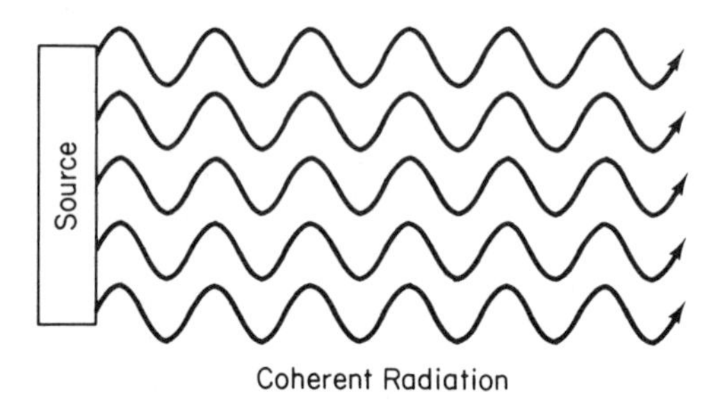

Fig. 17-27. Incoherent and coherent radiation (conceptual).

Atomic electrons can exist in a variety of energy states. In the simplified picture of the atom we have adopted, this means that each electron has a family of different orbits available to it. These are very specific and well-defined orbits for each kind of atom. Normally the electrons will be in their lowest energy states or *ground states*. Quantum mechanics tells us that if a photon of exactly the right energy (i.e., having a specific frequency) comes along, it can be absorbed by an electron, as in the left illustration of Fig. 17-28*a*, forcing the electron to jump into a higher energy orbit. This phenomenon is called *excitation*. The excited atom, if left to itself, will more or less randomly emit this acquired energy, and the electron returns to the ground state, as in the right half of Fig. 17-28*a*. If the electron has no other possible energy state between the ground state and the one to which it has been excited, it must emit a photon of the same frequency it absorbed

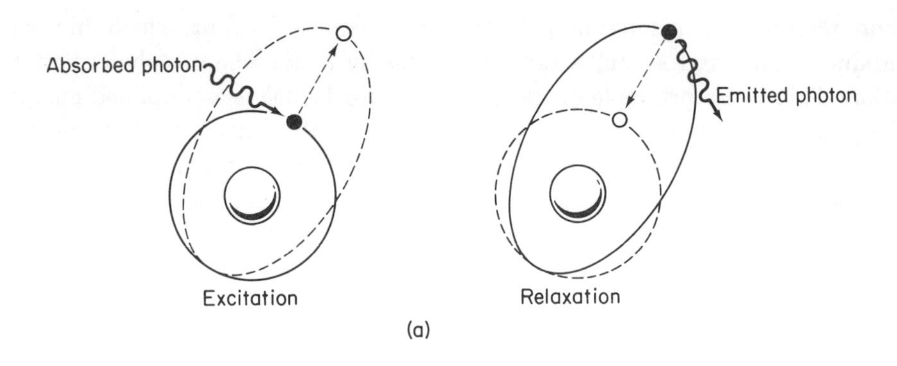

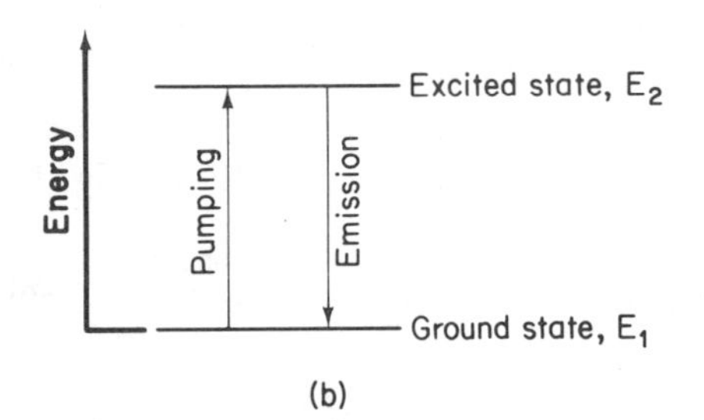

Fig. 17-28. The maser principle. (a) Excitation-relaxation phenomenon. (b) Energy level diagram.

to get back to the ground state. If an intermediate state is available, it may return by a two-step process, releasing first a photon of enough energy to get it down to that intermediate state and then emitting a second photon, whose energy makes up the difference between the intermediate and the ground states. This process of returning from an excited state to the ground state is called *relaxation*. The relaxation process can occur in two ways: it can occur haphazardly, if the atom is left alone, or it can be *forced* to occur by striking the atom with another photon of the same frequency. This second way is called *stimulated emission* and, in this case, both photons leave in the same direction and in phase.

Now, ordinarily, most atoms in a "chunk" of matter are in their lower energy states. If an incident photon happens to strike an excited atom, producing two photons by stimulated emission, both photons are quickly absorbed by the atoms in the low-energy states, and that is that. If, however, most of the atoms in the chunk of matter could be "pumped" up into the excited state, then the two photons produced by stimulated emission in one atom would, in most cases, each strike another excited atom, producing two more photons. In this manner, the two photons produce four, which, in turn, produce eight, and so on, a rapidly increasing avalanche of coherent radiation. This is the principle of the *maser*. Figure 17-28*b* is a simplified energy diagram for a two-state system. The forced excitation of the bulk of the atoms is called *pumping*. The word maser itself is an acronym from the first letters of "*m*icrowave *a*mplification by the *s*timulated *e*mission of *r*adiation." (Interestingly, although the name is an acronym, and does *not* mean "something that 'mases,'" the verb "to mase" is now a part of accepted technical language. It is also proper to say that lasers "lase.")

The first maser was developed by Charles H. Townes of Columbia University while studying gas absorption of microwaves. He found that the ammonia molecule (which consists of a nitrogen atom bound to three hydrogen atoms) had an excited state which differed from the ground state by an energy equivalent to a microwave frequency. That frequency is 23.87 GHz. It also happens that excited ammonia molecules are repelled by an intense electrostatic field, while nonexcited ones are not. These properties allowed Townes to build the maser shown schematically in Fig. 17-29. In the boiler,

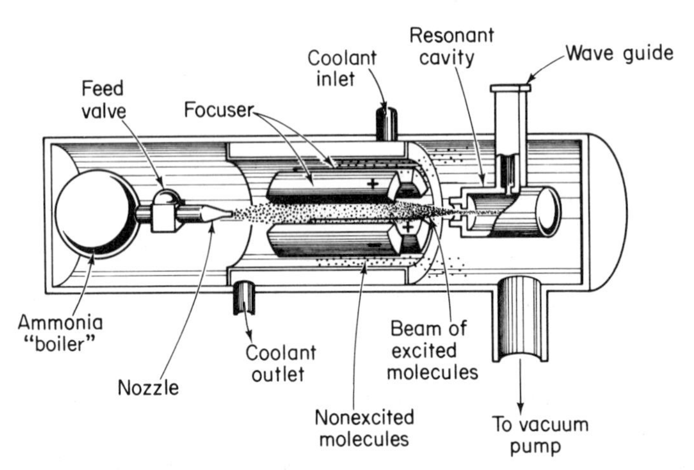

Fig. 17-29. The ammonia maser.

ammonia molecules are raised to excited states by heating to 300 °K (about 81 °F). They are injected through a nozzle into a cylindrical assembly of rods, and the rods are charged alternately positive and negative. The high-energy molecules are repelled toward the axis, forming a beam, while the low-energy molecules pass out between the rods and are eventually sucked out by the vacuum pump. The beam of excited molecules enters a cavity resonator tuned to 23.87 GHz, ready to be stimulated into emission. A weak signal of this frequency fed in by a waveguide can trigger the "masing" action, and this results in an enormously amplified output wave. Actually, even the incoming weak signal is not required, for the resonance of the cavity will cause the maser to trigger itself. The ammonia maser can thus serve either as a narrow-band amplifier or as a precise 23.87-GHz oscillator.

Other types of masers soon followed Townes' ammonia maser, which was not really practical as a microwave amplifier. Solid state masers are tunable over a range of frequencies and have better bandwidth characteristics.

Townes and A. L. Schawlow published a paper in 1958 describing a maser for "light" frequencies. Such a device is called either an *optical maser* or, more popularly, a *laser* (where the m for *m*icrowave has been replaced by l for *l*ight). The first successful laser was built by T. H. Maiman of the Hughes Aircraft Company in 1960. His active medium, which also formed the resonator, was a ruby rod.

Ruby is a transparent crystal of aluminum oxide that owes its charac-

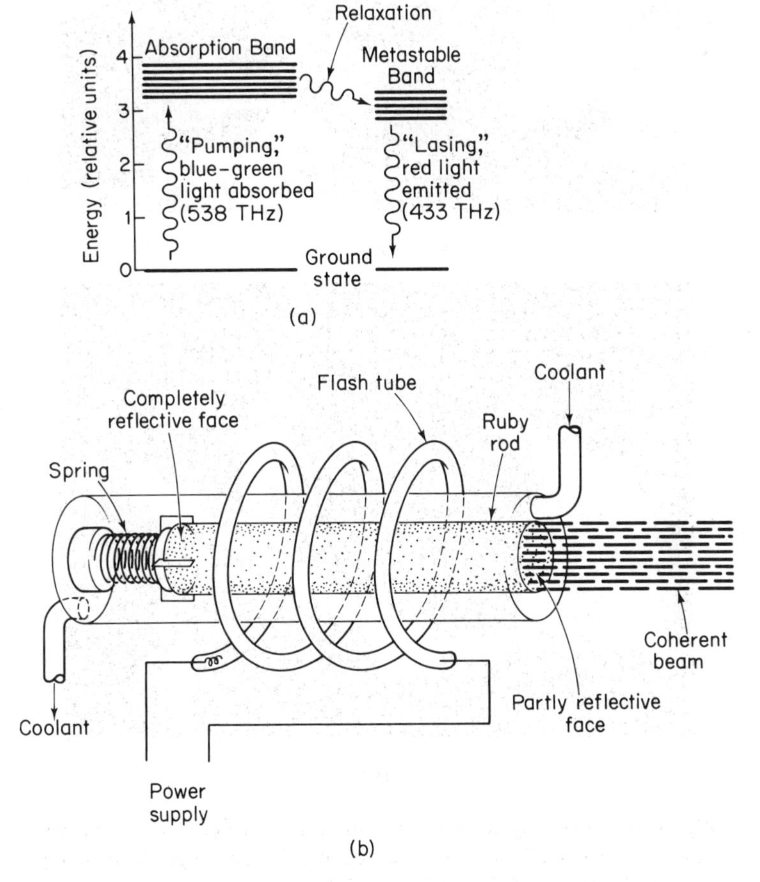

Fig. 17-30. Ruby laser. (a) Energy level diagram. (b) Structure.

teristic red color to chromium impurity atoms. It is the chromium atoms that also give the ruby its capacity to serve in a laser. The chromium atom is a three-state system, as shown in Fig. 17-30a, easily excited by photons of

(a)

(b)

Fig. 17-31. (a) Physicist makes final adjustment of argon gas laser. (Courtesy of the Aerospace Corp.) (b) Argon gas laser being tested. (Courtesy of Radio Corp. of America.)

blue-green light (at a frequency of about 538 terahertz). The excited atoms immediately relax to an intermediate energy level, in which state they can remain for several milliseconds. That is a long time on the atomic scale of events, and so this state is called *metastable*. The relaxation from the metastable state to the ground state is the useful one for laser action and is characterized by the emission of red light at about 433 terahertz.

The construction of a pulsed ruby laser is represented by Fig. 17-30*b*. The heart of this laser is a cylinder of ruby whose end faces are perfectly flat and parallel to each other. One face is completely silvered to be almost perfectly reflecting; the other face is incompletely silvered to permit the beam to leave the rod. This coated ruby rod is called a Fabry-Perot resonator. Pumping is accomplished with the coiled xenon flash tube, which is wrapped around the ruby rod. Initially, most of the chromium atoms are in their ground state. As the flash from the lamp begins, they are pumped to their higher energy state, from which they immediately relax to the metastable state. As the flash continues its pumping to the high-energy state, some of the chromium atoms already in the metastable state spontaneously emit photons and the rod begins to glow red. By the time the flash is completed, a point has been reached where the number of chromium atoms in the metastable state exceeds the number in the ground state. Under these circumstances, the randomly emitted photons can act as triggers, stimulating other atoms in the metastable state to emit. Those photons emitted parallel to the axis of the ruby rod bounce back and forth between the silvered reflecting faces, while those emitted in any other direction immediately leak out of the ruby. The beam moving back and forth parallel to the axis stimulates more and more photons to emit, always in the same direction as the beam and in phase with it. Finally, the intensity becomes so great that it "bursts" through the one silvered face that is slightly transparent, in a huge flash of coherent light.

Development of laser technology has been very rapid since Maiman's original ruby laser. Both gas lasers, such as the argon laser shown in Fig. 17-31, and solid state lasers have been proven practical. They have applications in welding, surgery, cancer treatment, guidance and detection, microbiology, metrology, and in a host of other areas as yet unimagined.

REVIEW QUESTIONS

1. Why are the imperfections of the human eye essential to the success of television?
2. What is a TV camera tube required to do?

3. Describe the process of video transmission.
4. Why do you suppose the TV video signal is amplitude modulated onto the carrier while the audio portion is frequency modulated?
5. Why is the sync generator considered the most important unit in the TV transmitter?
6. Color TV receivers have more control knobs than black-and-white sets. Why?
7. What characteristics are common to both monochrome and color TV?
8. How would you build a color TV camera without dichroic mirrors?
9. How does the monochrome receiver produce a black-and-white image from a color broadcast?
10. What is a subcarrier, and why is it used in color television?
11. Why is a shadow mask needed in a color kinescope?
12. What accounts for the unusual behavior of radio waves at microwave frequencies?
13. To what does the "skin effect" refer (besides burlesque shows)?
14. How is microwave energy conducted in a waveguide?
15. Why is an external magnetic field needed in the magnetron and TWT, but not in the klystron?
16. What are the major differences between an ordinary radio transceiver and a pulsed radar unit?
17. What properties must a substance have in order to be usable as a medium for a maser?
18. What characteristic of the maser makes it a suitable basis for an "atomic clock"?

18

ELECTRONIC DOCTOR, LAWYER, MERCHANT, CHIEF . . .

18-1 APPLICATIONS FILE

Surely the list of all possible applications of electronics is infinitely long! We could not even begin to enumerate them. However, it would be useful to have a reasonably small number of descriptive categories, in which we could group all the applications we come across. Even this more modest goal may not be completely attainable, but we can try. One such list of application categories might be:

1. Automation/Electromechanization/Automatic Control
2. Data Acquisition/Recording
3. Data Processing/Computation
4. Communication
5. Simulation
6. Electronic Phenomenon Itself/Miscellaneous

In this final chapter, we are going to flit like a hummingbird from one application of electronics to another, in fields as diverse as medicine, defense, and ladies' fashion. Hopefully, we will be able to fit all the applications we encounter into one of these six categories without overworking that catch-all sixth "miscellaneous" category.

In the first category, electromechanization, we will place applications where either (1) some process that previously used another source of energy was converted to operate from electric energy, or (2) electronic devices were incorporated in a process to make it self-controlling. We extend the word "process" here to include metabolic functions in a living organism.

The second category, data acquisition, is taken to include all applications where electronic sensors are used to acquire data that would otherwise be difficult or impossible to obtain and/or where electronic recording devices are used to preserve data. "Data" here include human speech and photographic images.

The next category, data processing, covers all phases of handling data from "shaping" the signal out of a particular transducer, so that it will suit a particualr display device, to mathematical calculations. This category, especially, is the domain of the electronic computer.

Most of the "hardware" associated with the fourth category, communication, is already familiar to us from previous chapters. Here we intend to look briefly at some perhaps unexpected uses of communications devices.

Simulation, the fifth category, describes a research or training process in which certain artificial (in this case, electronic) activities are used to represent activities in the real situation that are too dangerous, expensive, or otherwise cumbersome for that research or training process. Simulation is used extensively where human life and/or expensive equipment are at stake in the real situation.

The sixth category we reserve for cases where the electronic phenomenon is the objective in itself or any application that does not fit into at least one of the other categories. Many items in this category represent pure electronics research.

We now proceed into a dazzling kaleidoscope of electronic technology at work. To prepare for our encounter with the giant computers, we study first one of the major developments that made them possible, *microminiaturization.*

18-2 MAKING LITTLE ONES OUT OF BIG ONES

As electronic devices become more and more sophisticated, more and more components are similarly required to construct them. The difference in capability between the old crystal radio receiver and a big digital computer is of about the same order as the difference between an amoeba and a gorilla. Yet, if their sizes were also proportionate, the computer should weigh about 3 million tons and take up as much space as 2 million rooms, each the size of an average bedroom! The reason that computers, and other complex electronic devices, have not assumed such ridiculous proportions is that fantastic strides have been made in miniaturizing electronic circuitry. In spacecraft and missiles, in particular, where space and weight are always at a premium, the demand for light, compact electronics has spurred the development of *microminiature* semiconductor circuit technology.

The first steps in miniaturization consisted of trying to build little components just like the big ones. Fortunately, most electronic circuits do not consume or handle much power. Typical power levels are in the milliwatt range. This fact made it indeed feasible to build tiny resistors, capacitors, inductors, and so on. Such conventional *discrete components* were then tightly packed and encapsulated (sealed in a solid insulating material) in high-density modules, like the one illustrated in Fig. 18-1. This degree of size reduction is about as far as you can go with conventional components.

Fig. 18-1. High-density module using conventional miniature components. (Courtesy of Sprague Electric Co.)

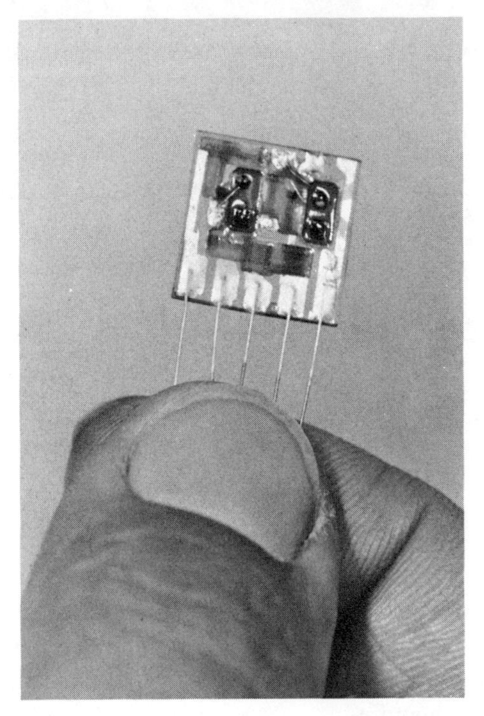

Fig. 18-2. Thin-film integrated circuit on ceramic substrate. (Courtesy of Sprague Electric Co.)

The next major step was the development of the solid state *integrated circuit*, or IC. In ICs, the individual components are not recognizable as such, but are functionally formed by the junctions of various layers of conductor and semiconductor materials. Figure 18-2 shows an IC formed by depositing thin films of such materials on a ceramic "substrate" (base material).

The present limit of miniaturization is represented by the *microcircuit*. Large numbers of identical microcircuits are "grown" into a small wafer of semiconductor material by a lengthy and ticklish process of masking, local doping, and plating, A typical finished wafer containing 36 microcircuits is shown in Fig. 18-3. Note the size of a standard paper clip by comparison.

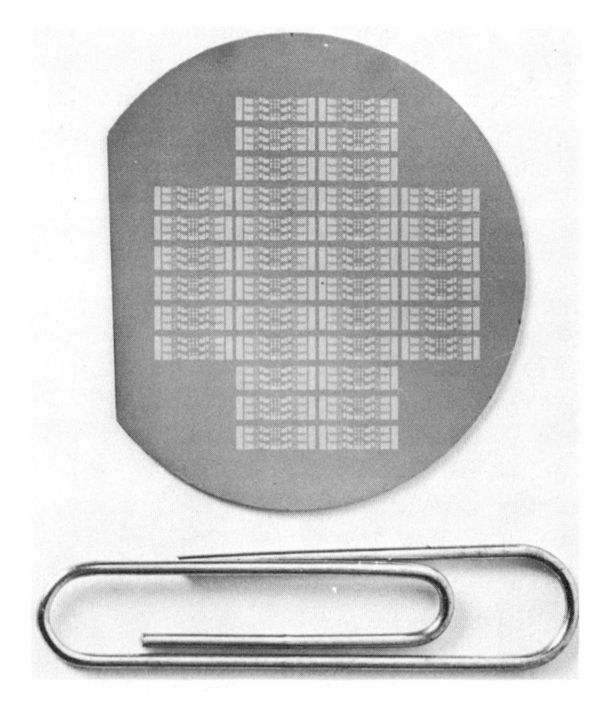

Fig. 18-3. Thirty-six microcircuits on a single silicon wafer. (Courtesy of Sprague Electric Co.)

Several such little ICs can, in turn, be combined into a module like that in Fig. 18-4*a*. At this degree of miniaturization, the size of contacts and lead wires becomes very significant. It is widely believed, in fact, that the miniaturization limit is rapidly being reached because of the minimum volume required by a bundle of connecting wires. A great deal of skill and patience is required in working with these tiny components, as depicted in Fig. 18-4 *b* and *c*.

The main structural features of the various types of component junctions used in integrated circuits are illustrated in Fig. 18-5*a*. A segment of pure semiconductor between two metallic contacts forms a resistor. Two broad segments of metallic film act as the plates of a capacitor with a P-N junction serving as the dielectric (in use, this junction must be reverse-biased, of course): By "cutting away" part of one plate, as in the third

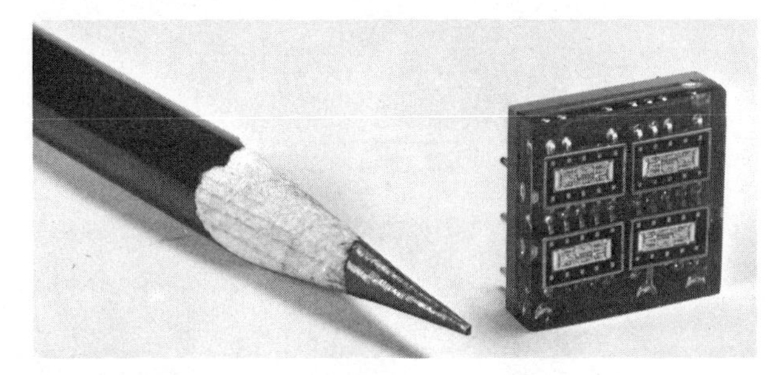

Fig. 18-4. (a) Completed integrated circuit module contains four IC's interconnected. (Courtesy of Sprague Electric Co.)

Fig. 18-4. (b) Assembling a subminiature telemetry transmitter. (Courtesy of the Aerospace Corp.)

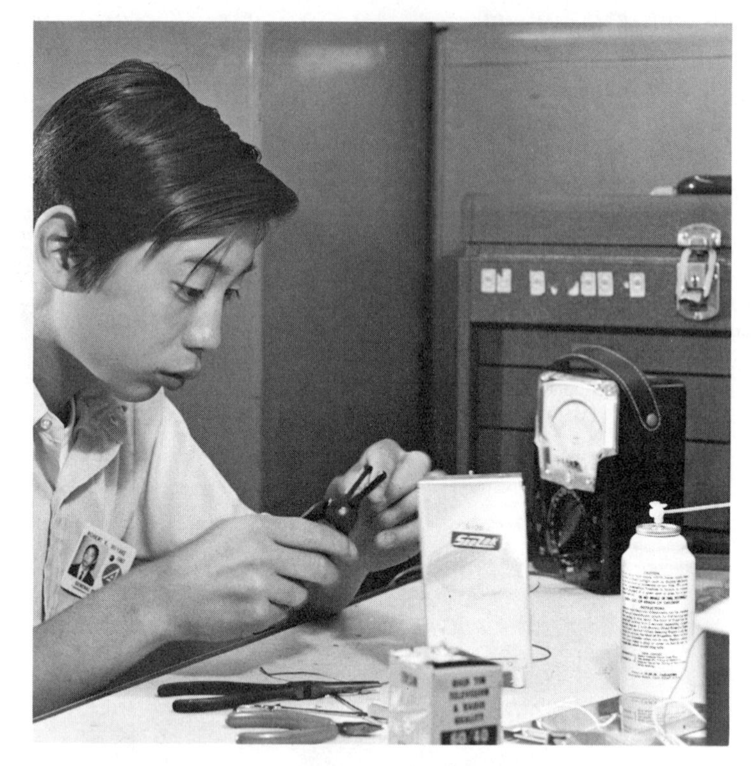

Fig. 18-4. (c) Completing a portable anemometer with IC amplifier. (Courtesy of the Aerospace Corp.)

illustration of Fig. 18-5*a*, we obtain an IC with the characteristics of a resistor-capacitor network. The fourth illustration shows a junction transistor formed by contact of two separated zones of one type of semiconductor with a common zone of the other type. A diode is, effectively, "half of a transistor." Figure 18-5*b* shows a sectioned view of the construction of a simple IC with

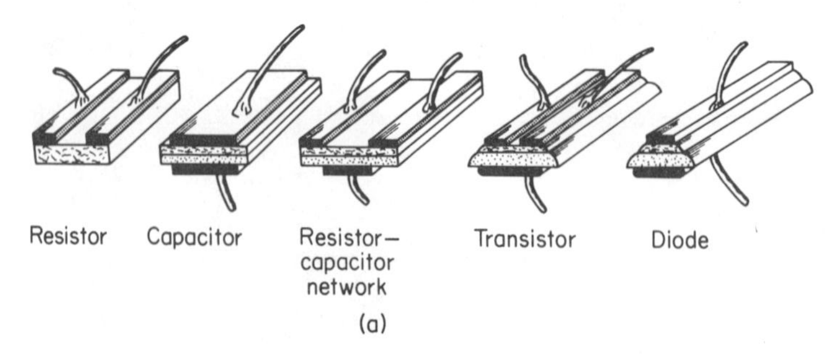

Fig. 18-5. The integrated circuit. (a) How various elements are formed.

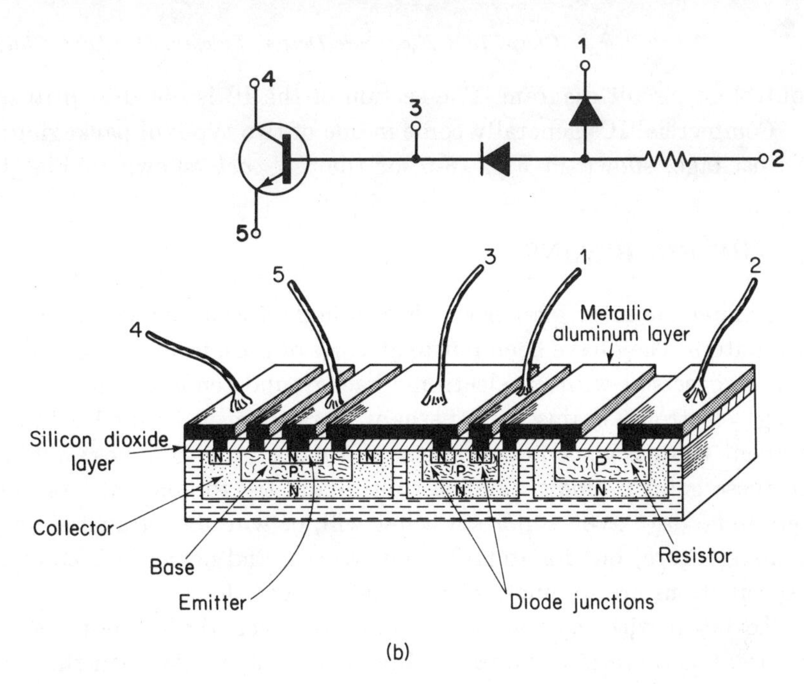

Fig. 18-5. (b) A simple integrated circuit (sectioned view) with its equivalent schematic.

Fig. 18-6. Integrated circuit, dual in-line package. (Courtesy of Radio Corp. of America.)

Fig. 18-7. Integrated circuit, 14-lead *flat pack*. (Courtesy of Radio Corp. of America.)

its equivalent circuit diagram. The section of the IC is not drawn to scale.

Commercial ICs generally come in one of two types of packaging: the *in-line* package, shown in Fig. 18-6, or the *flat pack*, shown in Fig. 18-7.

18-3 THE BIG BRAINS

Among the most glamorous electronic devices in the world today are the computers. They have been put to all sorts of uses, from rating to dating. The average person with no electronic background tends to consider them thinking "electronic brains" and perhaps fears them just a little. After all, what about those science-fiction stories of robots running amuck? Might not one of these giant brains get out of control and try to dominate its human masters some day? We cannot say what will, or will not, be a possibility in the remote future, but for today's, tomorrow's, and next year's computers such speculations are so much melodramatic hogwash.

To begin with, computers are morons. Even the lowliest mammal, the mouse, has a brain of immeasurably greater capability than the largest and most sophisticated computer. So, too, do many of the lower animals. The question of whether a computer can think at all is hotly debated, and in the final analysis depends on your definition of "think." A computer can determine that (1) given two collections of two each, when put together they "form a collection of four," or (2) given that all metals conduct electricity and iron is a metal "implies that iron conducts electricity." If arriving at these arithmetic and logical conclusions fits into your definition of "think," then computers can indeed think. There is not, however, today, nor will there be tomorrow, and possibly there will never be, a computer that can train a set of sensors on the world and, at its own volition compose "Macbeth," or "The Iliad," or "La Traviata," or "Basin Street Blues." Where, then, lies the wonder of the computer? The answer is its lightning *speed* and unerring *accuracy* in doing the simple things of which it is capable.

There are two general types of computer: the *analog computer* and the *digital computer*. Sometimes digital and analog equipment are used together and the resulting complex of equipment is called a *hybrid computer*. The two remain essentially separate, however, requiring special "translating" circuitry to interconnect them, for analog and digital computers do not speak each other's language.

In principle, they are greatly different. The digital computer *calulates the answer* to the problem; the analog computer *simulates the problem itself*. Stated in another way, the digital computer *does the arithmetic* called for by the mathematical equations that describe the problem, much as a man might do with paper and pencil; the analog computer, on the other hand, *is* the problem. Furthermore, the digital computer is capable of unlimited precision, while that of the analog computer is limited. Thus a digital computer can easily carry on calculations and print out an answer to 16 decimal places (the last 13 places of which may be worthless numbers because the input data we gave the machine to work with were only accurate to three places). The answers from an analog computer can only be read to three, or at best four, significant figures.

We illustrate the difference between analog and digital solutions of a problem with an extremely simplified example. Consider the situation shown in Fig. 18-8*a*. A cold liquid at 0 °C is separated from a hotter liquid at T °C

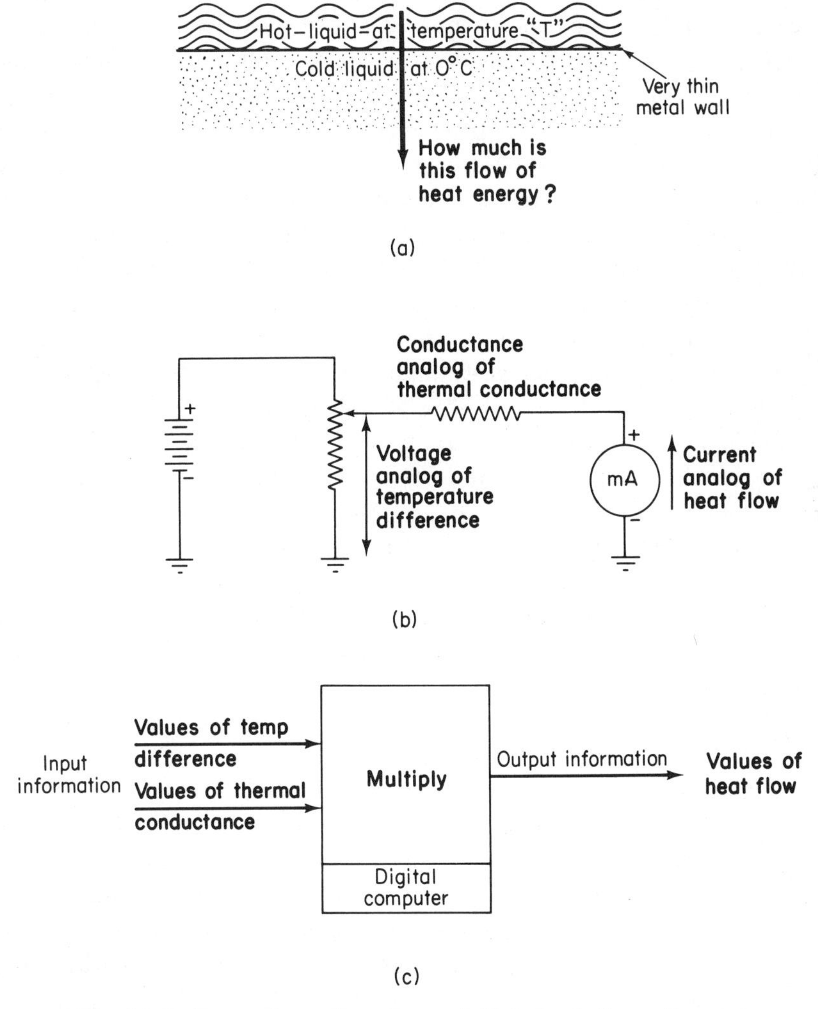

Fig. 18-8. Comparison of analog and digital solutions of a problem. (a) The problem. (b) Analog solution. (c) Digital solution.

(T is the temperature we are going to vary) by a thin wall of metal that is a good conductor of heat. The question to be answered is: What is the flow of heat energy through the wall when the hot liquid is at various temperatures represented by T?

For this simple problem, the laws of heat transfer tell us that flow of heat energy can be found by multiplying the temperature difference between the two liquids by a constant called the thermal conductance, which depends on the materials and conditions involved. The analog computer solution to the problem would go something like this: The heat transfer law, "heat flow = temperature difference × thermal conductance," has the same form as Ohm's law, "current flow = voltage difference × electric conductance." We can therefore build an electric circuit *model* of the heat transfer equation that looks like Fig. 18-8*b*. The voltage that is "picked off" at the potentiometer is a fraction of the supply voltage and is numerically equal to the desired temperature, T, of the hot liquid, in degrees Celsius. The effective thermal conductance for the real problem would be known from the materials and conditions of the problem. The series resistor in our analog model is chosen so that its electrical conductance (in millimhos) is the same numerically as that effective thermal conductance, say in units of Calories[1] per hour per square meter per degree Celsius. Ohm's law then tells us that the milliammeter will read a current equal to the voltage applied to the resistor times its conductance. This current is the electrical analog of the heat flow whose value we are trying to determine. The number of milliamperes of current flowing in our model is equal to the number of Calories per hour flowing through each square meter of the thin metal wall in our problem; we read the answer to the problem from the milliammeter. By adjusting the setting of the potentiometer, we are effectively changing the temperature, T, in our analog model, and we can read the corresponding heat flow from the milliammeter. The analog solution is therefore a "working model" of the problem in which measurable electrical variables take the place of the physical variables in the real problem.

The digital solution to the problem is represented in Fig. 18-8*c*. The digital "computer" indicated here can be any device that can have two numbers put into it and will then multiply them together and print out this number. Here, instead of simulating the problem, we take the mathematical equation that results from a theoretical analysis of the problem and put appropriate numbers into the equation to calculate the answer. The digital computer is thus just a calculating machine that requires two kinds of imputs: a set of instructions, called the *program*, which tell it what opera-

[1] A Calorie (or kilocalorie) is a unit of heat energy—enough to make one kilogram of water hotter by one degree Celsius.

tions to perform; and the *data*, upon which it is to perform these operations. In our simple problem, the program would be equivalent to

1. read a "temperature"
2. read a "thermal conductance"
3. multiply "temperature" times "thermal conductance"
4. print out answer
5. repeat the preceding steps until out of data

The computer could be just a desk calculator, your fingers performing the "program" of inputting the numbers and pressing the "multiply" button.

Either the analog or digital computer, by itself, deserves an entire textbook for its explanation. Yet, although our text is limited in space, it is well worth our while to delve just a little deeper, since so many of our applications employ computers. The next few sections are therefore devoted to these electronic "brains."

18–4 ANALOG COMPUTERS

The analog computer, as we have already seen, is not really a computer in the sense of a calculating machine. Rather, it is a kind of do-it-yourself simulation kit—a sort of Tinkertoy set of functional modules which you (the user) connect together to make a working model of the problem to be solved.

The analog computer is especially useful for solving one class of problems, *dynamics* problems. These are, roughly speaking, problems in which physical variables (like forces, distances, angles, illuminations) are *changing with time*. You can see that, although this is indeed only one class of problems, it is a very broad and important class with far-reaching applications. Mathematically speaking, these are problems that are described by a type of equation called a *differential equation*. A differential equation tells how the variables of interest are changing (usually, with time), rather than giving the values of those variables. The process of finding the instantaneous values of the variables themselves, knowing the equation for how they are changing, is called *integration*. Integration is the specialty of the analog computer, the capability that puts it miles ahead of the digital computer in solving dynamics problems or any other problem that mathematically has the same form as a dynamics problem.

The first workable analog computer was built by Vannevar Bush and his associates at MIT in 1925. Early analog computers were partly electric and partly mechanical, whereas modern machines are completely electronic. The variables studied are voltages and how they vary with time. In any particular problem, a specific voltage will represent a force, a distance, etc. The most fundamental device in the analog computer is the *operational amplifier*. This is a special kind of amplifier that must meet some rather amazing requirements:

1. Extremely high gain (ideally "infinite gain"), a typical value being 10^8.
2. High-input impedance and low-output impedance.
3. Good frequency response (ideally "flat" at all frequencies), typically ± 1 db and $1°$ phase shift from 0 to 10 kilohertz.
4. Minimum voltage drift (ideally, zero), and high stability against a wide range of loading conditions.
5. Adequate output power, typically 2 watts for a vacuum-tube amplifier.

Besides serving as a basic amplifier, the operational amplifier can be made to perform a variety other operations. With a suitable choice of input and feedback impedances, the operational amplifier becomes an *inverter*, which changes the sign of the input signal (mathematically equivalent to multiplying by -1); an *adder* or *summer*, whose output is the sum of the two or more input signals; or an *integrator*, to perform the integration process described previously. Since it is frequently required that a signal be changed in level, a number of attenuator potentiometers are provided. These are often called *coefficient potentiometers*, or simply "pots," and are used to reduce the signal by some factor from 0.0000 to 0.9999. In conjunction with an amplifier, the coefficient potentiometer can, in effect, multiply a signal by any desired constant value. Other devices commonly used with the basic analog computer are *multipliers*, to multiply two signals together; *resolvers*, to break up vectors into sine and cosine components; and *function generators*, electromechanical or photoelectric devices that can generate a signal approximating the behavior of any desired mathematical function over a portion of its range.

A modern analog computer is illustrated in Fig. 18-9. The amplifiers, adders, integrators, pots, etc., are connected to terminals in the *patchboard* shown in the center of the unit. The operator sets up his problem by interconnecting appropriate functional modules with plug-in connector wires, much like a telephone switchboard is operated. The potentiometers are set using the row of black knobs above the patchboard area. Instantaneous voltages (including those used to check pot settings) are read on the digital voltmeter (DVM) on the left above the potentiometers.

The output voltages, the "answers" to the problem, are usually recorded on some sort of a chart, such as a strip chart or an oscillograph like that shown in Fig. 18-10. Each oscillograph "channel" is essentially a mirror

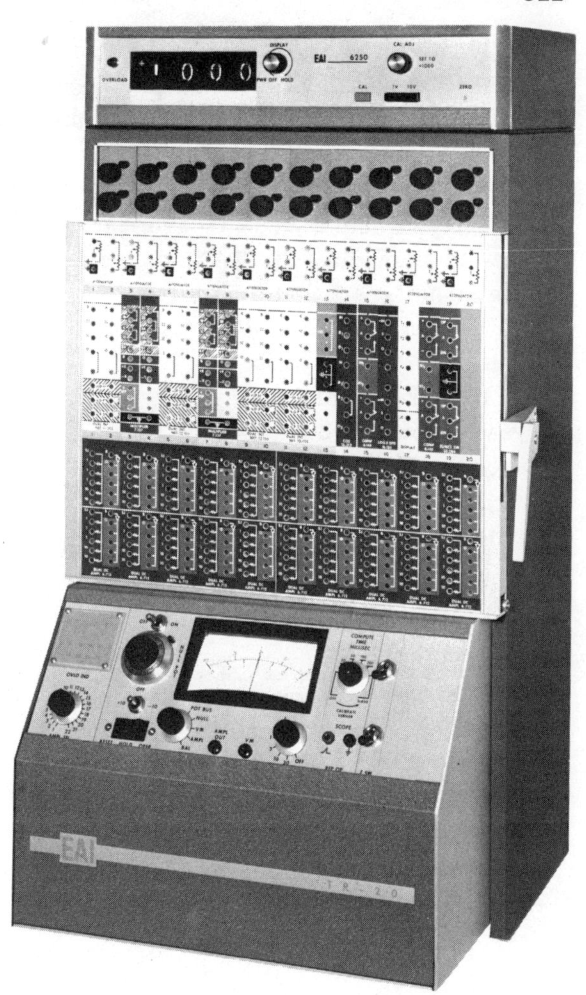

Fig. 18-9. PACE TR-20 analog computer. (Courtesy of Electronic Associates, Inc.)

galvanometer that uses a light beam to trace its deflection on a continuously moving belt of light sensitive paper. The oscillograph thus provides a graph called an *oscillogram* of how each output voltage varies with time. A grid of lines is also automatically printed on the oscillogram so that the deflections and time scale can be read off easily. Sometimes it is of greater interest to know how one voltage varies with another, say a force with a distance, instead of how either one varies individually with time. A useful read-out device for this purpose is an *x-y plotter*, such as the one shown in Fig. 18-11.

In addition to the fact that it is a dynamics problem specialist, the analog computer has certain other distinguishing characteristics. Its accuracy is limited, usually to something between 0.1 and 1 percent error. This accuracy limitation is inherent in using components that, although of the highest attainable quality, are not perfect. However, this degree of accuracy is generally more than adequate in typical dynamics problems. The analog computer is a poor device for doing arithmetic. Addition and subtraction are easy for it, but subject to the accuracy limit just discussed. Multiplication is more difficult, requiring servo multipliers that cannot operate at very high speed. Division is not done, unless absolutely unavoidable, and requires special accessory equipment. It is for this reason that *hybrid* computers, like the facility shown in Fig. 18-12, were developed. In these facilities, analog and digital units each perform the functions they do best, and then "communicate" the results to each other through analog-to-digital (AD) and digital-to-analog (DA) converters.

It is important to remember that the output of an analog computer is a reading on a *continuous* scale (like a pointer against a meter scale or a graph on an oscillogram) with a finite accuracy. One often talks about "analog variable" or "analog measurements" in this sense, even though no analog computer is involved.

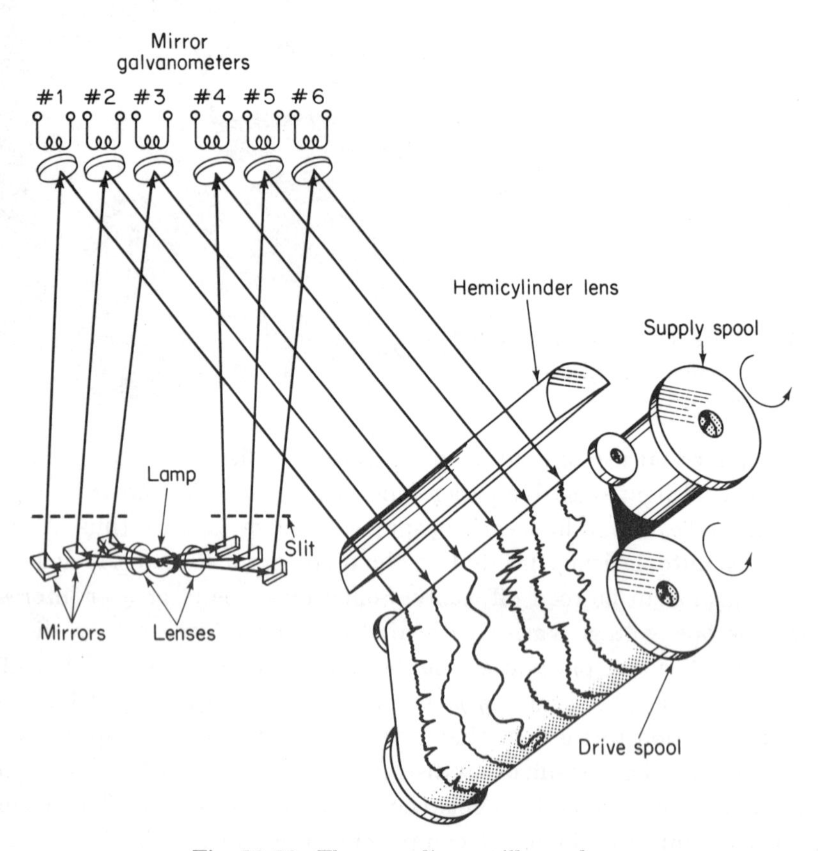

Fig. 18-10. The recording oscillograph.

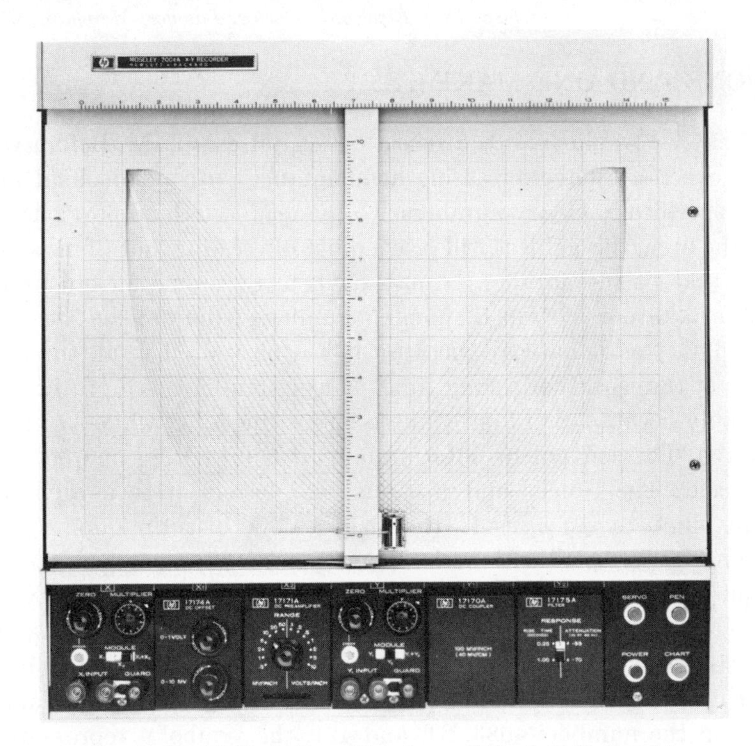

Fig. 18-11. X-Y plotter. (Courtesy of Hewlett-Packard.)

Fig. 18-12. Model 690 Analog/Hybrid Computing System. (Courtesy of Electronic Associates, Inc.)

18-5 ONE AND ONE MAKE TEN?

Next we would like to discuss the digital computer. Before we can, however, we must have a passing acquaintance with nondecimal number systems, specifically binary numbers. To the reader with no previous experience in the area, this subject will prove quite interesting, but (if his schooling was like that of the author) he may have to overcome an old prejudice.

We take our system of naming numbers—the *decimal* system—for granted. It is "most natural" because it has been used in all the civilized countries of the world for a long time. This system is based on the number 10, probably because man first counted on his fingers, of which he happened to have ten. The convenient notation now used in writing numbers is relatively recent. The Greeks and Romans used letters of their alphabets to represent numbers. As a result, they had a devil of a time doing what we would consider "simple" arithmetic. Thus a seasoned Roman merchant would spend quite some time with stylus and tablet to total up DLXII, CCXXXIX, and MCVIII, while any fifth grader can bat out the sum of 562, 239, and 1108 in a few seconds. The major "breakthroughs" in our number system notation were the inventions of the symbol *zero* and of assigning positional value. For instance, in the numbers 4083, 24, and 421, the symbol 4 represents three different quantities. In 4083, the 4 represents four thousands; in 24, it stands for just four units; and in 421, it means four hundreds. In other words, we have a position value system that uses the ten symbols 0, 1, 2, 3, 4, 5, 6, 7, 8, 9, and, depending on where they are placed with respect to the decimal point (whether it's shown or not), they represent

	(tens of thousands)	(thousands)	(hundreds)	(tens)	(units)	decimal point ↓
. . .	$\times 10^4$	$\times 10^3$	$\times 10^2$	$\times 10^1$	$\times 10^0 (= \times 1)$.	

and so on. Thus when we write the number symbol 4083, we mean

$4 \times 10^3 = 4 \times (10 \times 10 \times 10)$	= 4 thousands	=	4000
plus $0 \times 10^2 = 0 \times (10 \times 10)$	= no hundreds	=	000
plus $8 \times 10^1 = 8 \times (10)$	= 8 tens (eighty)	=	80
plus $3 \times 10^0 = 3 \times (1)$	= 3 units	=	3
	Totals		4083

Now, although we use ten symbols and the positions keep multiplying by ten as we go left from the decimal point, this is only a matter of historical usage and there is nothing sacred about ten. We could just as easily use the six symbols 0, 1, 2, 3, 4, 5, and have the places increase to the left by powers of 6. As a matter of fact, a number system can be built on any integral number greater than one. Several of these actually have been used at one time or another.

The number system that interests us now is the simplest of all, the *binary* system, based on the number two. It uses, of course, only the two symbols, 0 and 1. Successive positions in a written number therefore increase by powers of two (instead of ten) from the "binary" point

	(sixteens)	(eights)	(fours)	(twos)	(units) ↓ binary point
. . .	$\times 2^4$	$\times 2^3$	$\times 2^2$	$\times 2^1$	$\times 2^0 \; (= \times 1)$.

The binary number 110101 therefore means

	$1 \times (2^5) = 1 \times (2 \times 2 \times 2 \times 2 \times 2)$	$= 1 \times 32$	$=$	32
plus	$1 \times (2^4) = 1 \times (2 \times 2 \times 2 \times 2)$	$= 1 \times 16$	$=$	16
plus	$0 \times (2^3) = 0 \times (2 \times 2 \times 2)$	$= 0 \times 8$	$=$	0
plus	$1 \times (2^2) = 1 \times (2 \times 2)$	$= 1 \times 4$	$=$	4
plus	$0 \times (2^1) = 0 \times (2)$	$= 0 \times 2$	$=$	0
plus	$1 \times (2^0) = 1 \times (1)$	$= 1 \times 1$	$=$	1
		Totals		53

In other words, 110101 and 53 are two different ways of writing the same number. The decimal way, 53, is the one we are used to seeing, but the binary way, 110101, is just as "correct" and, in some cases, more useful. The first ten numbers are shown in Table 18-1 in both binary and decimal form. Note that the binary form of a number is longer than the decimal form. This is due to the fact that only the two symbols, 0 and 1, can be used in each position, instead of the ten symbols we use in our everyday decimal system. These symbols, 0 and 1, are called *bits*, a contraction of *bi*nary digi*ts*. A typical number in binary form contains between 3 and 4 times as many digits as it does in decimal form. This is a serious drawback to using the binary system, *for humans*, but it doesn't bother a large digital computer. One great advantage of the binary system is the simplicity of its arithmetic. There are only four basic additions to remember

$$\begin{array}{r}0\\+\,0\\\hline 0\end{array}\qquad\begin{array}{r}0\\+\,1\\\hline 1\end{array}\qquad\begin{array}{r}1\\+\,0\\\hline 1\end{array}\qquad\begin{array}{r}1\\+\,1\\\hline 10\end{array}$$

Table 18-1 Comparison of Some Binary and Decimal Numbers

Decimal	Binary
0	0
1	1
2	10
3	11
4	100
5	101
6	110
7	111
8	1000
9	1001
10	1010

+	0	1	2	3	4	5	6	7	8	9
0	0	1	2	3	4	5	6	7	8	9
1	1	2	3	4	5	6	7	8	9	10
2	2	3	4	5	6	7	8	9	10	11
3	3	4	5	6	7	8	9	10	11	12
4	4	5	6	7	8	9	10	11	12	13
5	5	6	7	8	9	10	11	12	13	14
6	6	7	8	9	10	11	12	13	14	15
7	7	8	9	10	11	12	13	14	15	16
8	8	9	10	11	12	13	14	15	16	17
9	9	10	11	12	13	14	15	16	17	18

Decimal

+	0	1
0	0	1
1	1	10

Binary

(a)

X	0	1	2	3	4	5	6	7	8	9
0	0	0	0	0	0	0	0	0	0	0
1	0	1	2	3	4	5	6	7	8	9
2	0	2	4	6	8	10	12	14	16	18
3	0	3	6	9	12	15	18	21	24	27
4	0	4	8	12	16	20	24	28	32	36
5	0	5	10	15	20	25	30	35	40	45
6	0	6	12	18	24	30	36	42	48	54
7	0	7	14	21	28	35	42	49	56	63
8	0	8	16	24	32	40	48	56	64	72
9	0	9	18	27	36	45	54	63	72	81

Decimal

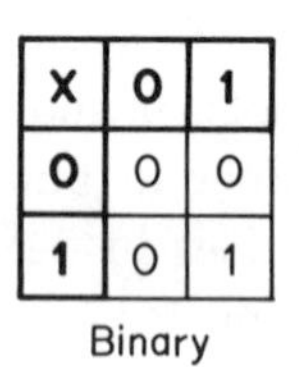

X	0	1
0	0	0
1	0	1

Binary

(b)

Fig. 18-13. Comparison of decimal and binary arithmetic tables. (a) Addition. (b) Multiplication.

The addition on the far right still says "one and one make two" (not "ten"). Remember that the symbol 2 does not exist in binary arithmetic and instead we write 1 two and 0 units, or 10. Binary multiplication is also very simple, there being only four basic products which look just like ordinary (decimal) multiplications

$$\begin{array}{r}0\\ \times\,0\\ \hline 0\end{array}\qquad\begin{array}{r}0\\ \times\,1\\ \hline 0\end{array}\qquad\begin{array}{r}1\\ \times\,0\\ \hline 0\end{array}\qquad\begin{array}{r}1\\ \times\,1\\ \hline 1\end{array}$$

Figure 18-13 shows a comparison of the addition and multiplication tables in the two systems. As children in grammar school, we were drilled in the decimal system tables on the left of Fig. 18-13. How much easier to remember the binary arithmetic tables at the right!

Another advantage of binary arithmetic is that it can be used in logic problems. The symbol 1 is then used to represent "true" and 0 to represent false." This arithmetic of logic is called *Boölean algebra*, after its inventor, George Boöle.

18-6 ELECTRONIC ARITHMETIC

The digital computer, as we have noted, is a calculating machine. It works directly with numbers. Perhaps the first digital calculator was the abacus, known in the Orient at least as early as 600 B.C. Designs for completely mechanical devices to perform arithmetic operations did not appear, however, until the late 1700s, and only modern precision manufacturing techniques made the desk calculator a reality. Finally, the expanding technology from World War II led to the development of the electronic digital computer.

The digital computer does binary arithmetic. This is most suitable because it requires only the two digits, 0 and 1, which can easily be represented electrically by any device having an "off" and an "on" state. Such a device, which has two stable states, is called *bistable*.

Why not, you ask, design the computer to operate in the decimal system? Couldn't we use ten different voltage levels to represent the digits from 0 to 9? For instance, let 0 volts represent zero, 1 volt represent one, 2 volts represent two, etc. In theory, such a ten-base computer could indeed be built. The problem is that electronic components are not perfect and tend to drift a little. Consequently, the voltage that should have been 7.0 volts, representing the number 7, may drift up to, say, 7.1 volts. If involved in a multiplication by 1000, the answer would come out 7100 instead of 7000, and this error would quickly propagate. Or, suppose that the voltage representing "7" dropped down to 6.5 volts. How would the operating circuit know whether that represented a "high six" or a "low seven"? From this short discussion, it should be evident that monumental problems are involved in designing a computer with thousands of reliable circuits, each having ten stable states. It is relatively easy, however, to distinguish reliably between the presence (1) or complete absence (0) of a signal. Hence the use of binary numbers and bistable circuits.

There are many types of bistable circuits. An important one, and one that is useful in gaining an elementary understanding of digital counting devices, is the Eccles-Jordan multivibrator circuit, more popularly called the *flip-flop*. Tube and transistor versions of the flip-flop are shown in Fig. 18-14. Both circuits operate in the same manner, except that the roles played by positive and negative pulses are reversed. We will describe the operation of the vacuum-tube circuit. The reader can easily reinterpret the explanation in terms of the transistor circuit.

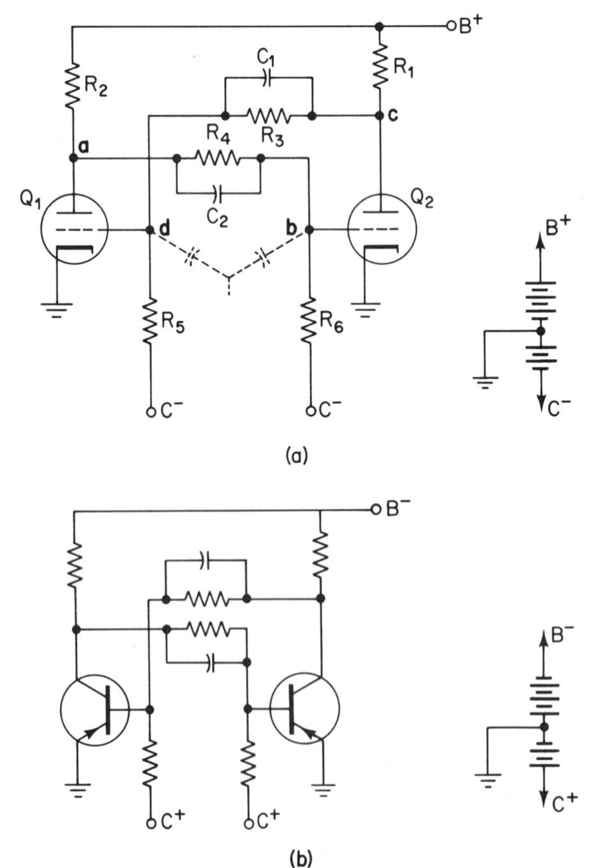

Fig. 18-14. *Flip-flop.* (Eccles-Jordan multivibrator circuit). (a) Vacuum tube version. (b) Transistor version.

Look at the circuit of Fig. 18-14*a*, and pretend for a moment that the tubes Q_1 and Q_2, and the capacitors, C_1 and C_2, are not there. We then have two groups of resistors in parallel, each group consisting of three resistors in series: the group R_1, R_3, R_5 and the group R_2, R_4. R_6, The operation of this circuit depends on the fact that no two vacuum tubes are perfectly identical. Consequently, when the voltage sources are connected, the plates of both tubes are more positive than their cathodes and both tubes will begin to conduct, but the current in one will rise faster than in the other. Let us assume that Q_1 conducts more heavily than Q_2. The heavy current results in a large voltage drop through R_2, which causes the point *b*, between R_4 and R_6, rapidly to become negatively biased. Point *b* is connected to the grid of Q_2 and the negative bias cuts off current through Q_2. This reduces the voltage drop through R_1, making points *c* and *d* more positive. Since the grid of Q_1 is connected to point *d*, this positive bias even further reinforces the conduction of Q_1, which is driven into saturation. This is one of the stable states of the flip-flop, and it will so remain indefinitely with Q_2 cut off and Q_1 conducting at saturation level.

Suppose, now, that a sharp negative pulse is applied to both grids simultaneously at points *b* and *d* (through the isolation capacitors shown dashed). Q_2 is already negatively biased to cut off so it is not immediately affected. For Q_1, however, the negative pulse momentarily biases the grid negative tending to cut Q_1 off. Point *a* then becomes more positive, as does point *b*, driving Q_2 into conduction, which makes points *c* and *d* more negative and completely cuts off Q_1. This is the other stable state of the flip-flop. Thus a single negative pulse has switched the flip-flop from one stable state to the other. A second pluse will return it to its original state. It should be noted that a positive pulse will work as well as a negative pulse, except that the positive pulse will affect the nonconducting tube first. Note, also, that when a tube is switched "on," its plate voltage suddenly drops. This is, in effect, another negative pulse which could be used to trigger another flip-flop. The capacitors, C_1 and C_2, are provided to make the switching time small, for they effectively provide a short-circuit path for the high-frequency components that make the peak of a switching pulse "sharp."

Let's put several flip-flops together to make a simple adding register, or accumulator. Figure 18-15*a* shows four flip-flops connected together to form a four-bit register. A neon lamp is connected across, let us say, the left-hand tube (or transistor) within each flip-flop, and its output is connected to trigger the next flip-flop to the left. Each lamp gives us a binary readout: lit, it's a 1; off, it's a 0. We are now going to illustrate how this counter does the simple addition

$$2 + 3 = 5$$

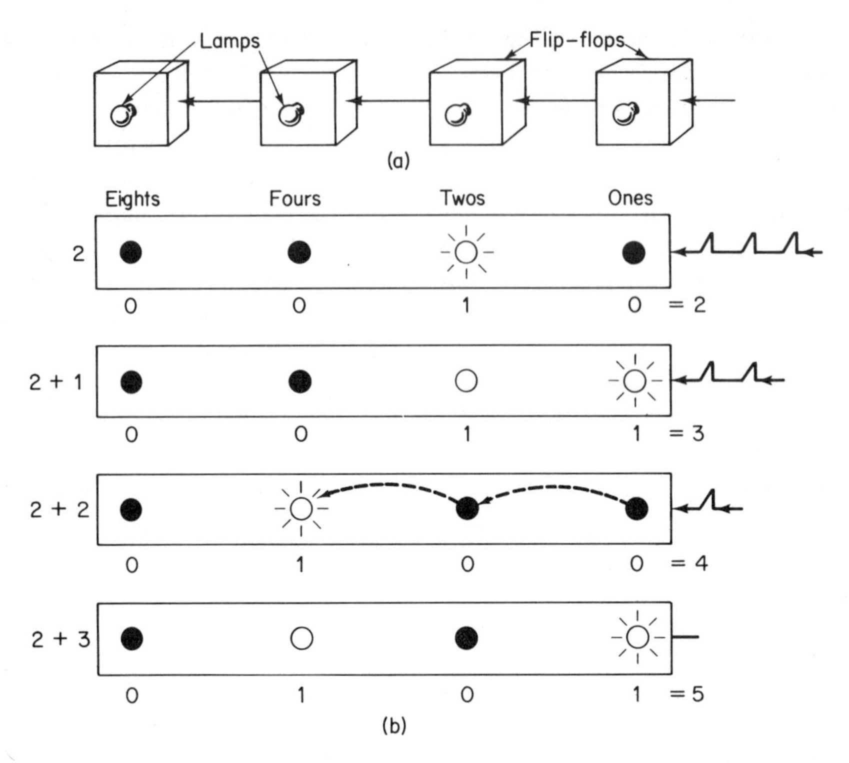

Fig. 18-15. Digital addition. (a) Counter schematic. (b) Adding 2 + 3.

Remember (consult Table 18-1 if necessary) that in binary notation, this is written

$$0010 + 0011 = 0101$$

Suppose that all four flip-flops are set to zero with their left-hand tubes (or transistors) conducting. All lamps are off. Let one pulse enter the right-hand (or "units") flip-flop. This cuts off the left tube inside this unit, lighting the lamp, and turns on the right tube. The other three flip-flops are not affected. The lights now show 0001, the symbol for 1. When the second pulse enters this flip-flop, it cuts off the right tube and turns on the left tube, extinguishing the light and sending a pulse to the second, or "twos," flip-flop. This pulse switches the state of the "twos" flip-flop, turning on its light. Thus we have "carried one" from the "units" place to the "twos" place, and the light pattern now shows 0010, the binary form of 2. This is the situation in the beginning of Fig. 18-15*b*. To this 2 we are going to add 3, so we send a train of 3 more pulses to the right end of the register. As the first of these pulses is received, the state of the "units" flip-flop switches to *on*, giving a 0011 (= 3) light pattern. The second of these pulses switches the "units" flip-flop *off* again, passing a pulse to the "twos" flip-flop, which, being *on*, is also switched *off* and passes a pulse to the "fours" flip-flop. The "fours" flip-flop is switched *on*, and the light pattern is now 0100, the binary 4. Finally, the third in the train of 3 pulses enters the "units" flip-flop, turning it *on*. This gives the light pattern 0101, the binary equivalent of 5 showing that three added to two makes five. This addition would take a large computer about two millionths of a second to perform; how long does it take us to say it or think it?

This simple example should convey an idea of how binary arithmetic is done by bistable circuits. Space does not permit a detailed discussion of all the arithmetic operations. Note that this register had a *memory;* it "remembered" the number 2 until the train of 3 pulses came along, then it performed the addition and "remembered" the total, 5.

18-7 DIGITAL COMPUTERS

With suitable combinations of large numbers of bistable circuits and other devices, we finally arrive at the giant digital computer. Such devices, like the Remington Rand Univac, the Control Data Corporation 3600 Computer, and the International Business Machines Corporation 360 System, are not simply machines, but a whole room full of interconnected equipment, as shown in Fig. 18-16. The tremendous complexity of these computers is apparent when you remember that extensive use is made of microcircuitry.

The major component units of such a digital computer are depicted in the block diagram of Fig. 18-17*a*. The *memory* unit, as the name implies, is a place for storage of both instructions and data to be worked with. Typical memory units are capable of storing from 4000 to 32,000 "words" of 18 bits each.

The *arithmetic* unit as its name implies, does arithmetic. It does far more, however. All machine instructions, including logic and calculations, except for a few special instructions to external equipment, are executed in the arithmetic unit. It contains small memory devices called *registers*, to contain the data being operated upon at the moment. All arithmeitc calculations are reduced to a series of additions, subtractions, multiplications, and divisions, The multiplications and divisions themselves, in turn, are usually done as a series of additions and subtractions, respectively. For instance, the multiplication 4×25 would be done (in binary arithmetic, of course) as the additions

Fig. 18-16. IBM 360 Digital Computer System. (Courtesy of International Business Machines.)

11001	(= 25)
11001	(add 25)
110010	(= 2 × 25 or 50)
11001	(add 25)
1001011	(= 3 × 25 or 75)
11001	(add 25)
1100100	(= 4 × 25 or 100, the answer)

The *control* unit is the computer's nerve center, which ties all the parts together and makes them function as a unified whole. It is something like a telephone exchange in that it directs which circuits are to be activated and deactivated, according to the programmed sequence of events. It is connected directly to the master control console, which contains the switches and display devices necessary to the operator.

The *clock* is a precision source of a sequence of pulses at some fixed frequency. Commonly used clock frequencies are between 100 kHz and 10

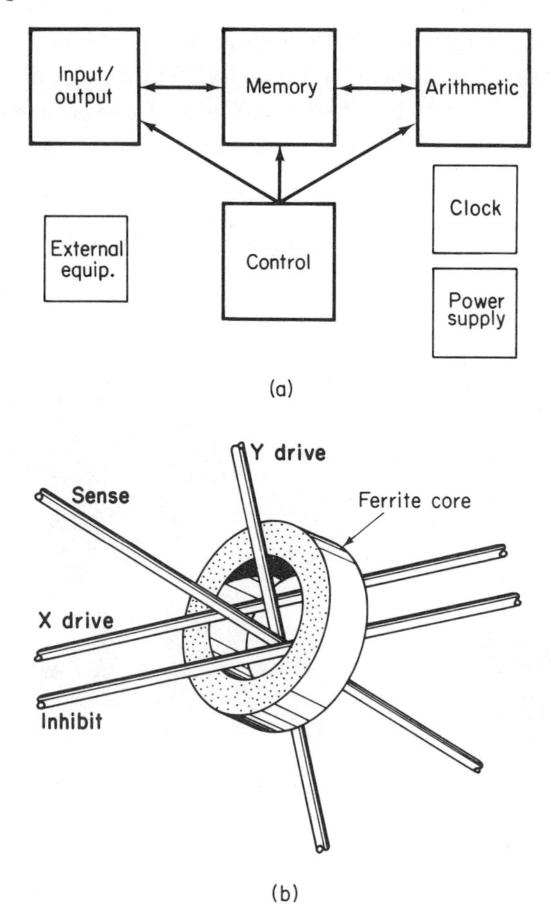

Fig. 18-17. (a) Major units in a digital computer. (b) Magnetic core.

MHz. The clock sets the pace for all operations internal to a *synchronous* computer. In some computers, called "asynchronous," the clock does not govern all operations.

The *external equipment*, like card readers, card punches, *x-y* plotters, typewriters, printers, etc., may be connected to the computer, but do not actually belong to the computer proper. Since their timing depends on mechanical devices used in their construction, like relays and motors, they cannot be synchronized directly with the computer clock. Consequently, *input/output* circuitry with its own temporary storage is used to handle the information transfer. Regulated voltages for all the operating units, and power for a cooling system (if one is required), are provided by the *power unit*.

Storage can be accomplished in several types of devices. We have already met the flip-flop. Another important device that can store one bit is the *magnetic core*, one form of which is shown in Fig. 18-17*b*. It is a little ring of *ferrite* (magnetizable iron oxide), typically 30 mils in inside diameter by 50 mils outside diameter by 15 mils thick. An idea of the actual size is given by Fig. 18-18. These little rings can be magnetized in either of two ways, that is,

Fig. 18-18. Individual cores for computer memory unit. (Courtesy of Ampex Corp.)

Fig. 18-19. Magnetic core assembly. (Courtesy of Ampex Corp.)

with the magentic flux clockwise or counterclockwise around the core, thus representing a binary 1 or 0. Many little cores are mounted on a frame to form a *matrix*. Wires are woven through the cores, four wires passing through each core as in Fig. 18-17*b*. Two of the wires, the x-drive and y-drive, are used to change the "state" (magnetization) of the core. The "sense" line is used to sample the changing flux when the state of the core is switched, while the "inhibit" prevents this occurrence. A typical matrix is shown in Fig. 18-19. This particular unit contains 1024 cores and can therefore store 1024 bits. To *address* a particular core in the matrix, a pulse is sent to the x-drive wire passing through the row in which it is located and another pulse to the y-drive wire passing through the column in which it is located. The core in question, located at the intersection of these two wires, feels both pulses at the same time, and if they are in the same direction, its state is switched. This bit of information shows up as a pulse on the sense line.

Note that a core-type memory unit has its information content destroyed by the act of "reading" it. This type of memory is called a *destructive-read-out* memory. There are also several types of nondestructive-read-out storage devices. These are the magnetic tape unit, the magnetic drum unit, and the magnetic disk unit, shown in Fig. 18-20, 18-21, and 18-22, respectively. The magnetic tape unit is similar to a commercial tape recorder except that it uses high-precision heads and exceptional quality tape. The tape used is

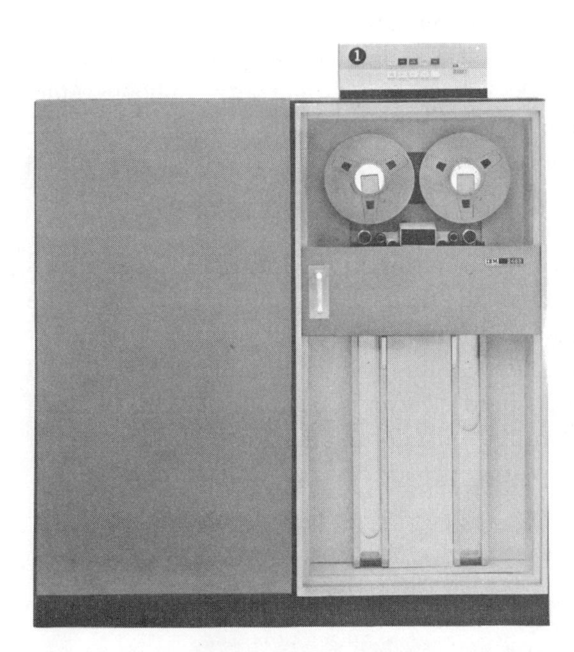

Fig. 18-20. Magnetic tape unit. (Courtesy of International Business Machines Corp.)

Fig. 18-21. Magnetic drum storage unit. (Courtesy of International Business Machines Corp.)

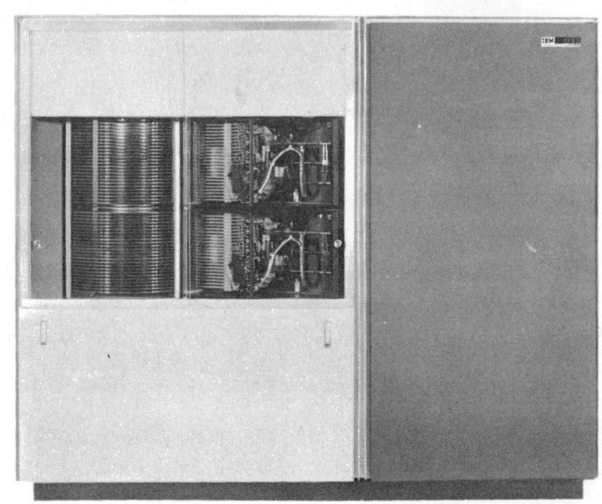

Fig. 18-22. Magnetic disk storage unit. (Courtesy of International Business Machines Corp.)

one-half inch wide, runs at speeds in excess of 240 ips, and uses seven channels. Bits are stored as magnetized spots on the tape with one of two magnetic polarities. Tape capacity is more than 500 bits per inch. The magnetic drum is essentially a continuous "short, fat tape" somewhere between 5 and 20 in. in diameter. The drum surface is subdivided into several tracks that run around the drum. "Read" and "write" heads pick up and imprint patterns in these channels, just like those made on tape. The disk storage device is also similar in principle, but uses flat disks rather than a cylindrical drum for the recording surface. These disks are in effect, magnetic "phonograph" records stacked in juke box fashion, with a pickup head between each pair. Some disk units can store millions of bits. In any case, most large computers use several different devices to make up the largest possible total memory, consistent with a reasonable cost.

Output from the computer can be either in the form of printed *alphanumeric* (alphabetical and numerical) data, or graphic, or both. The alphanumeric printer can be anything from a typewriter printing 10 characters per second, to an electromechanical line printer capable of about 1200 characters per second, to a high-speed electronic printer that can produce 10,000 characters per second. Graphic outputs require some DA conversion before being applied to an x-y plotter or cathode ray tube display device.

Fig. 18-23. CRT display unit. (Courtesy of International Business Machines Corp.)

The latter has been used very successfully in conjunction with a computer for visual solution of mechanical and structural design problems, as illustrated Fig. 18-23.

Digital computers can be categorized according to their intended applications. A *scientific* computer is a general purpose computer with a moderately large memory and great speed and flexibility in the types of computation and logic it can handle. Figure 18-24 shows such a computer in

Fig. 18-24. Using the IBM 360 Computer System. (Courtesy of International Business Machines Corp.)

use. A *business* computer is used primarily for bookkeeping and accounting-type problems. It is only required to do relatively simple arithmetic. The emphasis is on outputting directly on business forms (like insurance forms, account statements, paychecks) and in producing punched cards that can be used to control other business machines. For libraries, law enforcement, and other applications where large masses of data are to be stored, periodically updated, and easily retrieved, a *file* computer is more applicable. This type has little arithmetic capability, but very large memory and special facilities for rapid location and modification of single items of data. It is ideally suited to search and sort operations. Finally, there are the *special purpose* computers designed to handle only one problem or one type of problem. Typical examples are the guidance computers, used on ships, planes, and spacecraft, and instrumentation computers like the one in Fig. 18-25. The instrumentation computer takes raw data signals from various sensors

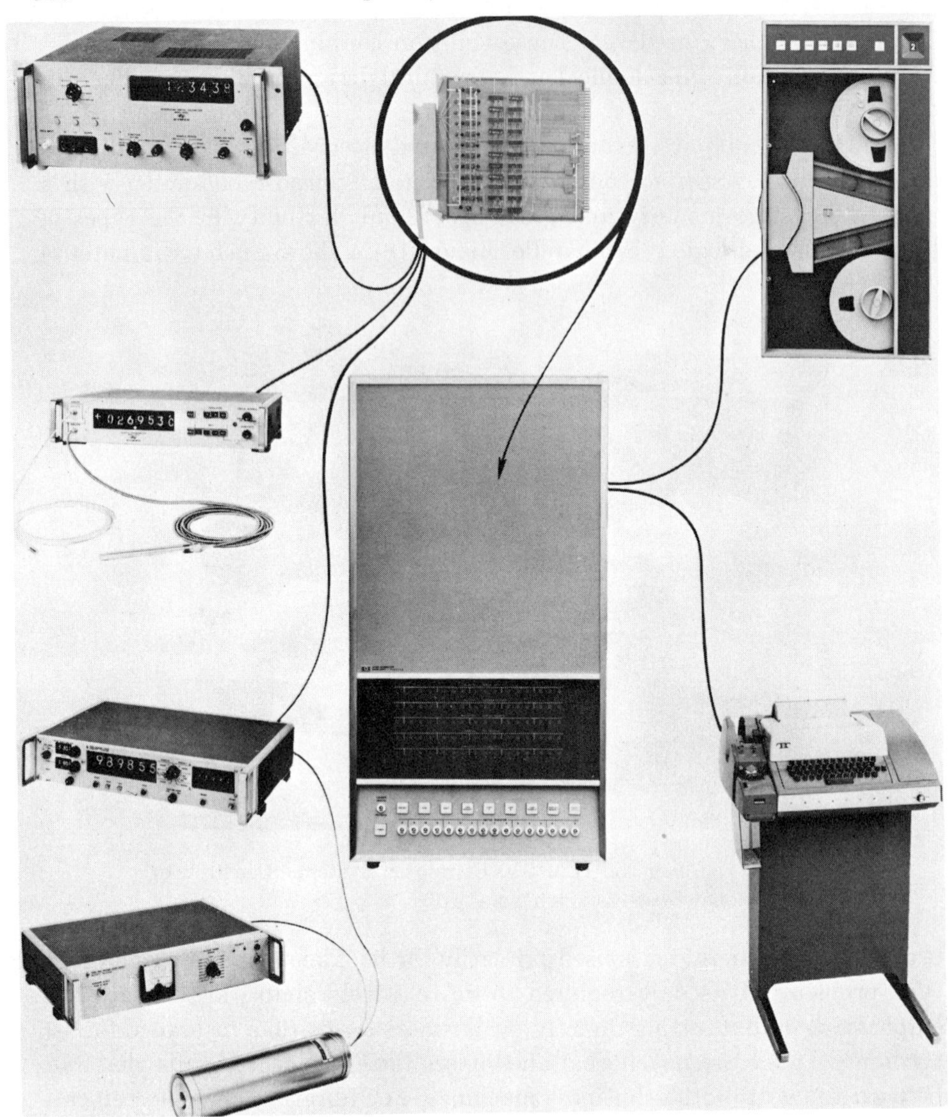

Fig. 18-25. Model 2116A Instrumentation Computer. (Courtesy of Hewlett-Packard.)

(which are analog devices), puts them through suitable AD converters, calculates and corrects as necessary, and displays the final "reduced" data in tabular and/or graphic form. Business, file, and instrumentation computers, as well as others like them, also come under the general heading of *data processors*.

18–8 ELECTRONIC MEDICINE

Medicine has traditionally been slow to adopt the technologies developed by her sister sciences. For a long while, only the stethoscope, otoscope (ear inspection device), sphygmomanometer (blood pressure checker), and occasionally the microscope and centrifiuge, were accepted by many doctors. Part of the reason lies in the fact that until recently physicians and engineers could not communicate effectively, due to the different technical jargon of their respective fields, and a lot of plain ingroup snobbishness on the part of both. Fortunately, this situation is changing. Electronics has had a special obstacle to overcome to gain medical respectability. Early in the century, charlatans posing as doctors foisted off quack electrical "cure" devices on the public, such as "healing rays" of colored light and sparks from electrostatic generators. These quacks gave a black eye to both professions, but now most of them have also faded into history.

Today, due to the impetus of wars, the manned space programs, and the emergence of a new breed of men called *biomedical engineers*, electronics has become at least the well-kept mistress, if not the handmaiden, of medicine. It has immediate applications in diagnosis, therapy, and artificial organs. In terms of the categorization scheme we set up at the beginning of this chapter, diagnosis comes under data acquisition and artificial organs under automation.

When we think of "electronics" and "diagnosis" in the same breath, we are almost invariably led to think *X ray*. Hand in hand with X-ray diagnosis goes fluoroscopy. A fluoroscope is simply an X-ray device in which the image is formed on a fluorescent screen rather than on a sheet of photographic film. Thus the fluoroscope permits the doctor to watch "X-ray movies" of the body, but with no permanent record. X-ray equipment has come a long way since the first Coolidge tubes. Figure 18-26 shows a remote-controlled diagnostic X-ray system that incorporates linear tomography (selectively X-raying sections of the body) as well as radiographic and fluoroscopic capabilities. A further application of electronics to X-ray technology is the radiograph intensifier of Fig. 18-27. With this device, one can view an X-ray while electronically adjusting the enlargement (magnification) and contrast, and also project the X-ray as either a negative or positive image. Finally, by combining the fluoroscope with television and recording technologies, we arrive at the recorded fluoroscopic examination, shown in Fig. 18-28. Here a vidicon camera monitors the fluoroscope images, which are then recorded on videotape. These videotapes can be played back by the doctors at their leisure, or duplicated and sent to various medical schools.

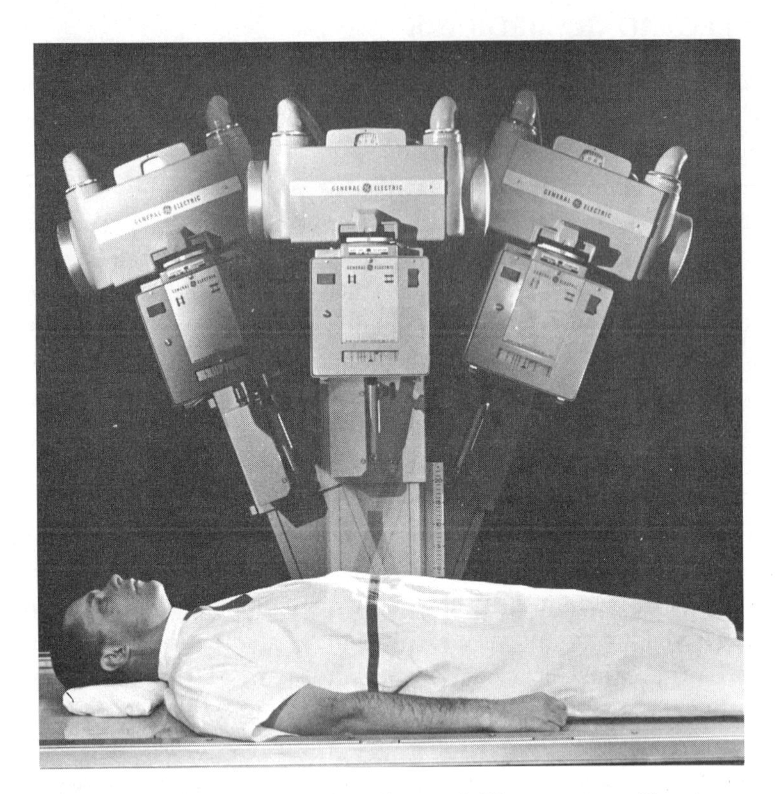

Fig. 18-26. Remote-controlled diagnostic X-ray system. (Courtesy of General Electric Co.)

While we are on the subject, many adaptations of closed-circuit television are found in medical practice. Coupling a TV camera to a microscope permits the display of microscopically examined material to a wide audience. A modified TV microscope called a *sanguinometer* automatically does blood counts as accurately as and much more rapidly than, a laboratory technician, freeing him to do less tedious work. A similar device, called a *cytoanalyzer*, does cancer cell counts electronically.

An interesting diagnostic tool is the application of the sonar principle to the human body. This technique, called *ultrasonic ranging*. consists of sending 1-μsec pulses of 15-MHz "sound" waves into the tissue to be tested every millisecond and observing the echoes. A quartz crystal transducer is used and coupled to the tissue with a column of water. The display devices are similar to conventional marine sonar displays. This device is useful diagnostically because sound waves are strongly reflected by interfaces between materials with different elastic properties (e.g., normal and malignant tissue).

Fig. 18-27. *Explorex* Radiograph Intensifier. (Courtesy of General Electric Co.)

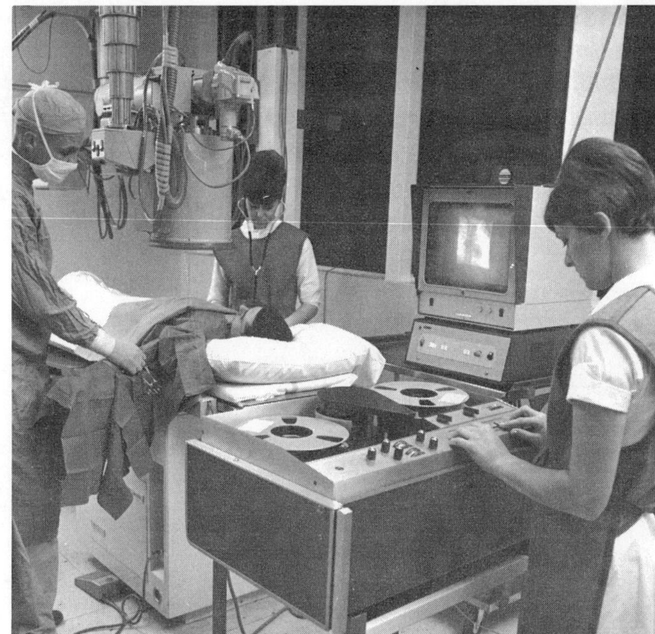

Fig. 18-28. Recording a fluoroscopic examination. (Courtesy of Ampex Corp.)

Two highly useful devices that are nothing more than very sophisticated voltmeters are the *electrocardiograph*, or ECG (sometimes EKG), and the *electroencepholegraph*, or EEG. The electrocardiograph measures the tiny potential changes at the surface of the body caused by contraction of the heart muscle. The ECG principle was discovered by the Dutch physiologist, Willem Einthoven, in 1903, eventually gaining him the Nobel Prize in medicine. In its present form, the ECG comprises a set of electrodes, connected to the arms, legs, and chest of the patient, whose output is amplified and displayed on a strip chart or oscillograph. The chart produced by the electrocardio*graph* is called an electrocardio*gram* (also abbreviated ECG). The need to monitor the heart rates of astronauts has led to the development of portable and rugged miniature ECGs. Along similar lines, the noted German psychiatrist, Hans Berger, devised a system of connecting electrodes to the scalp which could permit recording and study of the varying electric potentials produced by the brain, commonly called "brain waves." Thus, in 1929, was born the electroencephalgraph. Depending on the state of its activity, the brain produces several different types of rhythmic patterns termed alpha, beta, theta, etc. rhythms. It should be noted that these EEG patterns represent skin potentials, measured by conduction and *not* "radio broadcasts" from the brain. Although far too coarse in resolution to study the details of the thinking process, the electroencephal*gram* (the record produced) is a good indicator of certain major nervous afflictions, notably epilepsy.

Miniaturization and widespread use of solid state circuitry, has really accelerated the development of electromedical technology. Outputs from a number of tiny transducers, like the thermistor of Fig. 18-29, can be displayed

Fig. 18-29. Diamond thermistor, smaller than a preying mantis, can read temperatures from −325°F to 1200°F. (Courtesy of General Electric Co.)

on a central patient monitoring system, such as that in Fig. 18-30. Typical "intensive care" monitors continuously display the patient's ECG, blood pressure, temperature, respiration rate, and other critical data.

The therapeutic applications of electronics are fairly well known. X-ray emission (radiology) and rf heat treatment (diathermy) are the more obvious. In psychotherapy, electric shock treatment was once widely used but has now been largely supplanted by other techniques. The psychology department of San Jose State College has been experimenting with closed-circuit TV psychotherapy for mental patients at Agnew State Hospital in California.

Electronics has also contributed its share to the field of artificial organs, including the cumbersome kidney and heart-lung machines and the more widely publicized "iron lungs." In recent years, the electronic *pacemaker* has received a lot of attention. This device, shown next to a model of the

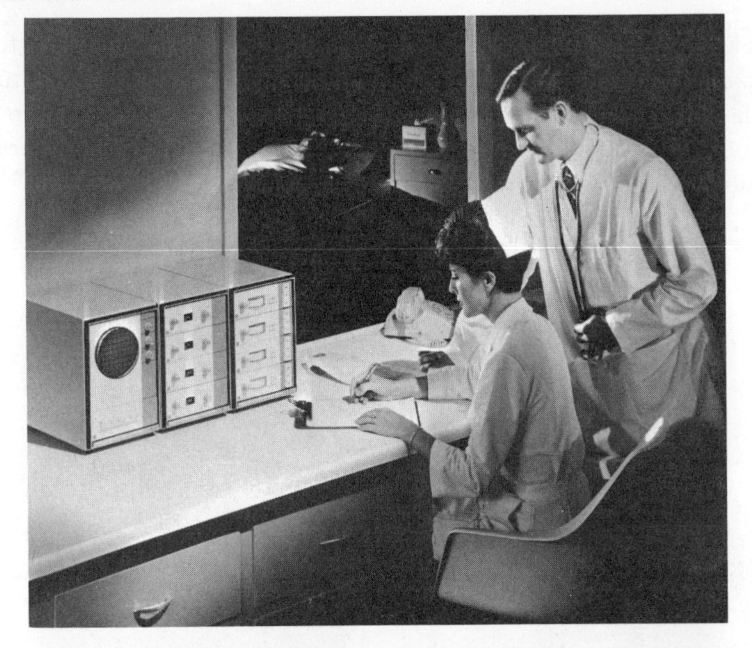

Fig. 18-30. Patient monitoring system for intensive care units. (Courtesy of General Electric Co.)

Fig. 18-31. Cardiac *Pacemaker* next to model of human heart. (Courtesy of General Electric Co.)

human heart in Fig. 18-31, is not a true "artificial heart" but a transistorized, battery-operated device used to stimulate, or "pace," the weakened muscles of the heart in a cardiac patient. It is implanted surgically, as shown in the X ray of Fig. 18-32. More than 10,000 Americans now have their hearts electronically paced.

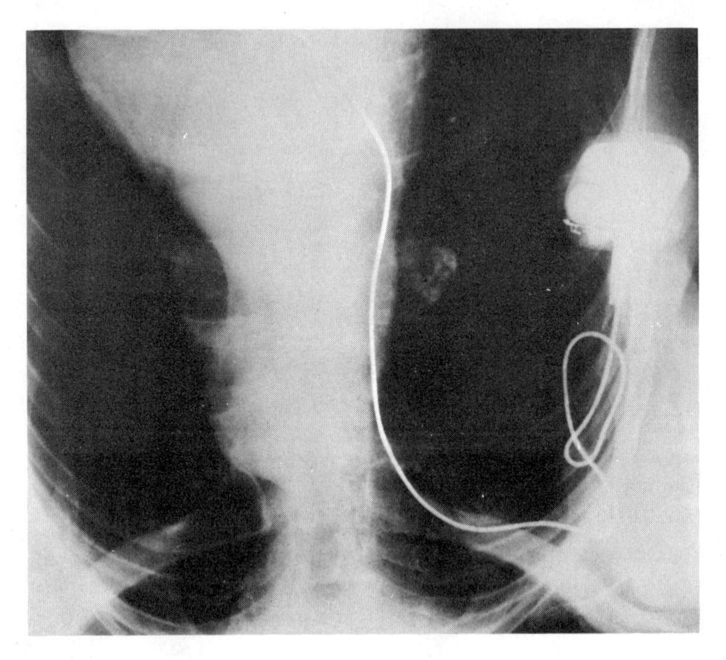

Fig. 18-32. Cardiac *Pacemaker* generator implanted in patient. (Courtesy of J. D. Morris, M. D. and General Electric Co.)

18-9 ELECTRONICS IN THE FACTORY

The applications of electronics in manufacturing are legion. They fall, predominantly, into the area of electromechanization and automatic control. Beyond such grass-roots items as replacing foot treadles or steam engines with electric motors, and hand levers with solenoids, one of the most widely implemented electronic ideas is the *closed-loop automatic control system*. Consider a process in which some variable quantity (like temperature, or pressure, or flow rate) is to be held at a particular value. An observer monitors the value of that variable and, if it drifts outside some allowable limits, he takes a corrective action. For example, he may open a valve wider if a steam pressure falls too low and close it part way if the pressure gets too high. Now, in any such industrial process, if the variable quantity can be measured by a transducer and the corrective action performed by an electrically operated device, then we have the makings of an automatic closed-loop control system. The controlled process may by purely mechanical, or it may involve the flow of a gas (pneumatic) or liquid (hydraulic). Whatever the case, the controlling device is driven by an electronic amplifier called the *servo amplifier*, which is fed a reference signal corresponding to the desired level of the quantity being controlled. A transducer senses the actual immediate value of that quantity and feeds its signal back to the input of the servo amplifier. There, the feedback signal from the transducer is compared with the reference signal to form an *error signal*. The further the controlled quantity tries to get away from the desired value, the larger the error signal, and it will be either positive or negative, depending on whether the actual value of the variable tries to go below or above the desired level. This error signal is amplified and used to drive the controlling device, thereby adjusting the controlled quantity in the proper direction toward its desired value. In other

Fig. 18-33. Closed-loop control system automatically controls moisture content in paper production. (Courtesy of General Electric Co.)

Fig. 18-34. Automatic X-ray inspection device checks for proper amount of soft drink in cans. (Courtesy of General Electric Co.)

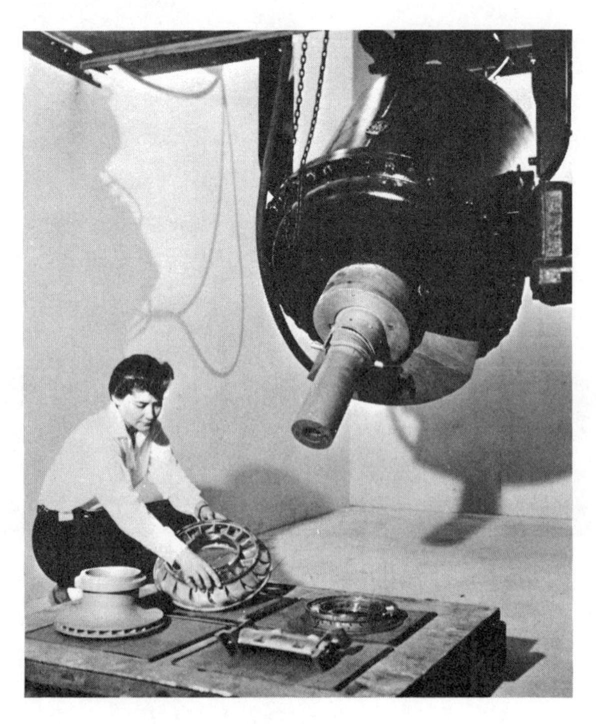

Fig. 18-35. High-voltage industrial X-ray machine used to inspect jet engine turbines. (Courtesy of General Electric Co.)

words, the closed-loop control system always drives itself in such a way as to make its own error signal zero. The term *servo* comes from the word *servomechanism*, a name given to devices in which a primary control is operated by a relatively weak electrical signal. Electromechanical and electrohydraulic servomechanisms first gained broad popularity in aircraft control systems. Figure 18-33 shows part of a typical automatic control system. Besides controlled flow processes, today's industry uses lathes and shapers controlled by a tape-recorded program, automatic electron-beam and laser-beam welding equipment, etc., in the mass production of intricate and delicate shapes.

TV and X-ray monitoring equipment make it possible to inspect areas normally inaccessible or hazardous to humans. Figure 18-34 shows an automatic device that uses X-rays to check the amount of soft drink in a sealed can, while a jet engine turbine receives a similar penetrating inspection in Fig. 18-35.

In many plants, the paychecks and group insurance forms for those working in the factories, and possibly also inventory and procurement records, are processed by computer. The same computer may be used, on a time-shared basis for engineering calculations, or even computerized design as illustrated in Fig. 18-36.

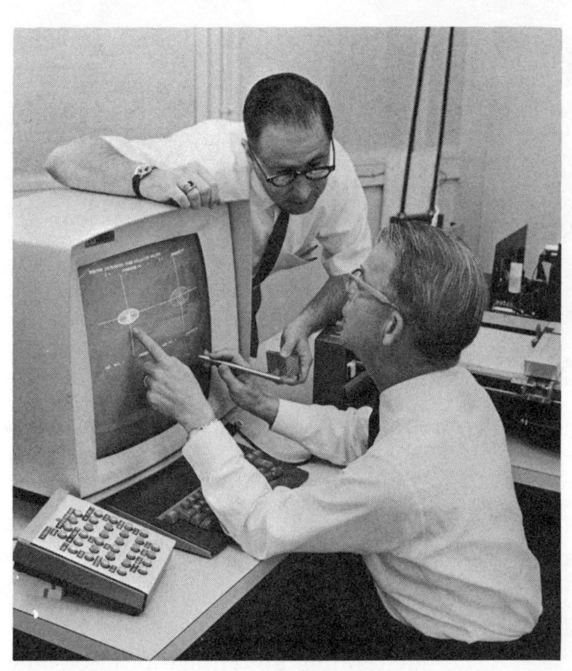

Fig. 18-36. Designing with the computer. (Courtesy of International Business Machines, Corp.)

18-10 ELECTRONIC TEACHERS

In both our formal schools and in industrial instruction programs, electronics is playing an ever-growing role. Most of these applications come under the headings of communication and data processing. From the communications standpoint, closed-circuit TV has gone a long way in bringing the instructor and his demonstrations closer to a large audience. Coupled with a videotape system, as shown in Fig. 18-37, the TV system also eliminates the

Fig. 18-37. Closed-circuit educational videotape system records chemistry lesson. (Courtesy of Ampex Corp.)

need for the instructor to perform the same demonstration over and over for class after class. Furthermore, the talents of professional filmmakers can be put to use, as in Fig. 18-38, to make certain subject matter more palatable to the younger set.

The outstanding advantage of electronic recording, however, is in its capacity for *instant replay*. Thus, whether it is an audio tape in the foreign-language lab feeding "parlay voo fransay mahn soor" back to the would-be

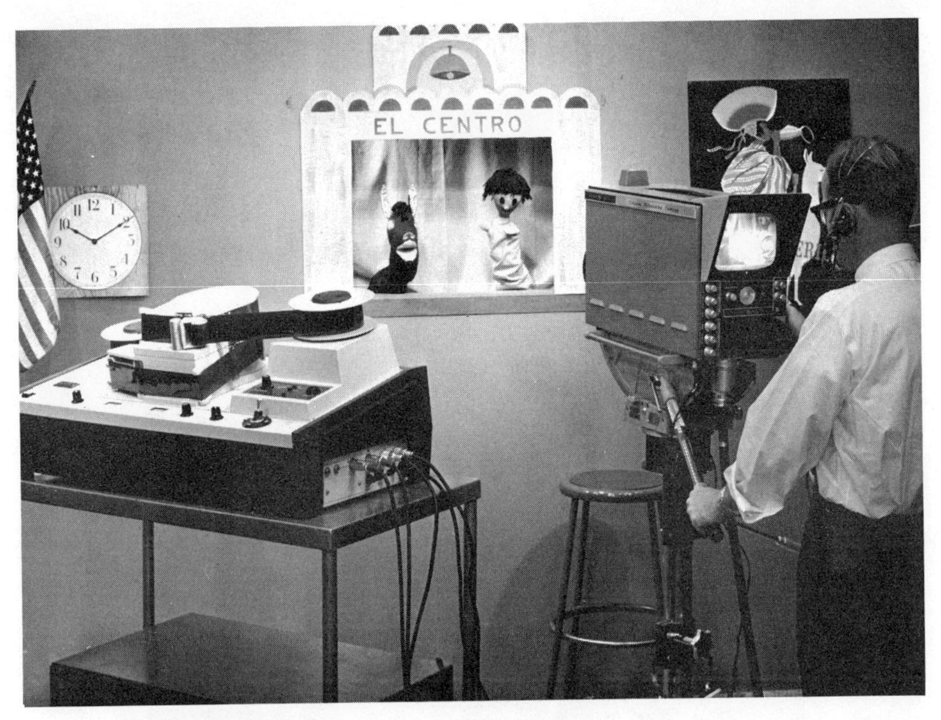

Fig. 18-38. Recording a "painless" Spanish lesson for fourth-graders. (Courtesy of Ampex Corp.)

Fig. 18-39. Instant replay shows "What'd we do wrong, coach?" (Courtesy of Ampex Corp.)

speaker of French or the videotape system of Fig. 18-39 showing the first-stringers how the second-stringers managed to get through their defense, the immediate playback of recorded data gives the student the maximum opportunity to profit by his mistakes.

Videotape training programs have proven invaluable in standardizing food-preparation techniques. Such programs, as illustrated in Fig. 18-40, are

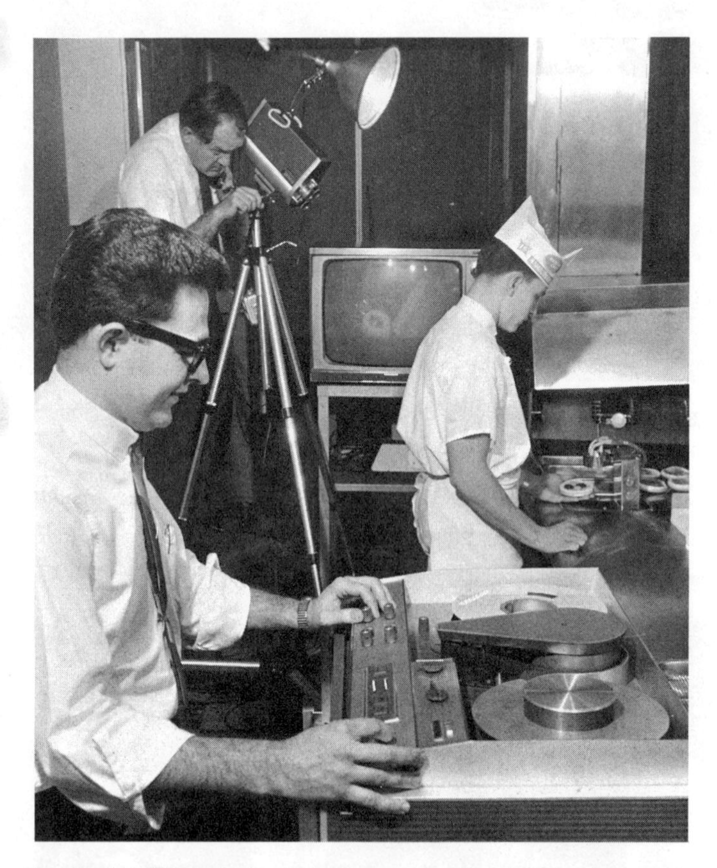

Fig. 18-40. Helping a trainee learn to make a hamburger. (Courtesy of Ampex Corp.)

rapidly gaining popularity with food-vending franchise organizations.

One of the most effective pedagogical applications of simulation devices is in driver training. In most high schools around the country, simulators like those in Fig. 18-41 subject novice drivers to all the psychological pressures of road and expressway without physically endangering learner or other drivers. The same type of analog simulator, with the human being as

Fig. 18-41. Electronic driver training simulator. (Courtesy of Raytheon Corp.)

one of its elements, is found in varying degrees of sophistication throughout the spectrum of manual-skills training programs. The most glamorous of these devices are used in training pilots and astronauts.

18–11 ELECTRONICS IN THE AIR AND IN SPACE

Today's aircraft, commercial and military, are bristling with electronic systems. Most of these are either control systems or communications systems. In particular, electronic navigation instruments, such as those displayed in Fig. 18-42, literally take the blinders off the pilot in heavy clouds or fog. While these radio and radar devices are busy "up front" making the passenger's flight safe, another type of device is entertaining him as in Fig. 18-43.

In space, electronics has reigned supreme since the beginning. Virtually every operating system in any spacecraft has at least some electronic aspect; many of them are completely electronic. Consider, for example, the TIROS meteorological satellites of Fig. 18-44*a* and *b*, whose purpose is acquisition of weather data. Their sensors, the signal-processing equipment that handles the sensors' output, the communication links that transmit it,

Fig. 18-42. Airborne bearing and distance indicators. (Courtesy of International Telephone and Telegraph Corp.)

and the satellite's power supply and control systems—all are electronic. Figure 18-45 shows the installation of an electronic X-ray detector in another scientific satellite.

Remember the wonderful moon pictures from RANGER? Electronic guidance got it there safely, and the combined communication/data processing links shown in Fig. 18-46 brought those pictures back to us.

The Apollo vehicles that ferry astronauts to the moon are massive assemblages of electronic systems. Figure 18-47 shows how even the countdown sequence in launching their Saturn V booster rocket is automatically checked by computer.

A relatively novel space application of electronics is electric propulsion, more specifically, electrostatic propulsion. A group of electrostatic propulsion devices, called *ion rocket engines,* is shown in Fig. 18-48. The principle of one variety of ion engines is illustrated in Fig. 18-49. In a conventional rocket engine, chemicals react to form a hot gas that is expanded

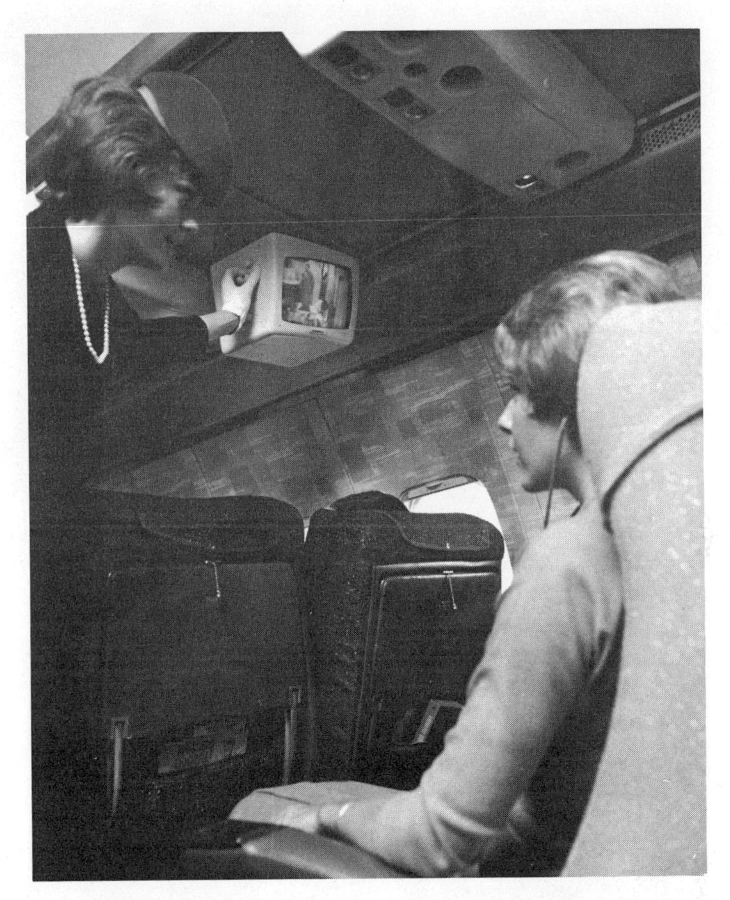

Fig. 18-43. In-flight entertainment system. (Courtesy of Ampex Corp.)

through a nozzle; the reaction force to the momentum of these escaping high-velocity gases is experienced as *thrust*. In the ion engine, a metal (usually cesium or mercury) is vaporized in a chamber. Then, under the bombardment of electrons moving from cathode to anode (in the device shown), some valence electrons are "knocked off" the vaporized metal atoms, making them positively charged ions. The magnetic field, from the permanent magnet sleeve, bends the paths of the electrons, increasing their chance of striking a metal atom. From this point on, the ion engine works like an electron gun, the ions being accelerated by high-voltage electrodes. The momentum of the ion beam produces a reaction thrust, just like the hot gas in an ordinary rocket. This positive ion beam must be neutralized before it leaves the engine, so that it doesn't cause a permanent charge on the spacecraft, and a little electron-

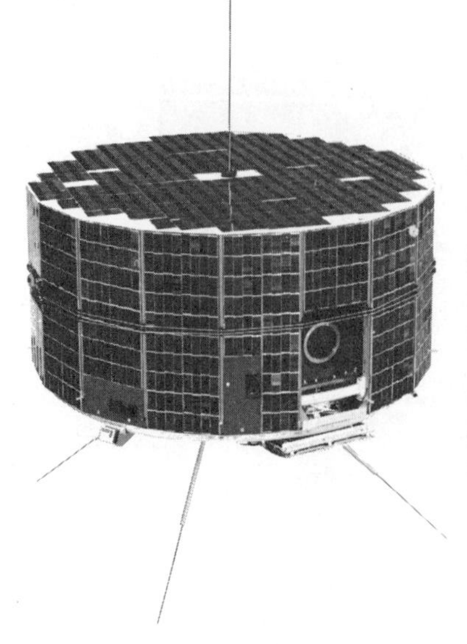

Fig. 18-44. (a) TIROS weather satelite. (Courtesy of Radio Coporation of America.)

(b) Advanced TIROS weather satellite. (Courtesy of Radio Corporation of America.)

Fig. 18-45. Electronic X-ray detection sensor being installed in a satellite. (Courtesy of the Aerospace Corp.)

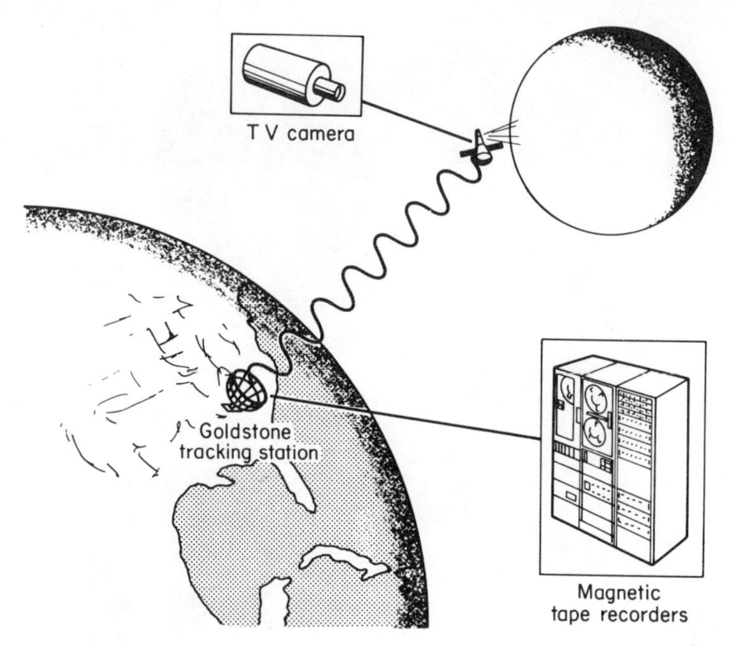

Fig. 18-46. Ranger 7 moon picture technique. (Courtesy of Ampex Corp.)

Fig. 18-47. Automatic countdown checkout computer for Saturn V rocket. (Courtesy of Radio Corporation of America.)

Fig. 18-48. Ion rocket engines of various sizes. (Courtesy of Electro-optical Systems, Inc.)

Fig. 18-49. Principle of ion rocket engine, cesium electron-bombardment type.

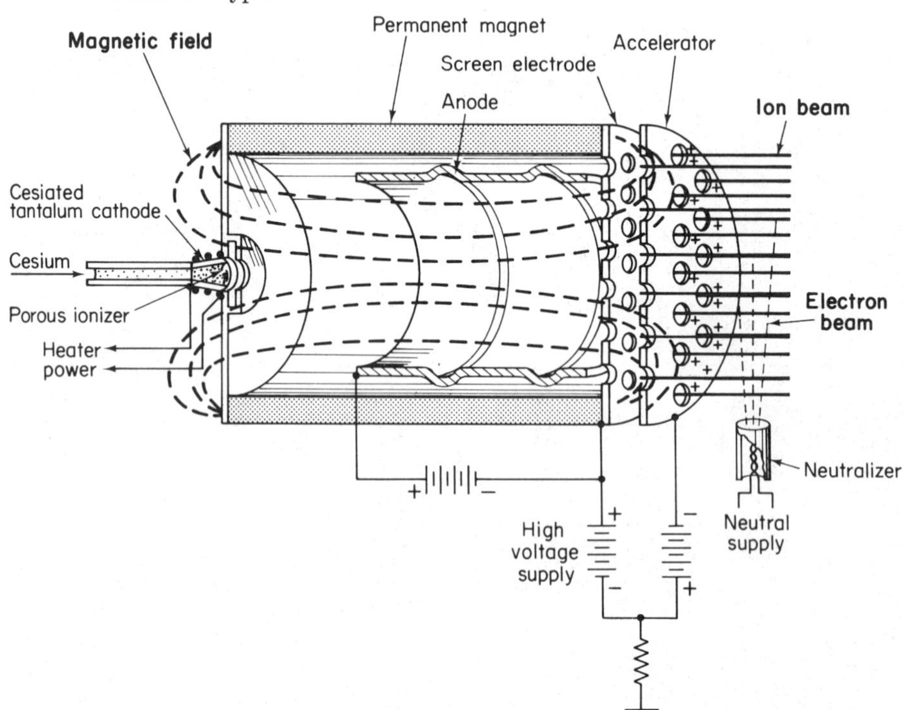

emitter, called the neutralizer, is used for this purpose. Ion engines are not widely used in spacecraft yet, for they are still somewhat experimental, and they require large amounts of power to produce a tiny thrust. However, their performance is (in terms of propellant consumption) more than ten times better than that of equivalent chemical rockets, and their modest thrust levels will undoubtedly find application in deep space probes.

18-12 ELECTRONICS IN THE LABORATORY

Electronics is busily at work in all phases of "pure" scientific research. We could not even begin to enumerate them. We may, however, take a quick look at a select sample. Data acquisition and data processing are the watchwords, with somewhat lesser emphasis on electromechanization.

Fig. 18-50. Scripps Institute of Oceanography shipboard computer. (Courtesy of International Business Machines Corp.)

The oceans are receiving increased research emphasis today. "Hydrospace," it seems, may hold the key to problems as diverse as the origin of the continents and what to do about the overpopulation problem. Research ships use seagoing computers, like the one in Fig. 18-50, to process the wealth of data they collect. Deep under the sea, exploration vessels of the type proposed in Fig. 18-51, probe the mysteries of the ocean's floor directly.

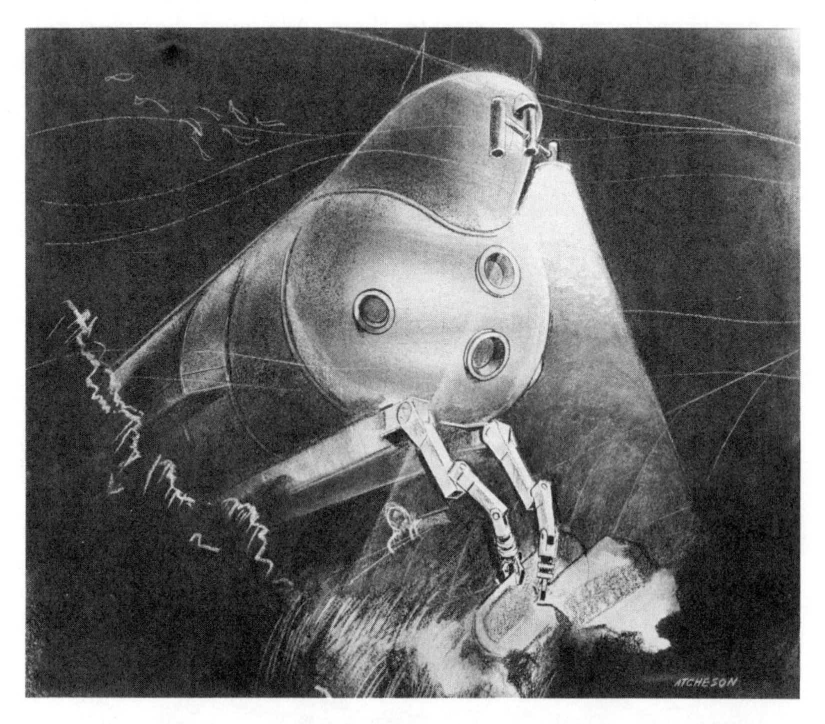

Fig. 18-51. *Aluminaut* 2500-fathom undersea Exploration Vessel with manipulator arms. (Courtesy of General Electric Co.)

Thanks to the work of Watson, Crick, Beadle, and others, the process of inheritance and the nature of the chromosome are becoming rapidly understood. Medical researchers are deeply interested in the genetic aspects of disease. Figure 18-52 shows that they have found a new partner to assist in the tedious correlation work required, the digital computer.

Chemical analysis has always been a time-consuming job, requiring a patient and meticulous technique. Now the "wet" analysis is being rapidly supplanted by absorption or emission spectrographic techniques with elec-

Fig. 18-52. Studying chromosome X-rays with a computer. (Courtesy of International Business Machines Corp.)

Fig. 18-53. Identification of micro-organisms by differences in their metabolic products. (Courtesy of General Electric Co.)

tronic data processing. In Fig. 18-53, we see an electronic analyzer at work in the biology laboratory, identifying various microorganisms by the differences in their metabolic waste products. Figure 18-54 shows a digitally controlled X-ray spectrometer, which can automatically analyze up to ten samples for as many as 11 different elements and print out the results as each analysis is completed.

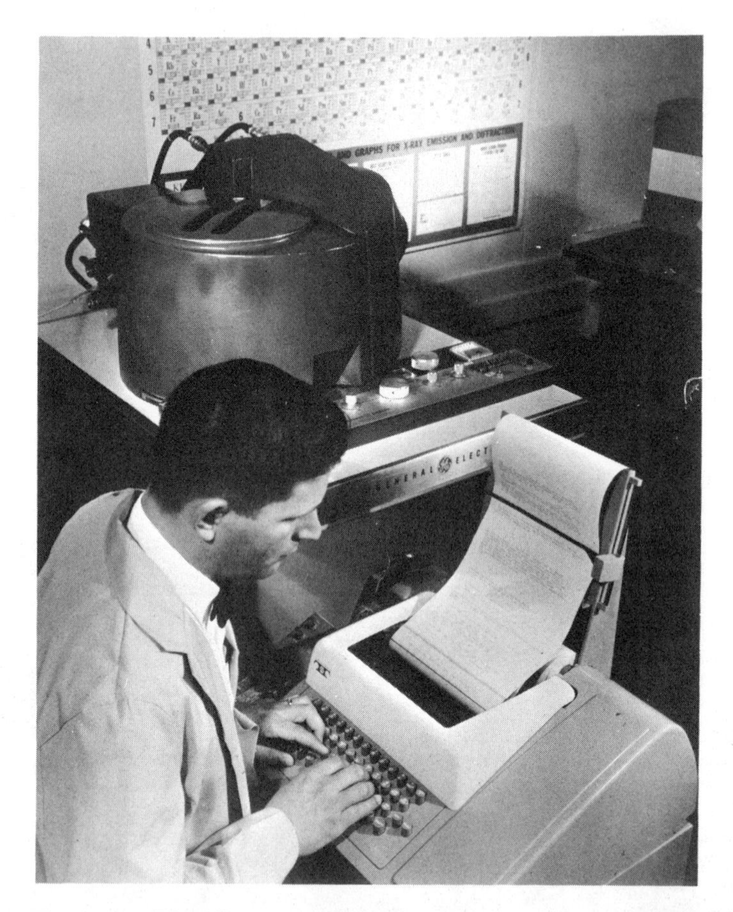

Fig. 18-54. Digitally controlled X-ray spectrometer performs rapid automatic chemical analysis. (Courtesy of General Electric Co.)

18-13 ELECTRONICS IN DEFENSE

Military applications of electronics cover our entire list of categories, for the Armed Forces, from the standpoint of the various functions they perform, are essentially a "world within a world." We have already touched on some defense electronics: widespread use of radar and sonar, microwave communications links, and the potential future development of a "death ray" out of laser technology.

Computers are heavily relied on in many strategic applications. Automatic fire-control systems are used to operate large naval artillery, like that in Fig. 18-55. Their huge shells, with the damage potential of a high

Fig. 18-55. Automatic fire control system directs huge naval guns. (Courtesy of General Electric Co.)

explosive bomb, must be fired over a long distance, through varying wind conditions, to hit, perhaps, a moving target, and this while the ship is bobbing and rolling in rough seas. The fire-control computer makes all the necessary adjustments while the guns are being aimed. Similar special-purpose computers are used, with gyroscopic reference devices, to make inertial guidance systems for strategic missiles. Figure 18-56 shows one such guidance package. In addition, large general-purpose digital computers are used by all branches of the service for technical computations and for administrative data processing. Simulators, embodying specialized analog computers, get extensive workouts in training personnel for a variety of jobs. They are particularly valuable in instructing pilots and flight crewmen, as previoulsy noted.

Television, often in conjunction with a videotape system, is also widely exploited. Tactical applications like communications, reconnaissance, and

Fig. 18-56. Missile guidance systems. (Courtesy of General Electric Co.)

visual weapon guidance have been virtually revolutionized by the development of rugged, lightweight vidicon camera tubes. TV is also a valuable training aid, of course. A Navy pilot landing aid videotape system called PLAT is shown in Fig. 18-57. This system enables the pilot to review the details of carrier landings by himself and others. Other TV application include closeup observation of dangerous processes and entertainment of personnel at remote duty stations.

18-14 ELECTRONIC LAW AND LAW ENFORCEMENT

The applications of electronics to law practice are just beginning to be realized. The rapid store-search-retrieve capability of the business computer is a "natural" in connection with the files of attorneys and the courts. Crim-

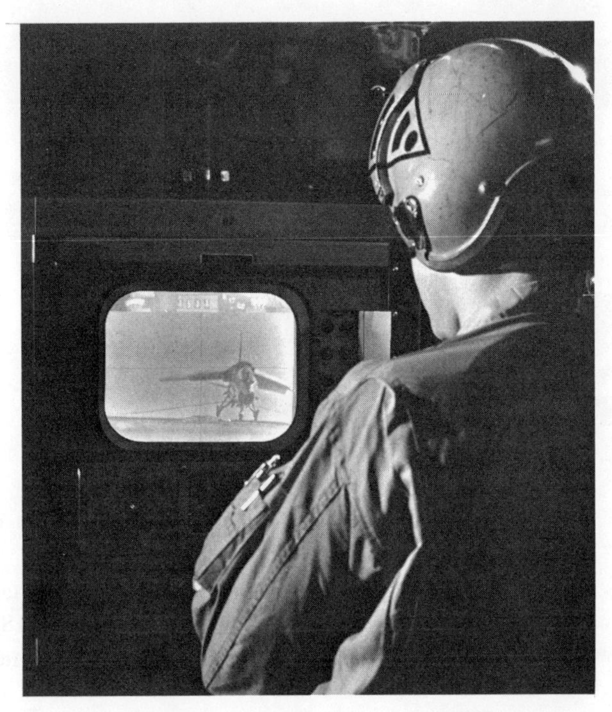

Fig. 18-57. Pilot landing aid television, a closed-circuit videotape system for U. S. aircraft carriers. (Courtesy of Ampex Corp.)

inological computerized files are already maintained by the Federal Bureau of Investigation and other major police organizations.

Banks and department stores have long since implemented closed-circuit TV surveillance systems and electronic alarms. Somewhat more controversial monitoring devices are the tiny electronic transmitters called *bugs*. Microminiaturization has made it possible to build these remote-listening devices small enough to be disguised as a man's tie clasp, a lady's brooch, a cocktail olive, etc. Such devices are available to anyone, easy to "plant," and difficult to detect. Legislation severely limits the acceptance of evidence so obtained in the courtroom, but unscrupulous persons outside the law enforcement profession are relatively free to use them, hence their controversial nature.

Radar has found some use in law enforcement, primarily in traffic control. The precision of radar Doppler techniques makes for easy entrapment of speed-limit violators on streets and highways. Patrol boats, such as the one in Fig. 18-58, can have as much as 1000 square miles of water and coastline under surveillance at one time. These little boats cruise more than 7000 miles each year enforcing the state's marine fishing and small boat laws, as well as aiding boats in distress.

Fig. 18-58. Radar-equipment patrol boat of Massachusetts Dept. of Natural Resources. (Courtesy of Raytheon Corp.)

18-15 ELECTRONICS IN BANKING AND COMMERCE

The major contribution of the electronic age to the commercial world, thus far, has been the digital computer. Figure 18-59 represents a computer installation at the headquarters of a major insurance company. This computer

Fig. 18-59. Insurance company computer. (Courtesy of International Business Machines Corp.)

is linked by telephone to nine smaller ones located in the company's offices in other cities. Daily policy information is fed into the small computers at the divisional offices, which check it for accuracy and then transmit it directly to update the company's master file, which is stored on tape.

Another interesting example of a commercial computer at work is the tribal records computer shown in Fig. 18-60, which is operated by the Navajo

Fig. 18-60. Navajo tribal records computer. (Courtesy of International Business Machines Corp.)

Tribe in Arizona. This computer is used to conduct more than 125 different projects for the tribe, which now numbers some 118,000 persons. The projects range from inventory of all livestock on the reservation to statistical analyses of the tribe's oil leases to various corporations.

Perhaps even more unusual is the computer application of Joseph Marshall Imports Co. in Los Angeles. The product here is ladies' wigs, and the firm presently handles some 30 different styles in 65 shades and variations, as indicated in Fig. 18-61. The computer keeps track of all data, from manufacturing to marketing, and generates fashion-trend reports.

Fig. 18-61. Computer processes wig styling data. (Courtesy of International Business Machines Corp.)

Modern banking is highly computer oriented. Checks are imprinted with electromagnetically susceptible ink to permit automatic processing of accounts. Several institutions have a central records room for verification of signatures, balances, etc. This facility communicates with individual tellers via two-way closed-circuit TV.

The same techniques are now employed by credit orgainzations in general. Credit cards, punched-card account statements, and one or more central data processors together make an almost foolproof automatic credit system. Furthermore, the quick sort and retrieve capability of the computerized file permits a business to obtain a rapid check on the credit references of a prospective client.

18–16 YET TO COME

As we come to the end of our study, it is only fair that we do a bit of fanciful crystal gazing. With just a little extrapolation of existing technology, we can see some distance down the road ahead.

Automation is certainly proceeding at a brisk pace. The automatic automobile (just tell it where you want to go and relax!) is not more than 50 years away. Neither is the fully automatic home—one that is self-cleaning, with an automatically controlled environment, and just tell it what you want to eat—more than half a century from realization.

Cybernetics, the science of automata founded by Norbert Wiener in 1948, will eventually give us the robot. Actualization of a versatile moderate-cost robot could provide a degree of prosperity and well-being, for all men, unknown in the world to date (providing we don't poison ourselves, suffocate ourselves, or blow ourselves up, first). The rudiments of robotics—microcircuits, ovonics (the new technology of amorphous semiconductors), computer technology, control theory, and information theory—are here even now.

Communication will progress by leaps and bounds. The video telephone is just a few years away. The "citizens band" craze a couple of decades from now will undoubtedly be small personal television transceivers. Commercial television itself should be three-dimensional, as well as in color, by the time the millenium arrives. Let us hope that the programming matures with the device.

Economically we will probably live in a "cashless" society long before the year 2000. This is a logical extension of the credit card system, whose details are being worked out even as these words are written. Perhaps that society can also be hungerless.

The same pre-2000 technology will put man on Mars, and improved communication coupled with central data processing may lead to rapid-response direct popular vote on political candidates and proposed legislation. There is no telling where electronics may ultimately lead in medicine, since all phenomena, including disease, are ultimately electronic at the molecular level. In the near future, however, electronics will at least offer improved weapons for the war on cancer, and more satisfactory artificial organs.

Beyond these almost obvious forecasts, electronic phenomena and applications, presently undreamed, remain to be discovered. Perhaps you, the reader, will be the first to discover a new application of electronics in our world.

REVIEW QUESTIONS

1. Under which of the six categories defined in Section 18-1 would the following fit: electric cake mixer, radio-controlled model airplane, buried metal locator, laser bombardment of cancer cells?
2. What were the major technological steps in attaining the present degree of miniaturization of electronic circuits?
3. Describe the fundamental differences between an analog computer and a digital computer. Which one is best suited to figure your income tax? Why?
4. What is a hybrid computer?
5. In an analog computer, describe what the following components do: summer, coefficient "pot," integrator, resolver, inverter.
6. Do these simple arithmetic problems, using binary numbers, and check your answers:

$$6 + 3 \qquad 2 \times 5 \qquad 4 + 7 + 12$$

7. Why are bistable devices used extensively in digital computers?
8. Briefly describe the major units that make up a digital computer.
9. In what areas does electronics seem to benefit medicine most at present?
10. Suppose you wanted to build a closed-loop control system to hold the speed of a motorboat at 15 knots. What sort of devices would you need?
11. Can you give an application for electronics in politics? In religion? Can you name an area to which electronic technology has not been applied?

APPENDIX

Table 1 Greek alphabet

Alpha	A	α	Nu	N	ν
Beta	B	β	Xi	Ξ	ξ
Gamma	Γ	γ	Omicron	O	o
Delta	Δ	δ	Pi	Π	π
Epsilon	E	ϵ	Rho	P	ρ
Zeta	Z	ζ	Sigma	Σ	σ
Eta	H	η	Tau	T	τ
Theta	Θ	θ	Upsilon	Υ	υ
Iota	I	ι	Phi	Φ	ϕ
Kappa	K	κ	Chi	X	χ
Lambda	Λ	λ	Psi	Ψ	ψ
Mu	M	μ	Omega	Ω	ω

THE INTERNATIONAL SYSTEM OF UNITS

All mechanical and electrical units in this system are derived from four basic units: the *meter*, the unit of length; the kilogram, the unit of mass; the *second*, the unit of time; and the *coulomb*, the unit of electric charge. The meter (m) is defined as an exact number of wavelengths of the reddish-orange light emitted in a particular electron orbital transition in the Krypton

Table 2 Practical MKS Units of Physical Quantities

Quantity	Nominal Unit and Abbreviation	Derived Unit	Fundamental Unit	Approximate Equivalent in English System
Acceleration	meter/second^2(m/s^2)	–	m/s^2	3.3 ft/sec^2
Area	meter^2 (m^2)	–	m^2	11 ft^2
Capacitance	farad (F)	coulomb/volt	C^2-s^2/kg-m^2	same
Conductance	mho (mho)	ampere/volt	C^2-s/kg-m^2	same
Current	ampere (a)	–	C/s	same
Elect. field strength	volt/meter (v/m)	newton/coulomb	kg-m/C-s^2	0.3 v/ft
Electromotive force, Potential, Voltage drop	volt (v)	joule/coulomb	kg-m^2/C-s^2	same
Energy, Work	joule (J)	newton-meter or watt-second	kg-m^2/s^2	0.74 lb-ft
Force	newton (N)	–	kg-m/s^2	0.22 lb$_f$
Frequency	hertz (Hz)	(cycles)*/second	(~)*/s	same
Inductance	henry (h)	weber/ampere	kg-m^2/C^2	same
Magnetic field strength	ampere(-turn)*/meter	–	C/m-s	0.3 a/ft
Magnetic flux, Pole strength	weber (Wb)	volt-second	kg-m^2/C-s	same
Magnetic flux density	tesla (T)	weber/meter^2	kg/C-s	0.009 Wb/ft^2
Magnetomotive force	ampere(-turn)*	–	C/s	same
Power	watt (W)	joule/second	kg-m^2/s^2	0.74 lb-ft/sec
Pressure, Stress	pascal	newton/meter^2	kg/m-s^2	0.002 lb/ft^2
Reluctance	ampere(-turn)*/weber(a/Wb)	1/henry	C^2/kg-m^2	same
Resistance	ohm (Ω)	volt/ampere	kg-m^2/C^2-s	same
Torque	newton-meter (N-m)	–	kg-m^2/s^2	0.74 ft-lb
Velocity	meter/second (m/s)	–	m/s	3.3 ft/sec
Volume	meter^3 (m^3)		m^3	35 ft^3

*Not a true physical unit, but usually written for clarity.

Errata for ELECTRONICS IN OUR WORLD by Gregory J. Nunz.

The following table accompanies the table title on page 561 (table 3).

TABLE 3 SCHEMATIC SYMBOLS USED IN CIRCUIT DIAGRAMS

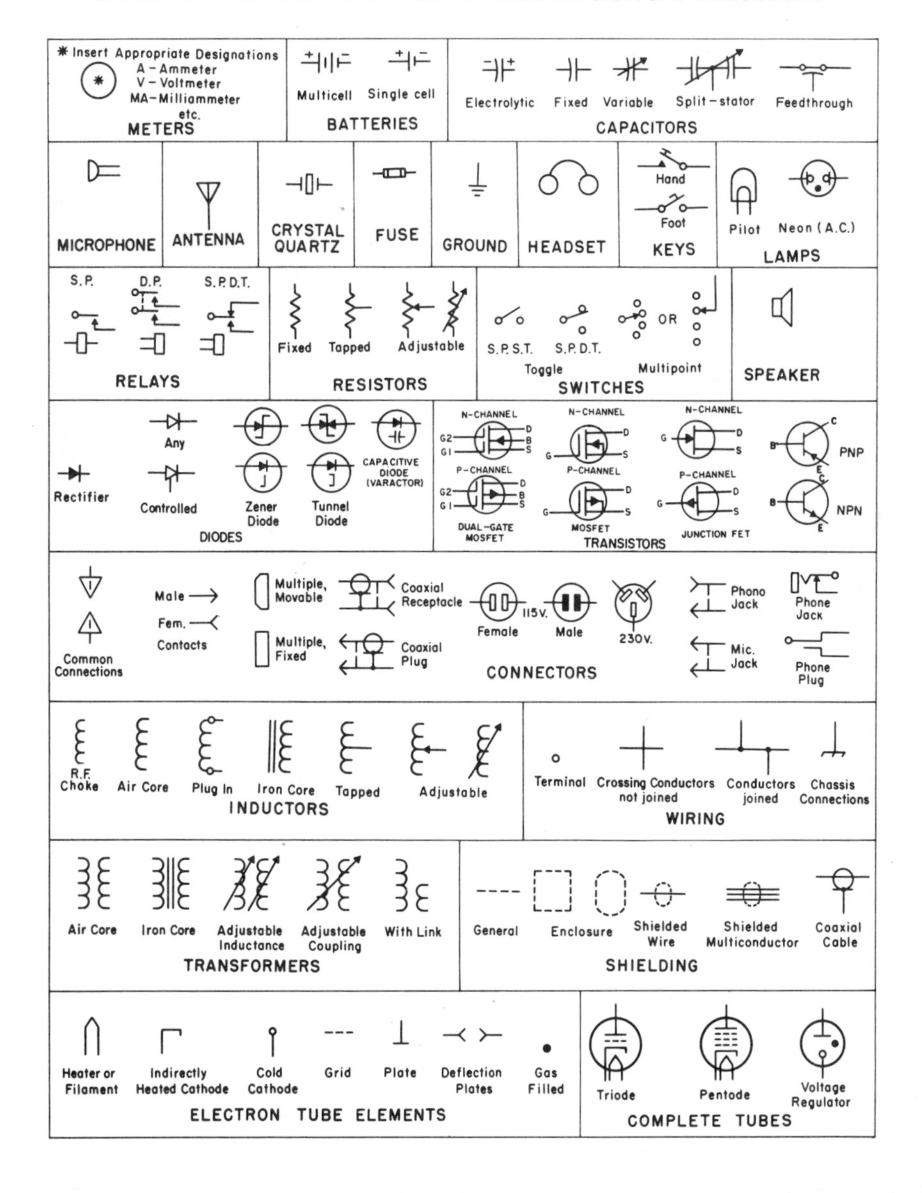

86 atom, and is about 3.3 feet. The kilogram (kg) is defined as the mass of a special cylindrical bar kept in the International Bureau of Weights and Measures vault in Sèvres, France, and is equivalent to a mass of about 2.2 pounds. The second (s) is defined as an exact fraction of the length of the tropical year 1900. The coulomb (C) is internationally defined on the basis of the ampere of current, which is in turn defined in terms of the force produced between two parallel conductors.

Given the four basic units—m, kg, s, C—the units of the other mechanical and electrical quantities are as on page 562.

Table 3 The Decimal Prefixes

Factor by which unit is multiplied	Prefix	Symbol	Factor by which unit is multiplied	Prefix	Symbol
10^{12}	tera	T	10^{-2}	centi	c
10^{9}	giga	G	10^{-3}	milli	m
10^{6}	mega	M	10^{-6}	micro	μ
10^{3}	kilo	k	10^{-9}	nano	n
10^{2}	hecto	h	10^{-12}	pico	p
10	deka	da	10^{-15}	femto	f
10^{-1}	deci	d	10^{-18}	atto	a

Table 4 Schematic Symbols Used in Circuit Diagrams

SUBJECT INDEX

N

O

P

Q

R

U

V

W

X

Z

PERSONALITY INDEX

M

N

O

P

R

S

T

V

W

Z